ORGANIC CHEMISTRY SERIES

Series Editor: J E Baldwin, FRS

Volume 2

Organotransition Metal Chemistry Applications to Organic Synthesis

Related Pergamon Titles of Interest

BOOKS

Organic Chemistry Series
Volume 1
DESLONGCHAMPS: Stereoelectronic Effects in Organic Chemistry

Also of Interest

BARTON *et al:* R B Woodward Remembered
BARTON & OLLIS: Comprehensive Organic Chemistry
FREIDLINA & SKOROVA: Organic Sulfur Chemistry
HERAS & VEGA: Medicinal Chemistry Advances
NOZAKI: 4th International Conference on Organic Synthesis
PERRIN *et al:* Purification of Laboratory Chemicals, 2nd Edition
RIGAUDY & KLESNEY: Nomenclature of Organic Chemistry, "The Blue Book"
TROST & HUTCHINSON: Organic Synthesis — Today and Tomorrow
WILKINSON *et al:* Comprehensive Organometallic Chemistry

JOURNALS

Chemistry International (IUPAC news magazine for all chemists)
Polyhedron (primary research and communication journal for inorganic and organometallic chemists)
Pure and Applied Chemistry (official IUPAC research journal for all chemists)
Tetrahedron (primary research journal for organic chemists)
Tetrahedron Letters (rapid publication preliminary communication journal for organic chemists)

Organotransition Metal Chemistry Applications to Organic Synthesis

STEPHEN G. DAVIES

Lecturer in Organic Chemistry, The Dyson Perrins Laboratory
University of Oxford, UK
and
Fellow of New College, Oxford

PERGAMON PRESS

OXFORD · NEW YORK · TORONTO · SYDNEY · PARIS · FRANKFURT

U.K.	Pergamon Press Ltd., Headington Hill Hall, Oxford OX3 0BW, England
U.S.A.	Pergamon Press Inc., Maxwell House, Fairview Park, Elmsford, New York 10523, U.S.A.
CANADA	Pergamon Press Canada Ltd., Suite 104, 150 Consumers Road, Willowdale, Ontario M2J 1P9, Canada
AUSTRALIA	Pergamon Press (Aust.) Pty. Ltd., P.O. Box 544, Potts Point, N.S.W. 2011, Australia
FRANCE	Pergamon Press SARL, 24 rue des Ecoles, 75240 Paris, Cedex 05, France
FEDERAL REPUBLIC OF GERMANY	Pergamon Press GmbH, Hammerweg 6, D-6242 Kronberg-Taunus, Federal Republic of Germany

First edition 1982
Reprinted, 1983, 1984

Library of Congress Cataloging in Publication Data

Davies, Stephen G.
Organotransition metal chemistry.
(Organic chemistry series; v. 2)
Includes bibliographies and index.
1. Organometallic compounds. 2. Transition metal compounds. 3. Chemistry, Organic—Synthesis.
I. Title. II. Series: Organic chemistry series (Pergamon Press); v. 2.
QD411.D36 1982 547'.056 82-16626

British Library Cataloguing in Publication Data

Davies, Stephen G.
Organotransition metal chemistry.—(Organic chemistry series; v. 2)
1. Chemistry, Organic—Synthesis
2. Organometallic compounds
I. Title II. Series
547'.2 QD262
ISBN 0-08-026202-3 (Hardcover)
ISBN 0-08-030714-0 (Flexicover)

Printed in Great Britain by A. Wheaton & Co. Ltd., Exeter

To Kay

FOREWORD

During the past two decades enormous advances have taken place in our understanding of the structure and reactivity of organotransition metal compounds. These insights have prepared the way for applications of these compounds to the ever burgeoning field of organic synthesis, both as stoichiometric reagents and as catalysts.

At this point in time it is probably true to say that the major themes of mechanism and reactivity for purely organic compounds are fairly well understood, at least in qualitative if not quantitative terms. These same themes place fairly substantial limits on what is possible in a synthetic sense, for example, displacement reactions at tetrahedral carbon by nucleophilic species invariably requires an inversion (S_N2) and this fact places certain well-known structural and stereochemical limitations in the application of this reaction type. The introduction of a transition metal completely changes the picture, and even at this early stage it is quite evident that the old "rules" no longer apply. These new opportunities are presently being explored in the synthetic area and in this book Dr. Davies has successfully combined a qualitative bonding picture with a description of reactivity which will be easily understood by practising organic chemists, and so enable them to see new synthetic opportunities in organotransition metal systems synthesis. Dr. Davies' book therefore is to be warmly recommended to all organic chemists who have any interest in synthetic methodology and its ultimate objective, total synthesis.

Professor J.E. Baldwin, FRS

University of Oxford,
Dyson Perrins Laboratory

PREFACE

Organotransition metal chemistry is rapidly becoming an important tool for organic synthesis. The aim of this monograph is to provide an introduction to organometallic chemistry for organic chemists interested in synthesis. Emphasis has been placed not only on reactions that are established already as synthetically useful but also on those that are potentially interesting. Both catalytic and stoicheiometric reactions are discussed although it is recognised that the former are likely to gain more rapid acceptance for synthesis.

It is impossible for a monograph of this size to be comprehensive, therefore topics that have been extensively reviewed elsewhere are only briefly discussed. The references cited cover the literature up to mid-1980.

I am indebted to Dr. Hugh Felkin for many helpful discussions. I also wish to express my gratitude to Steven Abbott, Dr. Nurgün Aktogu, Gordon Baird, Dr. Taquir Fillebeen-Khan, John Hibberd, Nicholas Holman, Charles Laughton, Dr. Stephen Simpson, Fatemeh Tadj, Susan Thomas, Dr. Oliver Watts and Dr. Susan Wollowitz, all of who read the complete manuscript.

Stephen G. Davies

The Dyson Perrins Laboratory,
University of Oxford

CONTENTS

ABBREVIATIONS xv

CHAPTER 1

General Introduction to Organometallic Chemistry 1

1.1 18-ELECTRON RULE 2

1.2 BONDING OF HYDROCARBON LIGANDS TO TRANSITION METALS 4

1.3 ELECTRONIC EFFECTS OF COORDINATION OF UNSATURATED HYDROCARBONS TO TRANSITION METALS 5

1.4 STEREOCHEMICAL AND STERIC EFFECTS 10

1.5 CATALYTIC CYCLES 13

1.5.1 Hydrogenations, hydrosilylations etc. 14

1.5.2 Hydroformylations 17

1.6 REFERENCES 18

CHAPTER 2

Complexation and Decomplexation Reactions 19

2.1 PREPARATION AND DECOMPLEXATION OF η^1-COMPLEXES 20

2.1.1 Nucleophilic metal and RX 20

2.1.2 Metal halide and nucleophilic alkyl 24

2.1.3 Nucleophilic addition to transition metal complexes 25

2.1.4 Metal hydride and olefins 26

2.1.5 Metal hydride and diazomethane 28

2.1.6 Interconversion of M-alkyl and M-acyl 28

2.1.7 Deprotonation of cationic olefin complexes 30

2.1.8 Decomplexation of η^1*-ligands* 30

2.2 PREPARATION AND DECOMPLEXATION OF η^2-OLEFIN AND ACETYLENE COMPLEXES 32

2.2.1 Ligand exchange reactions 32

2.2.2 Deoxygenation of epoxides 36

2.2.3 Other methods 37

2.2.4 Decomplexation of η^2*-olefin ligands* 40

OMC - A*

2.3 PREPARATION AND DECOMPLEXATION OF η^2-CARBENE COMPLEXES 40
2.3.1 From CO and RNC ligands 40
2.3.2 From M-acyl ligands 41
2.3.3 Miscellaneous methods 41
2.3.4 Decomplexation of carbene ligands 42
2.4 PREPARATION AND DECOMPLEXATION OF η^3-ALLYL COMPLEXES 44
2.4.1 From Dienes 44
2.4.2 From olefins with an allylic leaving group 46
2.4.3 From olefins 48
2.4.4 Miscellaneous methods 50
2.4.5 Decomplexation of η^3-allyl ligands 51
2.5 PREPARATION AND DECOMPLEXATION OF η^4-DIENE COMPLEXES 53
2.5.1 From metal carbonyls 53
2.5.2 From exchange reactions between dienes and other ligands 56
2.5.3 Other methods 59
2.5.4 Decomplexation of η^4-diene ligands 60
2.6 PREPARATION AND DECOMPLEXATION OF η^5-DIENYL COMPLEXES 62
2.6.1 Hydride abstraction from η^4-diene complexes 62
2.6.2 Reactions of acids with substituted η^4-diene complexes 64
2.6.3 From nucleophilic addition to η^6-arene and η^6-triene complexes 66
2.6.4 Decomplexation of η^5-dienyl ligands 67
2.7 PREPARATION AND DECOMPLEXATION OF η^6-ARENE AND RELATED COMPLEXES 68
2.7.1 Preparation 68
2.7.2 Decomplexation of η^6-arene ligands 73
2.9 REFERENCES 75

CHAPTER 3
Organometallics as Protecting and Stabilising Groups 83
3.1 THE USE OF ORGANOMETALLIC SPECIES AS PROTECTING GROUPS 83
3.1.1 Protection of olefins. 83
3.1.2 Protection of acetylenes. 86
3.1.3 Protection of dienes. 91
3.1.4 Protection of amines during peptide synthesis. 98

3.2 STABILISATION OF REACTIVE COMPOUNDS AND TRAPPING OF REACTION INTERMEDIATES 99

3.2.1 Stabilisation of reactive compounds 99

3.2.2 Stabilisation of thermodynamically disfavoured tautomers 109

3.2.3 Trapping of reaction intermediates 112

3.3 REFERENCES 113

CHAPTER 4

Organometallics as electrophiles 116

4.1 STOICHEIOMETRIC NUCLEOPHILIC ADDITIONS TO ORGANOTRANSITION METAL CATIONS 117

4.1.1 General rules governing the regioselectivity 117

4.1.2 Nucleophilic addition to cationic η^2-olefin complexes 128

4.1.3 Nucleophilic addition to cationic η^3-allyl complexes 135

4.1.4 Nucleophilic addition to cationic η^4-diene complexes 139

4.1.5 Nucleophilic addition to cationic η^5-dienyl complexes 140

4.1.6 Nucleophilic addition to cationic η^6-arene complexes 151

4.1.7 Cyclopropanation reactions 155

4.2. STOICHEIOMETRIC NUCLEOPHILIC ADDITION AND SUBSTITUTION REACTIONS INVOLVING NEUTRAL ORGANOTRANSITION METAL COMPLEXES 156

4.2.1 Nucleophilic addition to neutral η^2-olefin complexes 156

4.2.2 Nucleophilic addition to neutral η^2-carbene complexes 162

4.2.3 Nucleophilic addition to neutral η^3-allyl complexes 164

4.2.4 Nucleophilic addition to neutral η^6-arene complexes 166

4.3 CATALYTIC NUCLEOPHILIC ADDITION AND SUBSTITUTION REACTIONS 172

4.4 REFERENCES 181

CHAPTER 5

Organometallics as Nucleophiles 187

5.1 NEUTRAL COMPLEXES AS NUCLEOPHILES 187

5.2 ANIONIC COMPLEXES AS NUCLEOPHILES 201

5.2.1 Anions derived from carbene ligands 201

5.2.2 Anions derived from (η^6-arene)$Cr(CO)_3$ complexes 203

5.2.3 The anion derived from (η^4-cycloheptatriene)$Fe(CO)_3$ 213

5.3 REFERENCES 215

CHAPTER 6

Coupling and Cyclisation Reactions 218

6.1 COUPLING REACTIONS INVOLVING BIS-η^1-COMPLEXES 218

6.2 THE HECK REACTION AND RELATED TRANSFORMATIONS 224

6.2.1 Coupling of organic nucleophiles with olefins 224

6.2.2 Coupling of organic halides with olefins 231

6.2.3 Coupling reactions involving C-H bond activation 234

6.3 COUPLING REACTIONS INVOLVING η^3-ALLYL INTERMEDIATES 236

6.4 COUPLING REACTIONS INVOLVING OLEFINS AND ACETYLENES 250

6.5 REFERENCES 261

CHAPTER 7

Isomerisation Reactions 266

7.1 ISOMERISATION OF OLEFINS AND ACETYLENES 266

7.2 ISOMERISATION OF DIENES 273

7.3 ISOMERISATION OF ALLYLIC ALCOHOLS, ETHERS, AMINES, ETC 282

7.4 REARRANGEMENTS OF SMALL RING HYDROCARBONS 291

7.5 ISOMERISATION OF SMALL RING HETEROCYCLES 294

7.5.1 Oxygen heterocycles 294

7.5.2 Nitrogen heterocycles 299

7.6 REFERENCES 301

CHAPTER 8

Oxidation and Reduction 304

8.1 OXIDATION 304

8.1.1 The Wacker process and related reactions 304

8.1.2 Dehydrogenation reactions 308

8.1.3 Epoxidation reactions with ROOH 311

8.1.4 Oxidations with transition metal dioxygen and peroxy complexes 317

8.1.5 Oxidative decomplexation reactions 318

8.2 REDUCTIONS 320

8.2.1 Addition of H_2 and related reactions 320

8.2.2 Asymmetric hydrogenation and hydrosilylation reactions 326

8.2.3 Deoxygenation and related reactions 331

8.2.4 Miscellaneous reductions 338

8.3 REFERENCES 344

CHAPTER 9

Carbonylation and Related Reactions 348

9.1 CARBONYLATION REACTIONS WITH Zr COMPOUNDS 352

9.2 CARBONYLATION REACTIONS WITH Fe COMPOUNDS 353

9.3 CARBONYLATION REACTIONS WITH Co AND Rh COMPOUNDS 366

9.4 CARBONYLATION REACTIONS WITH Pd and Ni COMPOUNDS 378

9.6 INSERTION OF CO_2 393

9.7 DECARBONYLATION REACTIONS 394

9.8 REFERENCES 400

INDEX 405

ABBREVIATIONS

Ac	Acetyl
acac	Acetylacetonate
cat	Catalyst
COD	Cyclooctadiene
Cp	Cyclopentadienyl
DDQ	2,3-Dichloro-5,6-dicyano-1,4-benzoquinone
DIOP	O,O'-Isopropylidene-2,3-dihydroxy-1,4-bis(diphenylphosphino)butane
diphos	Bis(diphenylphosphino)ethane ($Ph_2PCH_2CH_2PPh_2$)
DMF	Dimethylformamide
e	Electron
e.e.	Enantiomeric excess
Fp	$(C_5H_5)Fe(CO)_2$
h.o.m.o.	Highest occupied molecular orbital
HMPT	Hexamethylphosphoramide
L	Two electron ligand
LDA	Lithium diethylamide
l.u.m.o.	lowest unoccupied molecular orbital
M	Transition metal
[M]	Transition metal complex
MCPBA	meta-Chloroperbenzoic acid.
NBS	N-bromosuccinimide
NCS	N-chlorosuccinimide
Ph	Phenyl
Py	Pyridine
R	Alkyl radical
THF	Tetrahydrofuran
THP	Tetrahydropyranyl
Ts	p-Toluene sulphonyl
X	One electron ligand
*	Optically active centre

CHAPTER 1

GENERAL INTRODUCTION TO ORGANOMETALLIC CHEMISTRY

A certain number of remarkable organic reactions only take place in the presence of transition metal complexes e.g.

Ziegler-Natta stereoselective polymerisations

$$[M]—CH_2CH_2R \xrightarrow{R\diagdown\!\!=} [M]—CH_2\overset{\displaystyle R}{\overset{|}{C}}H—CH_2—CH_2—R \qquad [M] = \text{metal}$$

Wacker Process

$$CH_2{=\!=}CH_2 + O_2 \xrightarrow[Cu^{II}]{Pd^{II}} CH_3CHO$$

Fischer-Tropsch synthesis

$$CO + H_2 \longrightarrow C_n \text{ compounds}$$

Homogeneous hydrogenations using Wilkinson's catalyst

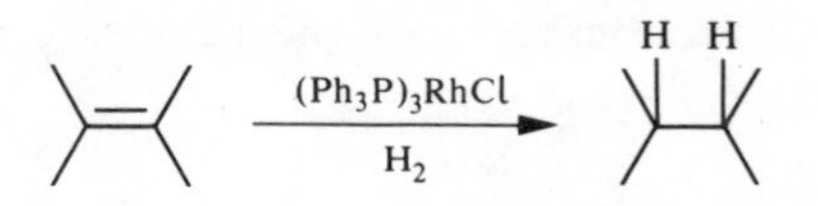

Also many highly selective enzymatic systems contain transition metals. The study of the preparation of transition metal complexes and their influence on the course of chemical reactions (organometallic chemistry) constitutes an area of chemistry that has been developing rapidly over the last 25 years. It is only recently, however, that this type of chemistry has begun to be exploited by organic chemists for synthesis.

Organometallic chemistry involves the interaction of an organic compound with a transition metal species to form an intermediate organometallic compound, which may or may not be stable. It is the reactions of these intermediate organometallic compounds that are of interest to the synthetic chemist. Generally, coordination of an organic compound to a transition metal markedly alters the properties of the compound such that it will undergo completely different types of reactions from the free molecule.

1.1 18-ELECTRON RULE

In order to achieve the stable inert gas configurations transition metal complexes need 18-electrons in their valence shells. That is to say they need to fill one s-orbital (2e), five d-orbitals (10e) and three p-orbitals (6e). The transition metals that will be considered are shown below together with the number of valence electrons associated with the metal atoms.

4e	5e	6e	7e	8e	9e	10e
Ti	V	Cr	Mn	Fe	Co	Ni
Zr	Nb	Mo	Tc	Ru	Rh	Pd
Hf	Ta	W	Re	Os	Ir	Pt

Many of these transition metals coordinate a variety of ligands to achieve the stable 18-electron configuration. However for Ti, Zr, Ni, Pd and Pt one of the orbitals has an energy unsuitable for ligand bonding and for these transition metals stable 16-electron complexes are generally found.[1] Steric problems associated with arranging a sufficient number of ligands around a metal can also lead to exceptions to the 18-electron rule.

Non-hydrocarbon ligands are classified below according to the number of electrons they contribute to the metal.

0e Lewis acids AlX_3, BX_3

1e -X, -H, $(NO)^*$

2e(L) Lewis bases. PR_3, $P(OR)_3$, CO, RCN, RNC, NR_3, R_2O, R_2S, etc.

3e NO^*

*NO normally acts as a 3e ligand but can also be a 1e ligand.

X^- = halide, CN^-, etc. R = alkyl, aryl.

Hydrocarbon ligands are classified according to their hapto (η) number.[2] η is the number of carbon atoms of the hydrocarbon bound to the metal. (Note that when η is odd the ligands are classified as *radicals*; this removes preconceived ideas about whether a ligand bears a positive or negative charge).

η^1	(1e)	alkyls, aryls, σ-allyls
η^2	(2e)	olefins (or polyolefins with only one double bond coordinated), carbenes.*
η^3	(3e)	π-allyls
η^4	(4e)	conjugated dienes
η^5	(5e)	dienyls and cyclopentadienyls
η^6	(6e)	trienes and arenes
η^7	(7e)	trienyls and cycloheptatrienyls
η^8	(8e)	cyclooctatetraene

It is important to consider only the number of carbon atoms bound to the metal. For example both σ, η^1 (1e) and π, η^3 (3e) allyl complexes are known, cyclooctatetraene can be η^8, η^6, η^4 or η^2 and cyclopentadienyl can be η^5, η^3 or η^1.

The number of electrons in a given complex is given by the sum:

number of electrons on free metal atom
+ (sum of the η numbers of all the hydrocarbon ligands)
+ [sum of the electrons donated by the other ligands]
+ number of negative charges on the metal in the complex
- number of positive charges on the metal in the complex

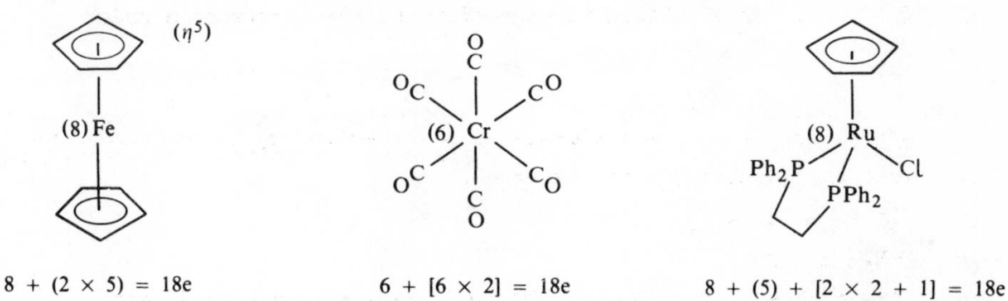

*Carbenes are classed as η^2-ligands even though bonding to only one carbon is involved since 2e rather than 1e are contributed.

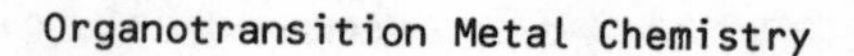

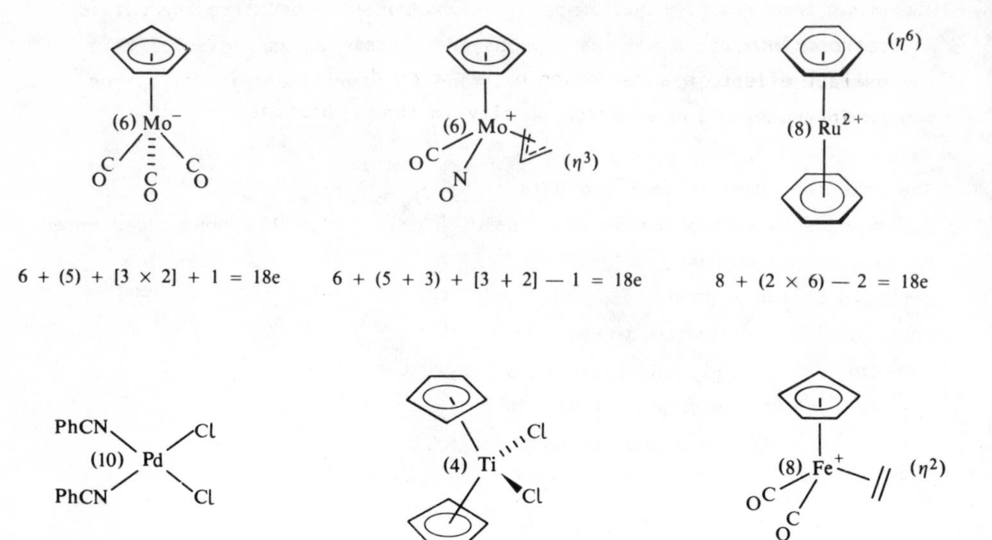

10 + [2 × 2 + 2 × 1] = 16e   4 + (2 × 5) + [2] = 16e   8 + (5 + 2) + [2 × 2] — 1 = 18e

The 18-electron rule is not absolute but it does serve as a useful guide to the types of compounds that can be expected to be stable and isolable. Also an understanding of how to count the number of electrons in a given complex or intermediate facilitates the understanding and description of reaction mechanisms. No organometallic complex with 20-electrons has been demonstrated.

1.2. BONDING OF HYDROCARBON LIGANDS TO TRANSITION METALS

The stability and reactivity of hydrocarbon complexes of transition metals can be attributed to bonding effects. According to the Chatt[3]-Dewar[4]-Duncanson[3] model for the bonding of ethylene to a transition metal

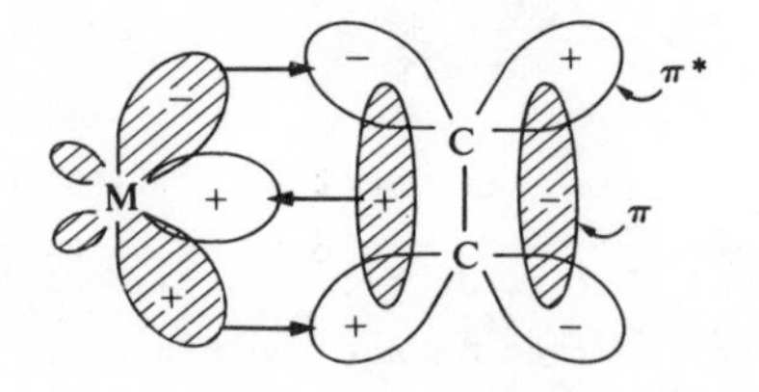

two types of bond are involved; (a) the unsaturated hydrocarbon donates electrons from its π-orbital to the metal (b) the metal donates its d-electrons into the antibonding π^*-orbital of the olefin (back-donation). The overall effect is a reduction of electron density in the π-orbital and an increase in the electron density in the π^*-orbital.

The same two types of bonding occur for all ligands i.e. donation of electron density from the filled ligand orbitals to the metal accompanied by back donation from the metal to the empty (normally antibonding) orbitals on the ligand. The relative sizes of the two types of bonding contribution depends on the actual complex. A bound hydrocarbon may be more or less electron rich than the free hydrocarbon.

1.3 ELECTRONIC EFFECTS OF COORDINATION OF UNSATURATED HYDROCARBONS TO TRANSITION METALS

The $Cr(CO)_3$ group influences the electronic properties of coordinated arenes by both inductive and resonance mechanisms. Various studies have demonstrated that the $Cr(CO)_3$ group exerts a net electron withdrawing inductive effect on coordinated η^6-arene ligands, i.e. the electron donation from the arene σ and π orbitals to the metal is greater than the back donation from the metal to the arene. For example (benzoic acid) $Cr(CO)_3$ and (phenylacetic acid)$Cr(CO)_3$ are stronger acids than the corresponding uncoordinated compounds.[5] The electron withdrawing effect of the $Cr(CO)_3$ group is very similar to that of a nitro substituent, for example, neither (arene)$Cr(CO)_3$ nor nitrobenzene are susceptible to Friedel Crafts acylation. Various dissociation constants are given below.[5]

	pKa		pKa
$C_6H_5CO_2H$	5.68	$C_6H_5CH_2CO_2H$	5.64
$(C_6H_5CO_2H)Cr(CO)_3$	4.77	$(C_6H_5CH_2CO_2H)Cr(CO)_3$	5.02
$p\text{-}NO_2C_6H_4CO_2H$	4.48	$p\text{-}NO_2C_6H_4CO_2H$	5.01

The inductive electron withdrawing effect of the $Cr(CO)_3$ group is also illustrated by the ready nucleophilic substitution by methoxide on (chlorobenzene)$Cr(CO)_3$ to give (anisole)$Cr(CO)_3$ under conditions where chlorobenzene itself is unreactive.

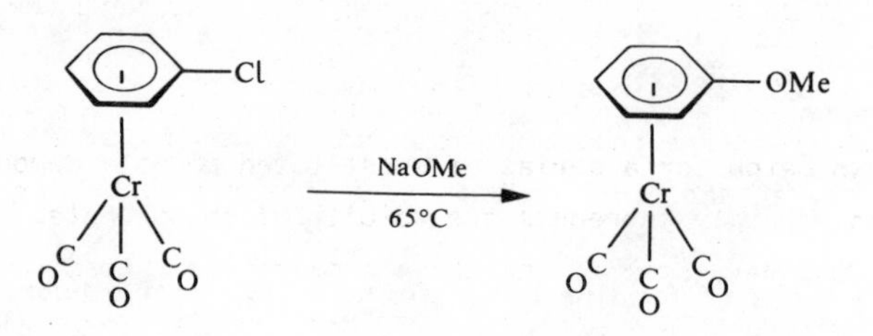

The relative magnitude of the forward and back donation components of the bonding of a ligand to a metal depends on the electron density on the metal. For (arene)$Cr(CO)_3$ complexes, the chromium is relatively electron poor because of the three CO ligands which are good π-acceptors and remove electron density from the metal. This results in the forward donation from the arene to the metal being of greater importance than the back donation. However, if one of the CO ligands is replaced by a donor ligand (e.g. phosphine or phosphite) then the chromium becomes relatively richer in electron density and back donation becomes more important. This can be seen from the acidities of the complexes (benzoic acid)$Cr(CO)_2L$ where the acidity of the acid complex decreases along the series L = CO > $P(OMe)_3$ > $P(OEt)_3$ > PPh_3.[6]

Compound	pKa
$(PhCO_2H)Cr(CO)_3$	4.77
$(PhCO_2H)Cr(CO)_2P(OMe)_3$	5.52
$(PhCO_2H)Cr(CO)_2P(OEt)_3$	5.62
$PhCO_2H$	5.68
$(PhCO_2H)Cr(CO)_2PPh_3$	6.15

Indeed the $Cr(CO)_2PPh_3$ unit has a net electron donating inductive effect. This series demonstrates that by changing the other ligands in a complex it is possible to control the electron density on a hydrocarbon ligand.

The resonance effect of a $Cr(CO)_3$ group in (arene)$Cr(CO)_3$ complexes is one of overall donation of electron density. However, in common with halogen substituents the magnitude of the resonance contribution depends markedly on the character of the reaction centre. Negative charges α to arene rings are stabilised by coordination of the arene to $Cr(CO)_3$. The resonance effects are relatively small in this case compared to the inductive effect. For example, (aniline)$Cr(CO)_3$ is a weaker base than aniline[7] (resonance effects cannot stabilise the anilinium cation). The results shown below for a series of substituted phenols demonstrate that complexation markedly increases the acidity of the phenols.[8] The $Cr(CO)_3$ group is inductively stabilising the phenoxide ion. Comparison of the pKa values of the complexed with the uncomplexed phenols leads to the conclusion that no distortion of the normal substituent effects occurs on complexation to $Cr(CO)_3$.

pKa values[8] for [OH / R-substituted phenol] and [OH / R-substituted phenol–$Cr(CO)_3$]

R =	Phenol	Phenol–$Cr(CO)_3$
p-Me	11.28	7.32
H	11.02	7.09
m-CO_2Me	10.05	6.77
m-COMe	9.99	6.82
p-CO_2Me	9.17	6.40
p-COMe	8.81	6.31

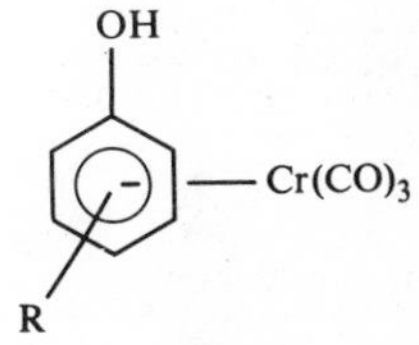

Similarly, coordination of arenes to $Cr(CO)_3$ leads to an increase in the acidity of benzylic protons, i.e. stabilisation of benzylic carbanions. Two examples are given below.[9]

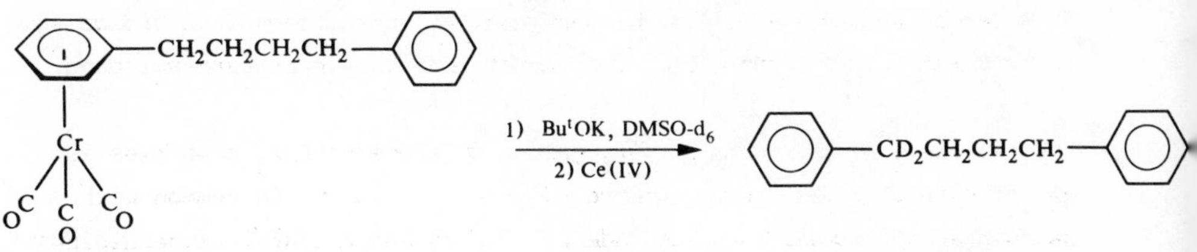

$(C_6H_5)Cr(CO)_3$–$CH_2CH_2CH_2CH_2$–C_6H_5 $\xrightarrow[\text{2) Ce(IV)}]{\text{1) Bu}^t\text{OK, DMSO-d}_6}$ C_6H_5–$CD_2CH_2CH_2CH_2$–C_6H_5

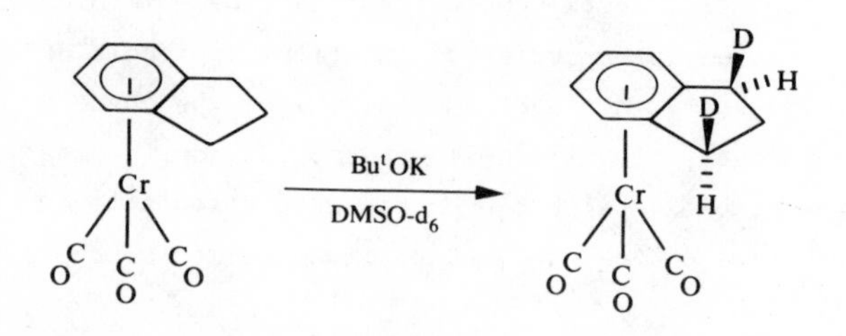

Nevertheless, the electron donating resonance effects of the $Cr(CO)_3$ group become important when positive charges are involved α to the arene ring. The first order rates of solvolysis of (benzyl chloride)$Cr(CO)_3$ and of (benzhydryl chloride)$Cr(CO)_3$ have been found to be 10^5 and $10^{3.3}$ times greater respectively than those of the uncomplexed chlorides.[10] This rate enhancement can be understood in terms of resonance stabilisation of the intermediate carbonium ion. A rate enhancement is also seen in the S_N1 solvolysis of (cumyl chloride)$Cr(CO)_3$.

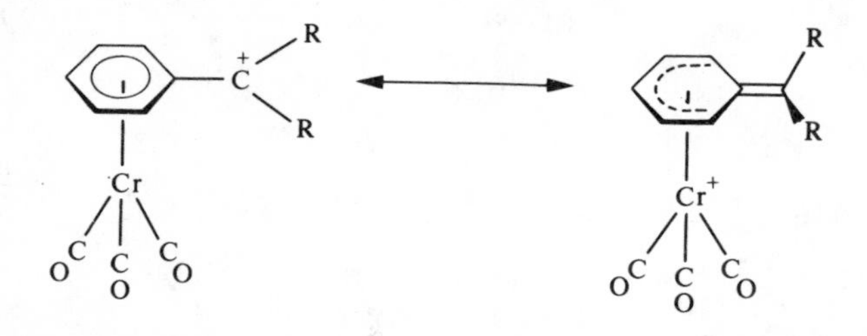

Similarly the acetolysis of (2-phenyl-2-methyl-1-propyl methanesulphonate) $Cr(CO)_3$ is 1.8 times faster than that of the uncomplexed compound.[11]

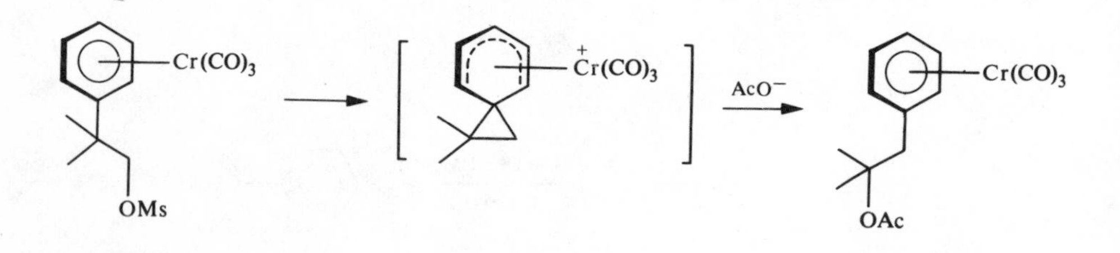

In contrast to the electron withdrawing effect of $Cr(CO)_3$ in (arene)$Cr(CO)_3$ complexes, the $Fe(CO)_3$ in (diene)$Fe(CO)_3$ complexes behaves as a net inductive electron donor. For example, the complexed acids shown below are weaker than the corresponding uncomplexed compounds. Also (1-phenylbutadiene)$Fe(CO)_3$ undergoes acylation in the para position indicating that the (diene)$Fe(CO)_3$ unit is electron donating.[12]

CO_2H $pK_a = 7.00$

CO_2H $pK_a = 7.00$

CO_2H Fe $(CO)_3$ $pK_a = 7.25$

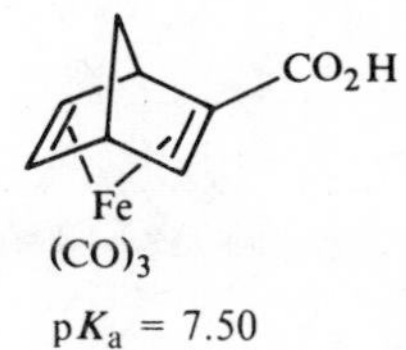

$pK_a = 7.50$

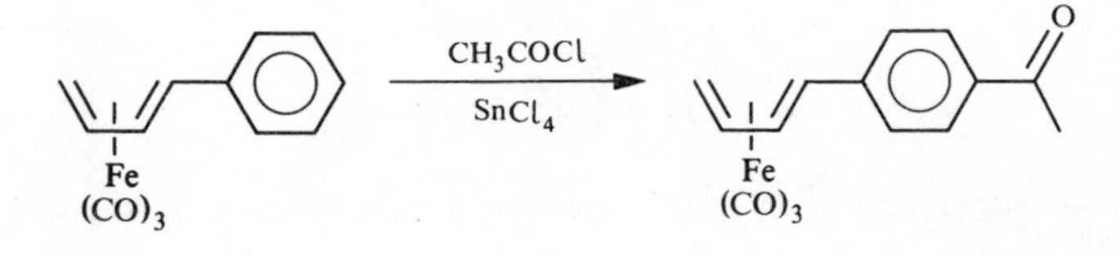

However negative charges adjacent to a (diene)$Fe(CO)_3$ unit can sometimes be stabilised by delocalisation of the charge onto the iron. The pKa of (cycloheptatriene)$Fe(CO)_3$ is 20 whereas for uncomplexed cycloheptatriene the pKa = 36.[13] The $Fe(CO)_3$ group effectively stabilises the antiaromatic cycloheptatrienyl anion.

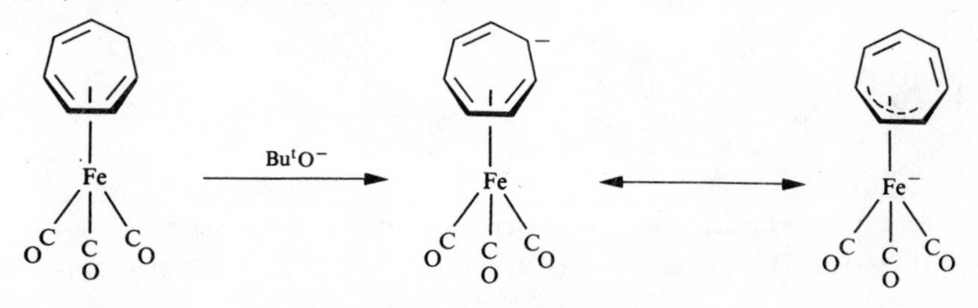

The solvolysis of (7-tosyloxynorbornadiene)$Fe(CO)_3$ is at least 10^6 times slower than that of the uncomplexed tosylate.[14] This is due to the π-electrons, which are involved in bonding to the endo $Fe(CO)_3$, being less available on the exo face for stabilisation of a C-7 carbonium ion.

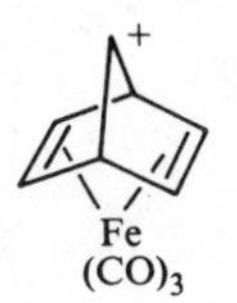

1.4 STEREOCHEMICAL AND STERIC EFFECTS

Many unsaturated hydrocarbon ligands are prochiral (substituted olefins, dienes, disubstituted arenes etc.) and thus form chiral transition metal complexes.

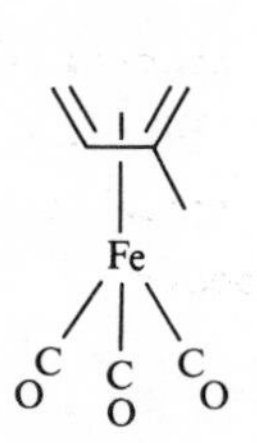

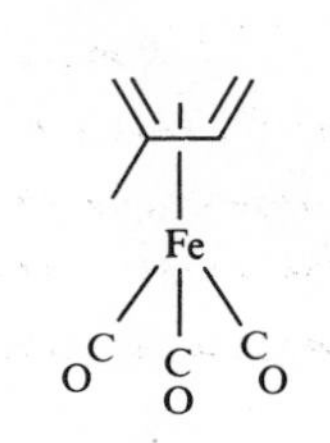

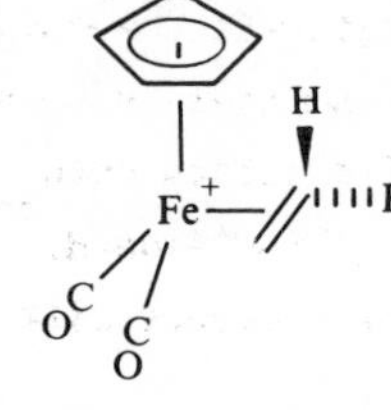

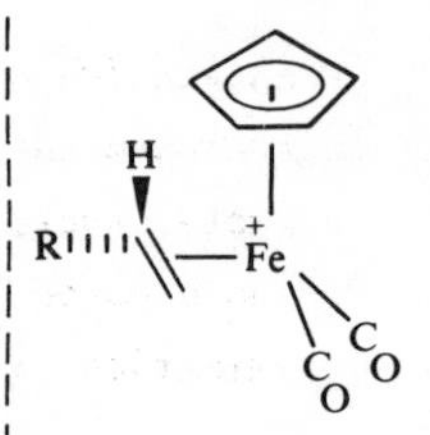

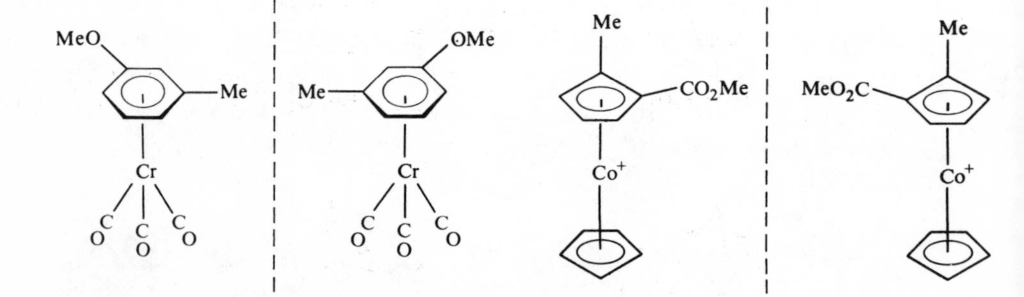

The resolution of such complexes generally may be achieved by standard organic methods if an appropriate functional group (e.g. CO_2H, CHO, NR_2, OH etc.)

is available on the ligand. For example, the acid 1 may be resolved as a salt with (-)-1-phenylethylamine.[15]

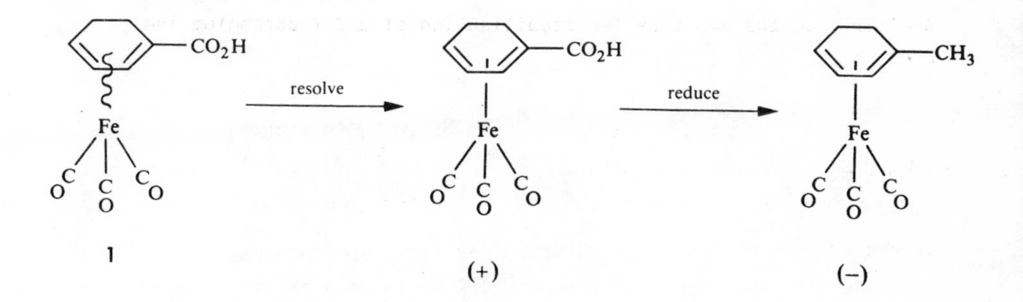

When either the metal centre or one of the ligands is chiral then the complexes with prochiral unsaturated hydrocarbons are diastereomeric. The use of chiral phosphine ligands, for example, allows the prochiral faces of olefins to be differentiated on complex formation with subsequent reactions leading to asymmetric syntheses.[16]

Complexation of unsaturated hydrocarbons to transition metal species has stereochemical consequences on subsequent reactions of these ligands. For example, nucleophilic attack onto 18e organometallic cations occurs on the uncoordinated face of the ligand.[17] The reverse reaction, namely elimination, similarly occurs with inversion at carbon.[18]

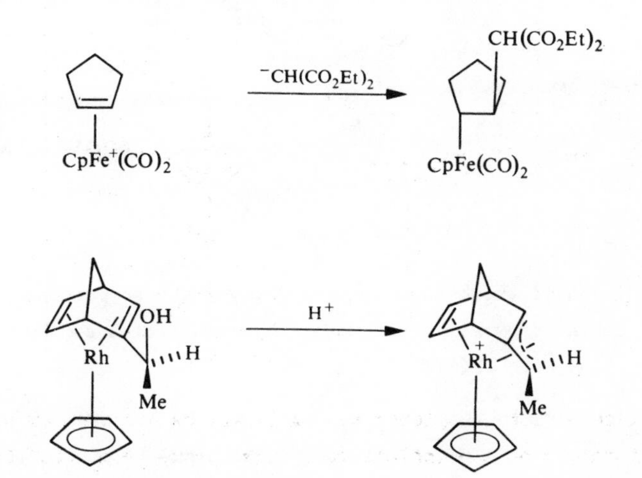

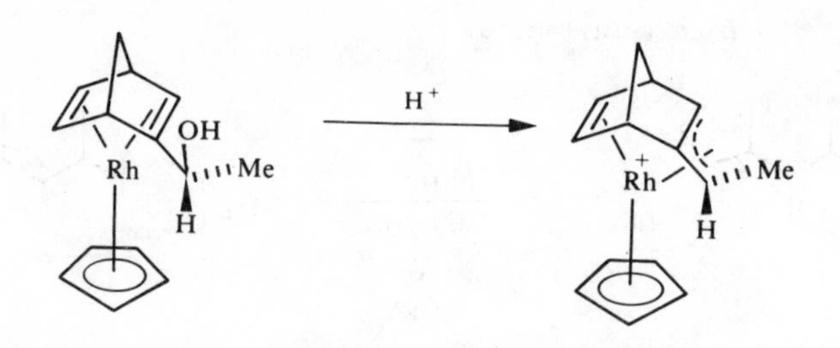

The steric bulk of $Cr(CO)_3$ results in reactants approaching (arene)$Cr(CO)_3$ complexes from the side away from the $Cr(CO)_3$. This is illustrated below for two reactions of (indanone)$Cr(CO)_3$.[19,20]

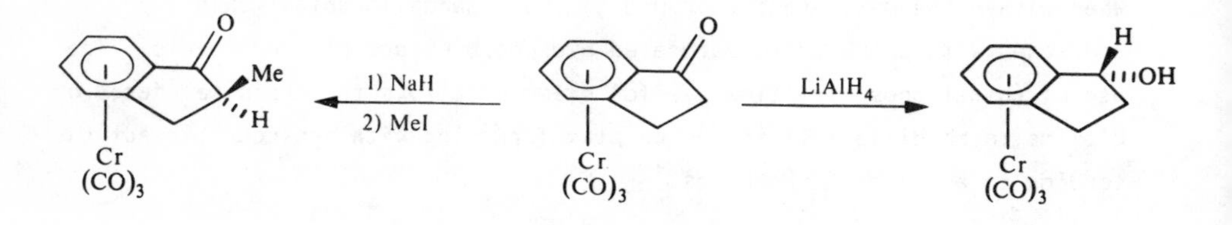

Starting with optically active (-)-(indanone)$Cr(CO)_3$ it is possible to prepare optically pure R(-)-2-methylindanone.[20]

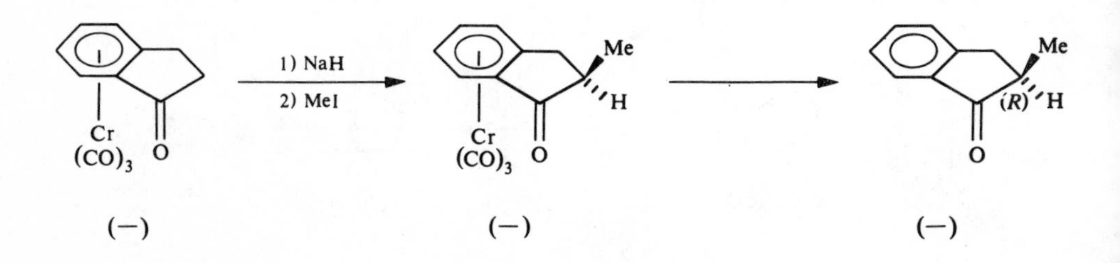

Hydride reduction of the ketone complex 2 gives the corresponding 3α-alcohol complex from which the novel epiergosterol 3 can be obtained by decomplexation with Fe(III).[21]

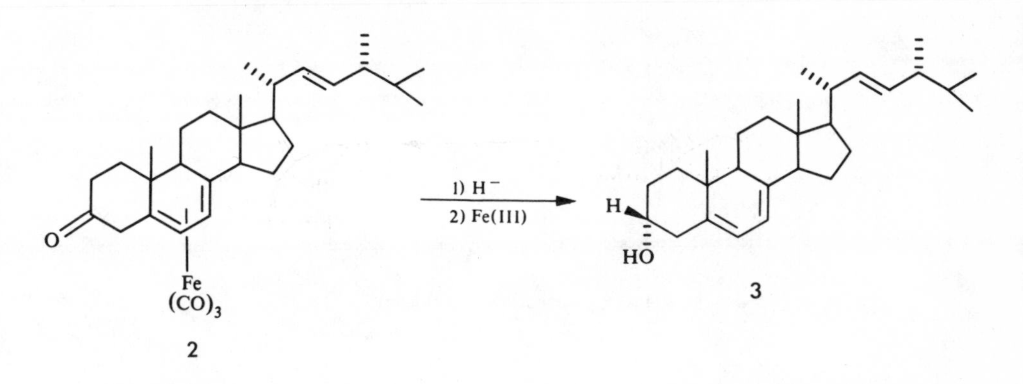

1.5 CATALYTIC CYCLES

One of the most important uses of transition metal complexes for synthesis is in catalytic reactions. In catalytic cycles the metal plays a much more active role than in the reactions described previously. To act as a catalyst the metal must possess the ability to change the number of electrons associated with it by 2 or 4 (2x2). Since the total number of electrons cannot exceed 18 most catalytic species possess 14- or 16-electrons. The type of reactions that are involved in most catalytic cycles are generalised below. Each reaction either increases or decreases the number of electrons on the metal by two.

(+2e) *coordination* *(−2e)* *dissociation*

$$[M] + L \longrightarrow [M]-L \longrightarrow [M] + L$$

(L = olefin, CO)

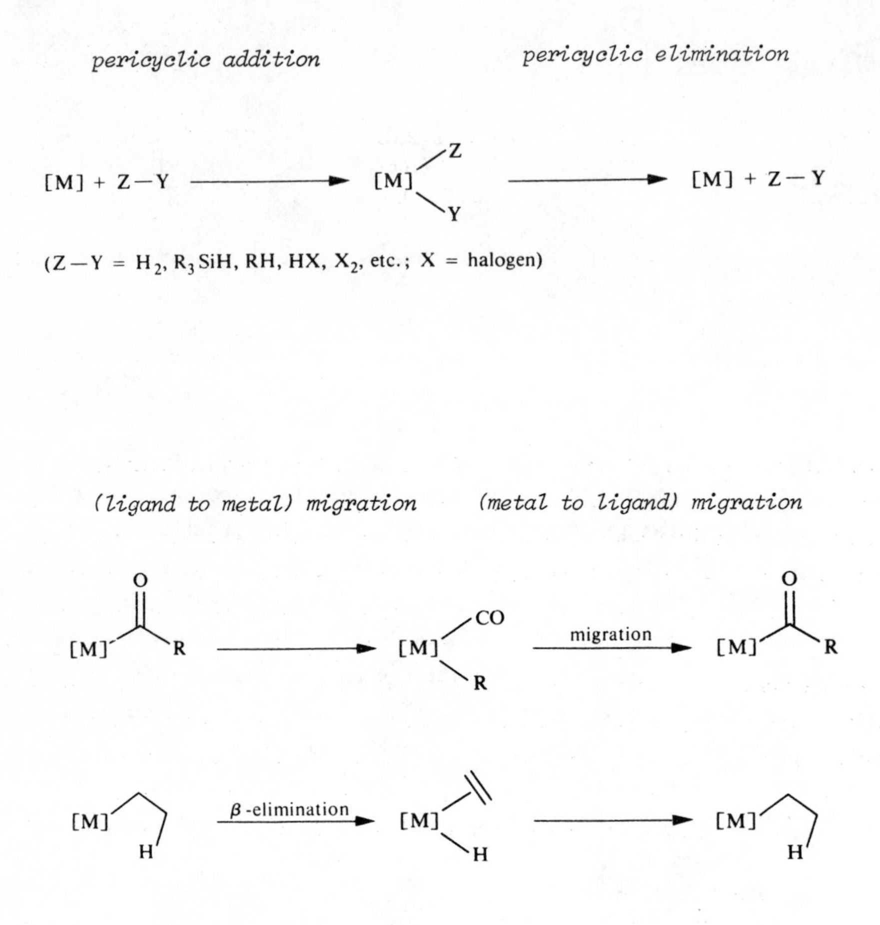

1.5.1 Hydrogenations, hydrosilylations etc. (see also chap. 8)

Two general types of catalysis are observed either involving a 14e intermediate or a 16e metal hydride intermediate. Wilkinson's catalyst, $(Ph_3P)_3RhCl$, is an example of the former type. A typical catalytic cycle is shown opposite.

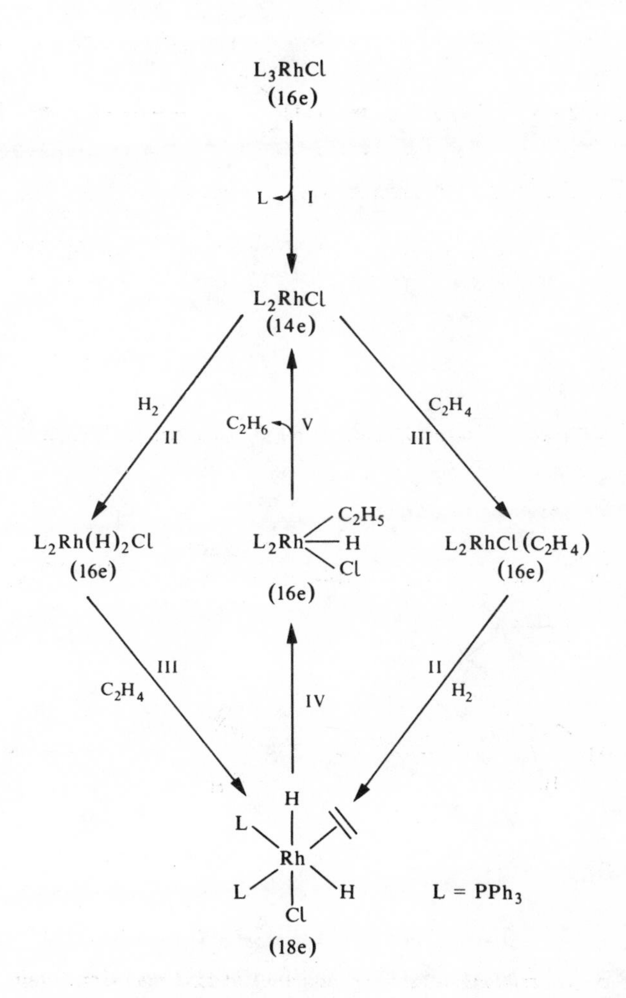

Dissociation of Ph_3P from the 16e $(Ph_3P)_3RhCl$ (step I) generates the 14e species $(Ph_3P)_2RhCl$ which may be stabilised by reversible coordination of solvent (alcohol). This intermediate reacts in two stages by coordination of ethylene (step II) and pericyclic addition of hydrogen (step III) to generate the 18e species $(Ph_3P)_2RhCl(H)_2(C_2H_4)$. This 18e complex undergoes a migration reaction (step IV) to the 16e ethyl hydride complex which undergoes pericyclic elimination (step V) of ethane and regenerates the starting 14e catalytic species. Many of these reactions are reversible.

The compound $(Ph_3P)_3RhH(CO)$ is an example of the second type of catalyst involving the steps shown below.

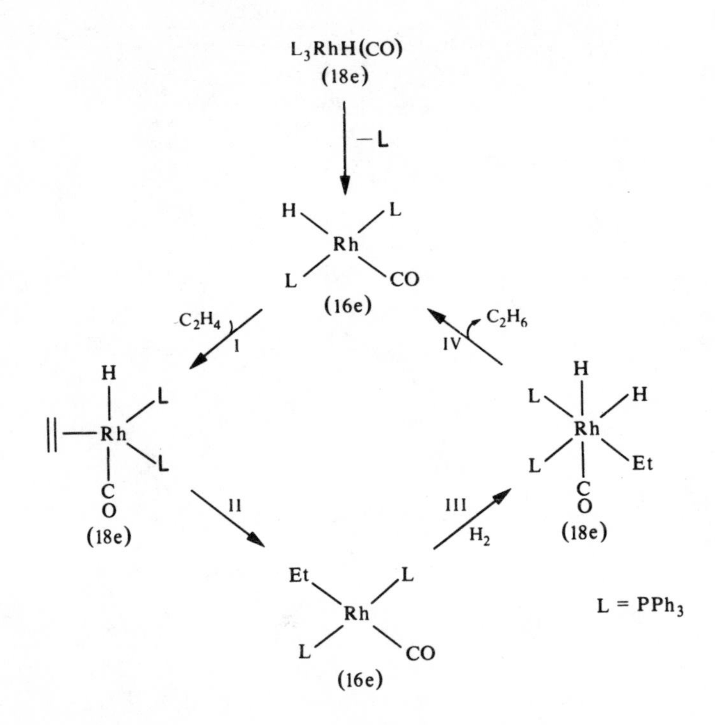

The same steps as before are involved namely coordination (I), migration (II), pericyclic addition (III) and pericyclic elimination (IV). Silanes may be used in place of H_2 with many hydrogenation catalysts.

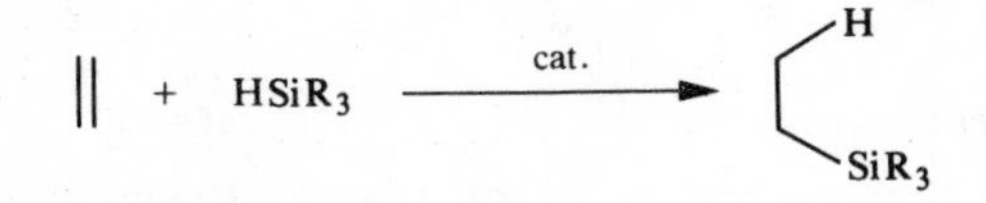

1.5.2 Hydroformylations (see also chap. 9)

Hydroformylation involves the same types of mechanism as the hydrogenation reactions described above with an additional step being involved, namely, the equilibrium between R[M]CO and [M]COR·(metal acyl). Catalytic hydroformylation is illustrated below for the catalyst precursor $HCo(CO)_4$.

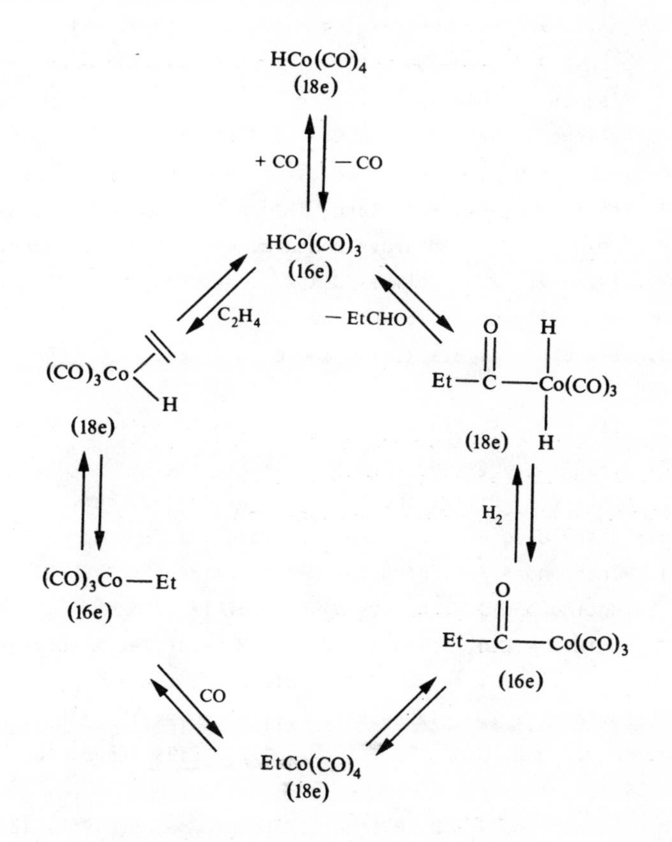

The similarities in mechanism between hydrogenation and hydroformylation reactions results in many catalysts being capable of performing both reactions. For example, $(Ph_3P)_3RhH(CO)$ acts as a hydrogenation catalyst in the presence of hydrogen and as a hydroformylation catalyst in the presence of a mixture of hydrogen and carbon monoxide.

1.6 REFERENCES

1. G.E. Coates, M.L.H. Green and K. Wade, Organometallic Compounds, Vol. II, Methuen and C^{o}, 1968.
2. F.A. Cotton, *J.Amer.Chem.Soc.*, 1968, *90*, 6230.
3. J. Chatt and L.A. Duncanson, *J.Chem.Soc.*, 1953, 2939.
4. M.J.S. Dewar, *Bull.Soc.chim.France*, 1951, C71.
5. B. Nicholls and M.C. Whiting, *J.Chem.Soc.*, 1959, 551.
6. G. Jaouen and R. Dabard, *J.Organometal.Chem.*, 1973, *61*, C36.
7. E.O. Fischer, K. Öfele, M. Essler, W. Fröhlich, J.P. Mortensen and W. Semmlinger, *Chem.Ber.*, 1958, *91*, 2763.
8. A. Wu, E.R. Biehl and P.C. Reeves, *J.Chem.Soc.Perkin II*, 1972, 449.
9. W.S. Trahanovsky and R.J. Card, *J.Amer.Chem.Soc.*, 1972, *94*, 2897.
10. S.P. Gupin, V.S. Khandkarova and A.Z. Kreindlin, *J.Organometal.Chem.*, 1974, *64*, 229; J.D. Holmes, D.A.K. Jones and R. Pettit, *J.Organometal.Chem.*, 1965, *4*, 324.
11. R.S. Bly, E.K. Ni, A.K.K. Tse and E. Wallace, *J.Org.Chem.*, 1980, *45*, 1362.
12. R. Pettit and G.F. Emerson, *Adv. in Organometallic Chem.*, 1964, *1*, 13.
13. G.A. Taylor, *J.Chem.Soc.Perkin I*, 1979, 1716.
14. D.F. Hunt, C.P. Lillya and M.D. Rausch, *J.Amer.Chem.Soc.*, 1968, *90*, 2561; (see also R.S. Bly and T.L. Maier, *J.Org.Chem.*, 1980, *45*, 980).
15. A.J. Birch and B.M.R. Bandara, *Tet. Letters*, 1980, 2981.
16. D. Valentine and J.W. Scott, *Synthesis*, 1978, 329.
17. S.G. Davies, M.L.H. Green and D.M.P. Mingos, Tetrahedron Report No.57. *Tetrahedron*, 1978, *34*, 3047 and references therein.
18. A.A. Koridze, I.T. Chizhevskii, P.V. Petrovskii and N.E. Kolobova, *Izv. Akad. Nauk. SSSR. Ser. Khim.*, 1979, 2395 (Chem. Abs., 1980, *92*, 129059e).
19. W.R. Jackson and T.R.B. Mitchell, *J.Chem.Soc. B*, 1969, 1228.
20. G. Jaouen and A. Meyer, *Tet. Letters*, 1976, 3547.
21. D.H.R. Barton and H. Patin, *J.Chem.Soc.Perkin I*, 1976, 829.

CHAPTER 2

COMPLEXATION AND DECOMPLEXATION REACTIONS

The use of organotransition metals for organic synthesis involves three stages; (i) the interaction of the organic compound with a transition metal species to form an intermediate transition metal complex; (ii) a chemical reaction on the coordinated ligand; (iii) recovery of the organic compound by decomplexation. This chapter describes the general methods available for the preparation of stable organometallic complexes and the types of reactions that lead to decomplexation.

Most of the transition metal complexes that are used as starting materials in the syntheses described below are commercially available. In many cases, however, it is economically profitable to undertake the normally short and well established syntheses of the starting complexes.

$ZrCl_4$ and $TiCl_4$ are readily available and reaction with NaCp leads to the formation of Cp_2ZrCl_2, Cp_2TiCl_2 or $CpTiCl_3$. Useful vanadium compounds include VCl_3, VCl_4, $CpV(CO)_4$ and $V(CO)_6$. Cr, Mo and W Halides are a convenient source of their complexes, as are the hexacarbonyls $M(CO)_6$ from which the dimers $[CpM(CO)_3]_2$ (M = Mo, W) are obtainable.

The carbonyl complexes $Mn_2(CO)_{10}$, $CpMn(CO)_3$ and $(MeC_5H_4)Mn(CO)_3$ are useful sources of Mn complexes. It is noteworthy that $(MeC_5H_4)Mn(CO)_3$, an anti-knock agent, is presently 100 times less expensive that $CpMn(CO)_3$. The inexpensive $Fe(CO)_5^*$, $Fe_2(CO)_9^*$, $Fe_3(CO)_{12}^*$ and $[CpFe(CO)_2]_2$ together with $FeCl_2$ and $FeCl_3$ are used to prepare the many different types of Fe complexes described below. $RuCl_3$ is the only reasonable source of Ru complexes. The Co complexes available include $Co_2(CO)_8$ and $CpCo(CO)_2$.

The dichlorides $NiCl_2$, $PdCl_2$ and $PtCl_2$ are the most used sources of these metals. Useful Ni(0) complexes include $Ni(CO)_4^*$ and $[(PhO)_3P]_4Ni$. $PdCl_2$ itself is not very useful for synthesis because it is insoluble in most noncoordinating solvents. $PdCl_2$ is usually first converted to the more soluble complexes $(RCN)_2PdCl_2$ with refluxing RCN (R = Me, Ph) or Li_2PdCl_4 by the addition of LiCl.

Transition metal cations are normally employed in conjunction with the large anions PF_6^- and BF_4^-. These anions are generally non-nucleophilic, non-coordinating and increase the solubility of the cations in organic solvents. ClO_4^- is rarely used as a general anion due to the explosive nature of its salts.

Many organometallic complexes and intermediates are sensitive to oxygen and therefore reactions are generally performed under an inert atmosphere (N_2 or Ar).

2.1 PREPARATION AND DECOMPLEXATION OF η^1-COMPLEXES

2.1.1 Nucleophilic metal + RX

A large number of anionic transition metal complexes are easily prepared

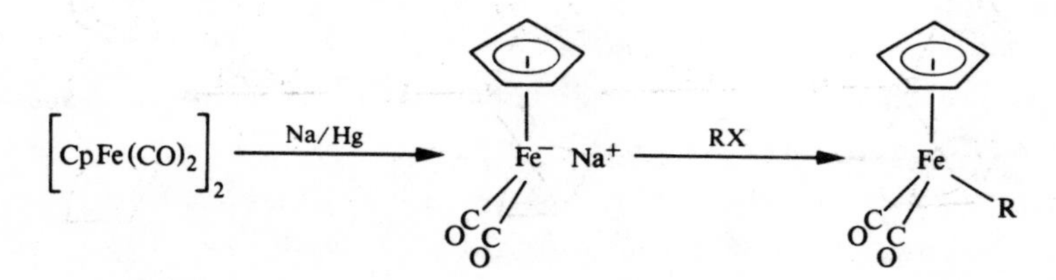

RX = Alkyl X, acyl X, allyl X, epoxides, etc.

---

*CAUTION: volatile metal carbonyls such as $Ni(CO)_4$ and $Fe(CO)_5$, and reagents that produce them e.g. $Fe_2(CO)_9$ and $Fe_3(CO)_{12}$ are extremely toxic and should be used with care.

by reduction of the corresponding metal halides or bimetallic complexes or by treatment of a transition metal hydride with base. These anions react with a variety of alkyl halides to give the corresponding neutral transition metal alkyl complexes.[1-4]

Examples of anionic transition metal complexes include:

$CpMo(CO)_3^-$ Cp_2Re^- $CpFe(CO)_2^-$ $Co(CO)_4^-$ $CpNi(CO)^-$

$CpW(CO)_3^-$ $(CO)_5Mn^-$ $CpRu(CO)_2^-$

Cp_2MoH^- $(CO)_5Re^-$ $CpFe(diphos)^-$

Cp_2WH^- $Fe(CO)_4^{2-}$

$Fe(CO)_4H^-$

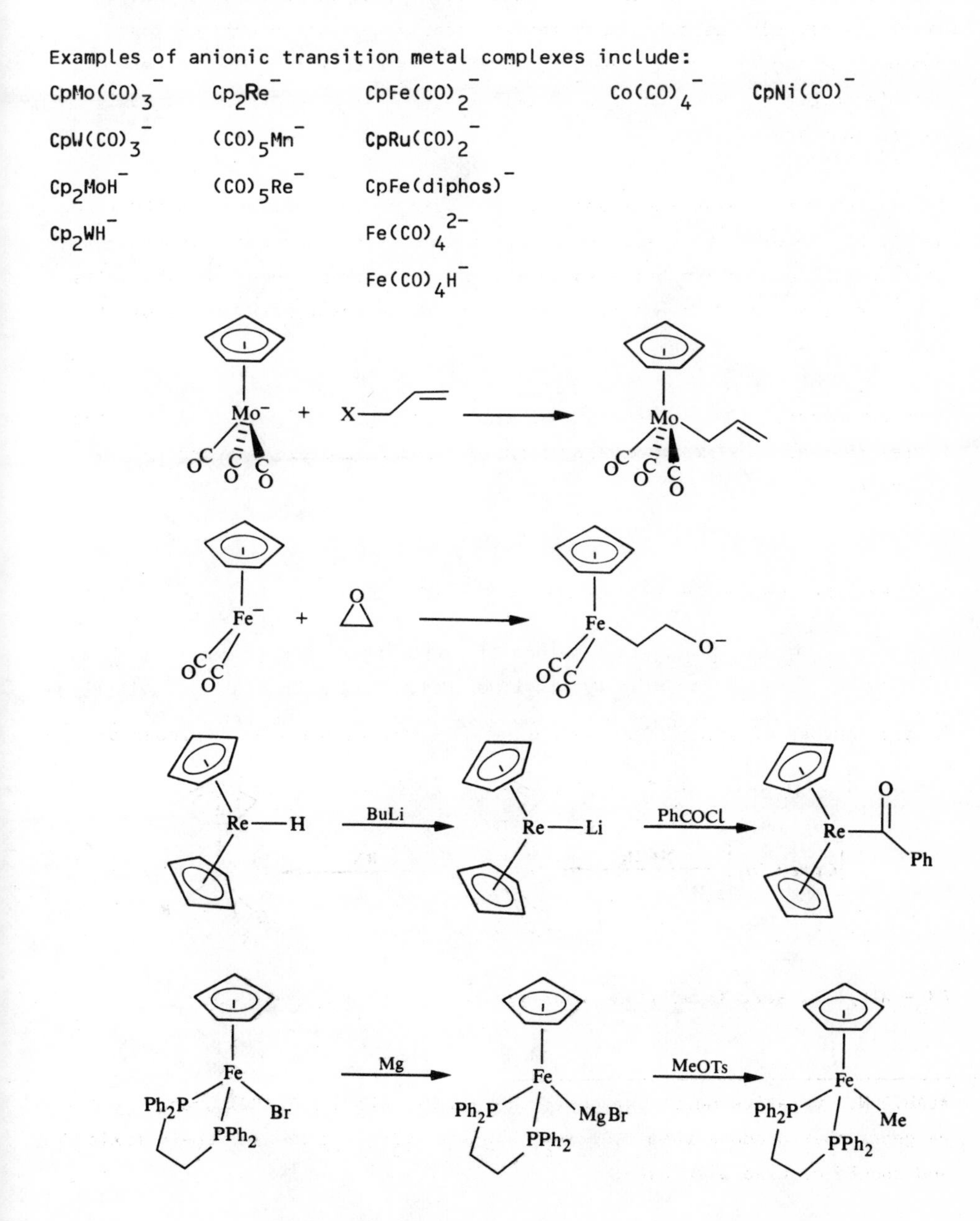

The anion $CpFe(CO)_2^-$ is one of the most studied with regard to organic synthesis. The reaction of $CpFe(CO)_2^-$ with alkyl halides has been shown to proceed by the expected S_N2 attack with inversion of configuration at carbon.[5]

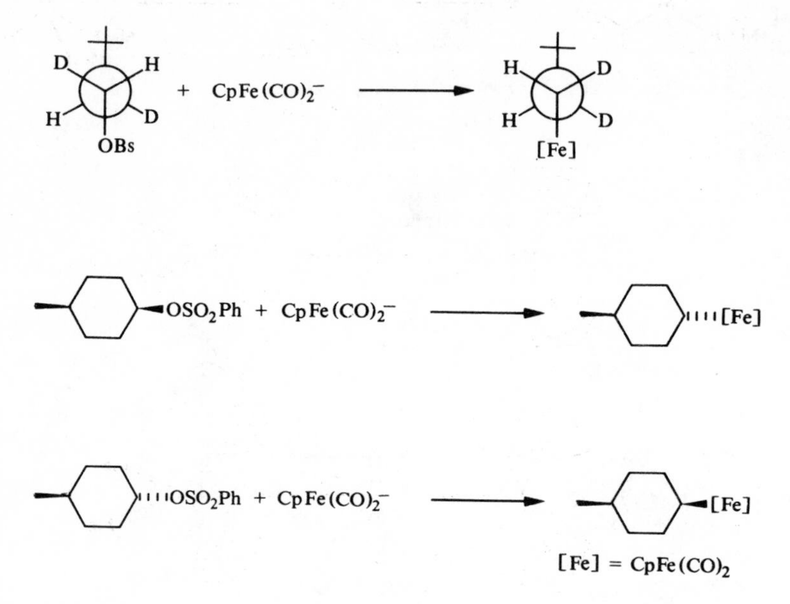

The main limitation for this method of preparing η^1-alkyl complexes is that it is only useful for primary alkyl halides. More substituted alkyl halides tend to undergo elimination to give metal hydride and olefin.

Many neutral organotransition metal complexes are also nucleophilic and react with unhindered alkyl halides to give metal alkyl cations.[3,6-8]

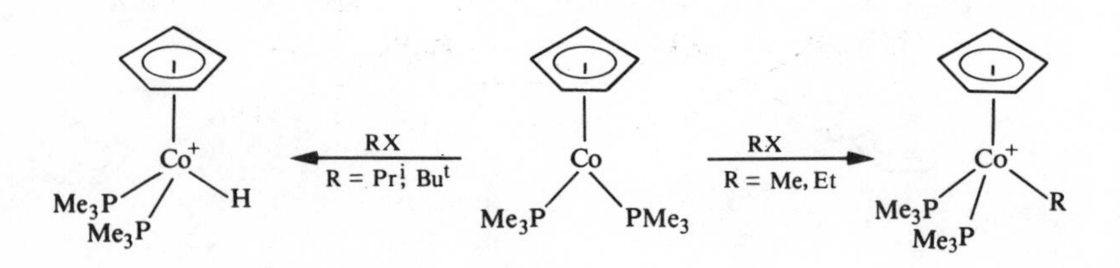

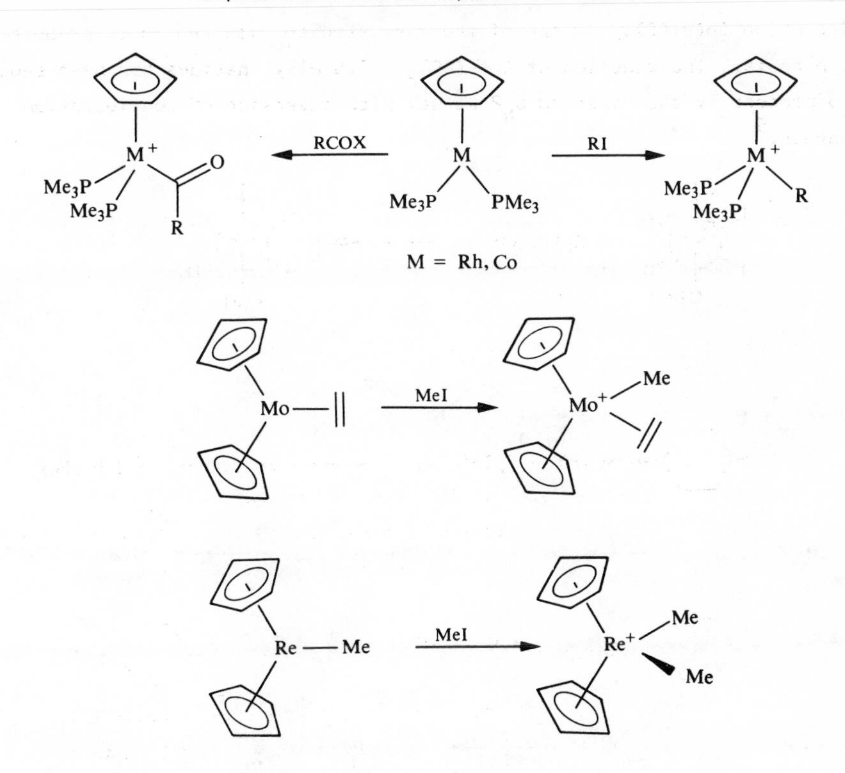

A related reaction is the addition of alkyl halides to transition metal complexes.[9-13]

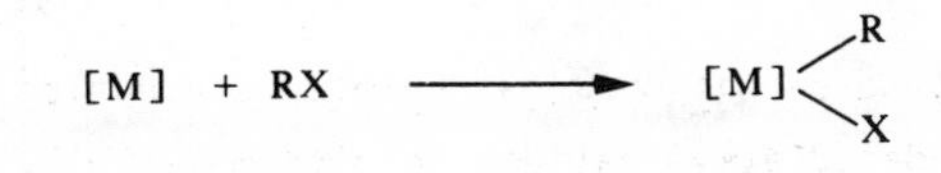

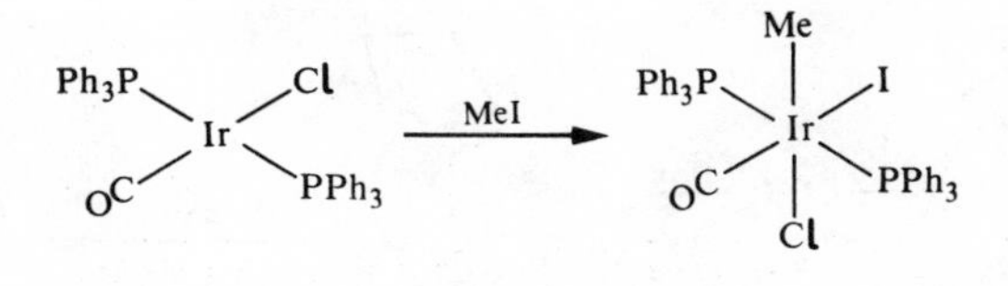

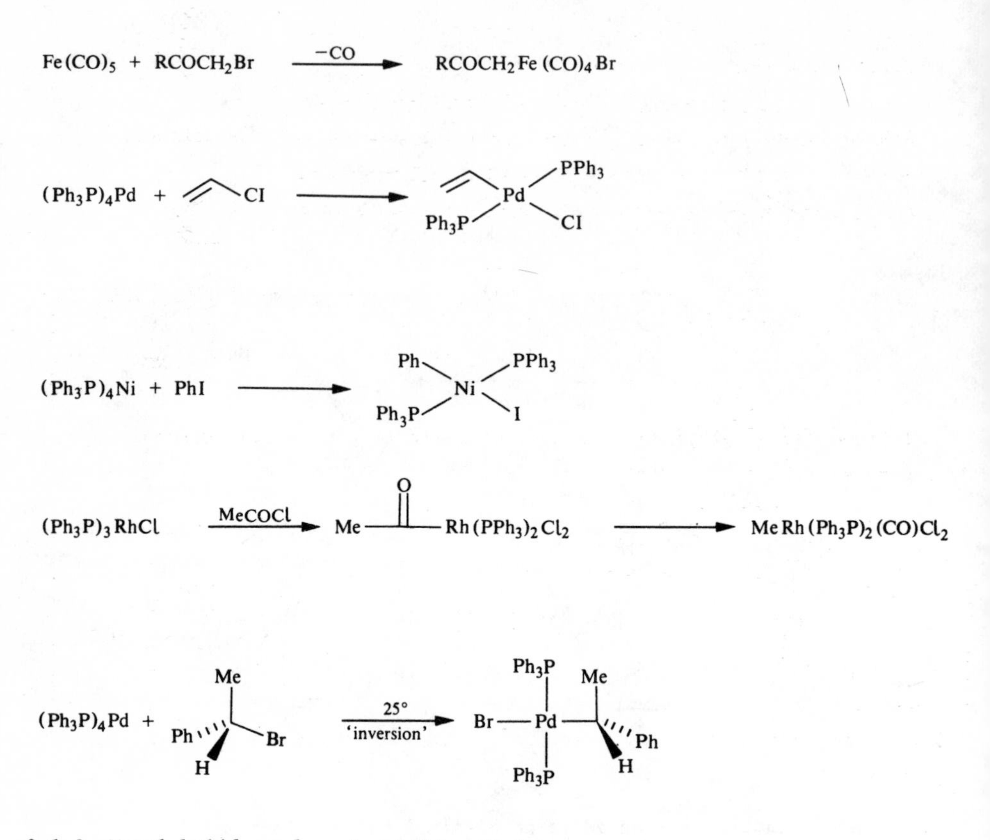

2.1.2 *Metal halide and nucleophilic alkyl*

Treatment of metal halides with primary alkyl derivatives of lithium, magnesium, zinc, aluminium and mercury generally results in metal alkyl derivatives being formed.[14-16]

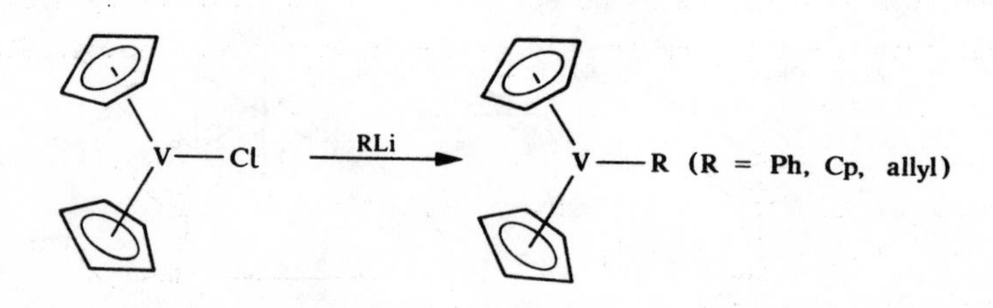

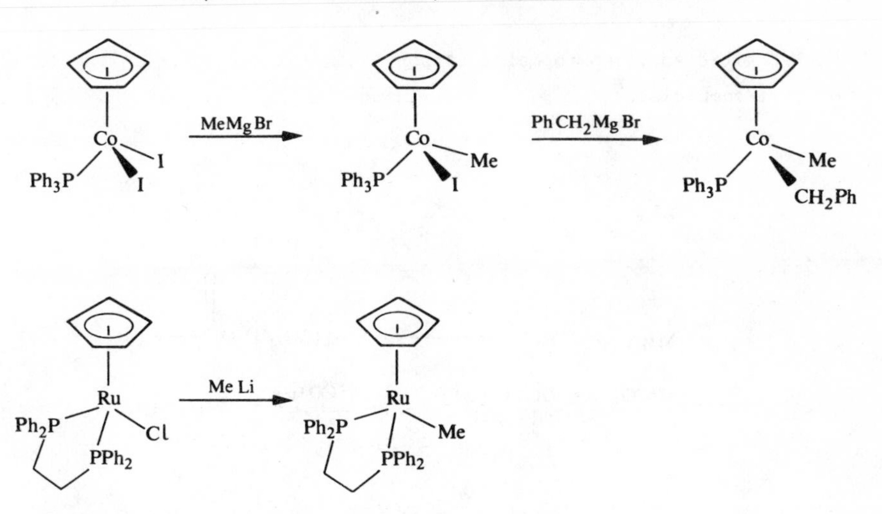

This method is limited to small primary alkyls. Secondary Grignard reagents, for example, give only metal hydrides.[17]

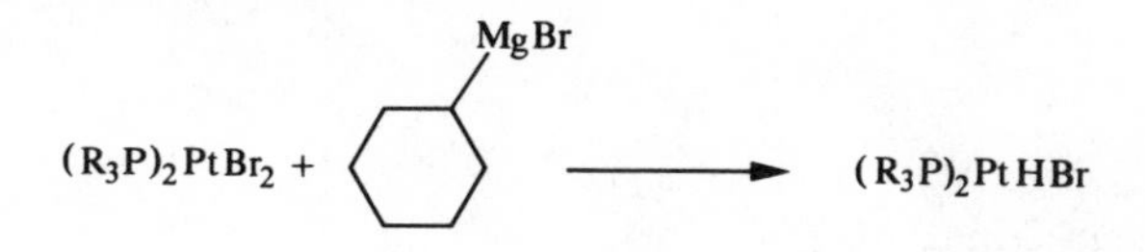

2.1.3 Nucleophilic addition to transition metal complexes

Nucleophilic addition to cationic transition metal olefin complexes is one of the most studied reactions in organometallic chemistry and is generalised below.[18]

$$[M]^{+}\text{—}\| \xrightarrow{Y^{-}} [M]\text{—CH}_2\text{CH}_2\text{—Y}$$

Many different types of cations containing a variety of olefins have been prepared and reacted with nucleophiles (e.g. R^-, H^-, CN^-, MeO^-, etc.). This reaction is dealt with much more fully in chapter 4.

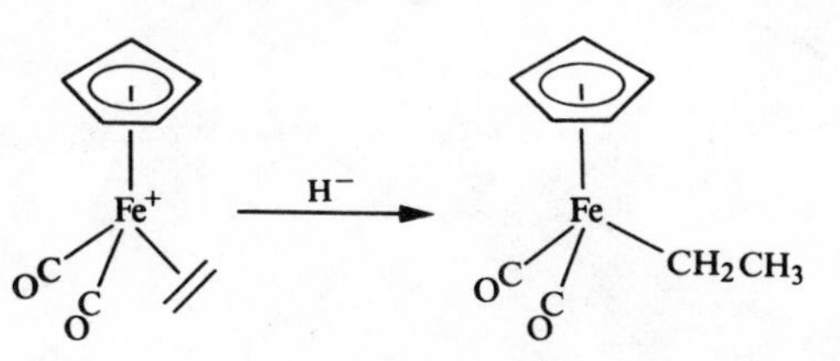

Nucleophilic addition to metal carbonyl complexes results in the formation of acyl complexes.[19]

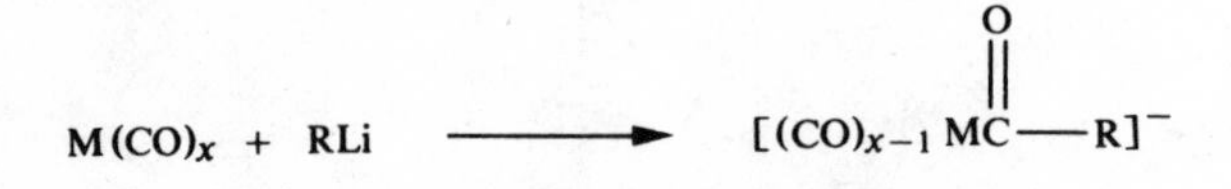

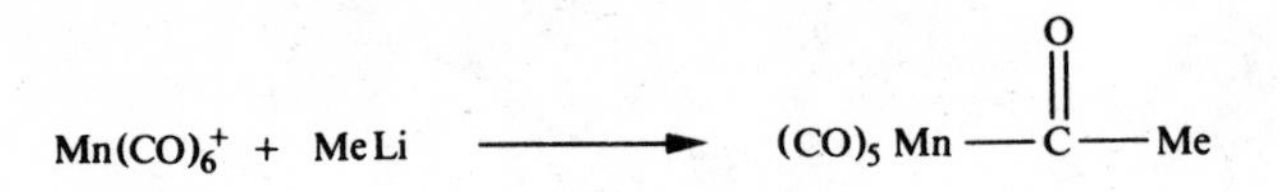

2.1.4 *Metal hydride and olefins*

Reaction of certain metal hydride complexes with olefins leads to η^1-alkyl complexes.[20,21]

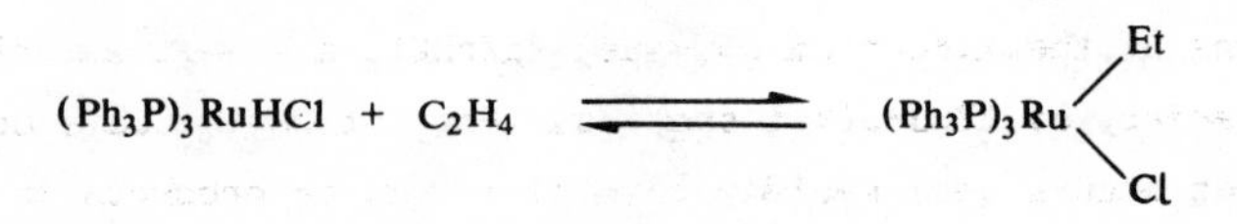

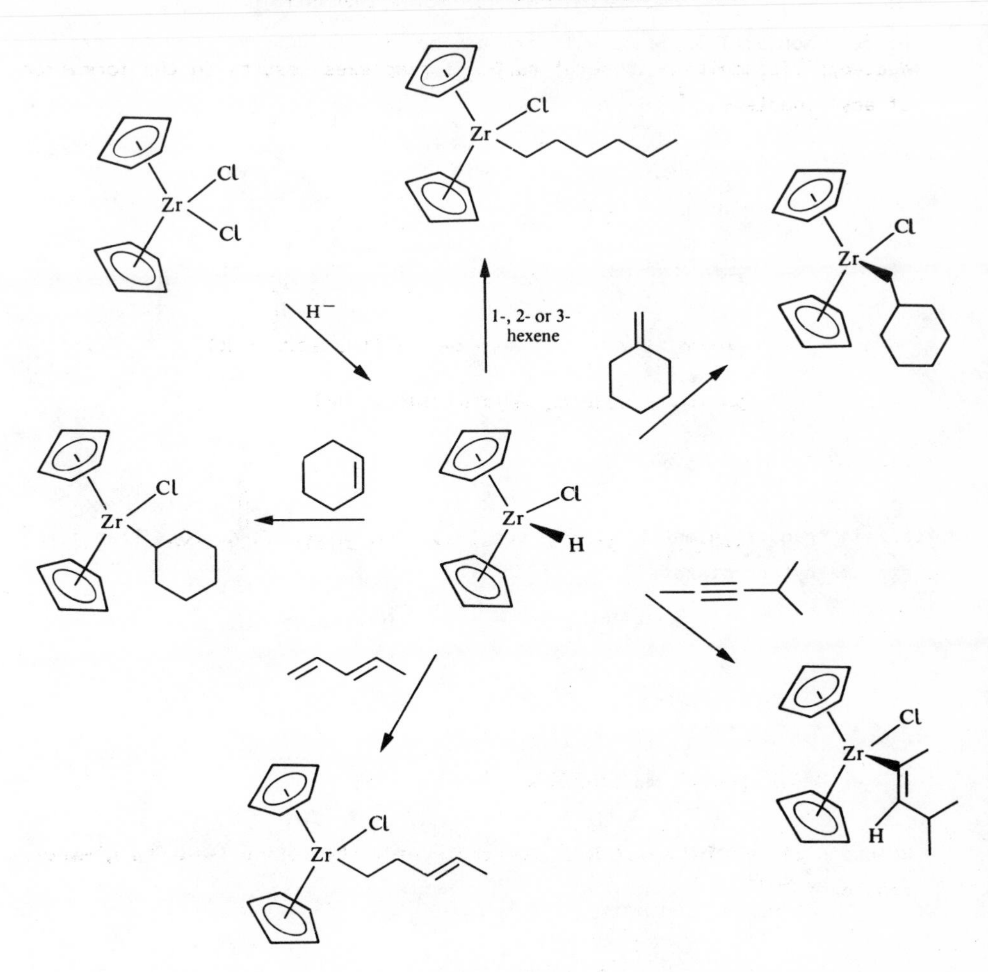

The reactions of the zirconium hydride, Cp_2ZrHCl, are particularly useful and lead to a variety of [Zr]-alkyl species. For internal double bonds, rearrangement occurs very rapidly resulting only in products in which the bulky Zr species is in a terminal position. The reactions of Cp_2ZrHCl are very much akin to the reactions of dialkylboranes.

$$R_2BH + \text{CH}_2{=}\text{CH}{-}R \longrightarrow R_2B{-}\text{CH}_2\text{CH}_2{-}R$$

The reaction of 1,3-dienes with certain metal hydrides can lead to the formation of η^1-allyl complexes.[22]

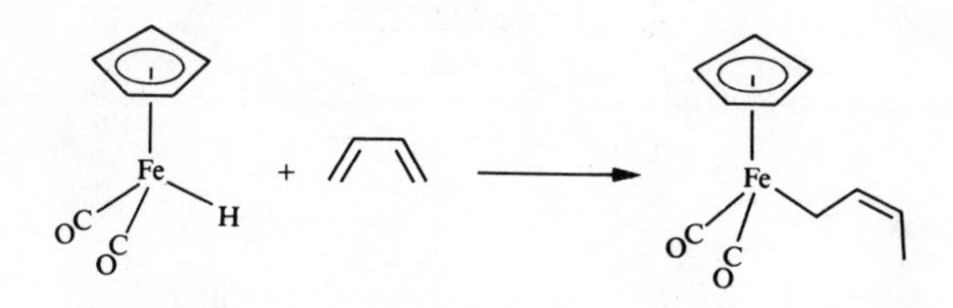

2.1.5 Metal hydride and diazomethane

. Certain transition metal hydride complexes and diazomethane form transition metal methyl complexes.[23]

$$(CO)_5MnH \xrightarrow{CH_2N_2} (CO)_5Mn—CH_3$$

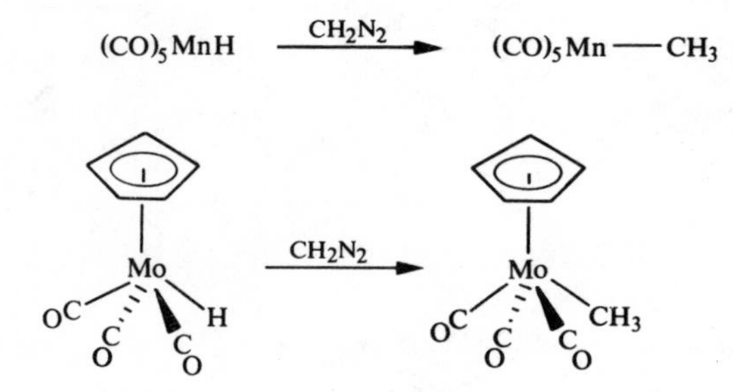

2.1.6 Interconversion of M-alkyl and M-acyl

Treatment of metal alkyl complexes with 2-electron ligands L (e.g. CO,Ph_3P) can cause the migration of an alkyl group from the metal to a carbonyl ligand. Similarly, removal of a 2-electron ligand can lead to the reverse reaction.[13,24-26]

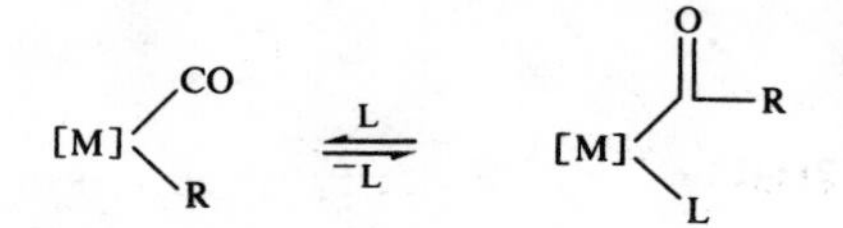

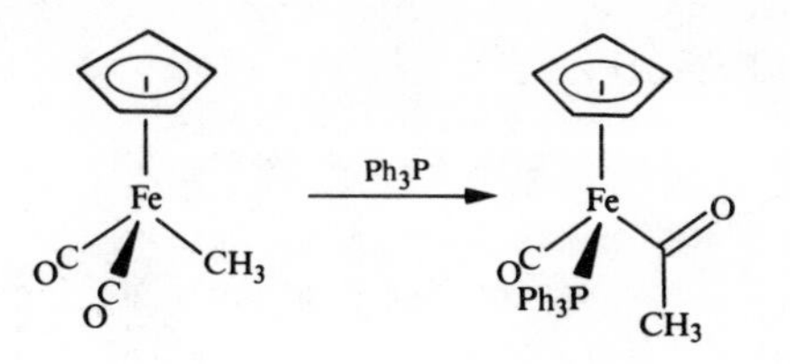
Fe
OC
CH3
Ph3P
Fe
O
OC
Ph3P
CH3

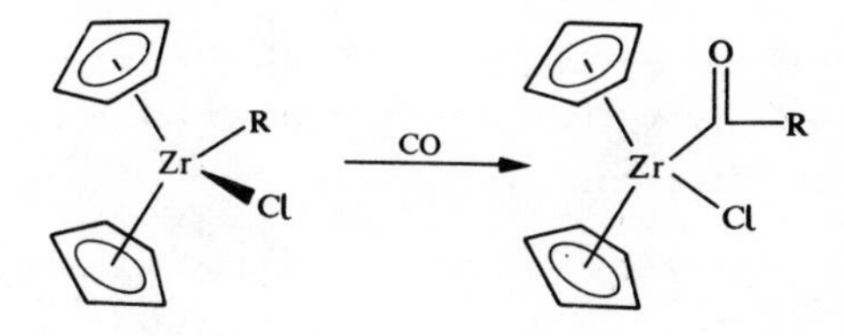
R
Zr
Cl
CO
O
R
Zr
Cl

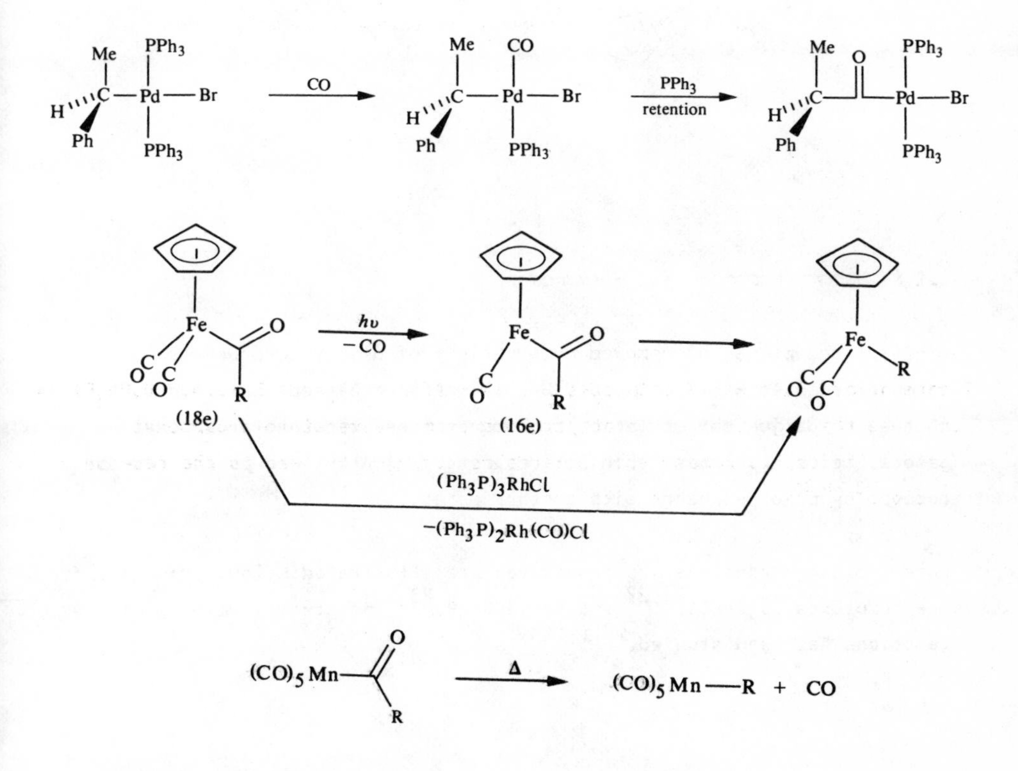
Me
PPh3
H
C
Pd
Br
Ph
PPh3
CO
Me
CO
H
C
Pd
Br
Ph
PPh3
PPh3
retention
Me
O
PPh3
H
C
Pd
Br
Ph
PPh3
Fe
O
OC
R
(18e)
hυ
− CO
Fe
O
R
(16e)
Fe
R
(Ph3P)3RhCl
−(Ph3P)2Rh(CO)Cl
O
(CO)5 Mn
R
Δ
(CO)5 Mn —R + CO

2.1.7 *Deprotonation of cationic olefin complexes*

The removal of an allylic proton from cationic olefin complexes can lead to the formation of η^1-allyl complexes.[26a] Deprotonation of $CpFe(CO)_2^+$(olefin) cations is stereospecific with the requirement that the C-H bond is *trans* to the Fe-olefin bond. For example it is possible to deprotonate the $CpFe(CO)_2^+$ cyclopentene and cyclohexene cations which satisfy this requirement but not the cycloheptene cation which does not contain a proton in the required orientation.

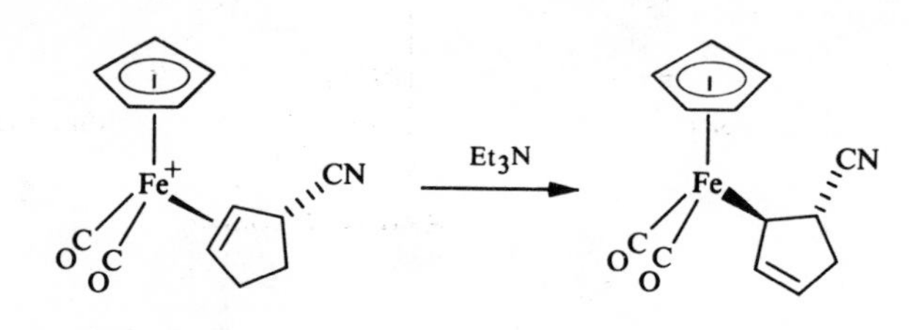

2.1.8 *Decomplexation of η^1-ligands*

η^1-Alkyl groups may be removed by a variety of oxidative procedures, to form derivatives RX (X = OH, Cl, Br, I). Acid treatment may produce alkanes whereas oxidative carbonylation procedures (see also Chapter 9) lead to esters, acids, lactones, acid halides, etc. η^1-Alkyl groups may also be removed by prior exchange with another metal.

Some typical decomplexation procedures are illustrated below, primarily for the complexes Cp_2ZrRCl[21,27] and $CpFe(CO)_2R$.[28] The stereochemistry of these reactions has been studied.[5,13,14,29]

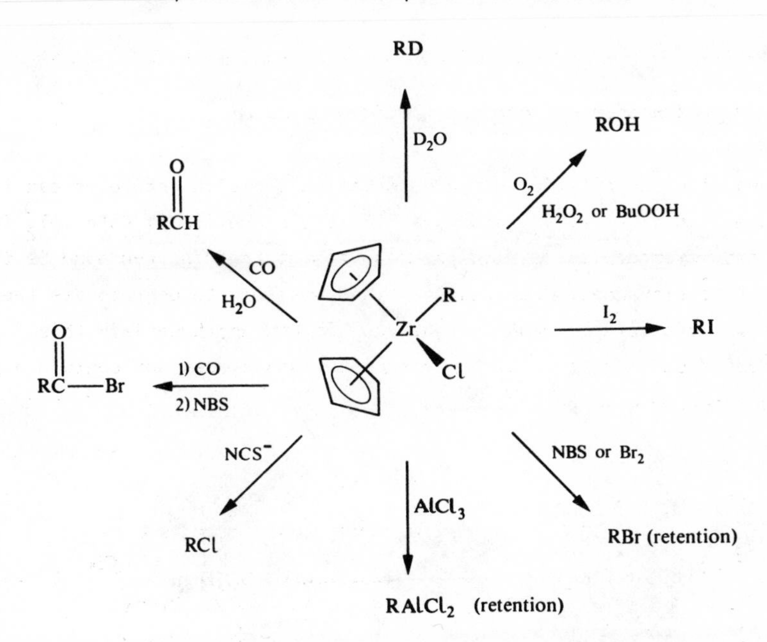
RD
D_2O
ROH
O_2
H_2O_2 or BuOOH
O
RCH
CO
H_2O
R
Zr
Cl
I_2
RI
O
RC—Br
1) CO
2) NBS
NCS⁻
NBS or Br_2
$AlCl_3$
RCl
RBr (retention)
$RAlCl_2$ (retention)

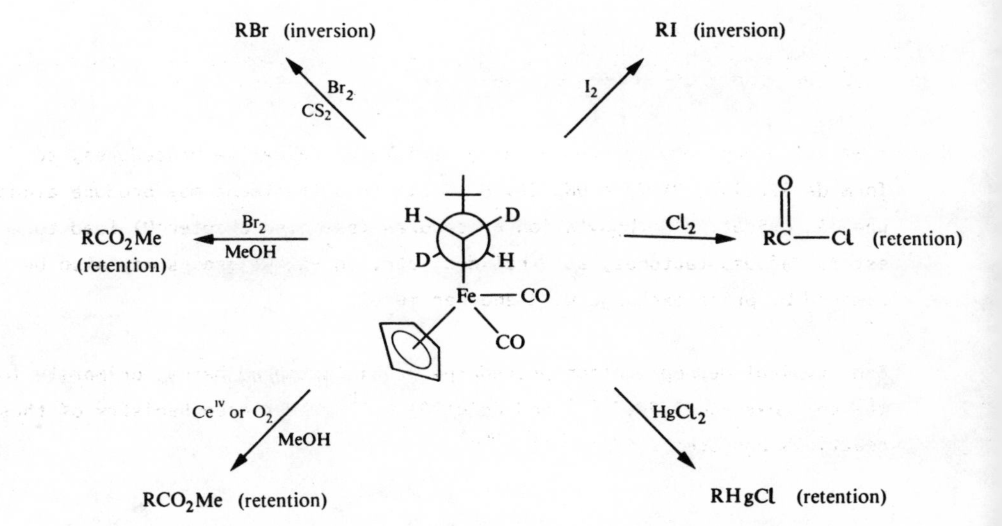
RBr (inversion)
Br_2
CS_2
RI (inversion)
I_2
H
D
D
H
Fe—CO
CO
Br_2
MeOH
RCO_2Me
(retention)
Cl_2
O
RC—Cl (retention)
Ce^{IV} or O_2
MeOH
$HgCl_2$
RCO_2Me (retention)
RHgCl (retention)

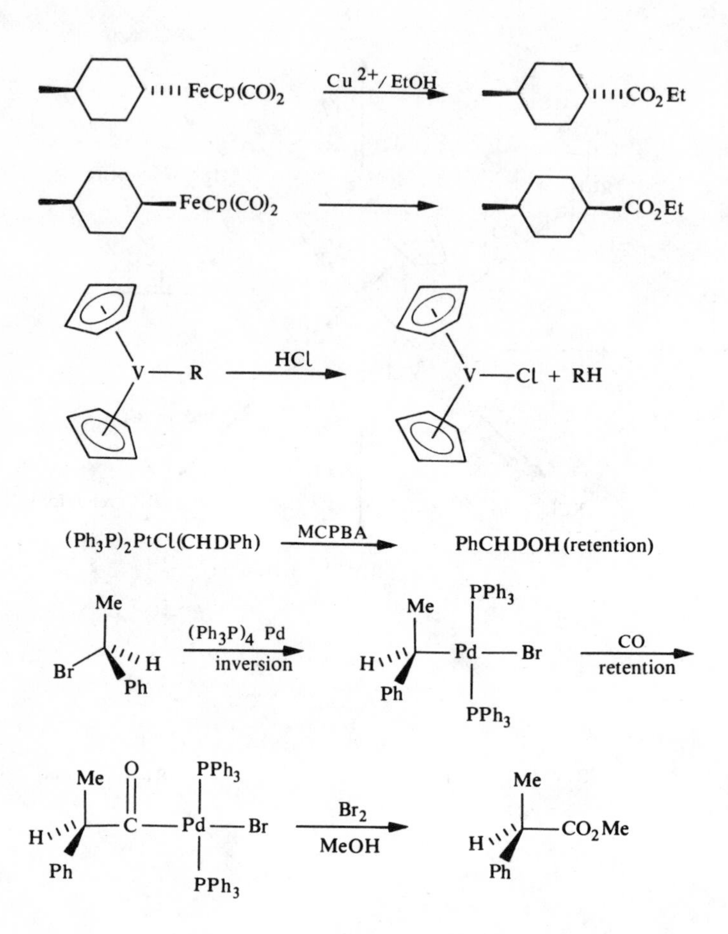

2.2 PREPARATION AND DECOMPLEXATION OF η^2-OLEFIN AND ACETYLENE COMPLEXES

2.2.1 *Ligand exchange reactions*

The reaction of K_2PtCl_4 with ethylene gave the first reported η^2-olefin complex $(C_2H_4)PtCl_3^-$, Zeise's salt.[30]

The exchange of chloride for olefin has been used to make many transition metal olefin complexes.[31-35]

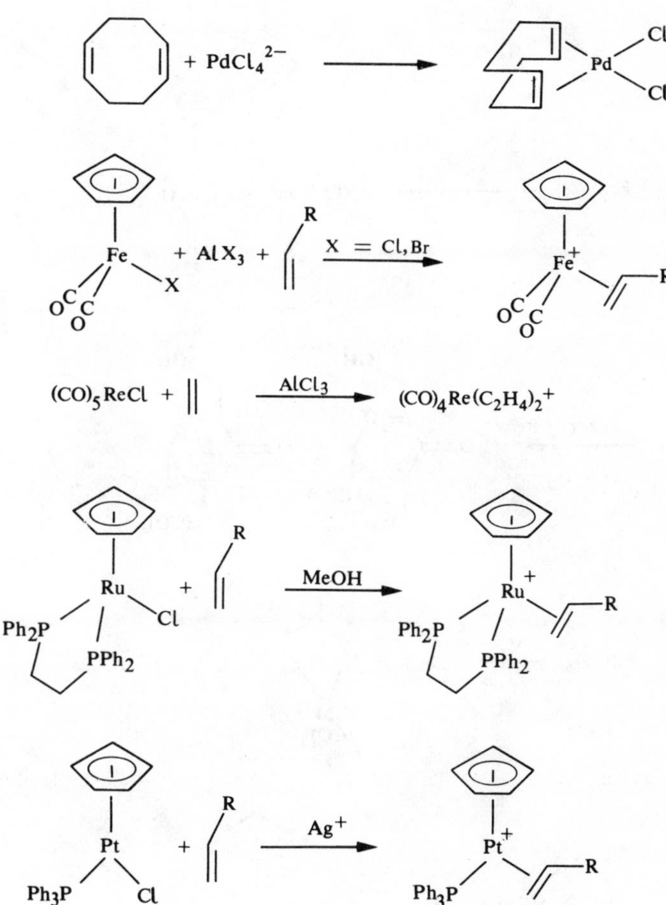

Transition metal olefin complexes may also be prepared by exchange of a CO ligand with an olefin. The ligand exchange reaction may occur either photochemically or thermally.[37-40]

$Fe(CO)_5$ + (olefin, X) $\xrightarrow{h\nu}$ (olefin, X)–$Fe(CO)_4$ $\xleftarrow{\Delta}$ (olefin, X) + $Fe_2(CO)_9$ ($Fe(CO)_4$ + $Fe(CO)_5$)

(X = CO_2H, CO_2Et, CHO, CN,etc.[36])

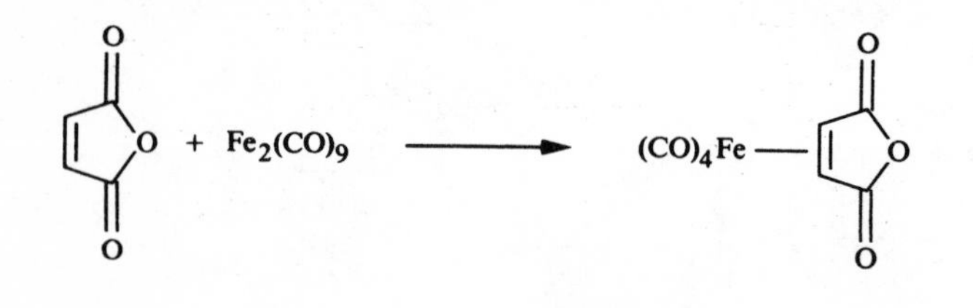

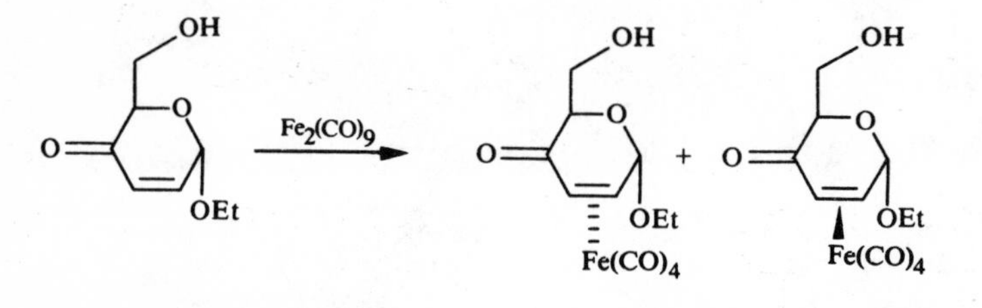

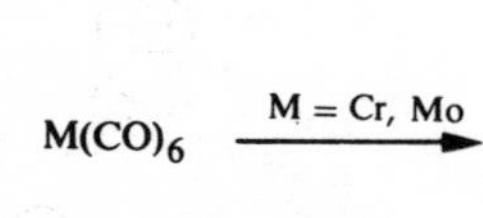

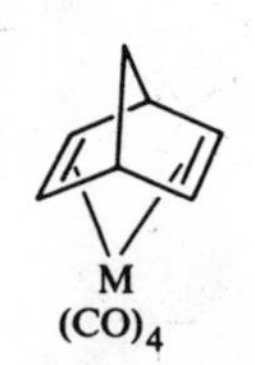

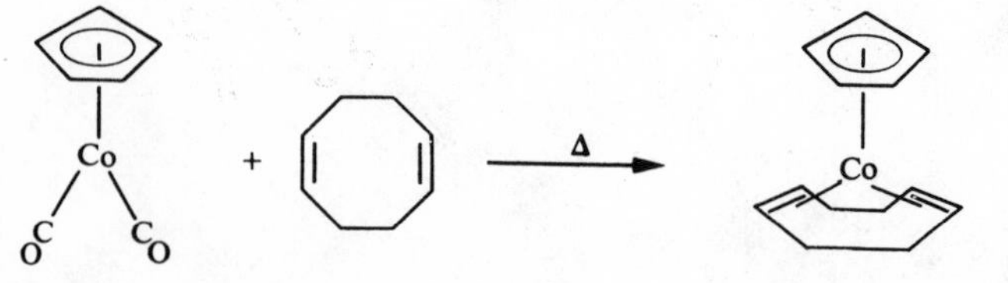

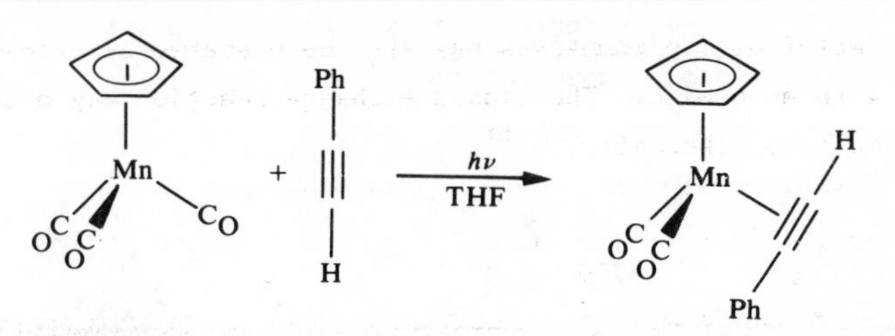

Hindered olefins such as isobutene may be thermally exchanged for mono- and 1,2-di-substituted olefins.[41] This is a particularly good method for preparing monosubstituted olefin complexes.[41-43]

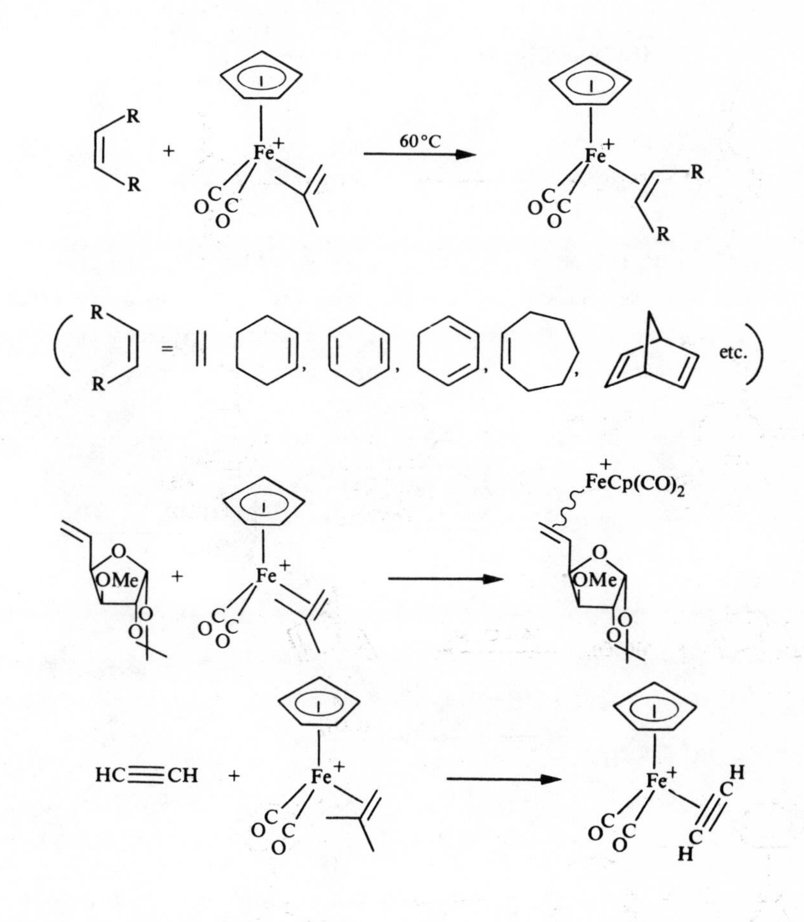

The resolution of *trans*-cyclooctene was accomplished using the Pt complex 1 prepared by an olefin exchange reaction.[44]

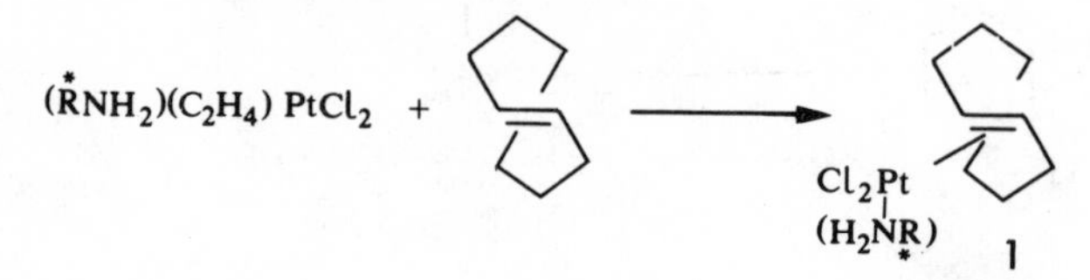

Olefinic palladium complexes may be made by exchange of PhCN in the complex $(PhCN)_2PdCl_2$.[45]

$$(PhCN)_2PdCl_2 + \text{RCH=CH}_2 \longrightarrow (\text{RCH=CH}_2)_2PdCl_2$$

2.2.2 *Deoxygenation of epoxides*

The reaction of the anion $CpFe(CO)_2^-$ with epoxides followed by treatment with acid provides an efficient method for the synthesis of $CpFe(CO)_2$ (olefin)$^+$ cations.[46] The stereochemistry of the complexed olefin is the same as the stereochemistry of the starting epoxide.

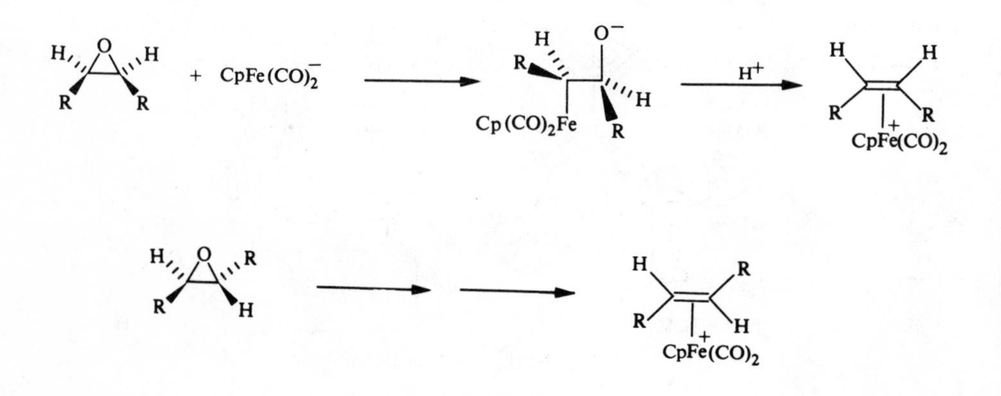

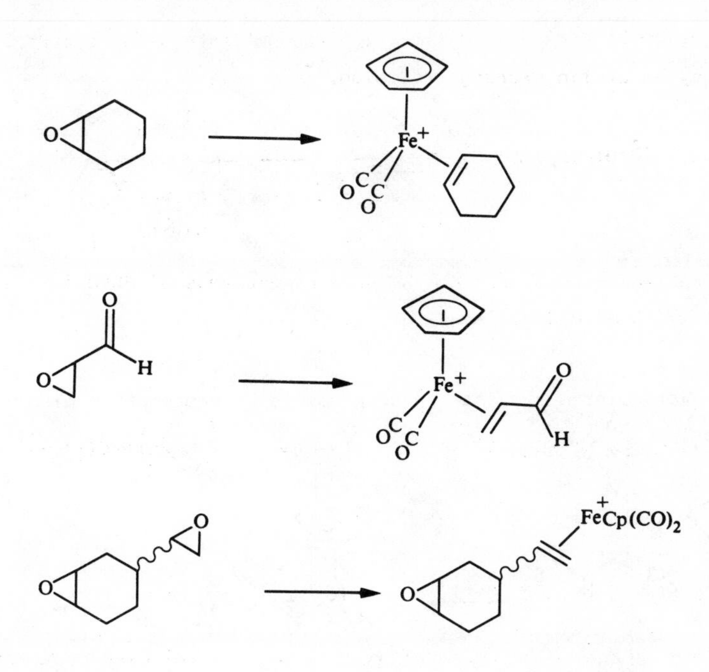

2.2.3 Other methods

Nucleophilic addition to the terminal carbon atom of η^3-allyl cations results in the formation of neutral η^2-olefin complexes (see section 4.1.3).[47,48]

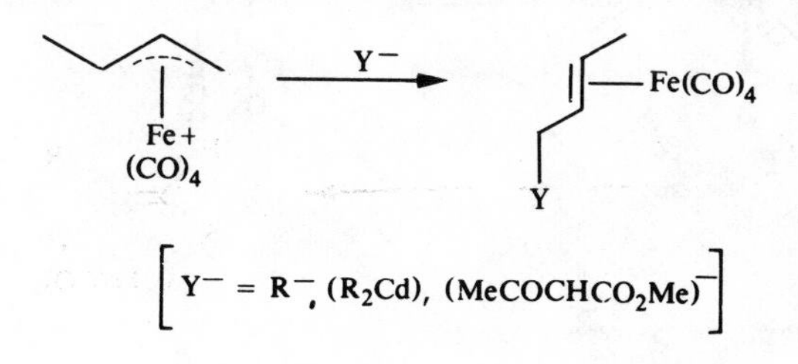

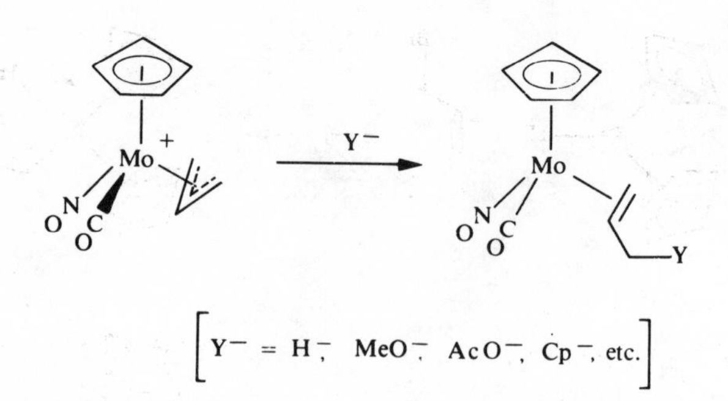

Alternatively olefin cations can be prepared by protonation of η^1-allyl complexes:[3,29,49-51]

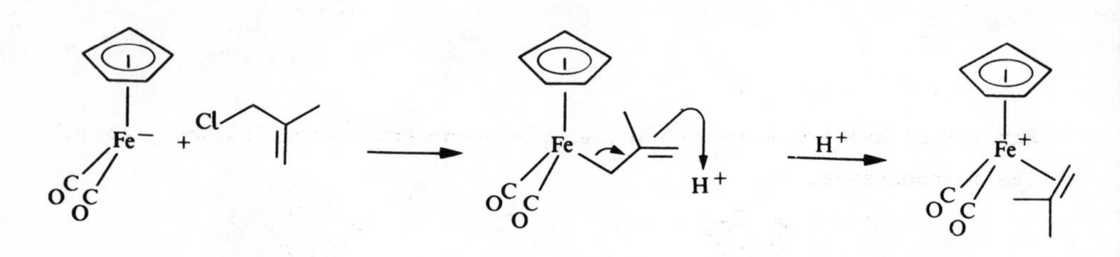

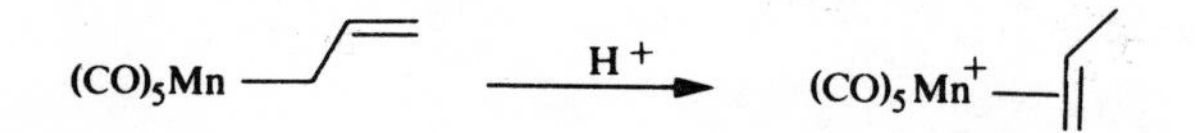

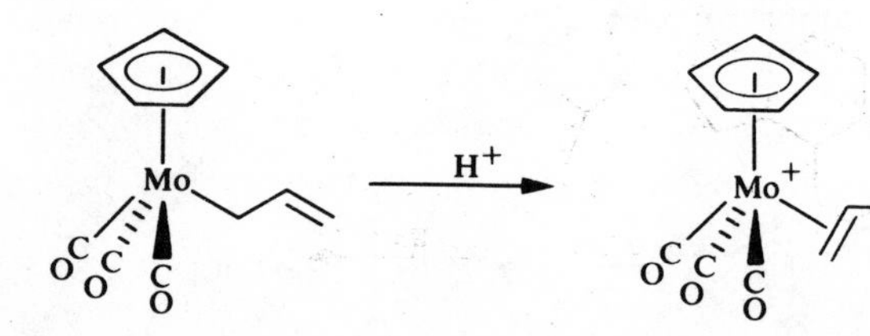

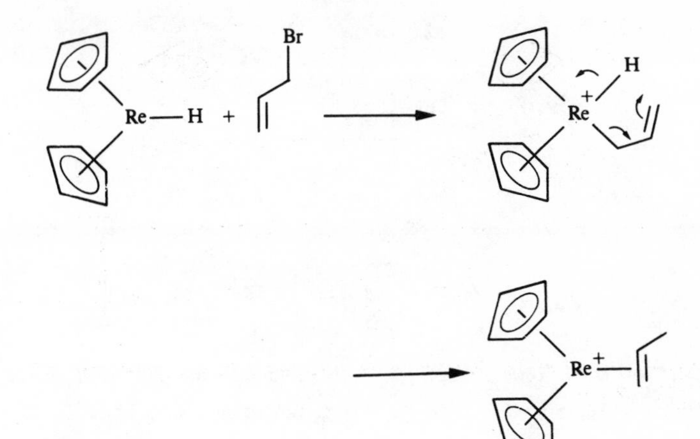

Removal of hydride from η^1-alkyl complexes by Ph_3C^+ can also give cationic olefin complexes.[52]

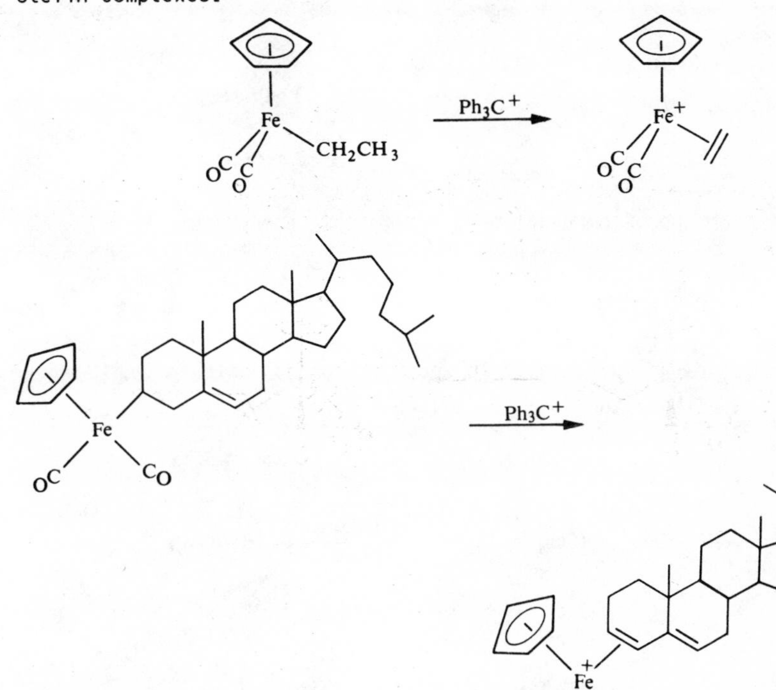

2.2.4 *Decomplexation of* η^2*-olefin ligands*

Olefins may be removed from $CpFe(CO)_2(olefin)^+$ cations either thermally or by iodide.[46]

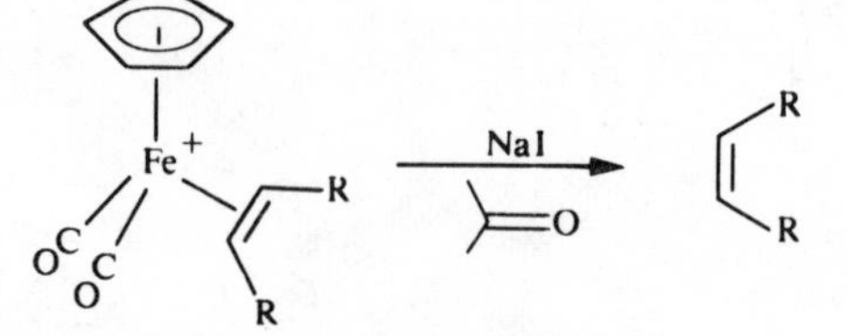

Decomplexation of $Fe(CO)_4(olefin)$ complexes may be achieved using a variety of 2-electron ligands (L) such as Ph_3P, pyridine, CO, etc.[53]

$$(CO)_4Fe(olefin) + L \longrightarrow (CO)_4FeL + olefin$$

Treatment of the complex (*trans*-cyclooctene) (amine)$PtCl_2$ with KCN releases the olefin.[44]

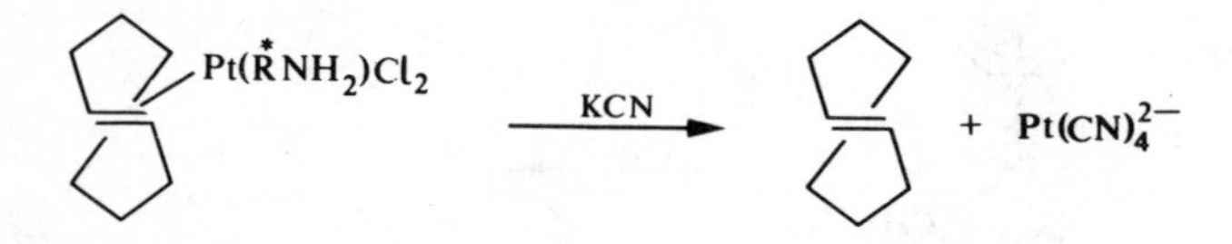

2.3. PREPARATION AND DECOMPLEXATION OF η^2-CARBENE COMPLEXES

2.3.1 *From CO and RNC ligands*

Nucleophilic attack on coordinated CO and RNC ligands followed by alkylation gives alkoxy and amino substituted carbene complexes respectively.[55]

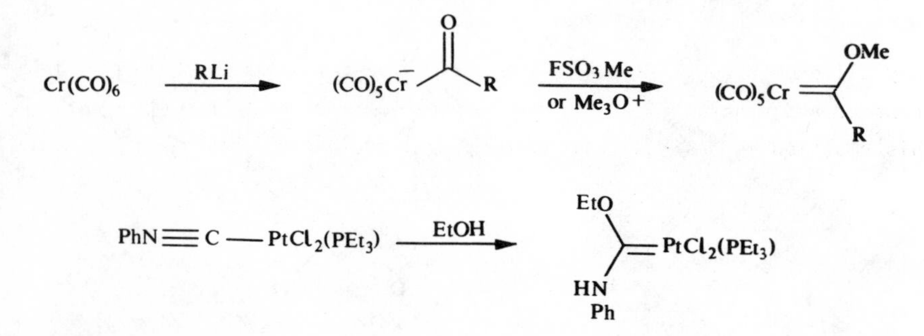

2.3.2 *From M-acyl ligands*

Alkylation of neutral metal acyl ligands produces carbene cations.[56]

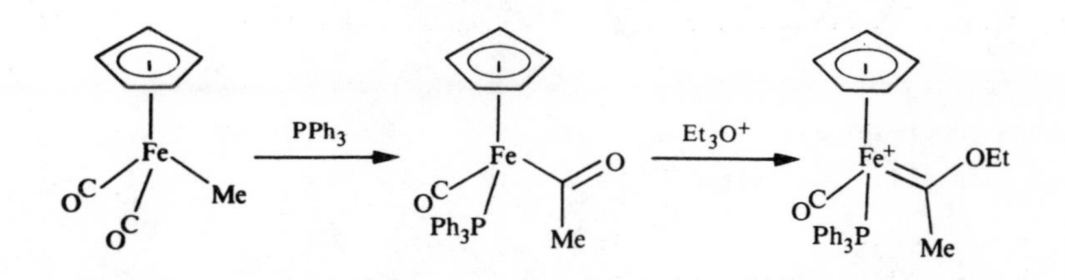

Intramolecular alkylation is also possible to give cyclic carbene complexes.[57]

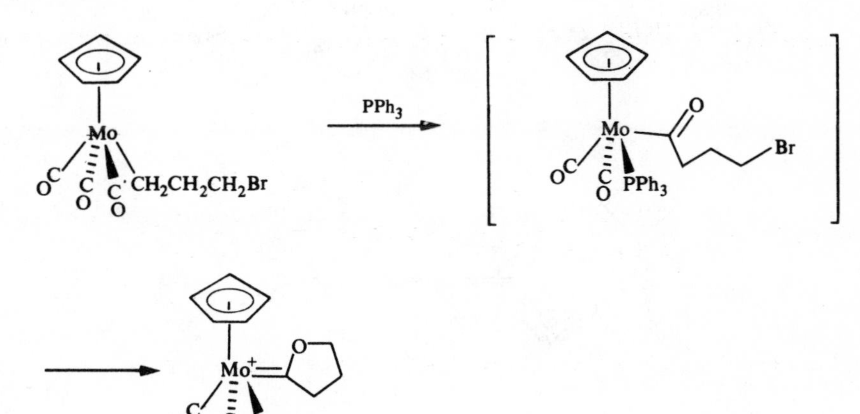

2.3.3 *Miscellaneous methods*

Carbene complexes may be prepared from diazo compounds or by hydride removal from η^1-alkyl complexes when β-elimination is unfavourable.[58,59]

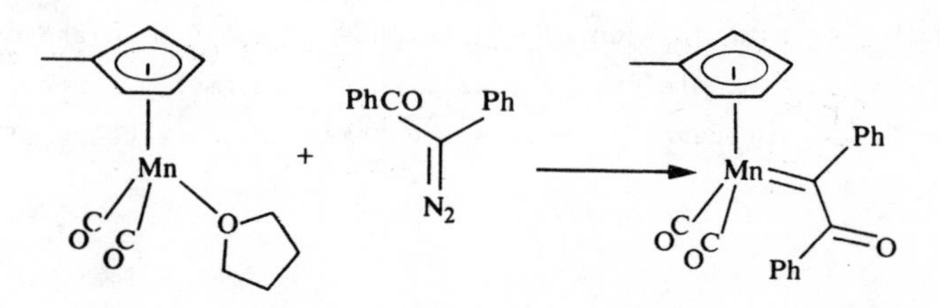

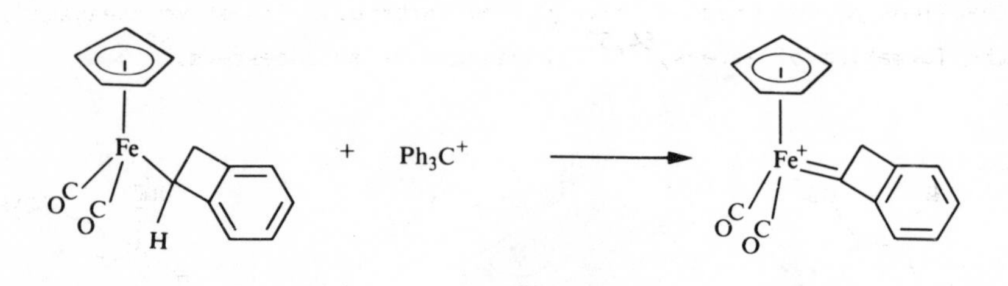

Carbene complexes may also be readily prepared from terminal acetylenes[34] and η^1-acetylide complexes.[60]

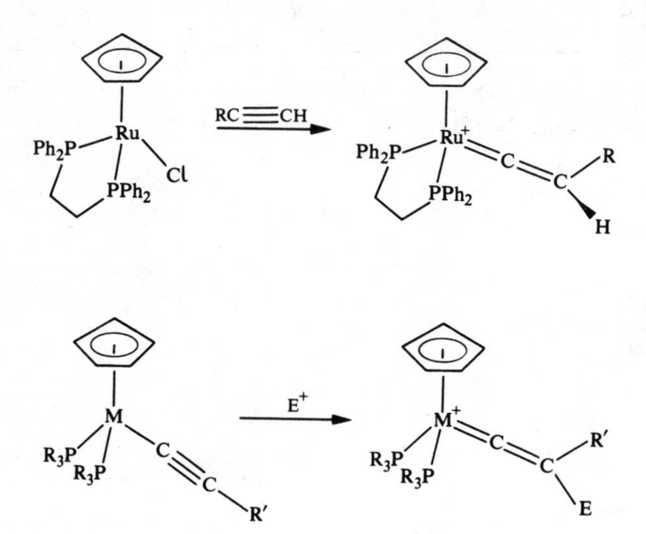

(M = Ru,Fe; E^+ = H^+, Me^+)

2.3.4 Decomplexation of carbene ligands

Alkoxycarbene ligands may be decomplexed thermally to give olefins corresponding to dimerization of the carbene ligand.[61] In the presence of electron deficient olefins or diazomethane, thermolysis results in the formation of cyclopropanes[62] and vinyl ethers[63] respectively. These reactions do not proceed through free carbenes. Oxidative cleavage allows the formation of esters,[64,57] thioesters or selenoesters.[64]

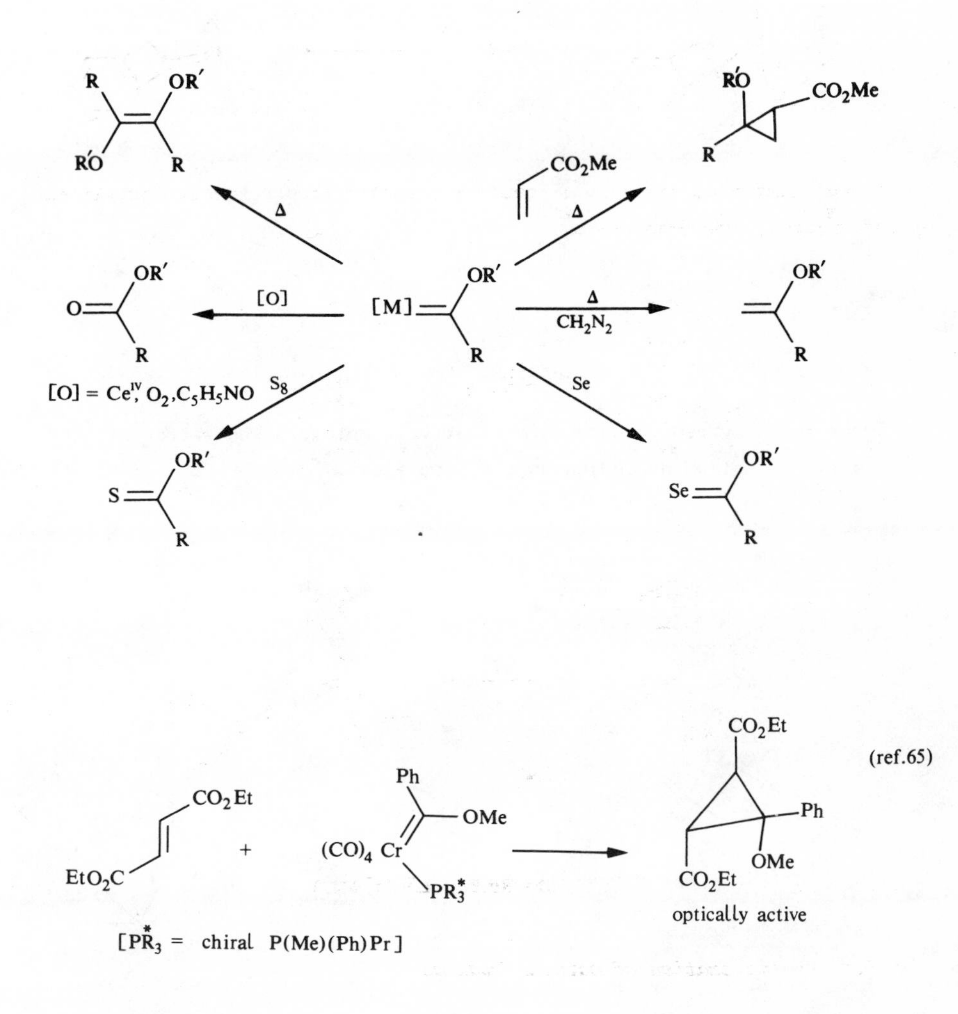

Carbene complexes containing α-hydrogens undergo pyridine catalysed decomposition to vinyl ethers.[65a]

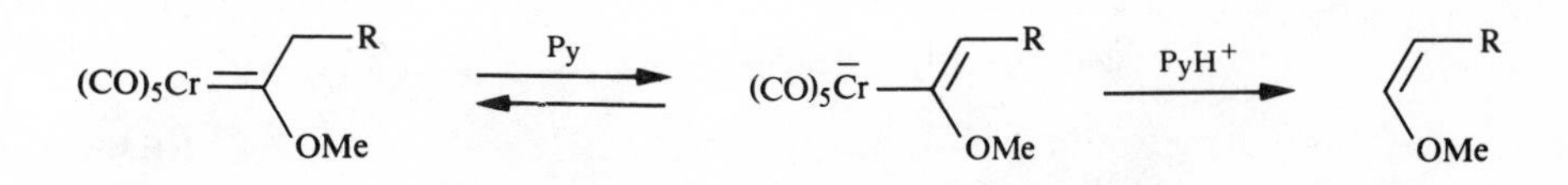

2.4 PREPARATION AND DECOMPLEXATION OF η^3-ALLYL COMPLEXES

2.4.1 From dienes

1,3-dienes react with $PdCl_2$ or $(PhCN)_2PdCl_2$ in the presence of nucleophiles to give η^3-allyl palladium complexes. Carbon nucleophiles may also be used in the form of RHgCl.[66]

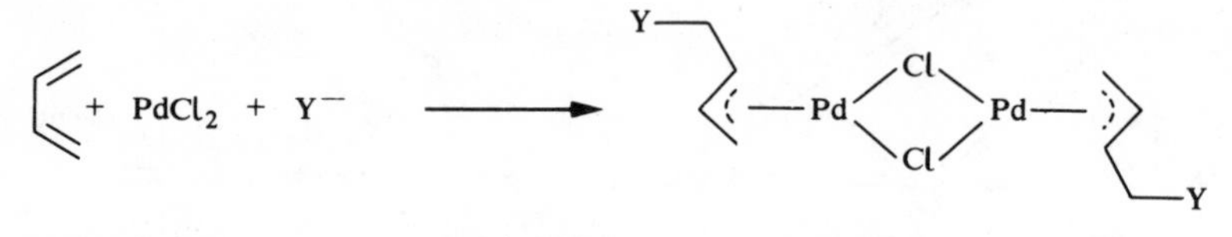

[$Y^- = Cl^-, RO^-, AcO^-, Ph^-$ (PhHgCl)]

Ocimene 2 and myrcene 3 both form η^3-allyl complexes, the latter with or without cyclisation depending on the conditions.[67]

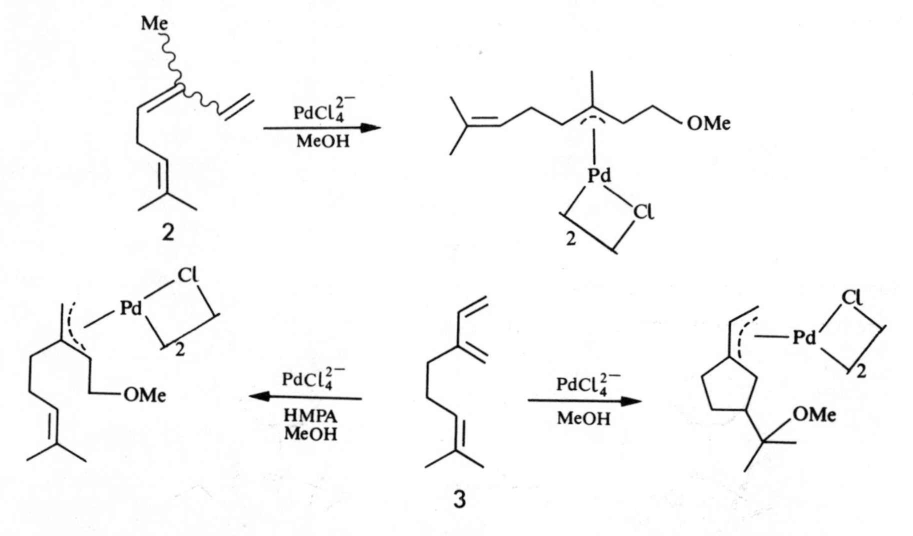

1,3-Dienes react with $(PhCN)_2PdCl_2$ in the presence of secondary amines to give monomeric η^3-allyl palladium complexes.[68]

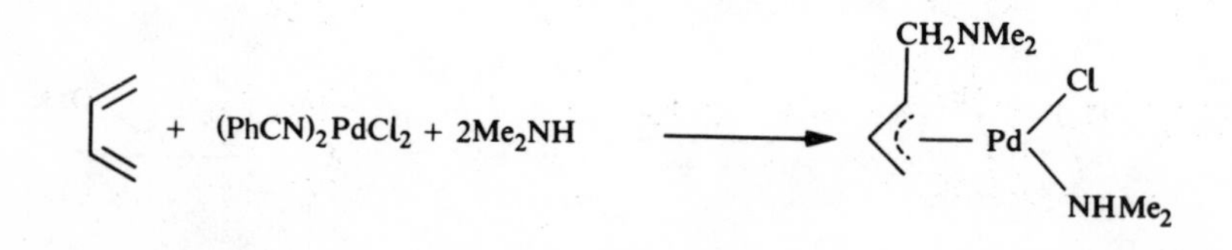

The addition of metal hydride complexes to 1,3-dienes may also produce η^3-allyl complexes.[69]

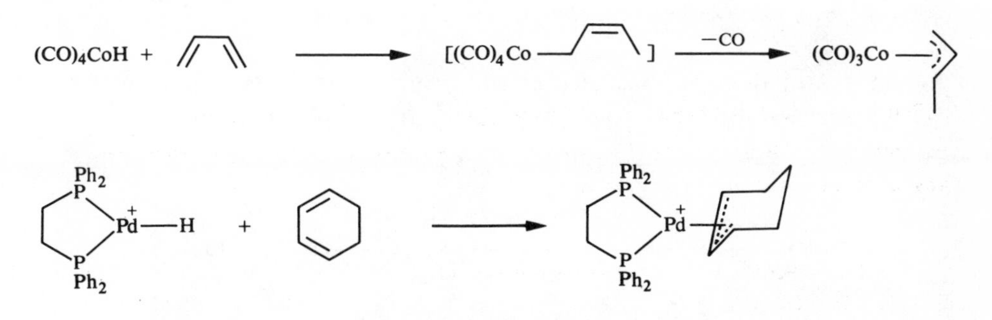

η^3-Allyl complexes may be prepared by protonation of the readily available η^4-diene or η^2-1,3-diene complexes.[70]

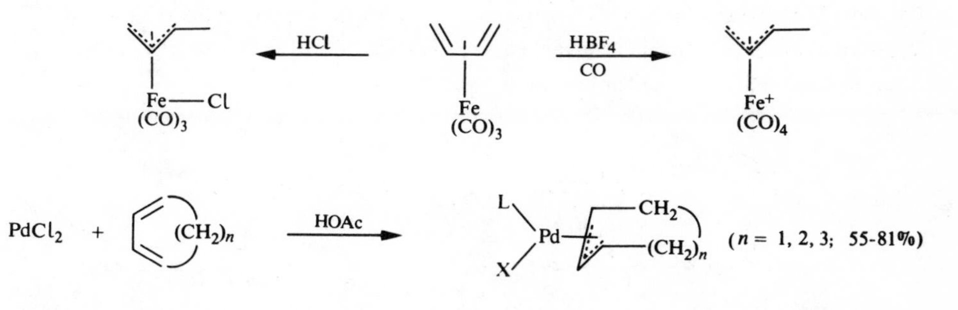

Nucleophilic addition to cationic η^4-diene complexes leads to neutral η^3-allyl complexes (see section 4.1.4).[17]

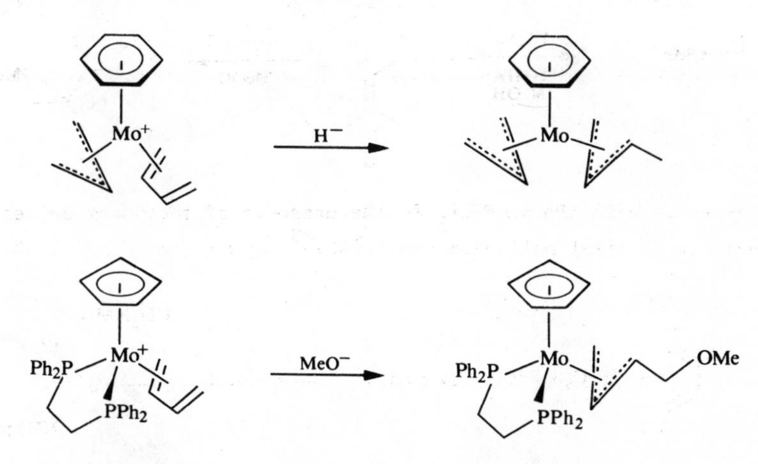

2.4.2 *From olefins with an allylic leaving group*

Pd(O) formed from the reduction of $PdCl_2$ by CO, $SnCl_2$, Fe, Cu, Zn or K undergoes addition with allyl chloride to give $[(allyl)PdCl]_2$.[66]

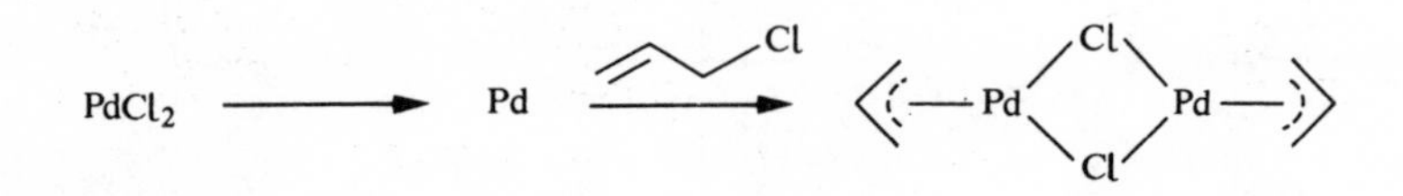

Terminal olefins may be converted to η^3-allyl complexes via the corresponding allyl chlorides.[71]

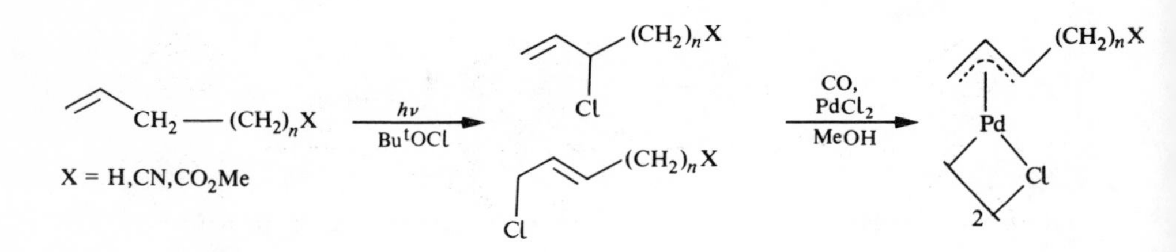

Allyl iodide reacts with $Fe(CO)_5$ or $Fe_2(CO)_9$ to afford the complex $(\eta^3\text{-allyl})Fe(CO)_3I$.[72]

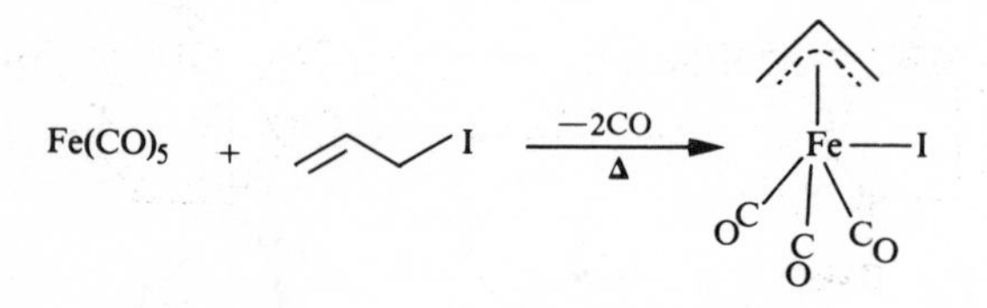

η^1-Allyl complexes prepared from the reaction of nucleophilic metals with allyl halides (see section 2.1.1) may be converted to η^3-allyl complexes by removal of a 2-electron ligand.[1,22, 73]

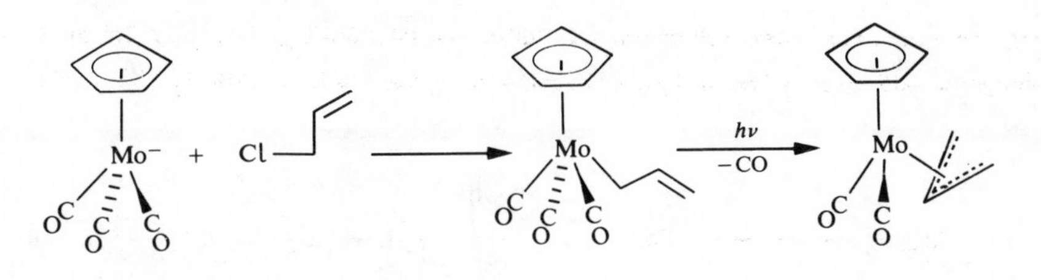

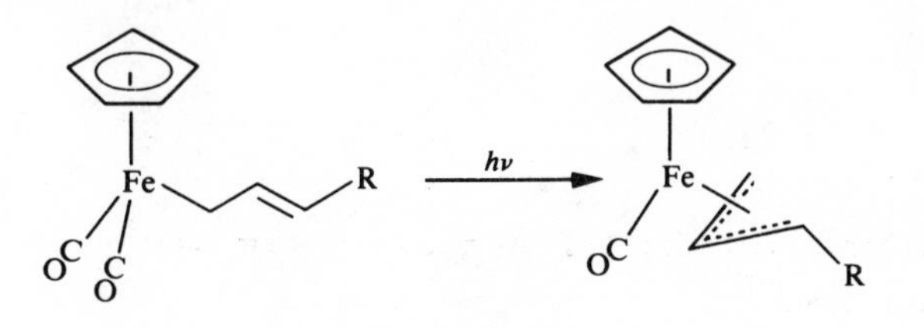

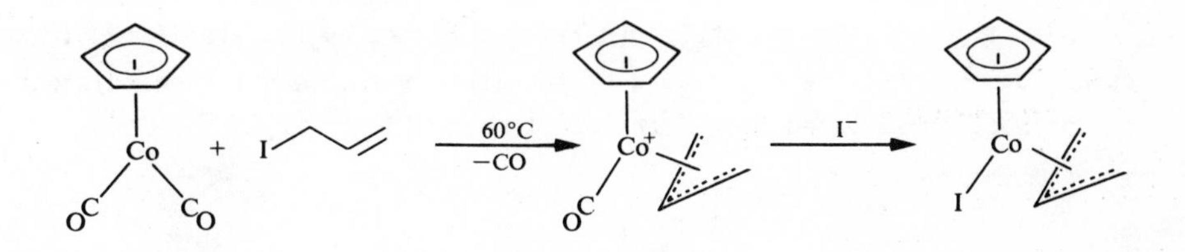

η^2-Olefin complexes of allyl ethers yield cationic η^3-allyl complexes with acid.[74]

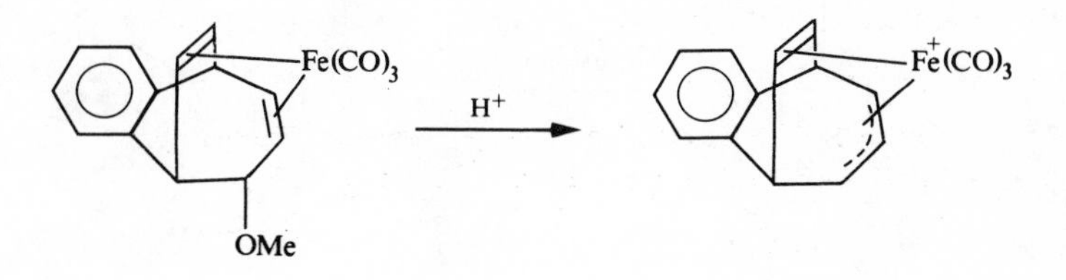

2.4.3 *From olefins*

For synthesis, the direct converion of olefins to η^3-allyl complexes is of interest since it allows the specific activation of an allylic position. There are many conditions available for the formation of η^3-allyl palladium complexes in particular.

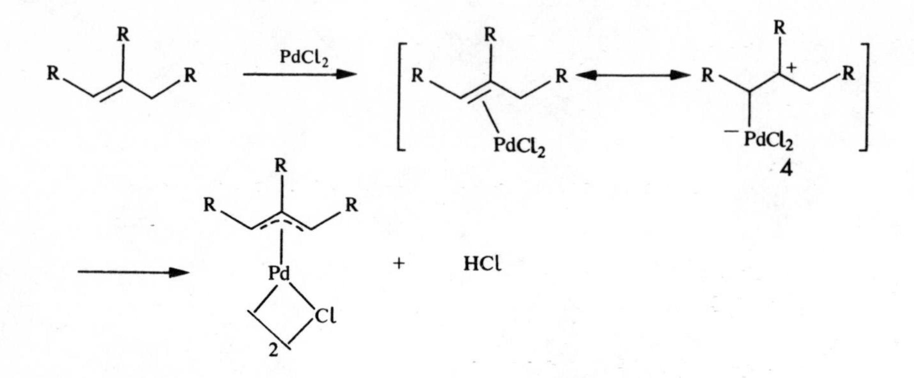

The direction of elimination is generally dictated by the stability of the carbonium intermediate **4** , i.e. a proton is lost from a position adjacent to the most substituted end of the double bond.[75]

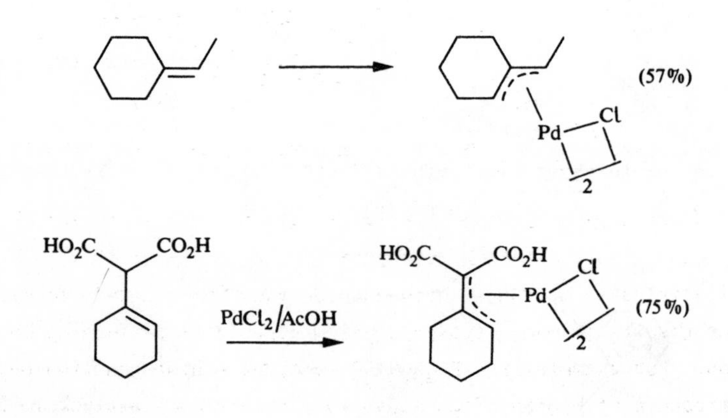

Various terpenes and steroids have been converted to η^3-allyl Pd complexes. *Syn-anti* rearrangements occur easily and generally the thermodynamically more stable *anti* isomer is formed.[76]

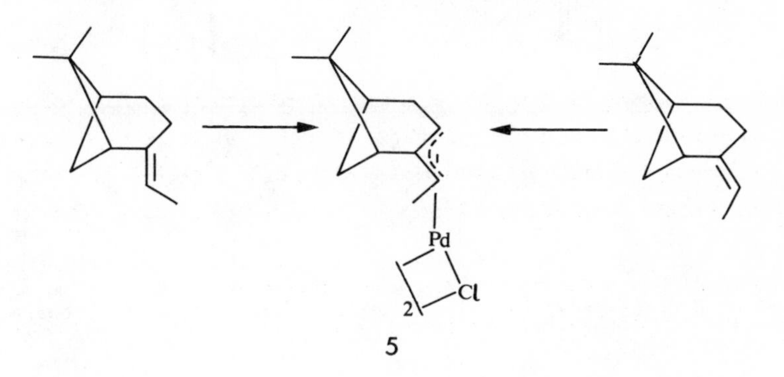

When the two faces of the double bonds are sterically non-equivalent the ratio of products depends on the relative stabilities of the initially formed olefin-$PdCl_2$ complexes. Thus only the complex 5 with the Pd *trans* to the dimethyl bridge is formed above. Similarly, cholest-4-ene and $(PhCN)_2PdCl_2$ give a mixture of the α- and β-complexes 6 and 7 whereas cholest-5-ene gives only the α- complex 6.[77] The β-face of both olefins is the more hindered with the β- face of cholest-5-ene being more hindered than the β-face of cholest-4-ene.

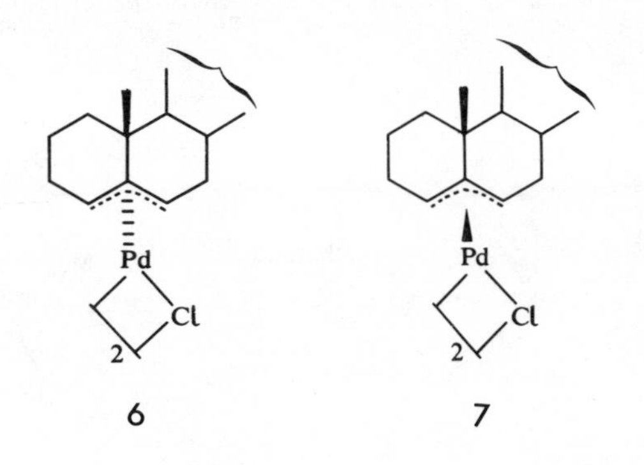

η^3-Allyl palladium complexes may be prepared from α,β-unsaturated ketones as well.[78]

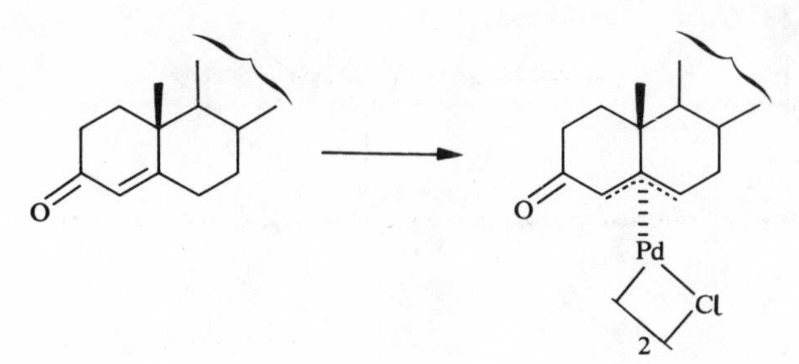

Removal of hydride from η^2-olefin complexes also allows the preparation of η^3-allyl complexes.[79]

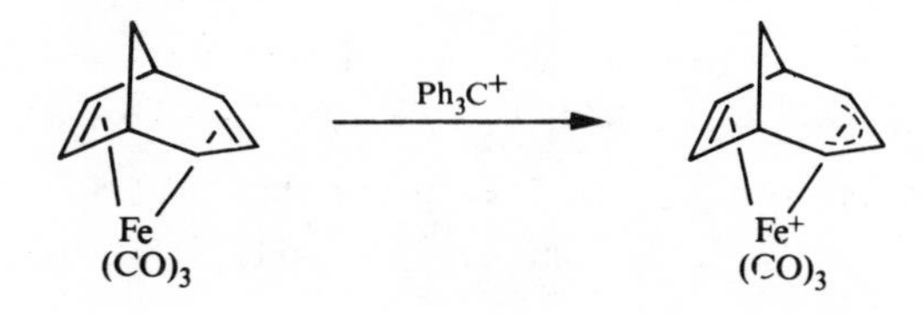

2.4.4. Miscellaneous methods

Treatment of the η^1-propargyl complexes **8** with methanol gives the η^3-allyl ester complexes **9**.[80]

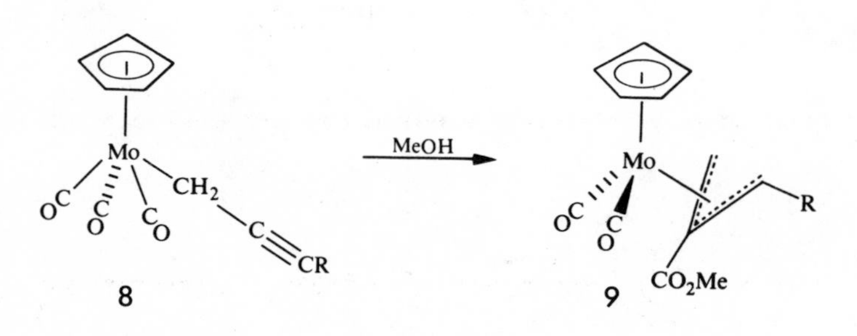

Vinyl cyclopropanes may be opened to η^3-allyl complexes.[81]

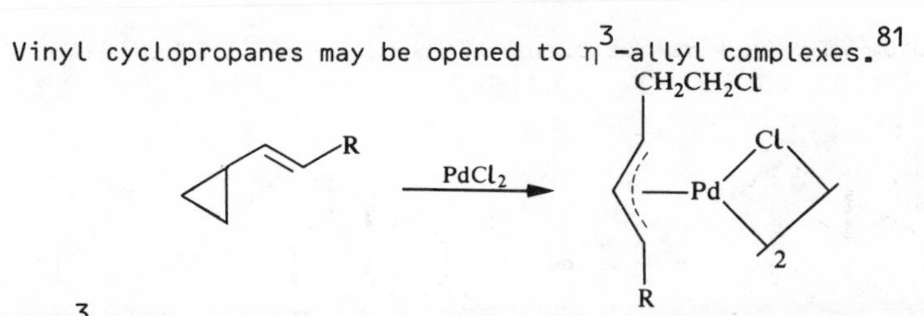

An η^3-allyl complex arises from the reaction of diketene with Na_2PdCl_4 in EtOH.[82]

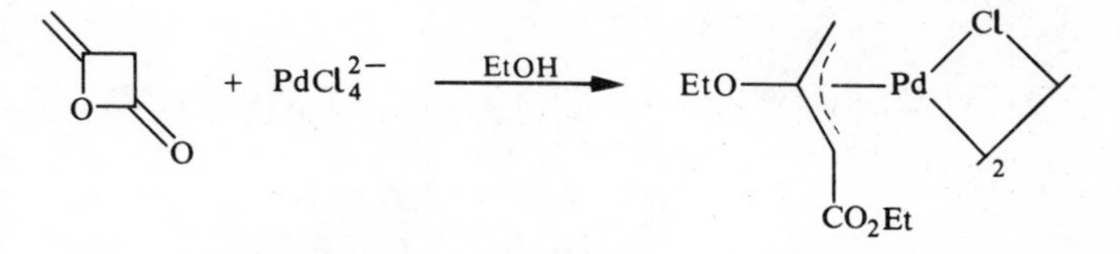

2.4.5 *Decomplexation of η^3-allyl ligands*

η^3-Allyl palladium chloride complexes react with a variety of nucleophiles to give η^2-olefin complexes which decompose with loss of olefin (see section 4.2.3).

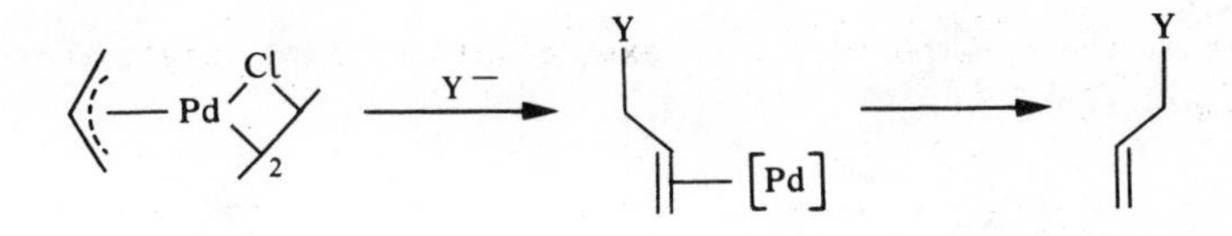

Decomplexation to give allyl acetates may be achieved with AgOAc and CO.[66]

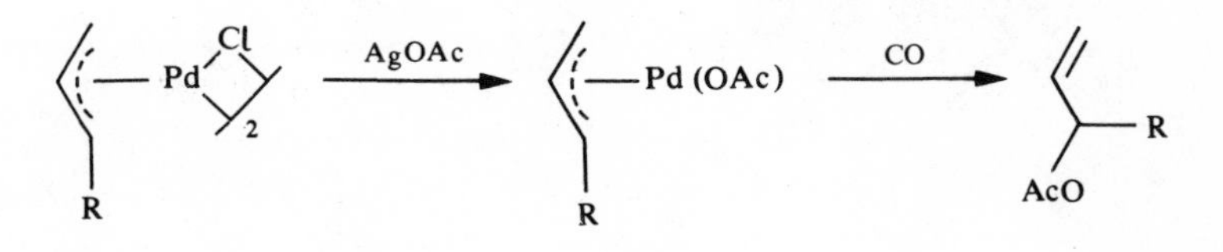

Decomplexation to give dienes may be achieved thermally or with KCN.[67,83]

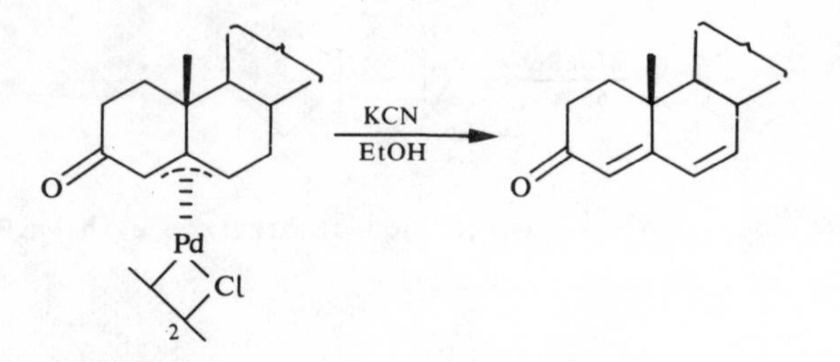

Carbonylation may lead to acid chlorides or esters (see Chapter 9).

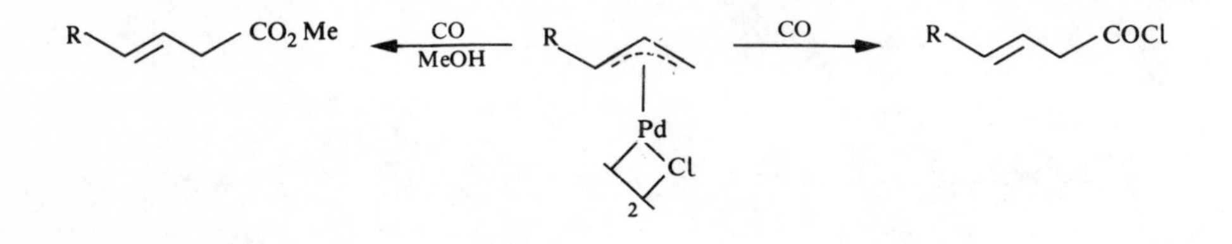

Oxidative decomplexation to α,β-unsaturated carbonyl compounds may be effected by MnO_2 or Collins' reagent.[84]

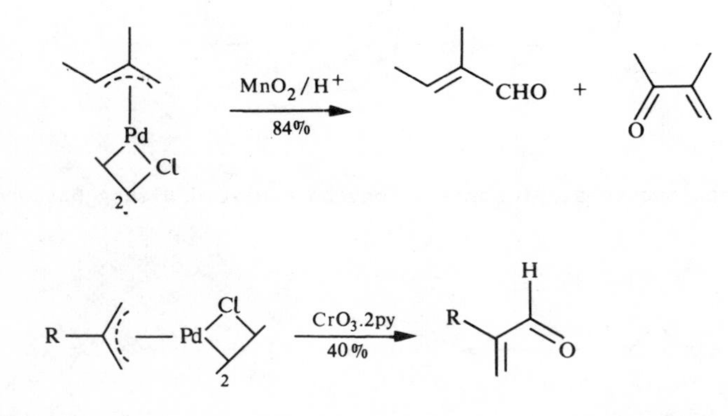

Allylic alcohols are produced by the action of MCPBA.[85]

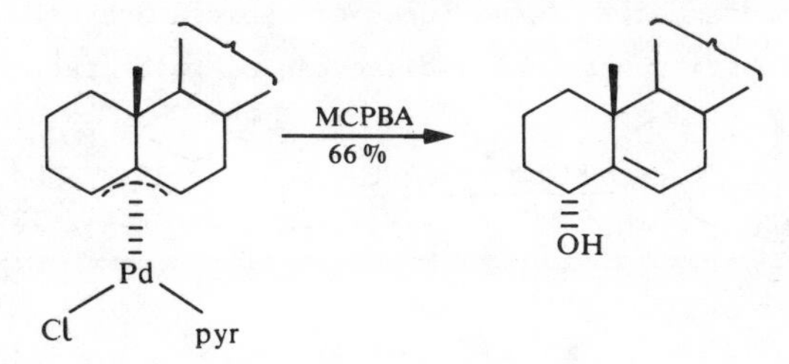

Photolysis of (η^3-allyl)Pd complexes leads to the formation of 1,5-dienes.[85a]

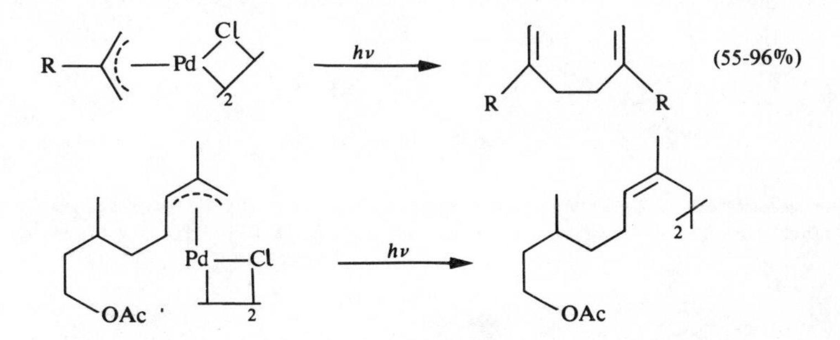

2.5 PREPARATION AND DECOMPLEXATION OF η^4-DIENE COMPLEXES

2.5.1 *From metal carbonyls*

A large number of η^4-1,3-diene $Fe(CO)_3$ complexes have been made from conjugated dienes and iron carbonyls. Non-conjugated dienes undergo rearrangement to give the 1,3-diene complexes where possible (see Chapter 7). The exchange of CO for diene may be achieved photochemically, thermally[86] or chemically.[87]

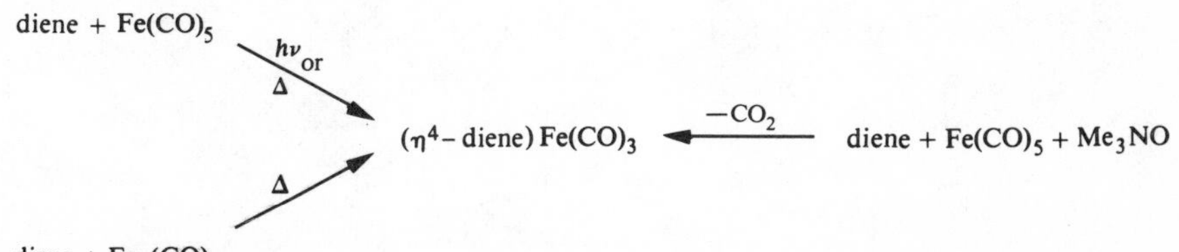

Both acyclic and cyclic dienes readily undergo complex formation. However, due to rearrangement reactions, mixtures of isomers are often formed. A wide variety of substituents on the diene can be tolerated.

Product mixtures may vary depending on the conditions. Thus 4-vinylcyclohexene gives different product ratios on thermolysis or photolysis with $Fe(CO)_5$.[88]

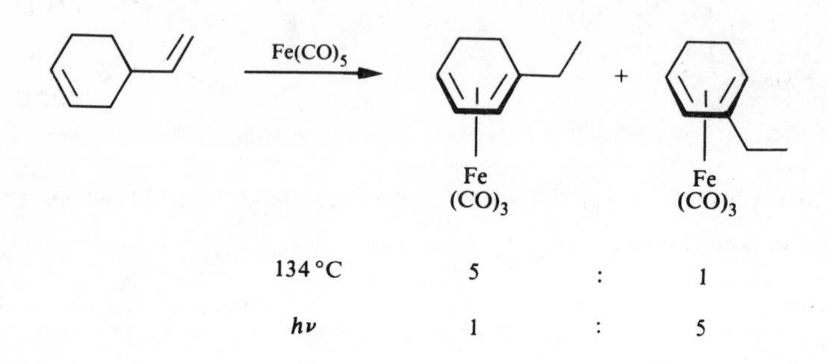

The reaction of 4-vinylcyclohexene with $Fe(CO)_4PPh_3$ is more selective.

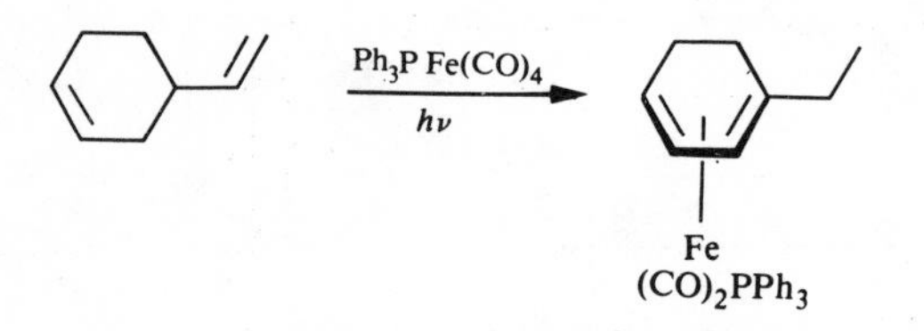

α-Phellandrene shows a similar change of product ratios with conditions. Photolysis gives the more reactive intermediate $[Fe(CO)_3]$ that is less selective, whereas thermolysis gives the less reactive intermediate $[Fe(CO)_4]$ that is more selective for the less-hindered face.

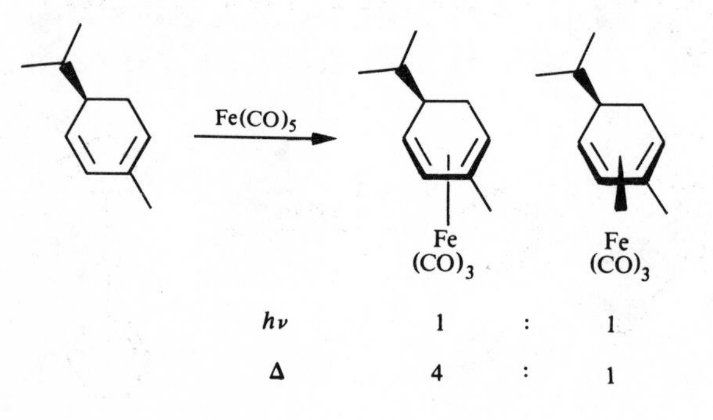

The iron tricarbonyl complex of thebaine may be obtained by reaction with $Fe_2(CO)_9$ or $Fe(CO)_5/h\nu$.[89] Owing to the steric requirements of the complexed diene, only one isomer is formed.

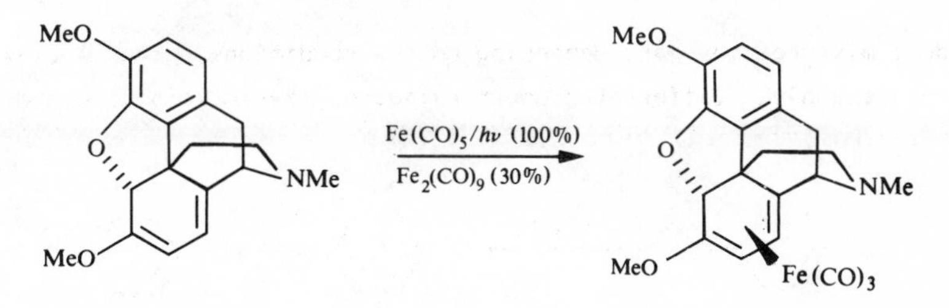

The sulphoxides 10a and 10b were separated and identified as their iron tricarbonyl complexes.[90]

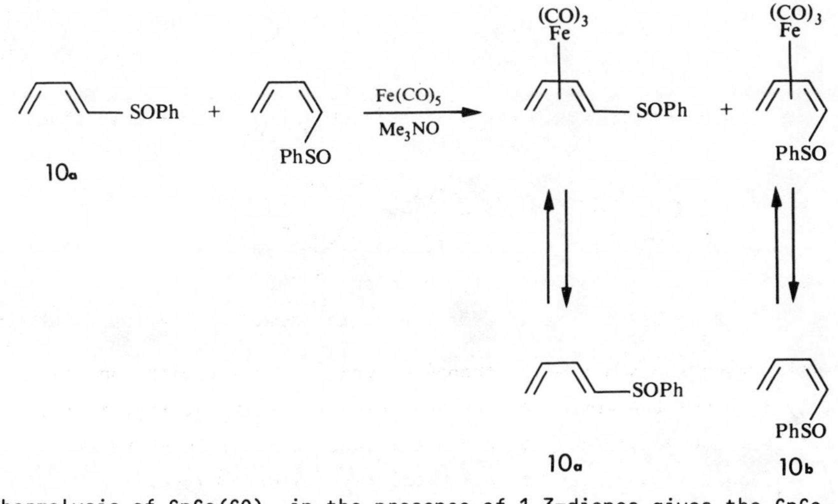

Thermolysis of $CpCo(CO)_2$ in the presence of 1,3-dienes gives the CpCo-(diene) complexes.[91]

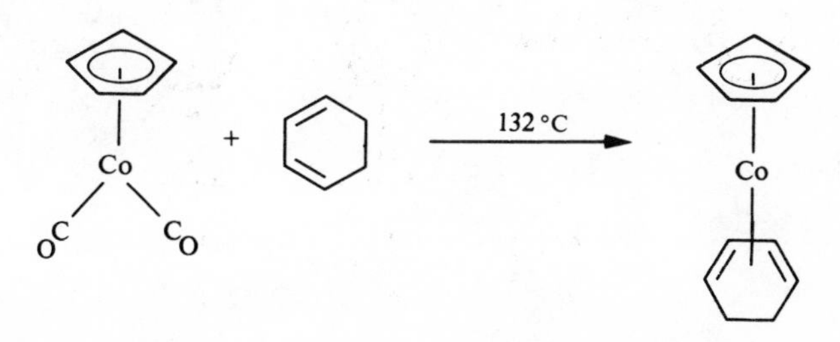

A variety of vanadium and manganese diene complexes have been prepared by photolytic removal of CO ligands.[92]

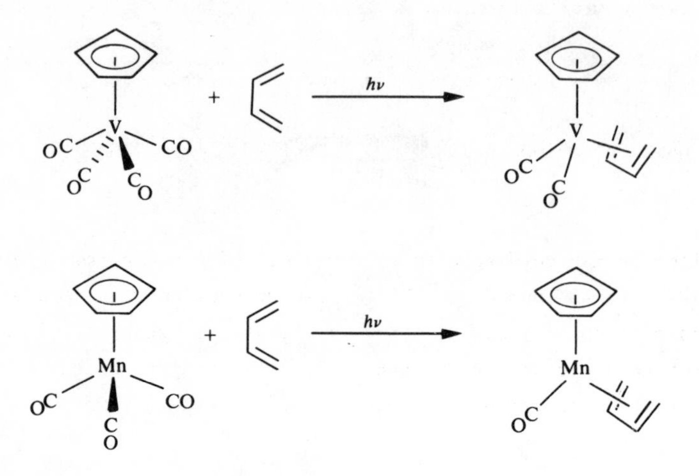

2.5.2 From exchange reactions between dienes and other ligands

An improved synthesis of (η^4- diene) $Fe(CO)_3$ complexes involves the reaction of benzylidene acetone $Fe(CO)_3$ with the diene. The benzylidene acetone $Fe(CO)_3$ acts as a transfer agent for the $Fe(CO)_3$ unit. It is particularly useful when complexing dienes that are sensitive to thermolysis and photolysis.[93]

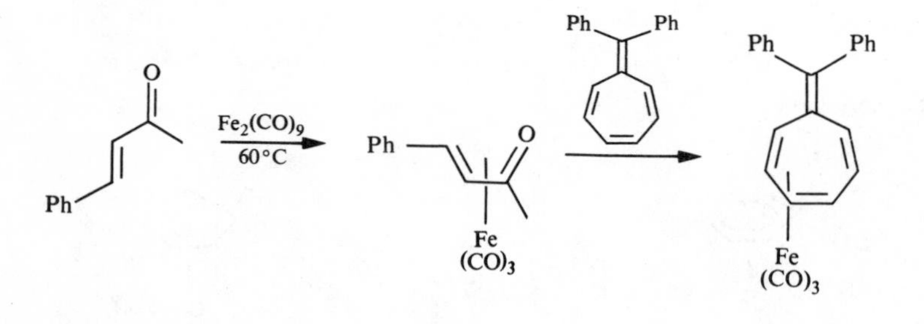

This procedure is the method of choice for the preparation of the iron tricarbonyl complex of ergosteryl acetate[94] and benzoate.[95] p-Methoxy benzylidene acetone can be used as a catalytic transfer agent in the reaction between ergosteryl benzoate and $Fe_2(CO)_9$.[95]

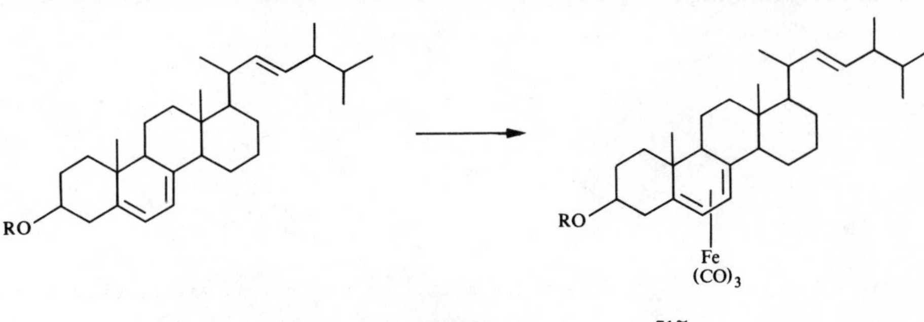

115 °C (benzylideneacetone)$Fe(CO)_3$ 71%
60 °C (*p*-methoxybenzylideneacetone)$Fe(CO)_3$ 66%
55 °C *p*-methoxybenzylideneacetone + $Fe_2(CO)_9$ 80%

R = Ac, PhCO

The mechanism of this exchange is believed to proceed through an intermediate with both the benzylideneacetone and the new ligand coordinated.[96]

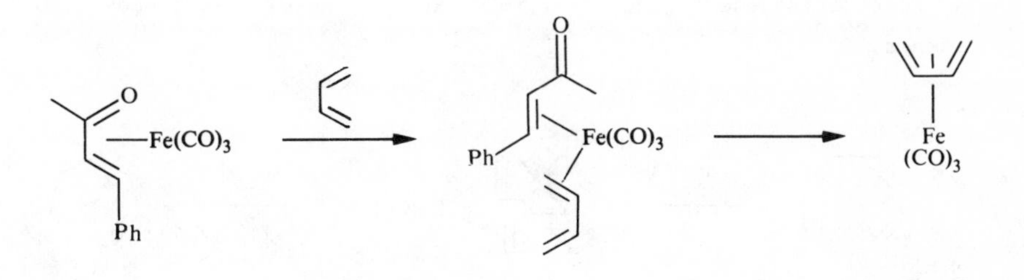

Since both the α,β-unsaturated ketone and the diene are coordinated to the metal at the same time, the use of chiral enone iron tricarbonyl complexes allows the synthesis of optically active substituted diene iron tricarbonyl complexes. This process distinguishes between the two faces of prochiral dienes. The iron tricarbonyl complexes of the enones 11 and 12 have been used to synthesise chiral iron tricarbonyl complexes of various dienes. Enantiomeric excesses are less than 20%.[97]

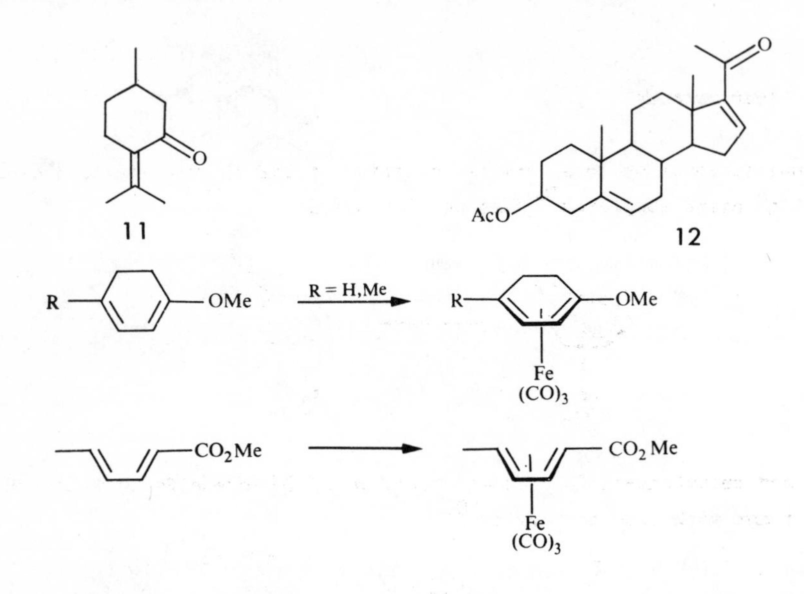

The 1,5-cyclooctadiene ligand in (1,5-COD)$Ru(CO)_3$ readily exchanges with 1,3-dienes.[98]

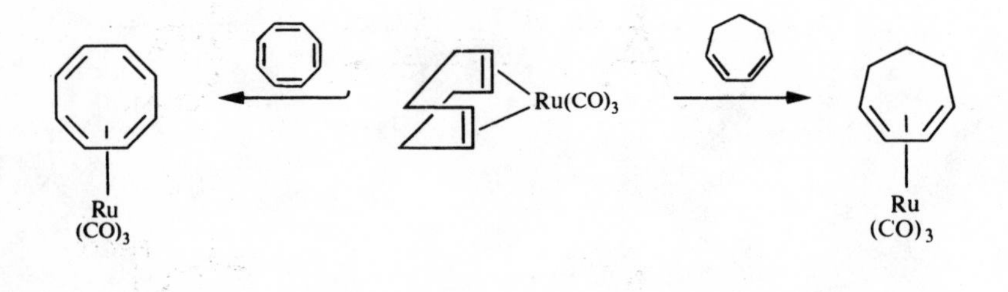

Cyclohexadiene or butadiene replaces one diphos ligand in the cation CpMo $(diphos)_2^+$.[99]

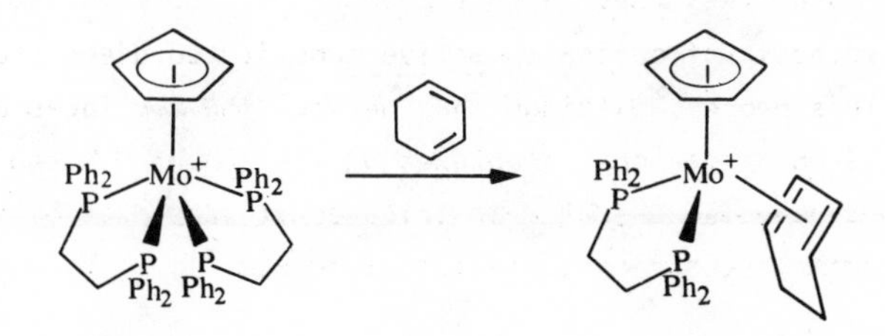

2.5.3 Other methods

Nucleophilic addition to a terminal position of dienyl metal cations gives neutral η^4-diene complexes[18] (see Section 4.1.5).

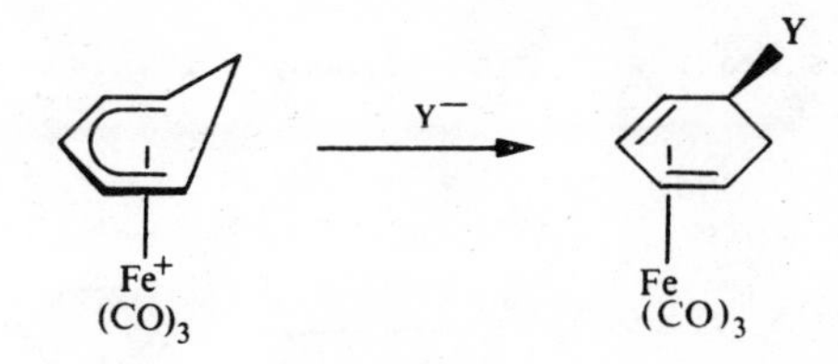

Vinyl- and methylene-cyclopropanes rearrange to (η^4-diene)$Fe(CO_3)$ complexes on treatment with iron carbonyls.[100]

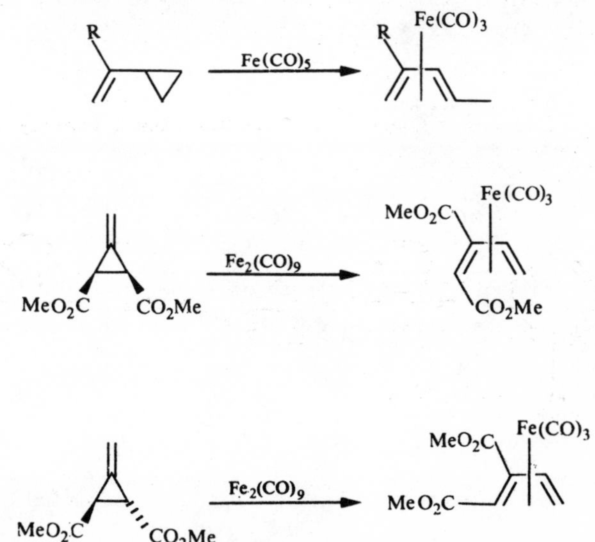

Vinyl epoxides also undergo rearrangement to diene complexes.[101]

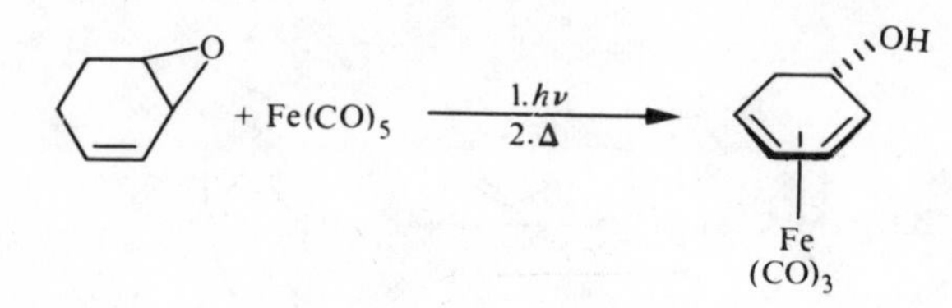

The stable o-quinodimethane iron tricarbonyl complex can be prepared from α,α'-dibromo-o-xylene and $Fe_2(CO)_9$ or $Fe(CO)_4^{2-}$.[102]

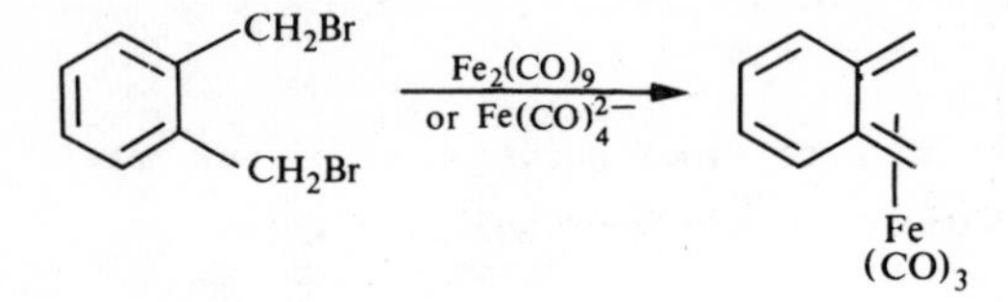

Cyclohexenyl bromides react with iron carbonyls to give cyclohexadiene $Fe(CO)_3$ complexes.[103]

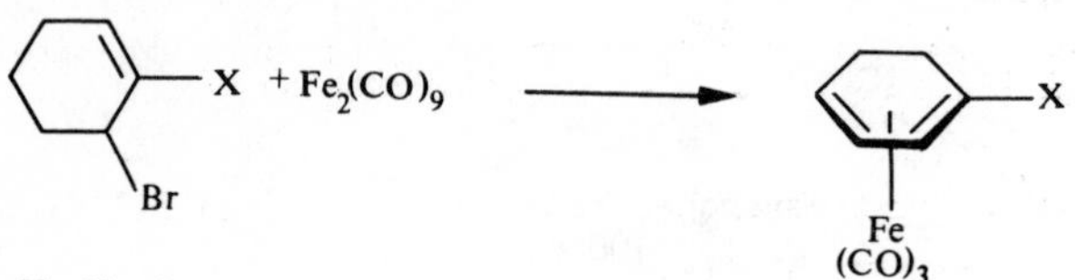

X = H, Cl

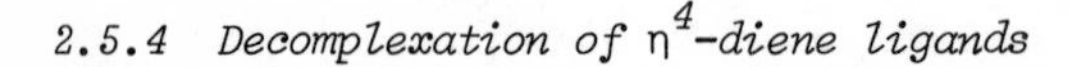

2.5.4 *Decomplexation of η^4-diene ligands*

The methods available for the release of dienes from their iron tricarbonyl complexes are generally oxidative. The oxidants include Me_3NO,[104]

$FeCl_3$,[95,105] Ce^{IV} [105], Collins reagent[106], $CuCl_2$[106a] and Ag^{I}[106b].

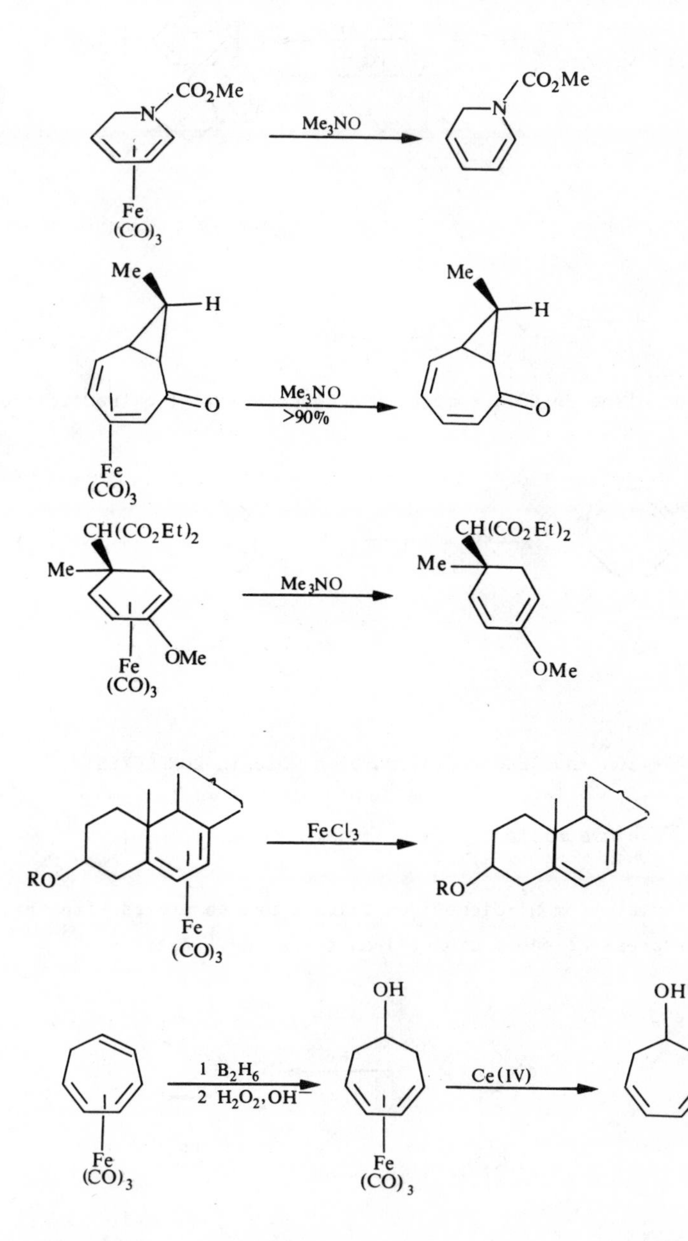

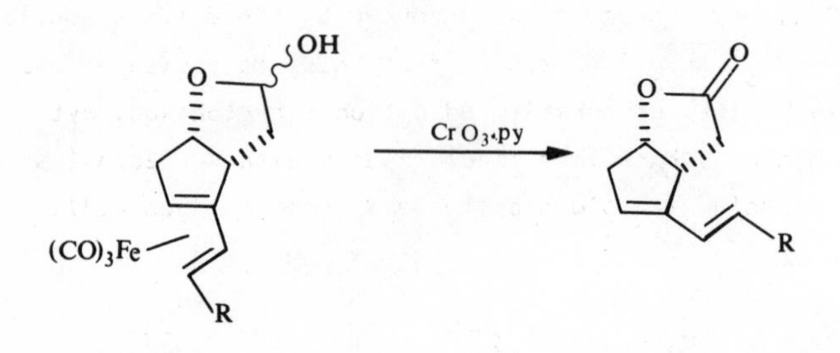

Reaction of diene $Fe(CO)_3$ complexes with Ph_3P may also lead to free diene.[107]

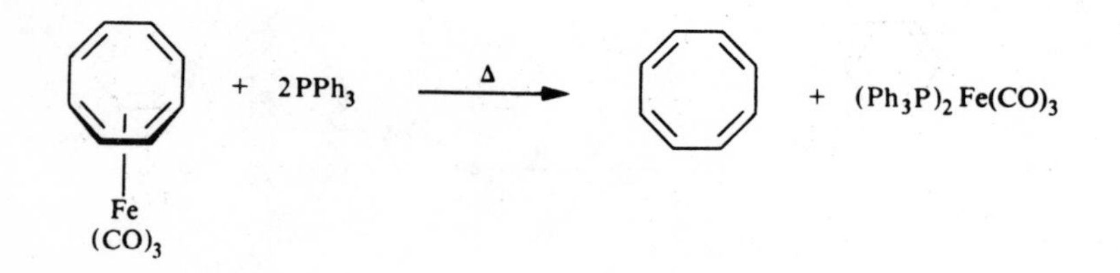

2.6 PREPARATION AND DECOMPLEXATION OF η^5-DIENYL COMPLEXES

2.6.1 Hydride abstraction from η^4-diene complexes

Hydride removal from η^4-diene iron tricarbonyl complexes with trityl cation allows the preparation of dienyl iron tricarbonyl cations.[108]

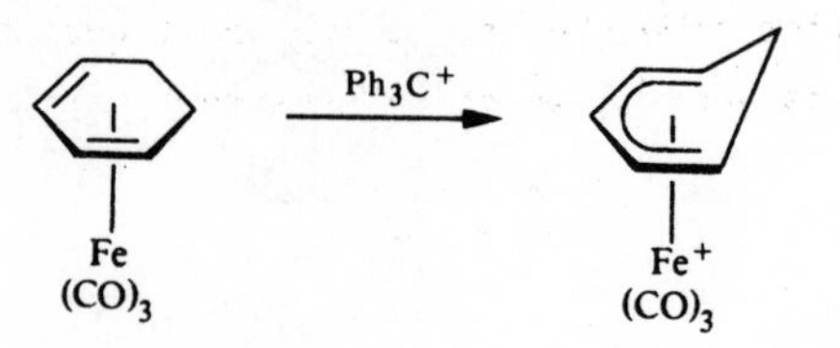

1,3-Cyclohexadiene iron tricarbonyl complexes are readily available from the reaction of 1,4-cyclohexadienes, produced by the Birch reduction of arenes, with $Fe(CO)_5$. Hydride removal from these complexes allows the preparation of a variety of substituted cationic cyclohexadienyl complexes.[109] Since such cations react readily with nucleophiles they have proved very useful for organic synthesis (see Section 4.1).

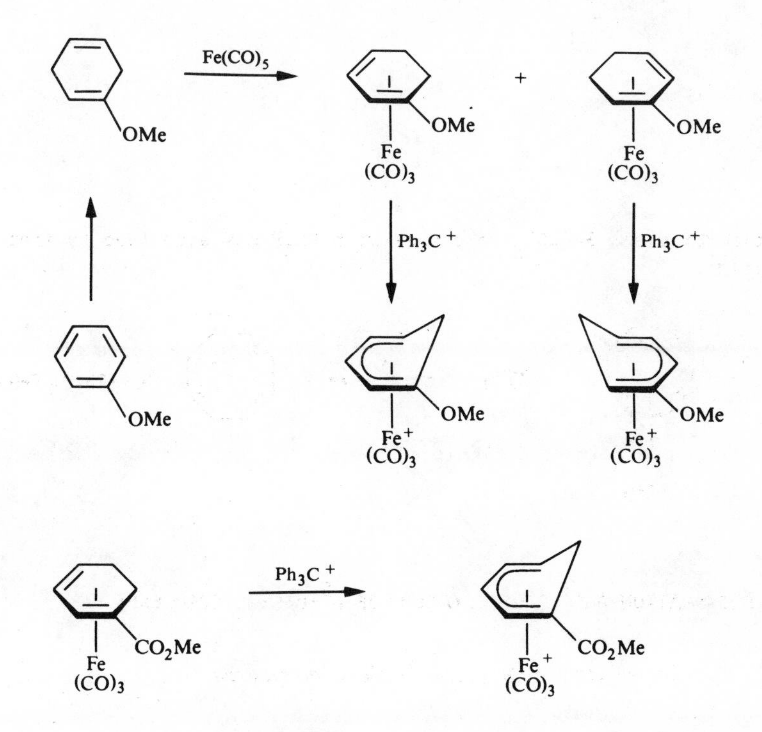

Hydride abstraction from the iron tricarbonyl complex of cholesta- 1,3-diene gives the cation 13 which on treatment with base gives the

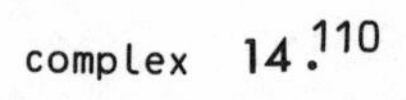
complex **14**.[110]

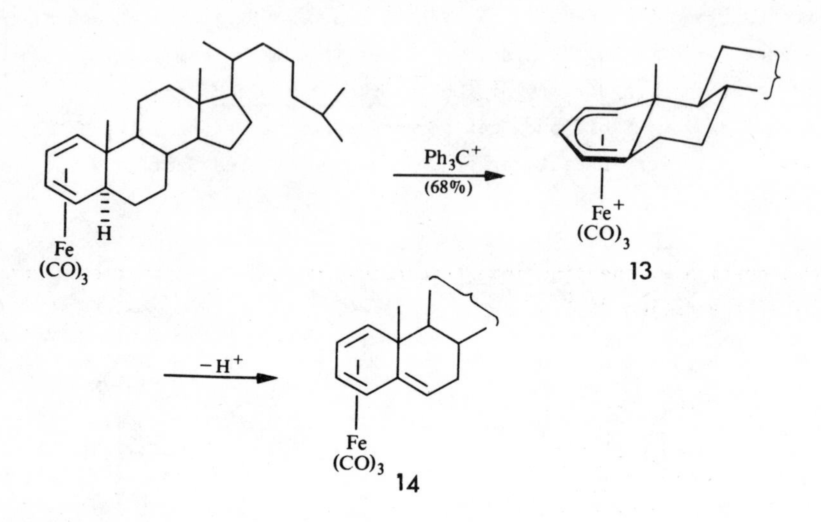

2.6.2 Reactions of acids with substituted η^4*-diene complexes*

Protonolysis of transition metal complexes of 1,3-dien-5-ols results in the formation of pentadienyl cationic complexes.[111]

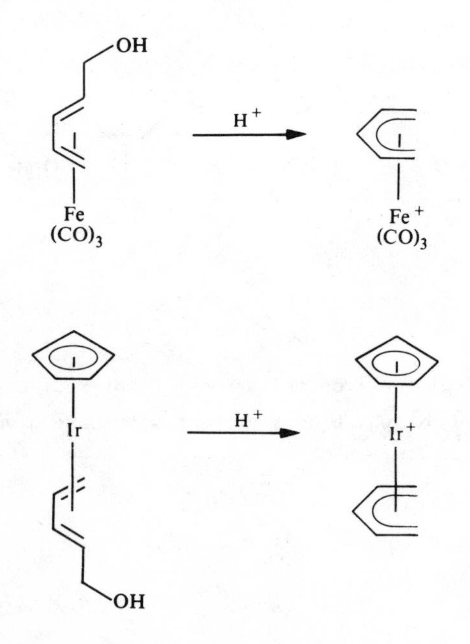

(3-Methoxycyclohexadienyl)$Fe(CO)_3^+$ may be prepared by treatment of (1,3-dimethoxycyclohexadiene)$Fe(CO)_3$ with acid.[112]

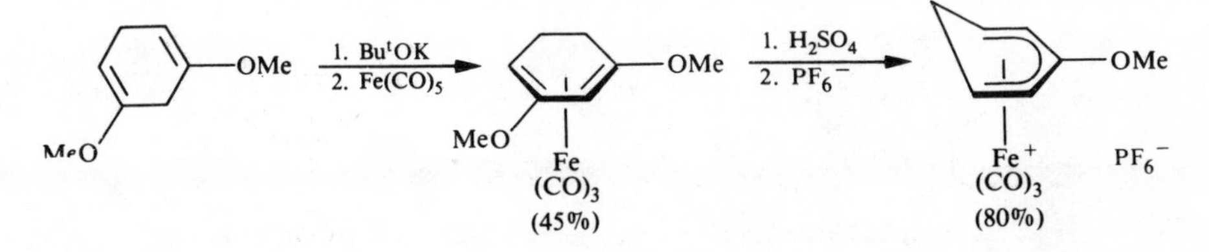

Protonation of thebaine iron tricarbonyl with HBF_4 in acetic anhydride gives the cation 15 [89]

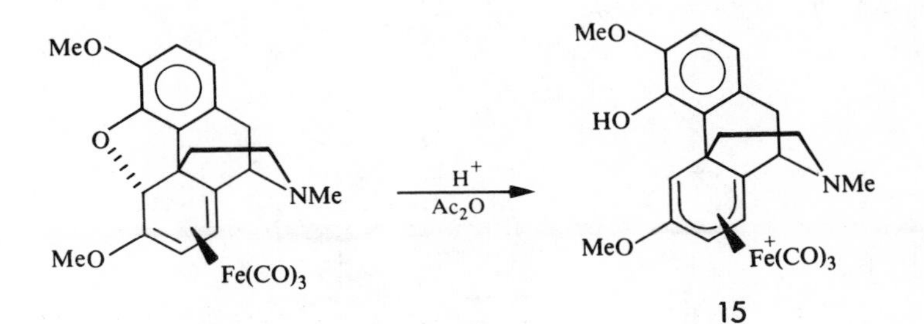

Protonation of non-complexed double bonds in η^4-diene iron tricarbonyl complexes also gives (dienyl)$Fe(CO)_3$ cations.[113]

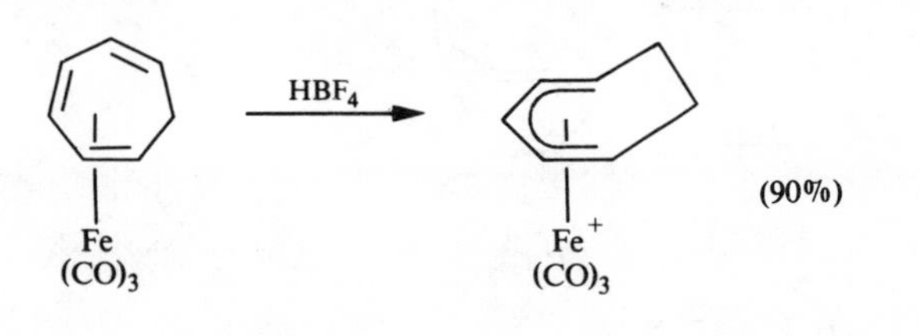

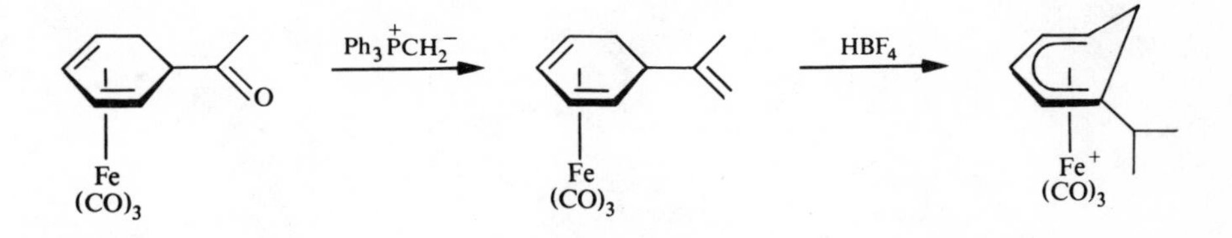

Protonation of the tosylate **16** gives the cation **17** after rearrangement.[114]

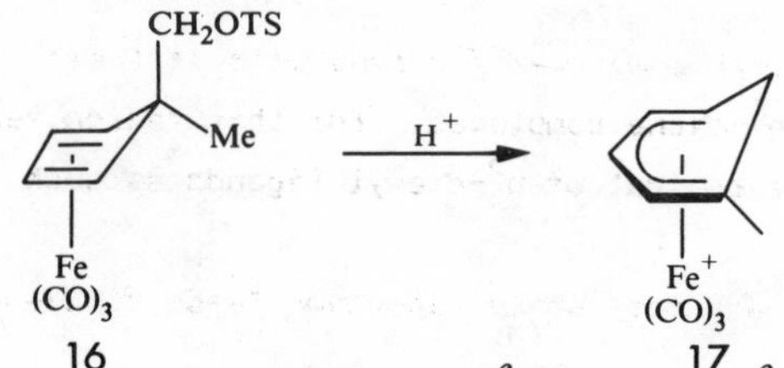

2.6.3 From nucleophilic addition to η⁶-arene and η⁶-triene complexes

Neutral η^5-dienyl complexes may be readily formed from the reaction of cationic η^6-arene and η^6-triene complexes with nucleophiles[18] (see section 4.1.6).

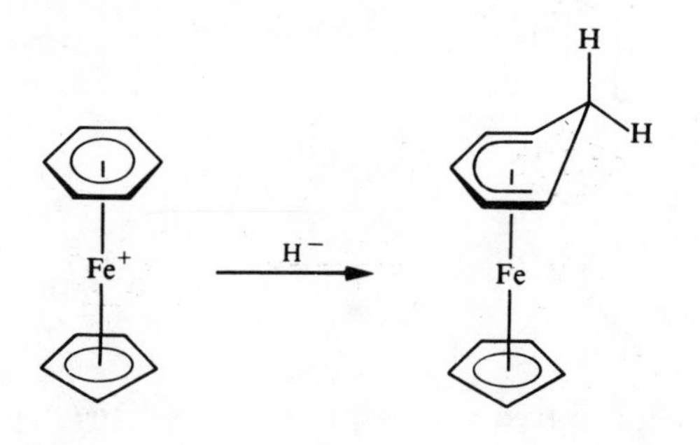

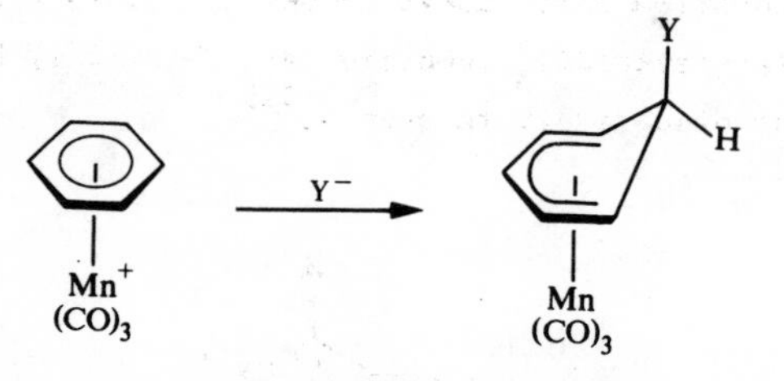

Y = H, R, MeO, N_3, etc.

2.6.4 *Decomplexation of η^5-dienyl ligands*

The main use of η^5-dienyl complexes for synthesis is their reaction with nucleophiles to give η^4-diene complexes. For this reason few methods have been developed for the removal of η^5-dienyl ligands as such.

Oxidation can lead to dienones while reduction leads to diene formation.

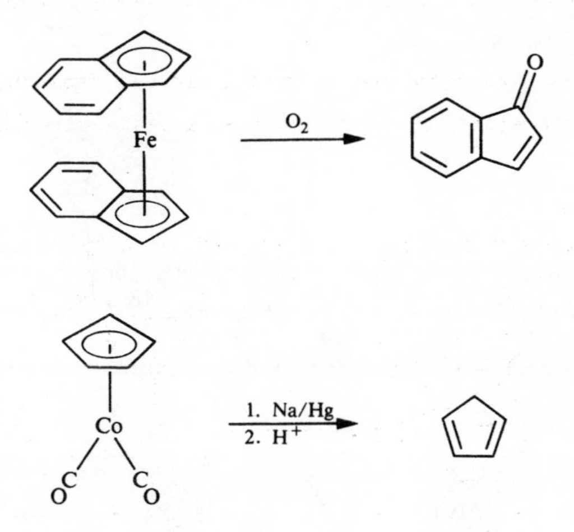

Cyclohexadienyl chromium tricarbonyl anions formed by the action of nucleophiles on (arene)$Cr(CO)_3$ complexes are decomposed by acid to cyclohexadienes or oxidatively to arenes.[115]

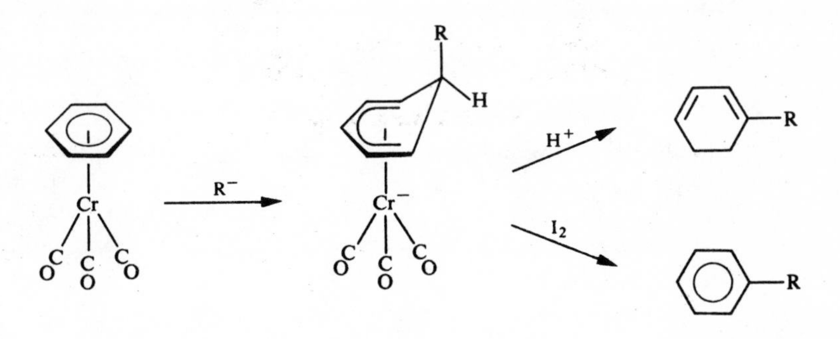

2.7 PREPARATION AND DECOMPLEXATION OF η^6-ARENE AND RELATED COMPLEXES.[116]

2.7.1 *Preparation*

Bis-arene iron dications may be prepared by the reaction of iron halides and arenes in the presence of aluminium trihalides.[117]

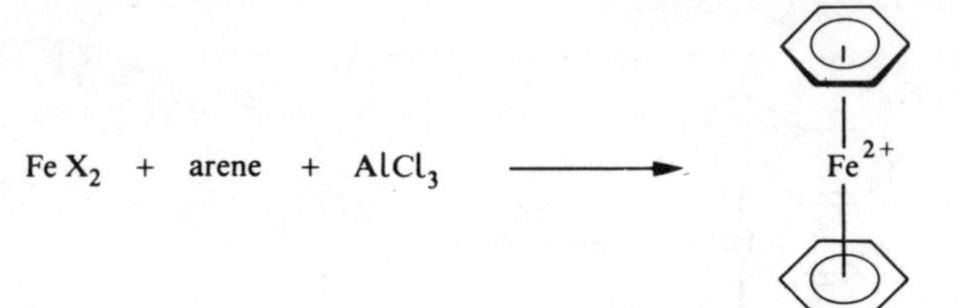

The readily available $CpFe(CO)_2Cl$ yields the cations $[(arene)FeCp]^+$ on treatment with $AlCl_3$ in the presence of arene.[118]

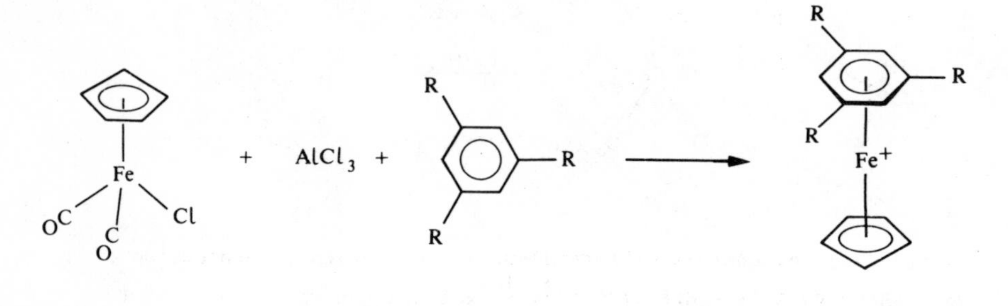

The bromide $(CO)_5MnBr$ behaves similarly.[119]

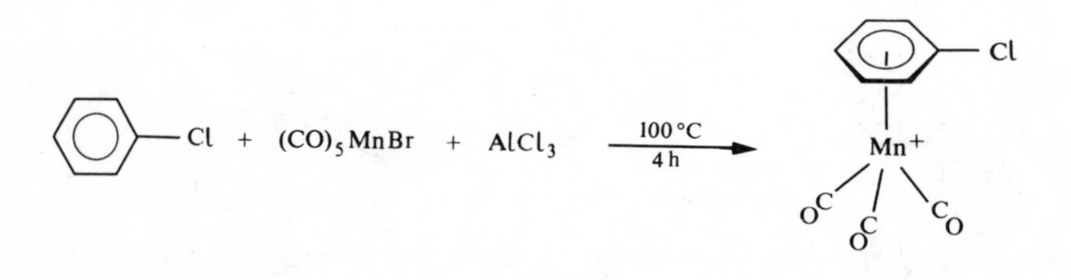

Treatment of the complex $[\{(PhO)_3P\}_2RhCl]_2$ with $AgPF_6$ generates the solvated cation $\{(PhO)_3P\}_2Rh^+$ which reacts with arenes to give the cationic (arene)$Rh\{P(OPh)_3\}_2^+$ complexes.[120]

Reaction of $RuCl_3$ with 1,3- or 1,4-cyclohexadiene generates the species $[(C_6H_6)RuCl_2]_2$. Other arene complexes may be prepared either from the appropriate substituted cyclohexadiene (often readily available from the Birch reduction of arenes) or by arene exchange reactions on thermolysis or photolysis.[121,122]

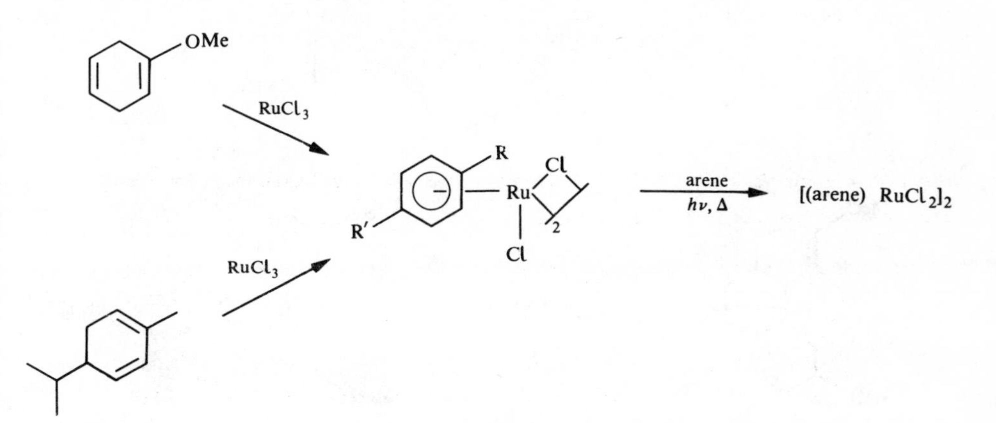

Treatment of the complexes $[(arene)RuCl_2]_2$ with TlCp generated the 18-electron cations $[(arene)RuCp]^+$.[121]

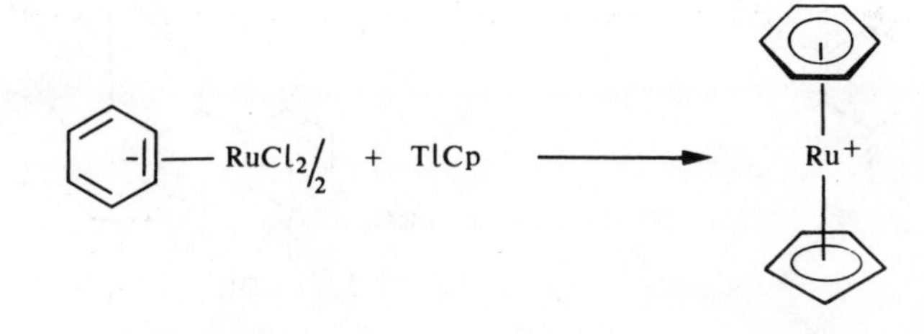

Thermolysis of the hexacarbonyls of chromium, molybdenum and tungsten in the presence of arenes leads to the formation of (arene)$M(CO)_3$ derivatives.[116] The preparations of (arene)$Cr(CO)_3$ complexes have been particularly well studied.[116,123] A wide variety of complexes have been made by this method.

Reaction times are generally long, however, and high temperatures are needed. Therefore, for many unstable aromatic molecules this method is unsuitable. It is also incompatible with certain functional groups such as CO_2H, CHO, CN and NO_2.

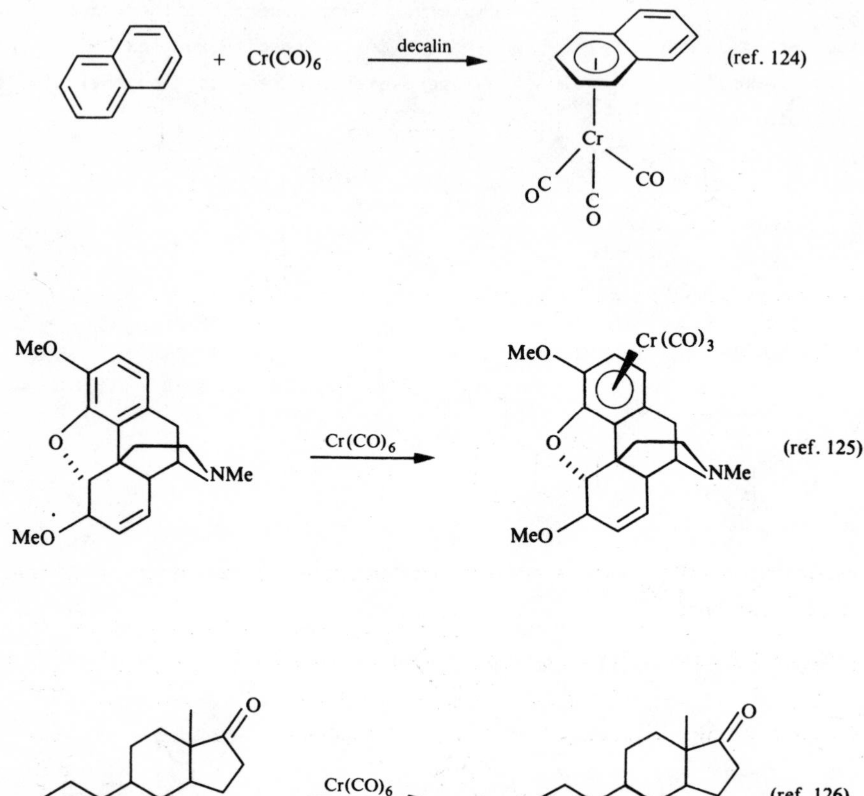

Benzene chromium tricarbonyl may be prepared in high yield from the reaction of $Cr(CO)_6$ with 1-methoxycyclohexa-1,4-diene.[127] This reaction has been shown to be useful for the synthesis of substituted (arene)$Cr(CO)_3$ complexes and provides a mechanism for the removal of an oxygen substituent from the arene ring of ring A aromatic steroids.[126]

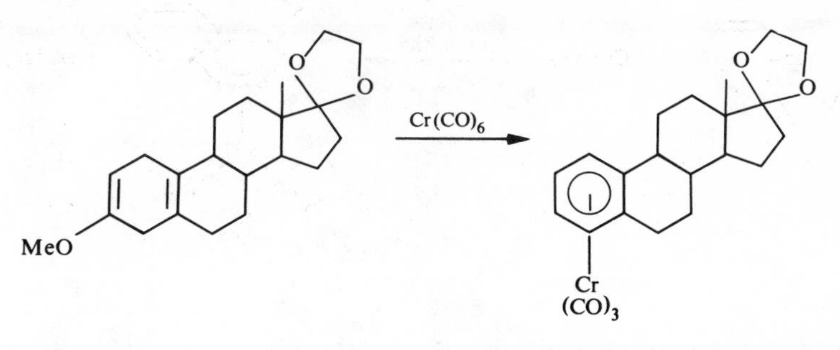

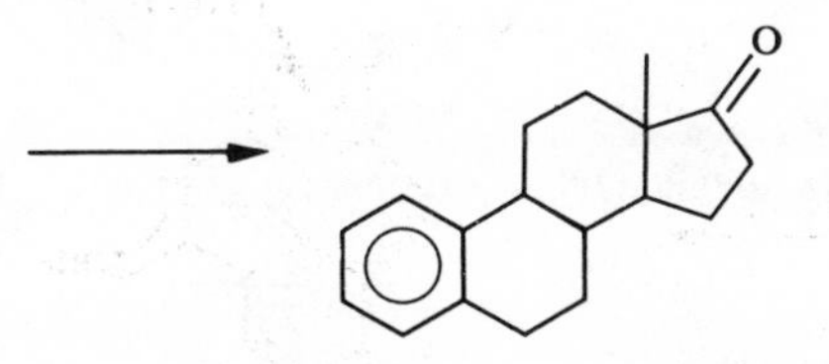

Reaction of $V(CO)_6$ with arenes allows the direct preparation of (arene)V$(CO)_4^+$ cations.[128]

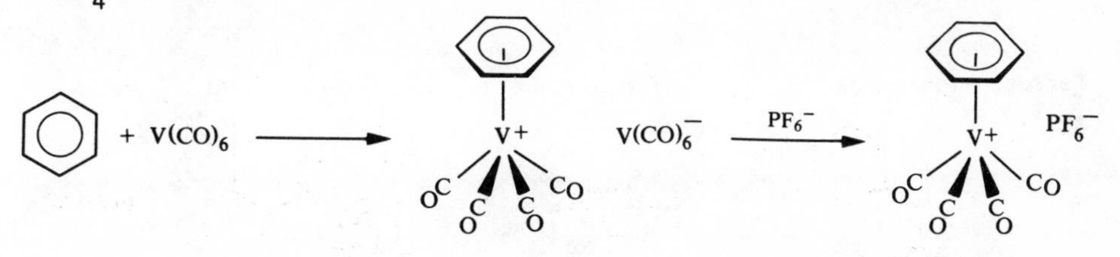

Arene/CO exchange occurs on thermolysis of the cation [(cycloheptatriene)$Mo(CO)_3]^+$ in the presence of arene.[129]

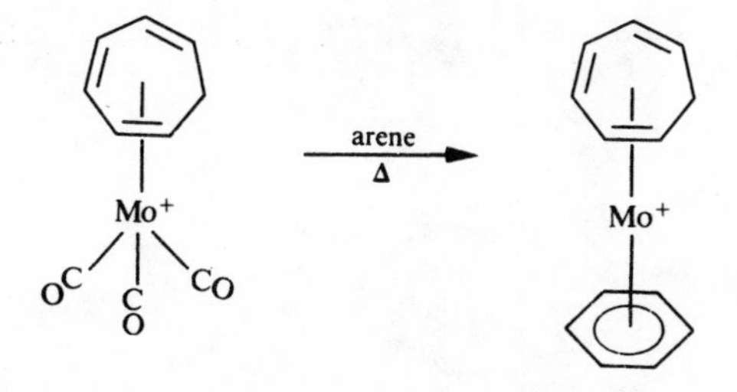

Pyrroles may also act as 6e ligands to transition metals.[130]

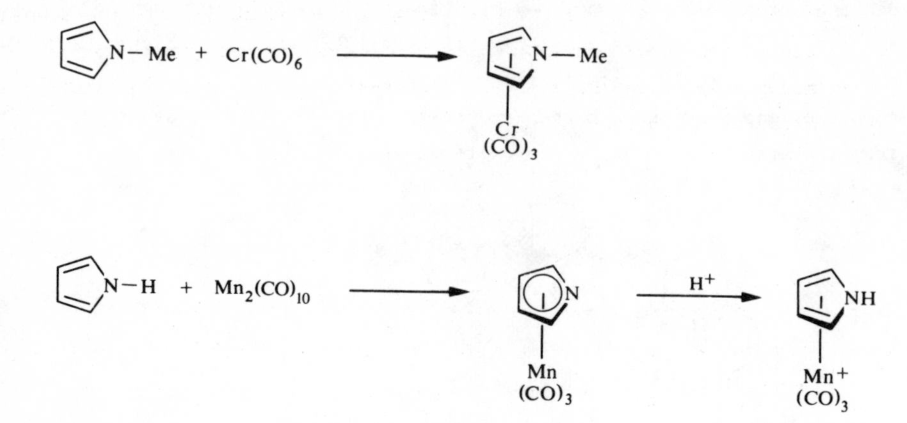

The problem of the high temperatures and long reaction times necessary to form (arene)$M(CO)_3$ complexes from $M(CO)_6$ (M = Cr, Mo, W) may be circumvented by prior formation of intermediate species of the type $L_3M(CO)_3$, which can then undergo arene/L_3 exchange under milder conditions.

$$M(CO)_6 + 3L \longrightarrow L_3M(CO)_3 \xrightarrow{\text{arene}} (\text{arene})M(CO)_3$$

Many different ligands (L) have been employed for this reaction, e.g. L = CH_3CN, THF, diglyme, pyridine, 4-methylpyridine, 2-methylpyridine, NH_3).[116]

Ferrocene exchanges one Cp ring for arene in the presence of $AlCl_3$ and Al.[131]

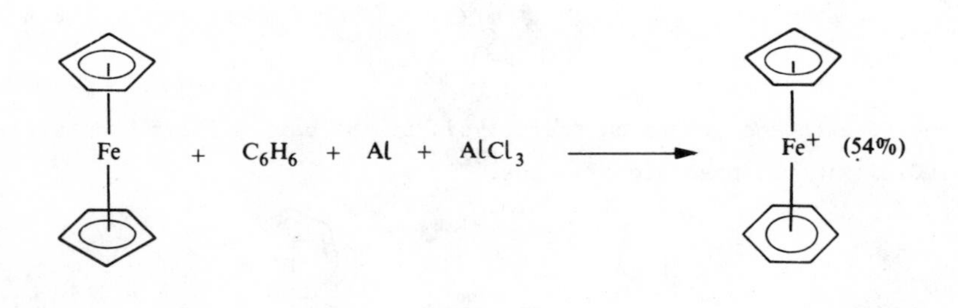

2.7.2 *Decomplexation of η^6-arene ligands*

There are two general methods for the decomplexation of η^6-arenes from transition metals. η^6-Arenes may be removed by oxidative processes or by displacement reactions with other ligands.

Oxidative methods include $\underline{h}\nu/O_2$,[126,132] Ce(IV),[133] I_2,[133] $KMnO_4$[123] and MnO_2.

η^6-Arene ligands may be displaced by a variety of two-electron ligands such as R_3P, R_3As and pyridine.[123,134] Decomplexation of η^6-arenes by pyridine is particularly useful since it allows the isolation of $(C_5H_5N)_3Cr(CO)_3$ in high yield as the inorganic product which can be subsequently recycled for the preparation of other (η^6-arene)$Cr(CO)_3$ complexes.[135]

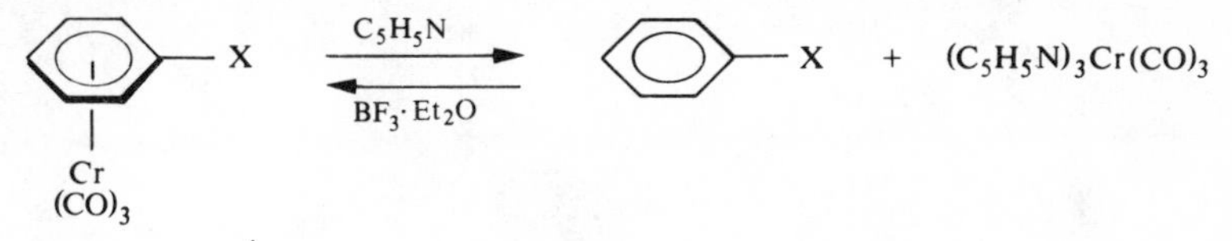

Decomplexation of η^6-arene ligands may also be achieved by ligand exchange with other arenes or fulvenes.[136]

2.8 PREPARATION OF η^7-TRIENYL COMPLEXES

η^7-Cycloheptatrienyl complexes may be formed either directly from cycloheptatriene[137] or by the action of trityl cation on cycloheptatriene complexes.[138]

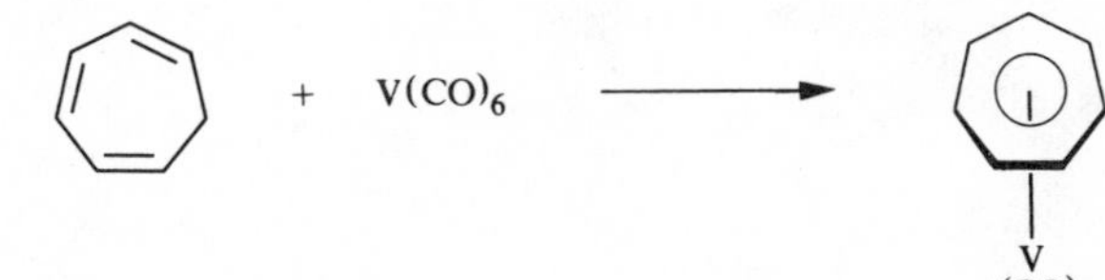

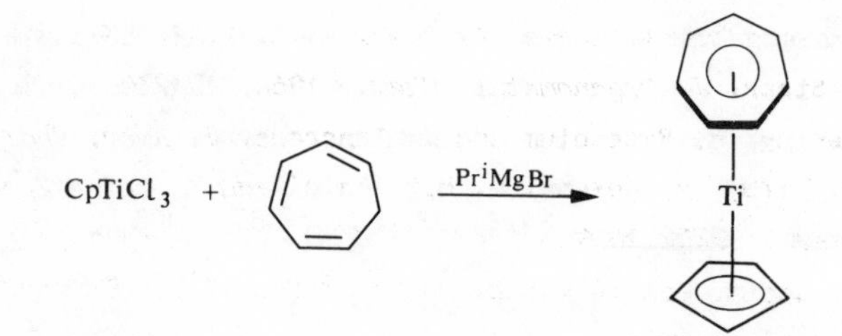

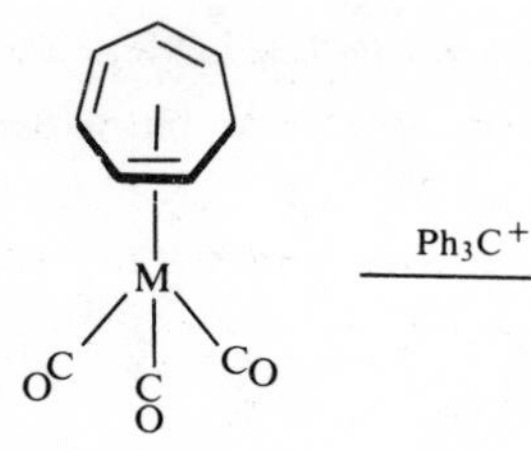

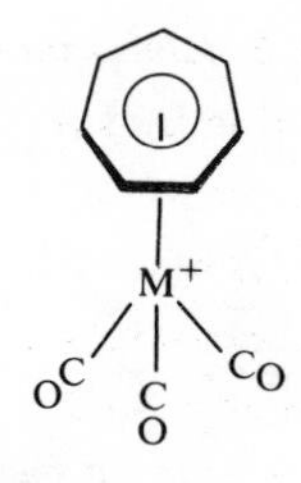

M = Mo, W

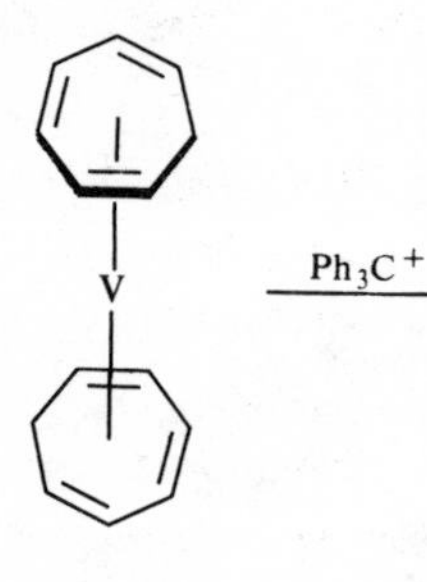

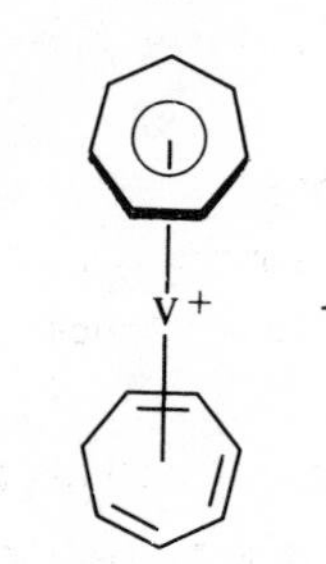

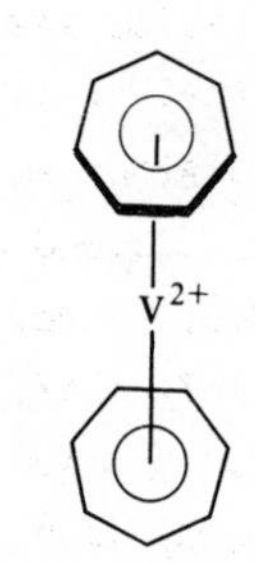

2.9 REFERENCES

1. M. Cousins and M.L.H. Green, *J. Chem. Soc.*, 1963, 889; M.L.H. Green and A.N. Stear, *J. Organometal. Chem.*, 1964, *1*, 230.
2. W.P. Griering, M. Rosenblum and J. Tancrede, *J. Amer. Chem. Soc.*, 1972, *94*, 7170; M. Rosenblum, M.R. Saidi and M. Madhavarao, *Tet. Letters*, 1975, 4009.
3. D. Baudry and M. Ephritikhine, *Chem. Comm.*, 1979, 895, and personal communication.
4. H. Felkin, P.J. Knowles, B. Meunier, A. Mitschler, L. Ricard and R. Weiss, *Chem. Comm.*, 1974, 44.
5. G.M. Whitesides and D.J. Boschetto, *J. Amer. Chem. Soc.*, 1971, *93*, 1529; K.M. Nicholas and M. Rosenblum, *J. Amer. Chem. Soc.*, 1973, *95*, 4449.
6. H. Werner and W. Hofmann, *Ang. Chem. Int. Ed.*, 1979, *18*, 158.
7. H. Werner and W. Hofmann, *Chem. Ber.*, 1977, *100*, 3481; H. Werner, R. Feser and W. Buchner, *Chem. Ber.*, 1979, *112*, 834.
8. F.W.S. Benfield, N.J. Cooper and M.L.H. Green, *J. Organometal. Chem.*, 1974, *76*, 49.
9. J.P. Collman, *Acc. Chem. Res.*, 1968, *1*, 136; L. Vaska, *Acc. Chem. Res.*, 1968, *1*, 335.
10. H. Alper and E.C.H. Keung, *J. Org. Chem.*, 1972, *37*, 2566.
11. W.J. Bland and R.D.W. Kemmitt, *J. Chem. Soc. A*, 1968, 1278.
12. M.C. Baird, J.T. Mague, J.A. Osborn and G. Wilkinson, *J. Chem. Soc.A*, 1967, 1347.
13. K.S.Y. Lau, P.K. Wong and J.K. Stille, *J. Amer. Chem. Soc.*, 1976, *98*, 5832; J.K. Stille and K.S.Y. Lau, *J. Amer. Chem. Soc.*, 1976, *98*, 5841.
14. F.W. Siegert and H.J.D.L. Meijer, *J. Organometal. Chem.*, 1968, *15*, 131.
15. H. Yamazaki and N. Hagihara, *J. Organometal. Chem.*, 1970, *21*, 431.
16. S.G. Davies and F. Scott,unpublished results.
17. R.J. Cross and F. Glockling, *J. Organometal. Chem.*, 1965, *3*, 253.
18. S.G. Davies, M.L.H. Green and D.M.P. Mingos, *Tetrahedron*, 1978, *34*, 3047.
19. T.L. Brown and P.A. Bellus, *Inorg. Chem.*, 1978, *17*, 3726.
20. P.S. Hallman, B.R. McGarvey and G. Wilkinson, *J. Chem. Soc. A*, 1968, 3143.

21. P.C. Wailes, H. Weigold and A.P. Bell, *J. Organometal Chem.*, 1971, *27*, 373; 1972, *43*, C32; J. Schwartz and D.W. Hart, *J. Amer. Chem. Soc.*, 1974, *96*, 8115; D.W. Hart, T.F. Blackburn and J. Schwartz, *J. Amer. Chem. Soc.*, 1975, *97*, 679; J. Schwartz and J.A. Labinger, *Ang. Chem. Int. Ed.*, 1976, *15*, 333; C.A. Bertelo and J. Schwartz, *J. Amer. Chem. Soc.*, 1976, *98*, 262.

22. M.L.H. Green and P.L.I. Nagy, *J. Chem. Soc.*, 1963, 189.

23. W. Hieber and G. Wagner, *Ann. Chem.*, 1958, *618*, 24.

24. M.L.H. Green, C.R. Hurley, *J. Organometal. Chem.*, 1967, *10*, 188; S.R. Su and A. Wojcicki, *J. Organometal. Chem.*, 1971, *27*, 231.

25. G. Fachinetti and C. Floriani, *Chem. Comm.*, 1975, 578; G. Fachinetti, G. Fochi and C. Floriani, *J.C.S. Dalton*, 1977, 1946; G.Erker and F. Rosenfeldt, *Ang. Chem. Int. Ed.*, 1978, *17*, 605.

26. J.J. Alexander and A. Wojcicki, *Inorg. Chem.*, 1973, *12*, 74.

26a. W.P. Giering, S. Raghu, M. Rosenblum, A. Cutler, D. Ehntholt and R.W. Fish, *J. Amer. Chem. Soc.*, 1972, *94*, 8251.

27. C.A. Bertelo and J. Schwartz, *J. Amer. Chem. Soc.*, 1975, *97*, 228; J.A. Labinger, D.W. Hart, W.E. Seibert III and J. Schwartz, *J. Amer. Chem. Soc.*, 1975, *97*, 3851; T.F. Blackburn, J.A. Labinger and J. Schwartz, *Tet. Letters*, 1975, 3041; J. Schwartz, *Pure and Applied Chem.*, 1980, *52*, 733.

28. P.L. Bock, D.J. Boschetto, J.R. Rasmussen, J.P. Demers and G.M. Whitesides, *J. Amer. Chem. Soc.*, 1974, *96*, 2814; M. Rosenblum, *Acc. Chem. Res.*, 1974, *7*, 122; P.L. Bock and G.M. Whitesides, *J. Amer. Chem. Soc.*, 1974, *96*, 2826.

29. I.J. Harvie and F.J. McQuillin, *Chem. Comm.*, 1977, 241.

30. R. Cramer, *Inorg. Chem.*, 1965, *4*, 445 and references therein.

31. R.G. Schultz, *J. Organometal. Chem.*, 1966, *6*, 435.

32. E.O. Fischer and E. Moser, *J. Organometal. Chem.*, 1965, *3*, 16; E.O. Fischer and K. Fichtel, *Chem. Ber.*, 1961, *94*, 1200; 1962, *95*, 2063.

33. E.O. Fischer and K. Ofele, *Ang. Chem.*, 1962, *74*, 76.

34. S.G. Davies and F. Scott, *J. Organometal. Chem.*, 1980, *188*, C41.

35. T. Majima and H. Kurosawa, *J. Organometal. Chem.*, 1977, *134*, C45.

36. G.O. Schenck, E. Koerner von Gustorf and M.J. Jun, *Tet. Letters*, 1962, 1059; E. Weiss, K. Stark, J.E. Lancaster and H.D. Murdoch, *Helv. Chim. Acta*, 1963, *46*, 288.

37. M.B. Yunker and B. Fraser-Reid, *J. Org. Chem*, 1979, *44*, 2742.
38. M.A. Bennett, L. Pratt and G. Wilkinson, *J. Chem. Soc.*, 1961, 2037.
39. R.B. King, P.M. Treichel and F.G.A. Stone, *J. Amer. Chem. Soc.*, 1961, *83*, 3593.
40. W. Barthelt, M. Herberhold and E.O. Fischer, *J. Organometal. Chem.*, 1970, *21*, 395; H. Alt, M. Herberhold, C.G. Kreiter and H. Strack, *J. Organometal. Chem.*, 1975, *102*, 491; M. Giffard, E. Gentric, D. Trouchard and P. Dixneuf, *J. Organometal. Chem.*, 1977, *129*, 371.
41. A. Cutler, D. Ehntholt, W.P. Giering, P. Lennon, S. Raghu, A. Rosan, M. Rosenblum, J. Tancrede and D. Wells, *J. Amer. Chem. Soc.*, 1976, *98*, 3495; W.P. Giering and M. Rosenblum, *Chem. Comm.*, 1971, 441.
42. S.G. Davies,unpublished results.
43. S. Samuels, S.R. Berryhill and M. Rosenblum, *J. Organometal. Chem.*, 1979, *166*, C9; D.L. Reger, C.J. Coleman and P.J. McElligott, *J. Organometal. Chem.*, 1979, *171*, 73.
44. A.C. Cope, C.R. Ganelli, H.W. Johnson, T.V. Von Auken and H.J.S. Winkler, *J. Amer. Chem. Soc.*, 1963, *85*, 3276.
45. M.S. Kharasch, R.C. Seyler and F.R. Mayo, *J. Amer. Chem. Soc.*, 1938, *60*, 882.
46. W.P. Giering, M. Rosenblum and J. Tancrede, *J. Amer. Chem. Soc.*, 1972, *94*, 7170; M. Rosenblum, M.R. Saidi and M. Madhavarao, *Tet. Letters*, 1975, 4009.
47. T.H. Whitesides, R.W. Arhart and R.W. Slaven, *J. Amer. Chem. Soc.*, 1973, *95*, 5792.
48. N.A. Bailey, W.G. Kita, J.A. McCleverty, A.J. Murray, B.E. Mann, and N.W.J. Walker, *Chem. Comm.*, 1974, 592.
49. M.L.H. Green and P.L.I. Nagy, *J. Chem. Soc.*, 1963, 189.
50. M.L.H. Green, A.G. Massey, J-T. Moelwyn-Hughes and P.L.I. Nagy, *J. Organometal. Chem.*, 1967, *8*, 511.
51. M. Cousins and M.L.H. Green, *J. Chem. Soc.*, 1963, 889.
52. M.L.H. Green and P.L.I. Nagy, *J. Organometal. Chem.*, 1963, *1*, 58; D.E. Laycock, J. Hartgerink and M.C. Baird, *J. Org. Chem.*, 1980, *45*, 291.
53. G. Cardaci, *Inorg. Chem.*, 1974, *13*, 368.
54. C.P. Casey, Transition Metal Organometallics in Organic Synthesis, Ed. H. Alper, Vol. I, p. 189, Academic Press, New York 1976.

55. R. Aumann and E.O. Fischer, *Chem. Ber.*, 1969, *102*, 1495; C.P. Casey, C.R. Cyr and R.A. Boggs, *Synth. Inorg. Met. Org. Chem.*, 1973, *3*, 249; E.M. Badley, J. Chatt and R.L. Richards, *J. Chem. Soc. A*, 1971, 21.
56. M.L.H. Green, L.C. Mitchard and M.G. Swanwick, *J. Chem. Soc. A*, 1971, 794.
57. F.A. Cotton and C.M. Lukehart, *J. Amer. Chem. Soc.*, 1971, *93*, 2672.
58. W.A. Herrmann, *Ang. Chem. Int. Ed.*, 1974, *13*, 599.
59. A. Sanders, L. Cohen, W.P. Giering, D. Kenedy and C.V. Magatti, *J. Amer. Chem. Soc.*, 1973, *95*, 5430.
60. R.A. Bell, M.H. Chisholm, D.A. Couch and L.A. Rankel, *Inorg. Chem.*, 1977, *16*, 677; M.I. Bruce and R.C. Wallis, *J. Organometal. Chem.*, 1978, *161*, C1; A. Davison and J.P. Selegue, *J. Amer. Chem. Soc.*, 1978, *100*, 7763; M.I. Bruce and R.C. Wallis, *Aust. J. Chem.*, 1979, *32*, 1471.
61. E.O. Fischer and C.G. Kreiter, *Ang. Chem. Int. Ed.*, 1969, *8*, 761.
62. K.H. Dötz and E.O. Fischer, *Chem. Ber.*, 1970, *103*, 1273; 1972, *105*, 1356.
63. C.P. Casey, S.H. Bertz and T.J. Burkhardt, *Tet. Letters*, 1973, 1421.
64. C.P. Casey, R.A. Boggs and R.L. Anderson, *J. Amer. Chem. Soc.*, 1972, *94*, 8947; E.O. Fischer and S. Riedmüller, *Chem. Ber.*, 1974, *107*, 915.
65. M.D. Cooke and E.O. Fischer, *J. Organometal. Chem.*,1973, *56*, 279.
65a. C.P. Casey and W.R. Brunsvold, *Inorg. Chem.*, 1977, *16*, 391.
66. B.M. Trost, Tetrahedron Report No. 32, *Tetrahedron*, 1977, *33*, 2615; J. Tsuji, "Organic Synthesis with Palladium Compounds" Springer-Verlag, Berlin 1980 and references therein.
67. K. Dunne, F.J. McQuillin, *J. Chem. Soc. C*, 1970, 2196, 2200; M. Takahashi, M. Suzuki, Y. Morooka and T. Ikawa, *Chem. Letters*, 1979, 53.
68. B. Akermark, J.E. Bäckvall, A. Löwenborg and K. Zetterberg, *J. Organometal. Chem.*, 1979, *166*, C33.
69. D.W. Moore,H.B. Jonassen and T.B. Joyner, *Chem. and Ind.*, 1960, 1304; D.J. Abbott and P.M. Maitlis, *J.C.S. Dalton*, 1976, 2156.
70. R. Hüttel, H. Dietl and H. Christ, *Chem. Ber.*, 1964, *97*, 2037; T.M. Whitesides and R.W. Arhart, *Inorg. Chem.*, 1975, *14*, 209; M. Brookhart, T.M. Whitesides and J.M. Crockett, *Inorg. Chem.*, 1976, *15*, 1550; T.M. Whitesides, R.W. Arhart and R.W. Slaven,

J. Amer. Chem. Soc., 1973, *95*, 5792; F.J. Impastato and K.G. Ihrman, *J. Amer. Chem. Soc.*, 1961, *83*, 3726.

71. R.C. Larock and J.P. Burkhart, *Synth. Commun.*, 1979, *9*, 659.
72. R.A. Plowman and F.G.A. Stone, *Z. Naturforsch*, 1962, *17b*, 575; H.D. Murdock and E. Weiss, *Helv. Chim. Acta*, 1962, *45*, 1927.
73. C.G. Hull and M.H.B. Stiddard, *J. Organometal. Chem.*, 1967, *9*, 519; R.F. Heck, *J. Org. Chem.*, 1963, *28*, 604.
74. A. Eisenstadt, *J. Organometal. Chem.*, 1972, *38*, C32.
75. B.M. Trost and P.E. Strege, *Tet. Letters*, 1974, 2603; R. Hüttel and H. Schmidt, *Chem. Ber.*, 1968, *101*, 252.
76. B.M. Trost and L. Weber, *J. Amer. Chem. Soc.*, 1975, *97*, 1611.
77. D.N. Jones and S.D. Knox, *Chem. Comm.*, 1975, 165.
78. R.W. Howsam and F.J. McQuillin, *Tet. Letters*, 1968, 3667; K. Henderson and F.J. McQuillin, *Chem. Comm.*, 1978, 15.
79. T.N. Margulis, L. Schiff and M. Rosenblum, *J. Amer. Chem. Soc.*, 1965, *87*, 3269.
80. J.L. Roustan, C. Charrier, J.Y. Mérour, J. Bénaim and C. Giannotti, *J. Organometal Chem.*, 1972, *38*, C37.
81. T. Shono, T. Yoshimura, Y. Matsumura and R. Oda, *J. Org. Chem.*, 1968, *33*, 876; A.D. Ketley and J. Braatz, *Chem. Comm.*, 1968, 959.
82. S. Baba, T. Sobata, T. Ogura and S. Kawaguchi, *Bull. Chem. Soc. Japan*, 1974, *47*, 2792.
83. I.T. Harrison, E. Kimura, E. Bohme and J.H. Fried, *Tet. Letters*, 1969, 1589.
84. E. Vedejs M.F. Salomon and P.D. Weeks, *J. Organometal. Chem.*, 1972, *40*, 221; R. Hüttel and M. Christ, *Chem. Ber.*, 1964, *97*, 1439.
85. D.N. Jones and S.D. Knox, *Chem. Comm.*, 1975, 166.

85a. J. Muzart and J-P. Pete, *Chem. Comm.*, 1980, 257.

86. R.B. King, "The Organic Chemistry of Iron", 1978, *1*, 525, Academic Press, New York.
87. Y. Shvo and E. Hazum, *Chem. Comm.*, 1975, 829.
88. P. McArdle and T. Higgins, *Inorg. Chim. Acta*, 1978, *30*, L303.
89. A.J. Birch and H. Fitton, *Aust. J. Chem.*, 1969, *22*, 971.
90. Y. Gaoni, *Tet. Letters*, 1977, 4521.
91. R.B. King, P.M. Treichel and F.G.A. Stone, *J. Amer. Chem. Soc.*, 1961, *83*, 3593.

92. E.O. Fischer, H.P. Kögler and P. Kuzel, *Chem. Ber.*, 1960, *93*, 3006.
93. J.A.S. Howell, B.F.G. Johnson, P.L. Josty and J. Lewis, *J. Organometal. Chem.*, 1972, *39*, 329.
94. G. Evans, B.F.G. Johnson and J. Lewis, *J. Organometal. Chem.*, 1975, *102*, 507.
95. D.H.R. Barton, A.A.L. Gunatilaka, T. Nakanishi, H. Patin, D.A. Widdowson and B.R. Worth, *J.C.S. Perkin 1*, 1976, 821.
96. G. Cardaci and G. Bellachioma, *Inorg. Chem.*, 1977, *16*, 3099; M. Brookhart and G.O. Nelson, *J. Organometal. Chem.*, 1979, *164*, 193.
97. A.J. Birch, W.D. Raverly and G.R. Stephenson, *Tet. Letters*, 1980, 197.
98. A.J.P. Domingos, B.F.G. Johnson and J. Lewis, *J. Organometal. Chem.*, 1973, *49*, C33.
99. M.L.H. Green, J. Knight and J.A. Segal, *J.C.S. Dalton*, 1977, 2189.
100. S. Sarel, R. Ben-Shoshan and B. Kirson, *Israel J. Chem.*, 1972, *10*, 787; T.M. Whitesides and R.W. Slaven, *J. Organometal. Chem.*, 1974, *67*, 99.
101. E.H. Braye and W. Hübel, *J. Organometal. Chem.*, 1965, *3*, 38.
102. W.R. Roth and J.D. Meier, *Tet. Letters*, 1967, 2053; B.F.G. Johnson, J. Lewis and D.J. Thompson, *Tet. Letters*, 1974, 3789.
103. M. Supozynski, I. Wolszczak and P. Kosztoowicz, *Inorg. Chim. Acta*, 1979, *33*, L97.
104. Y. Shvo and E.Hazum, *Chem. Comm.*, 1974, 336; H. Alper, *J. Organometal. Chem.*, 1975, *96*, 95; M. Franck-Neumann and D. Martina, *Tet. Letters*, 1975, 1759; A.J. Pearson, *J.C.S. Perkin 1*, 1977, 2069.
105. G.F. Emerson, J.E. Mahler, R. Kochhar and R. Pettit, *J. Org. Chem.*, 1964, *29*, 3620; C.H. Mauldin, E.R. Biehl and P.C. Reeves, *Tet. Letters*, 1972, 2955.
106. E.J. Corey and G. Moinet, *J. Amer. Chem. Soc.*, 1973, *95*, 7185.
106a. D.J. Thompson, *J. Organometal. Chem.*, 1976, *108*, 381.
106b. B.F.G. Johnson, J. Lewis, D.J. Thompson and B. Heil, *J.C.S. Dalton*, 1975, 567.
107. T.A. Manuel and F.G.A. Stone, *J. Amer. Chem. Soc.*, 1960, *82*, 366.
108. E.O. Fischer and R.D. Fischer, *Ang. Chem.*, 1960, *72*, 919.
109. A.J. Birch and K.B. Chamberlain, *Org. Synth.*, 1977, *57*, 107; A.J. Birch, P.E. Cross, J. Lewis, D.A. White and S.B. Wild, *J. Chem. Soc. A*, 1968, 332; A.J. Birch and D.H. Williamson, *J.C.S. Perkin I*, 1973, 1892.

110. H. Alper and C.C. Huang, *J. Organometal. Chem.*, 1973, *50*, 213.
111. J.E. Mahler and R. Pettit, *J. Amer. Chem. Soc.*, 1963, *85*, 3955, 3959: T.S. Sorensen and C.R. Jablonski, *J. Organometal. Chem.*, 1970, *25*, C62; P. Powell, *J. Organometal Chem.*, 1979, *165*, C43.
112. L.F. Kelly, A.S. Narula and A.J. Birch, *Tet. Letters*, 1980, 871.
113. H.J. Dauben and D.J. Bertelli, *J. Amer. Chem. Soc.*, 1961, *83*, 497; B.F.G. Johnson, J. Lewis and G.R. Stephenson, *Tet. Letters*, 1980, 1995.
114. G.E. Herberich and H. Müller, *Chem. Ber.*, 1971, *104*, 2781.
115. G. Jaouen, "Transition Metal Organometallics in Organic Synthesis," *II*, 64, Ed. H. Alper, Academic Press, 1978, New York.
116. W.E. Silverthorn, *Adv. in Organometal. Chem.*, 1975, *13*, 47.
117. E.O. Fischer and R. Bottcher, *Z. Anorg. Allgem. Chem.*, 1957, *291*, 305.
118. M.L.H. Green, L. Pratt and G. Wilkinson, *J. Chem. Soc.*, 1960, 989; T.H. Coffield, N. Sandel and R.D. Closson, *J. Amer. Chem. Soc.*, 1957, *79*, 5826.
119. P.L. Pauson and J.A. Segal, *J.C.S. Dalton*, 1975, 1677.
120. R.R. Schrock and J.A. Osborn, *J. Amer. Chem. Soc.*, 1971, *93*, 3089.
121. R.A. Zelonka and M.C. Baird, *J. Organometal. Chem.*, 1972, *35*, C43, *44*, 383.
122. M.A. Bennett and A.K. Smith, *J.C.S. Dalton*, 1974, 233.
123. B. Nicholls and M.C. Whiting, *J. Chem. Soc.*, 1959, 551.
124. E.O. Fischer, K. Öfele, M. Essler, W. Fröhlich, J.P. Mortensen and W. Semmlinger, *Chem. Ber.*, 1958, *91*, 2763.
125. H.B. Arzeno, D.H.R. Barton, S.G. Davies, X. Lusinchi, B. Meunier and C. Pascard, *Nouveau J. Chimie*, 1980, *4*, 369.
126. A.J. Birch, P.E. Cross, D.T. Connor and G.S.R. Subba Rao, *J. Chem. Soc. C*, 1966, 54.
127. A.J. Birch, P.E. Cross and H. Fitton, *Chem. Comm.*, 1965, 366.
128. F. Calderazzo, *Inorg. Chem.*, 1965, *4*, 223.
129. M. Bochmann, M. Cooke, M. Green, H.P. Kirsch, F.G.A. Stone and A.J. Welch, *Chem. Comm.*, 1976, 381.
130. K.K. Joshi, P.L. Pauson, A.R. Qazi and W.H. Stubbs, *J. Organometal. Chem.*, 1964, *1*, 471; K. Ofele and E. Dotzauer, *J. Organometal. Chem.*, 1971, *30*, 211.
131. A.N. Nesmeyanov, N.A. Vol'Kenau and I.N. Bolesova, *Tet. Letters*, 1963, 1725.

132. G. Jaouen and R. Dabard, *Tet. Letters*, 1971, 1015; D.A. Brown, D. Cunningham and W.K. Glass, *Chem. Comm.*, 1966, 306; G. Jaouen and A. Meyer, *J. Amer. Chem. Soc.*, 1975, *97*, 4667; G. Jaouen, A. Meyer and G. Simmonneaux, *Chem. Comm.*, 1975, 813.

133. L. Watts, J.D. Fitzpatrick and R. Pettit, *J. Amer. Chem. Soc.*, 1965, *87*, 3253; W.S. Trahanovsky and R.J. Card, *J. Amer. Chem. Soc.*, 1972, *94*, 2897; R.J. Card and W.S. Trahanovsky, *Tet. Letters*, 1973, 3823.

133b.M.F. Semmelhack and H.T. Hall, *J. Amer. Chem. Soc.*, 1974, *96*, 7091; G. Jaouen and A. Meyer, *J. Amer. Chem. Soc.*, 1975, *97*, 4667.

134. C.N. Matthews, T.A. Magee and J.H. Wotiz, *J. Amer. Chem. Soc.*, 1959, *81*, 2273; T. Kruck, *Chem. Ber.*, 1964, *97*, 2018; F. Zingales, A. Chiesa and F. Basolo, *J. Amer. Chem. Soc.*, 1966, *88*, 2707; J.A. Segal, M.L.H. Green, J.C. Daran and K. Prout, *Chem. Comm.*, 1976, 766; A. Pidcock, J.D. Smith and B.W. Taylor, *J. Chem. Soc. A*, 1969, 1604; J.F. White and M.F. Farona, *J. Organometal. Chem.*, 1972, *37*, 119.

135. G. Carganico, P. Del Buttero, S. Maiorana and G. Riccardi, *Chem. Comm.*, 1978, 989 and references therein.

136. C.A.L. Mahaffy and P.L. Pauson, *J. Chem. Res. S*, 1979, 126; F. Edelman, D. Wormsbächer and U. Behrens, *Chem. Ber.*, 1978, *111*, 817.

137. K.M. Sharma, S.K. Anand, R.K. Multani and B.D. Jain, *J. Organometal. Chem.*, 1970, *21*, 389; H.O. Van Oven, H.J. De Liefde Meijer, *J. Organometal. Chem.*, 1970, *23*, 159.

138. J.D. Munro and P.L. Pauson, *J. Chem. Soc.*,1961, 3475; H.J. Dauben and L.R. Honnen, *J. Amer. Chem. Soc.*, 1958, *80*, 5570; J. Müller and B. Mertschenk, *J. Organometal. Chem.*, 1972, *34*, C41.

CHAPTER 3

ORGANOMETALLICS AS PROTECTING AND STABILISING GROUPS

As discussed in Chapter 1, coordination of an olefin to a transition metal leads to a change in the electron density on the olefin. This results in the coordinated olefin being able to undergo reactions under conditions where the free olefin is inert. Alternatively, complexation of an olefin can lead to non-reactivity towards reagents that normally react with uncoordinated olefins. This latter property allows organometallic species to be used as protecting groups in organic synthesis.

3.1 THE USE OF ORGANOMETALLIC SPECIES AS PROTECTING GROUPS

3.1.1. Protection of olefins:

Coordination of an olefin to iron in the cations Fp(olefin)$^+$ [Fp = $(C_5H_5)Fe(CO)_2$] increases the reactivity of the olefin towards nucleophilic attack and reduces its reactivity towards electrophilic attack (see chaps. 4 and 5). This latter property allows coordination to Fp$^+$ to be used as a convenient protection method for olefins against electrophilic attack. Fp(olefin)$^+$ cations may be prepared by a variety of methods (see chap. 2) and release of the olefin is effected simply by treatment with NaI in acetone.[1] Using the exchange reaction between Fp(isobutylene)$^+$ BF_4^- and polyenes, selective coordination to the least substituted or most strained double bond in polyenes has been demonstrated.[2] Cations 1, 2, 3 and 4 react with H_2 over 10% Pd/C in CF_3CO_2H to afford the monoene complexes 5, 6, 7 and 8 respectively.[3] Hydrogenation of the free diene corresponding to 3, for example leads to the other monoene being produced. This protection method often complements the usual bromination-debromination and epoxidation-deoxygenation methods which protect the most substituted double bonds.

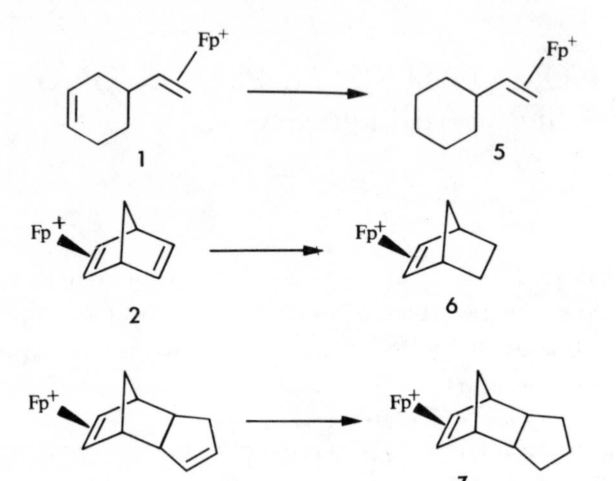

Coordination of one of the double bonds of a conjugated diene system allows selective reduction of the other.[4]

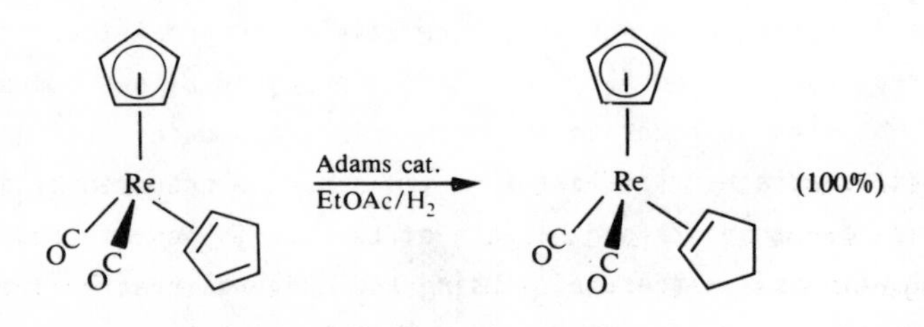

The protected norbornadiene derivative 2 undergoes electrophilic addition reactions without the usual homoallylic rearrangements occurring.[3]

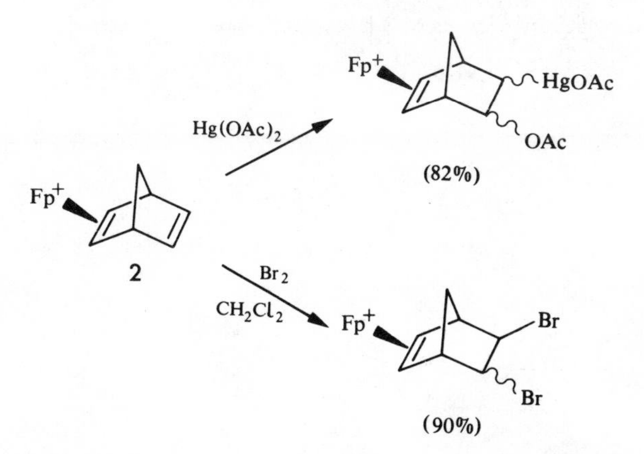

The stabilised and protected cation Fp(η^2-cyclobutadiene)$^+$ is constrained to act simply as a dienophile and allows the generation of cyclobutadiene-diene adducts.[5]

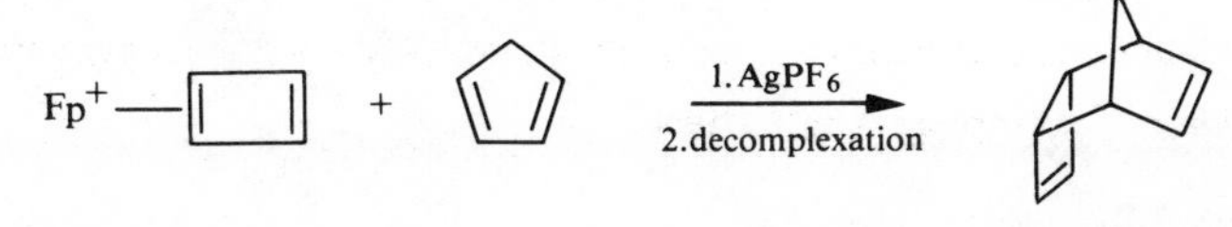

Electrophilic addition of bromine to double bonds is usually faster than electrophilic aromatic substitution. However protection of the double bond in eugenol 9 as the Fp(olefin)$^+$ cation leads to selective aromatic substitution to afford, after decomplexation, 10 .[5]

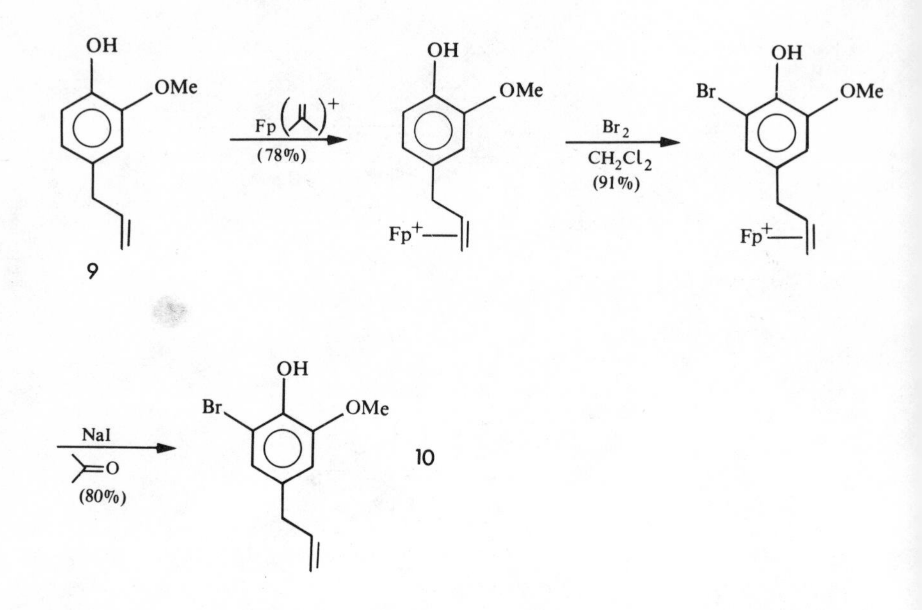

3.1.2 Protection of acetylenes:

Reactions of ene-ynes with $Co_2(CO)_8$ lead to selective coordination of the acetylene group. The high yields of complex formation and the efficient removal/recovery of the acetylene makes these complexes useful acetylene protecting groups.[6,7] Selective reaction of the double bond in ene-ynes with diimide or "BH_3" has been achieved in this way.[8]

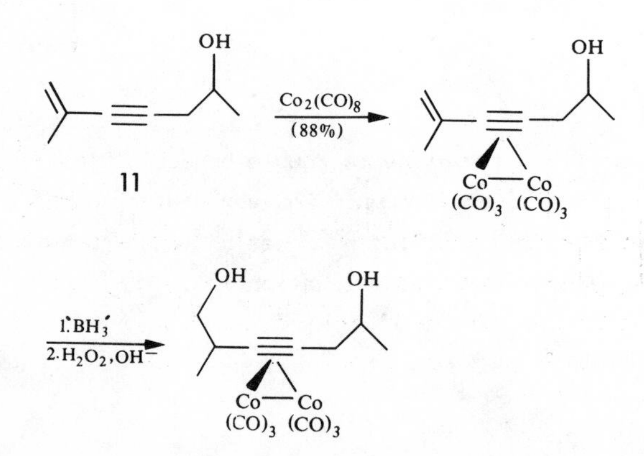

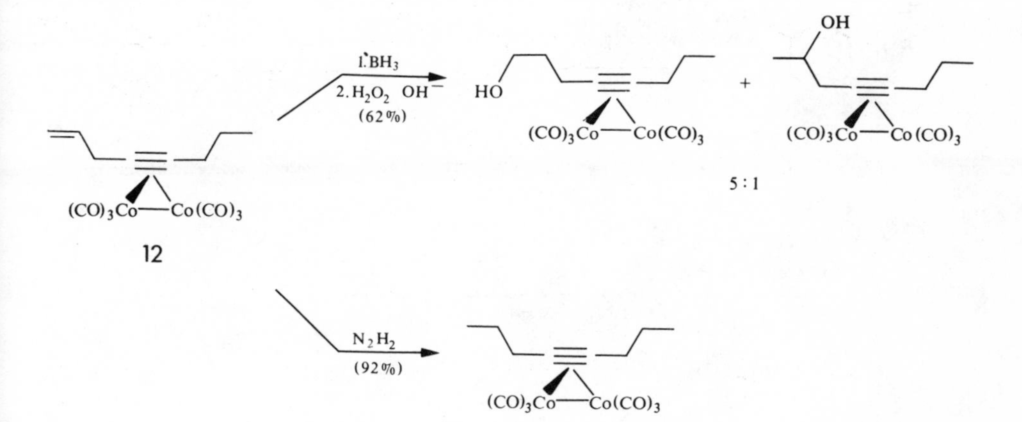
1. BH3
2. H2O2 OH−
(62%)
HO
OH
(CO)3Co
Co(CO)3
5 : 1
12
N2H2
(92%)

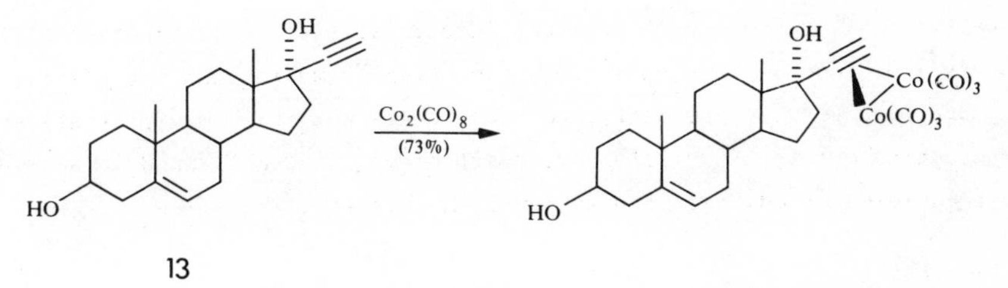
OH
HO
13
Co2(CO)8
(73%)
Co(CO)3

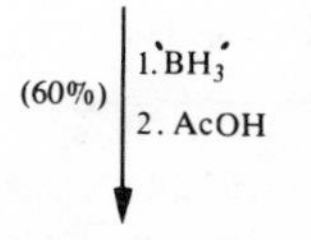
(60%)
1. BH3
2. AcOH

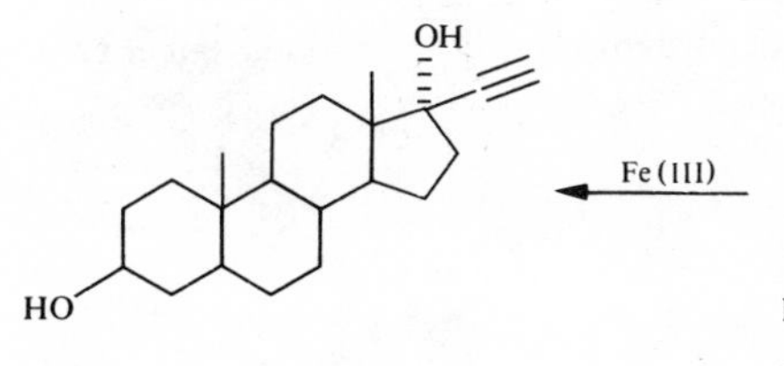
OH
HO
Fe(III)

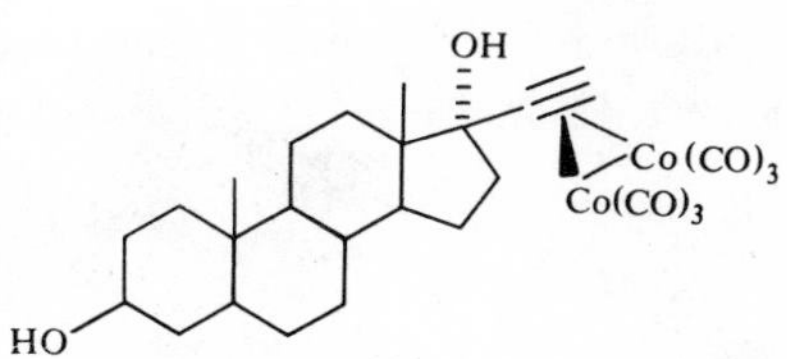
OH
HO
Co(CO)3

Compound 11 reacts with strong acid to give an intractable mixture of products. However its $Co_2(CO)_6$ complex reacts to give the diol 14 in high yield presumably via the stabilised carbonium ion shown.

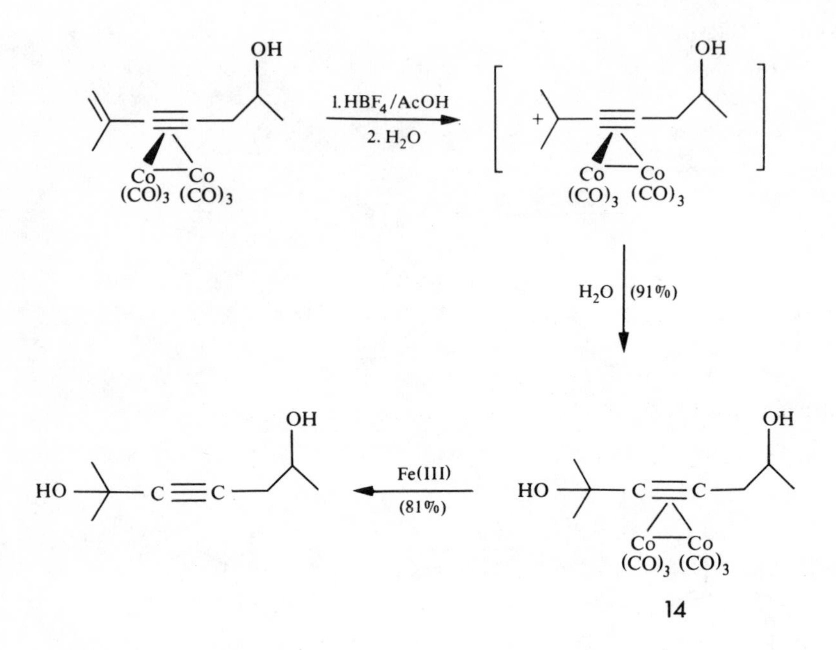

The isolable cationic complexes 15 react with ketones, trimethylsilyl enol ethers, enol acetates and allyl trimethylsilanes to yield after decomplexation a variety of substituted acetylenic compounds uncontaminated with allenic isomers.[9]

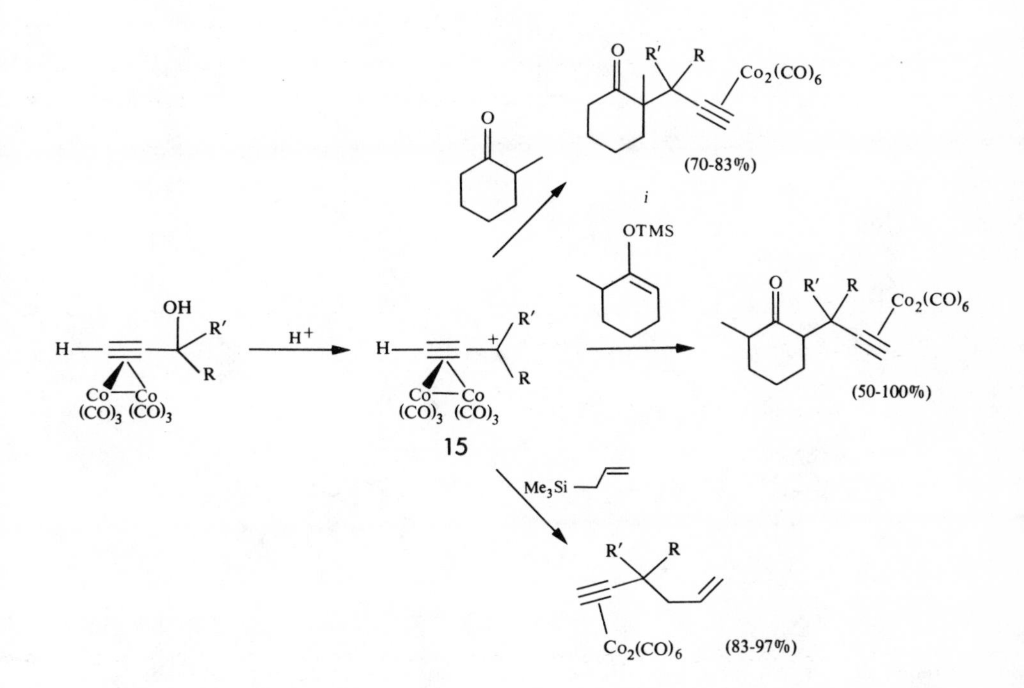

The increased stability of complexed acetylenic alcohols towards acid has been used in the synthesis of insect sex pheromones.[10] Treatment of the alcohol 16 with $HBr/ZnBr_2$ gives a mixture of the Z and E bromides shown in the ratio 98:2. Prior complexation with $Co_2(CO)_8$ leads to the opposite Z to E ratio of 2:98.

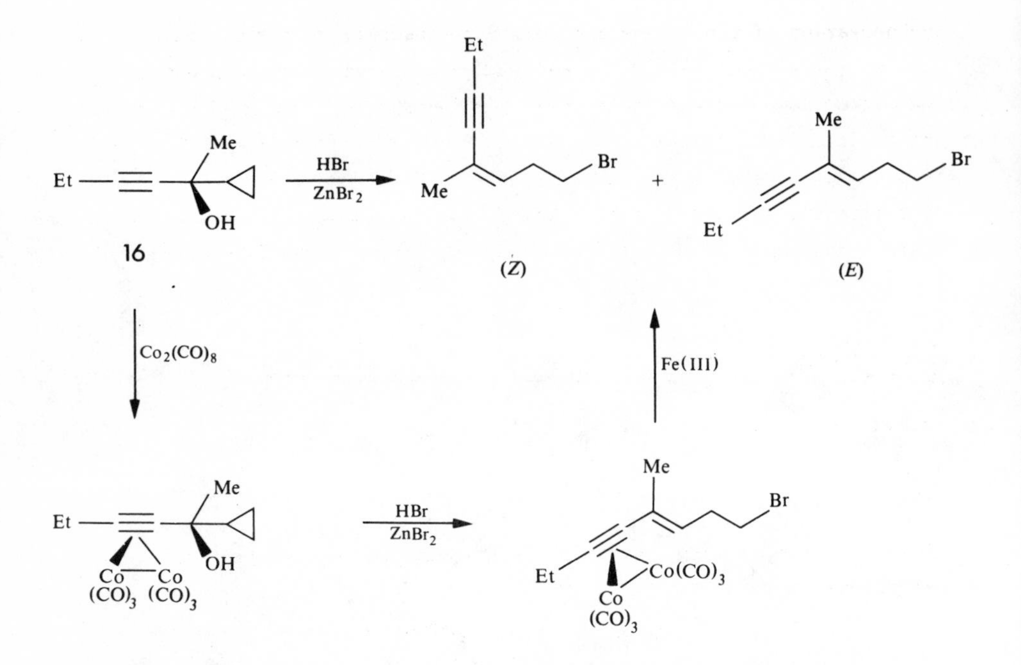

Friedel Crafts acylation of diphenyl acetylene cannot be accomplished directly. However para-substitution of the $Co_2(CO)_6$ complex 17 occurs efficiently.[6]

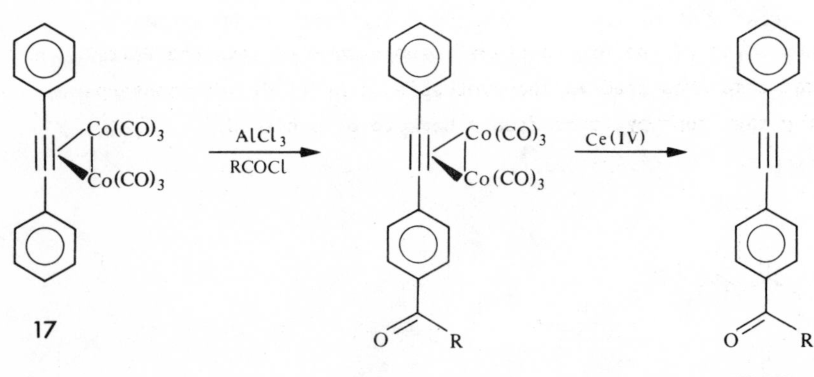

3.1.3. Protection of dienes:

Hydroboration of the triene complex 18 followed by oxidation leads only to the alcohol 19.[11] The $Fe(CO)_3$ group serves both to protect the diene component and to increase the regioselectivity of the reaction. The regioselectivity may be due either to the steric bulk of the diene iron tricarbonyl or to the ability of the diene iron tricarbonyl to stabilise the adjacent positive charge in the hydroboration transition state. Treatment of 19 with $FeCl_3$ releases *trans trans* 3,5-heptadien-1-ol.

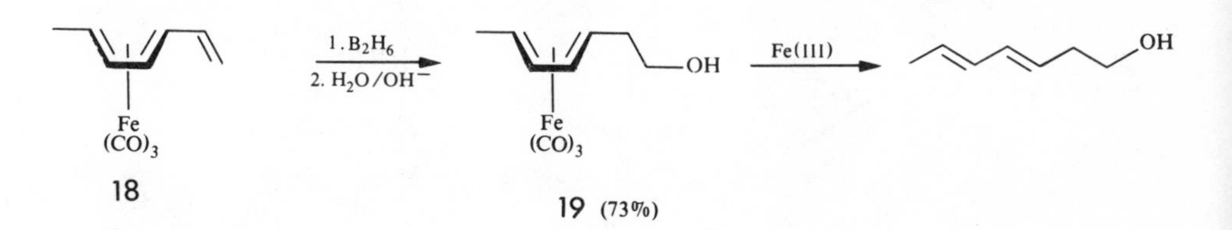

Similarly hydroboration-oxidation of 20 produces the alcohol 21 after decomplexation.

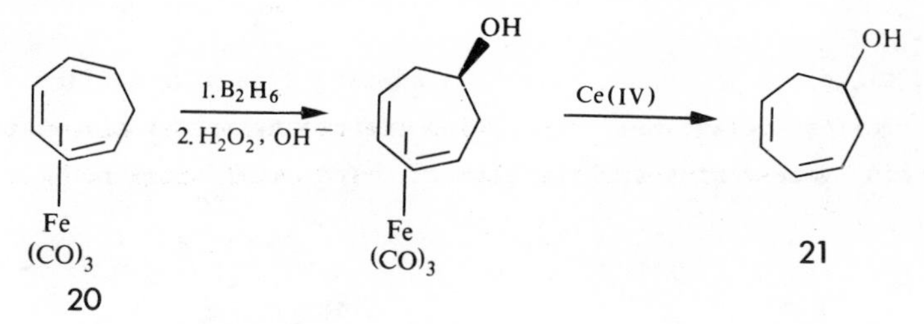

Hydroboration of the non-complexed double bond of (myrcene)$Fe(CO)_3$ and related complexes enables the synthesis of novel dihydromonoterpenes lacking the isopropylidene double bond to be achieved.[12]

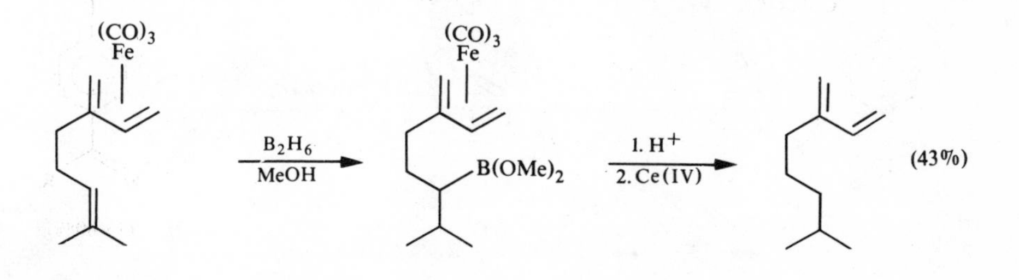

The diene system in bicyclo[4.2.1]nonatrienone 22 can be protected against catalytic hydrogenation by complexation to $Fe(CO)_3$.[13]

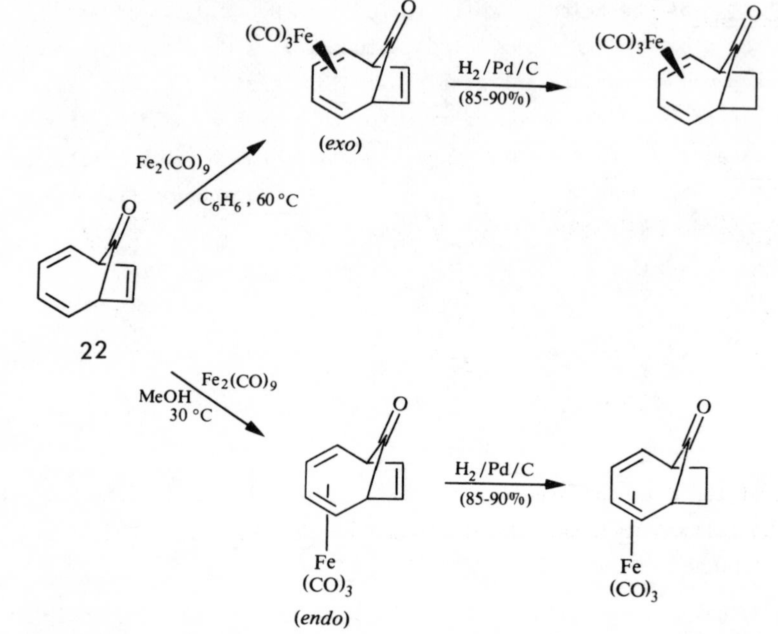

The $Fe(CO)_3$ group in the *exo* isomer also serves as a protecting group for the carbonyl group against borohydride reduction, presumably because of steric hindrance in the product alcohol. The *endo* isomer however reduces smoothly.

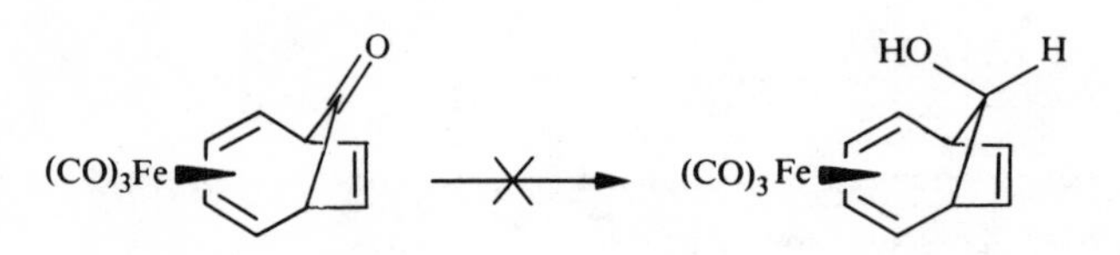

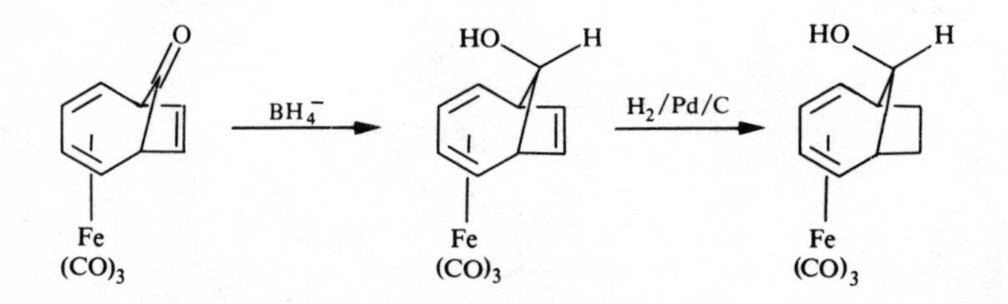

The diene groups of the complexes 23 and 24 are protected against isomerisation to the fully conjugated isomers during Grignard and Wittig reactions.[14]

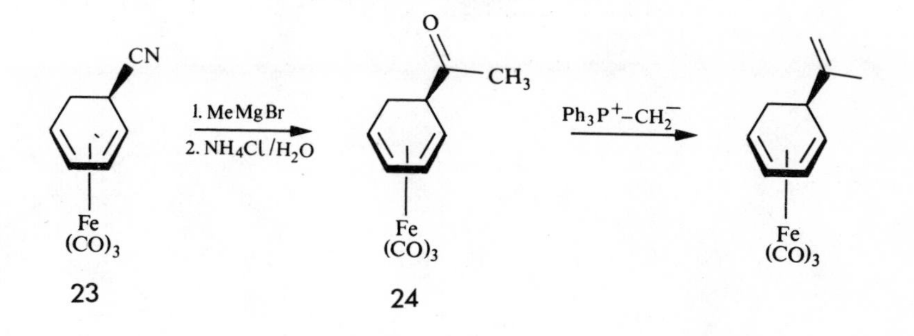

Protection of the diene of ergosterol esters 25 with $Fe(CO)_3$ in 26 allows the manipulation of the side chain double bond.[15]

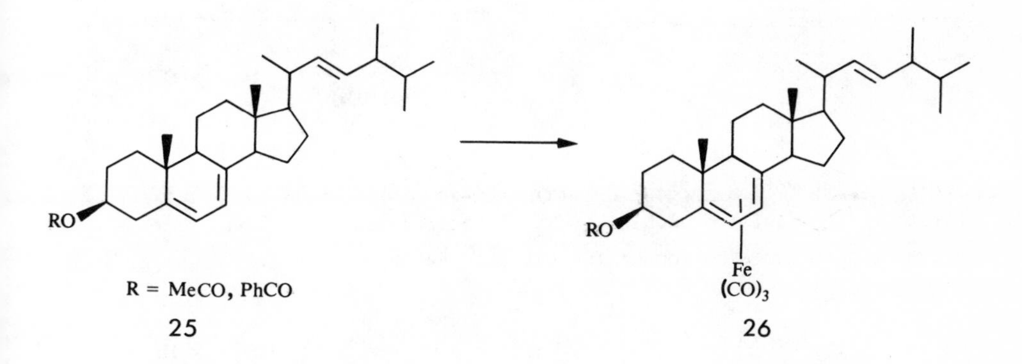

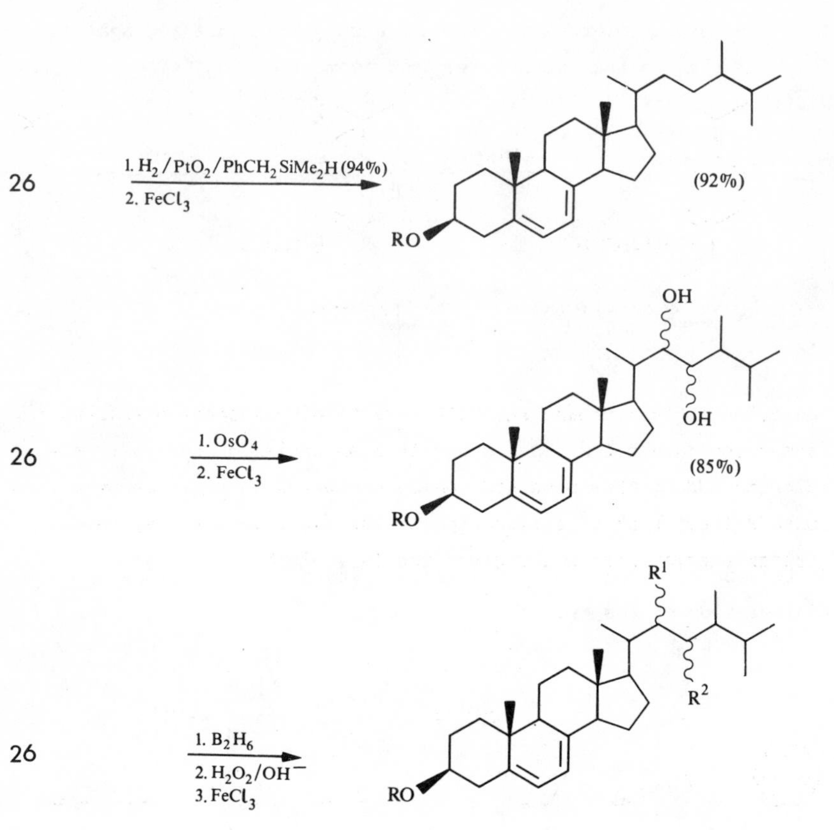

Dichlorocarbene is expected to add preferentially to conjugated dienes rather than isolated double bonds. This preference can however be overcome on coordination of the diene to $Fe(CO)_3$.[16] Some examples are shown below.

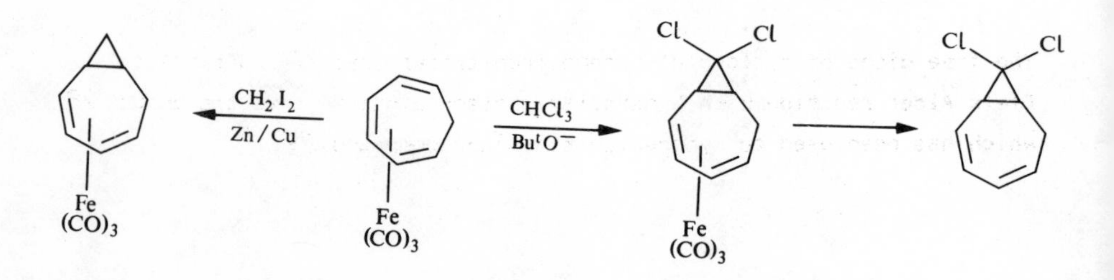

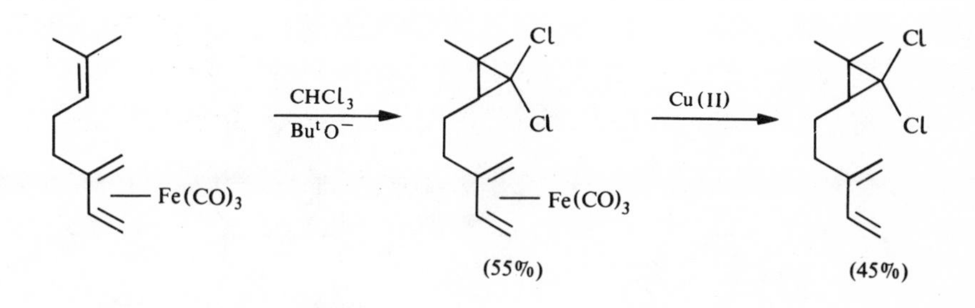

Reaction of (cyclohexadiene)$Fe(CO)_3$ with dibromocarbene results in insertion into a C-H bond rather than an addition reaction. The dibromo adduct thus produced can be converted to aldehyde by acid or base without risk of conjugation of the diene with the aldehyde. Decomplexation with Me_3NO gives the free aldehyde.[17]

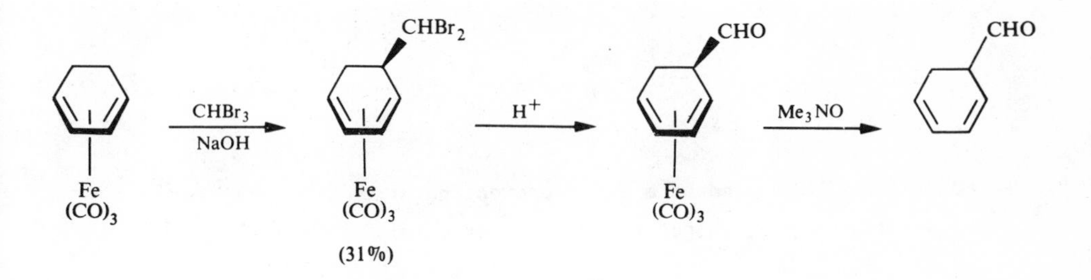

The ergosteryl acetate iron tricarbonyl complex 26 is inert to dichloro-carbene addition presumably because of steric hindrance around the 22,23-double bond.[16]

The free diene of cyclooctatetraene iron tricarbonyl 27 undergoes a Diels Alder reaction with N-methyltriazolene dione to give the adduct 28 which has been used to synthesise the diaza compound 29.[18]

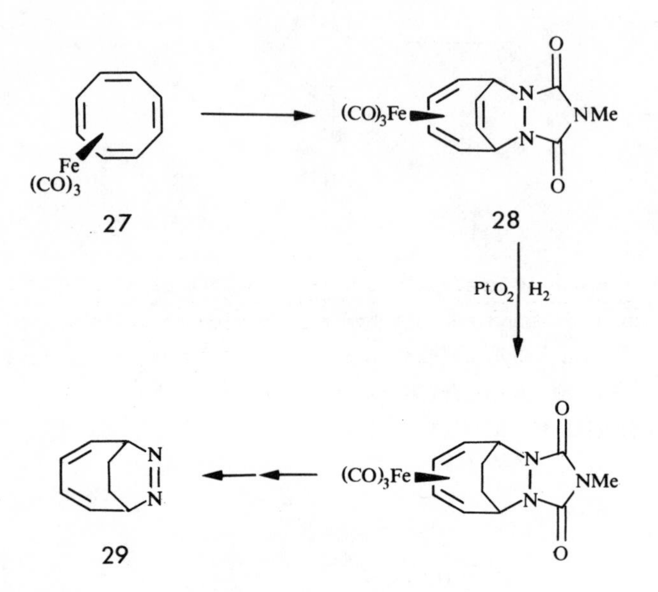

$Fe(CO)_3$ may also be used as a diene protecting group during acetyl chloride/$AlCl_3$ acylation reaction,[19] for example:

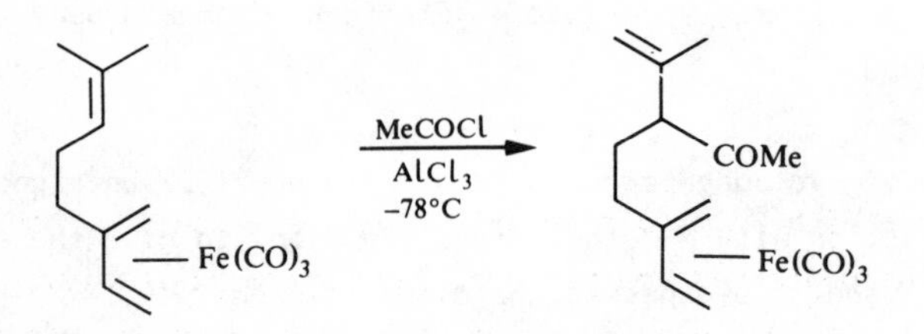

The reaction of tropones with diazoalkanes normally gives rise to ring-expanded products. Formation of their $Fe(CO)_3$ complexes, however, allows the cycloaddition reaction onto the uncomplexed double bond to proceed efficiently.[20]

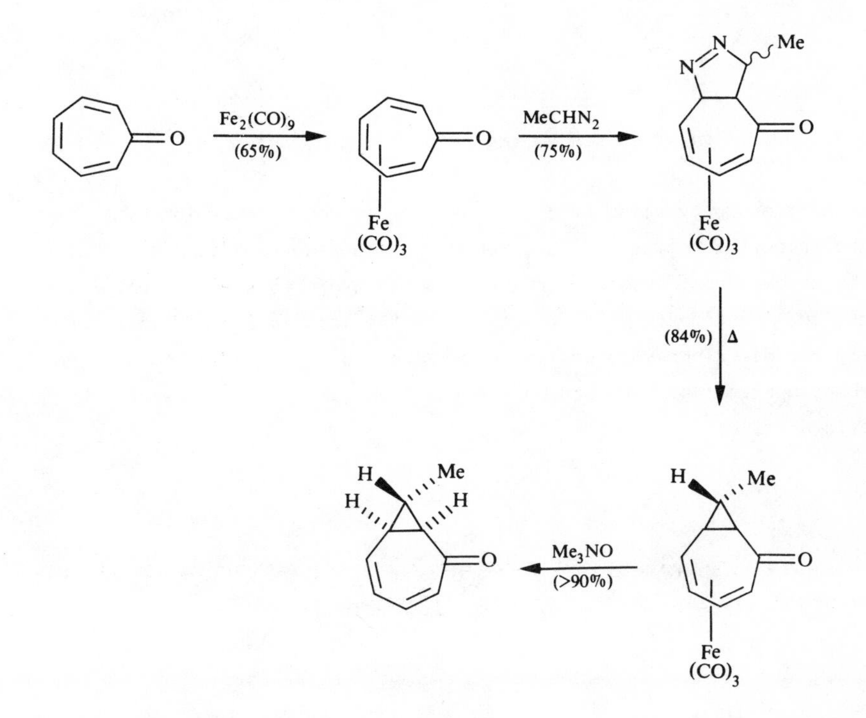

Thebaine reacts with BrCN to give extensively rearranged products. Thebaine can be protected against such rearrangements, however, by complexation to $Fe(CO)_3$ when reaction with BrCN leads to smooth conversion of NMe to NCN.[21]

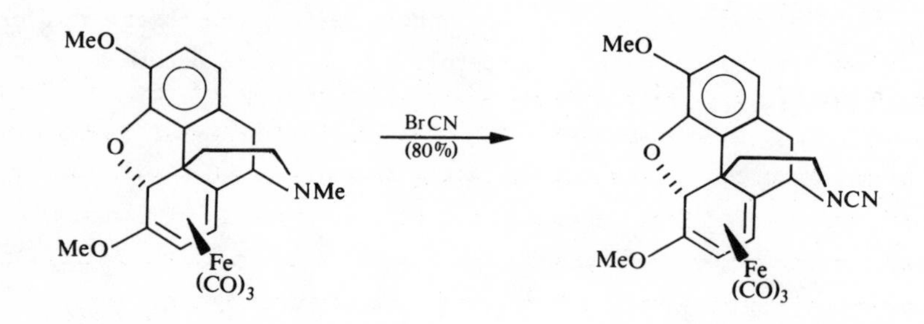

3.1.4 Protection of amines during peptide synthesis:

The amine group of amino acids has been protected by reaction with pentacarbonyl chromium and tungsten methoxycarbene complexes 30.[22] The amino carbene complexes 31 survive alkaline hydrolysis of the ester function. Peptide coupling reactions at the free acid can be performed using the dicyclohexylcarbodiimide/N-hydroxysuccinimide method and the carbene protecting group can be removed from the resulting peptide by 80% acetic acid.

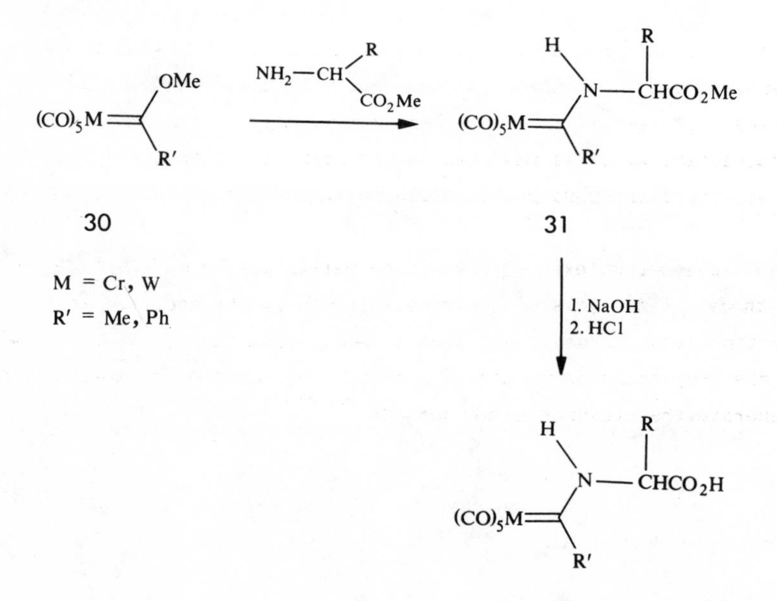

3.2 STABILISATION OF REACTIVE COMPOUNDS AND TRAPPING OF REACTION INTERMEDIATES

The change of reactivity that occurs on coordination of compounds to transition metals allows the isolation of transition metal complexes of highly reactive species which in the free state would have very short life times. Some examples of compounds which form stable transition metal complexes but are unstable in the free state are cyclobutadiene, 7-norbornadienone, strained acetylenes and dihydropyridines.

Transition metal complexes may also be used as a convenient form of storage of highly reactive species. The reactive species can then be generated by controlled release from the transition metal. Coodination to a transition metal may also be used for the stabilisation of thermodynamically disfavoured tautomers (e.g. enols) and the trapping of reaction intermediates that would otherwise react further.

3.2.1 Stabilisation of reactive compounds

Cyclobutadiene

It has not proved possible to isolate the highly reactive species cyclobutadiene.[43] However its complexes with transition metals are relatively stable and such complexes can be used as a source of cyclobutadiene which is released in the presence of oxidising agents. The cyclobutadiene thus produced can be trapped by suitable dienophiles.

Cyclobutadiene complexes of transition metals may be made by a variety of methods. Photolysis of α-pyrone followed by the addition of a transition metal carbonyl and further photolysis removes carbon dioxide from the α-pyrone and two carbon monoxide ligands from the metal carbonyl to generate the cyclobutadiene complex.[24-28]

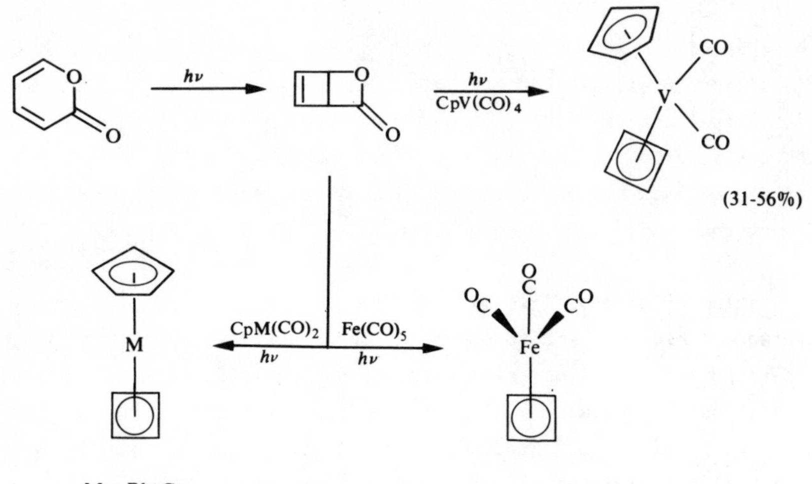

Substituted cyclobutadiene complexes may be made from the appropriate α-pyrone[29] or by modification of the cyclobutadiene complex.[26,28,30]

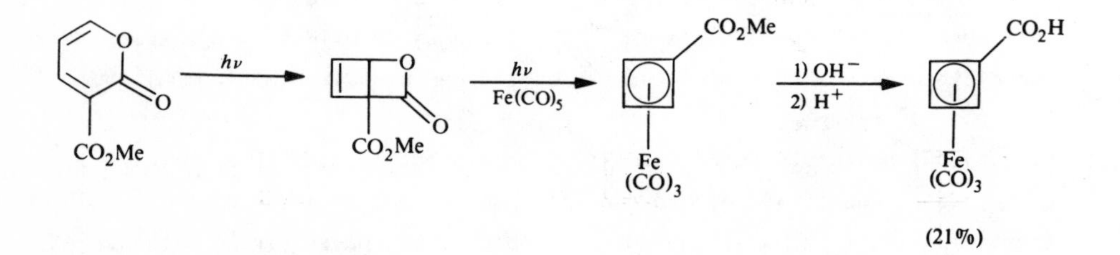

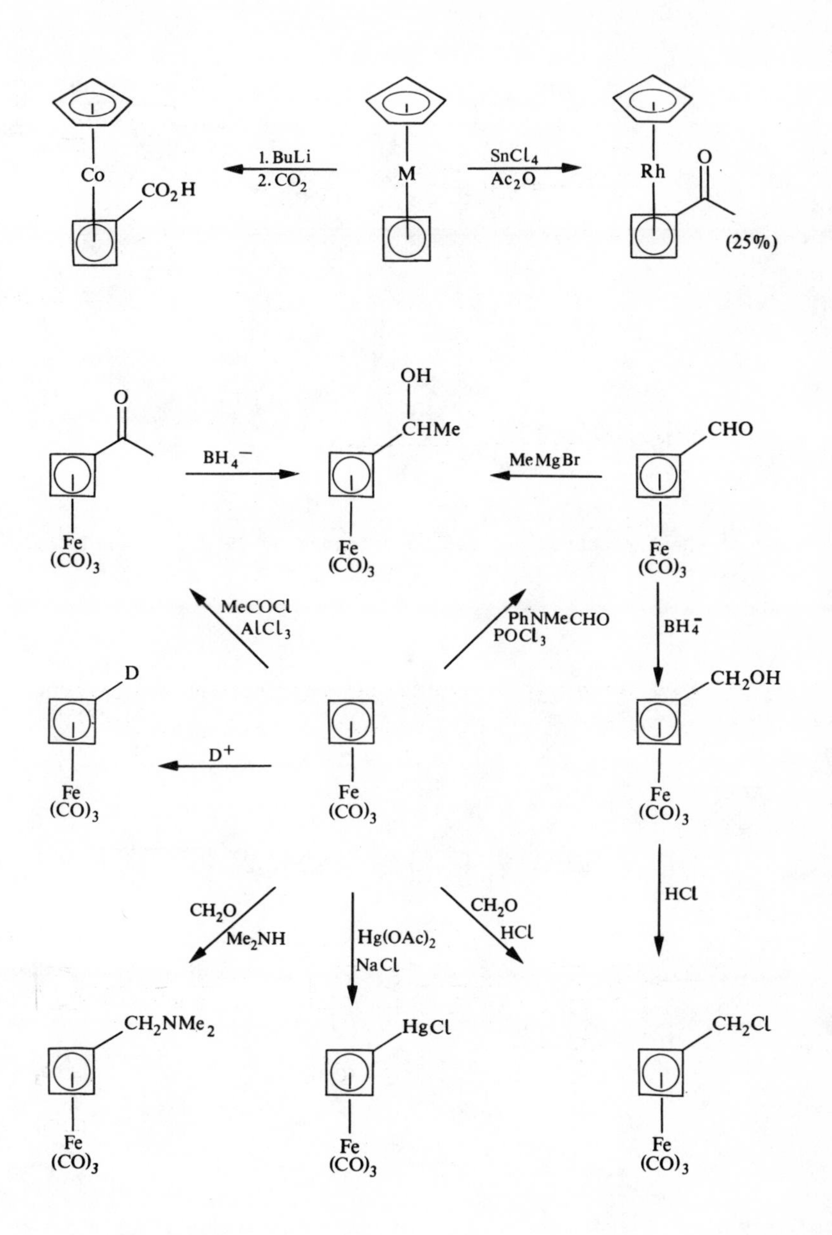

Cyclobutadiene complexes may also be prepared from dihalocyclobutenes.[31,32]

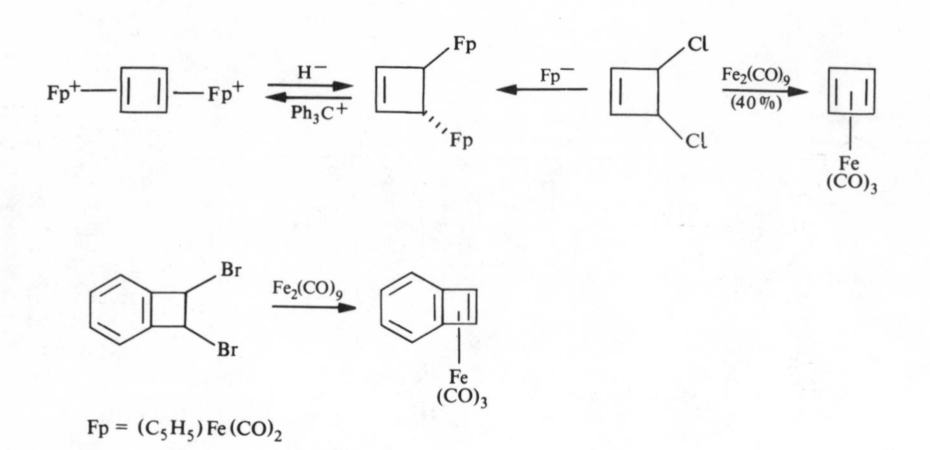

Release of the cyclobutadiene can be effected by oxidation with Ce(IV) or Fe(III) salts.[33]

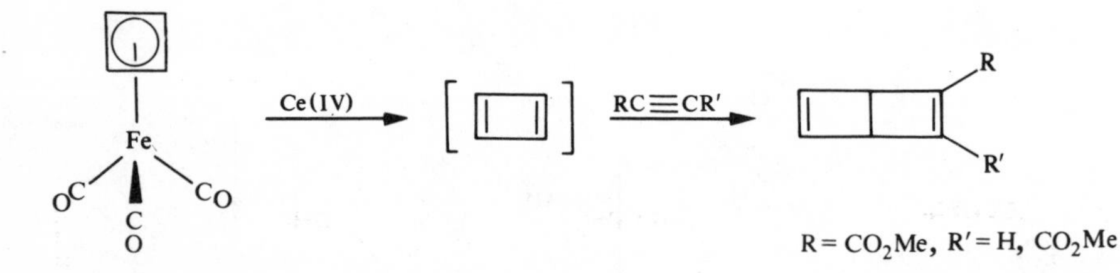

Some very elegant experiments using an optically active cyclobutadiene complex 32 [34] and polymer supported complexes[35] have demonstrated that free cyclobutadiene is indeed produced.

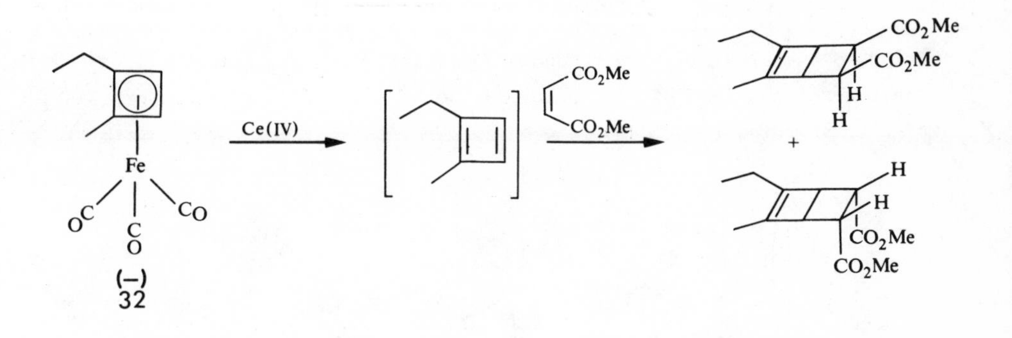

7-Norbornadienone

Complexation to iron tricarbonyl has been used to stabilise 7-norbornadienone.[36] 7-Norbornadienone-iron tricarbonyl 33 was prepared by oxidation of the 7-norbornadienol complex.

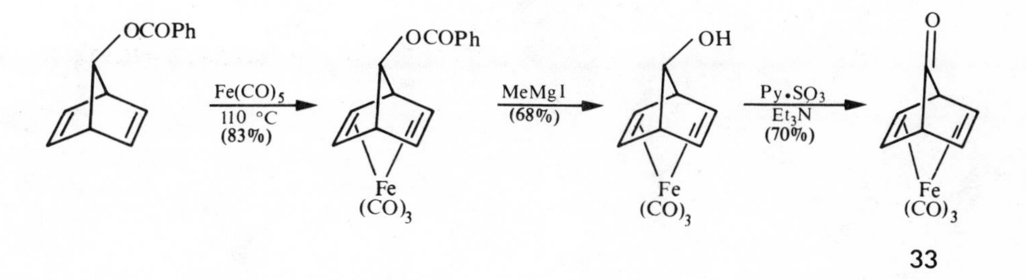

Decomplexation of the 7-norbornadienone by Ce(IV), Fe(III), thermolysis or photolysis led to the formation of carbon monoxide and benzene. 7-Norbornadienone iron tricarbonyl behaves as a normal ketone towards Grignard and alkyl lithium reagents and sulphur and phosphorus ylides. These reactions lead after decomplexation to a relatively easy synthesis of 7-substituted-7-norbornadienols.

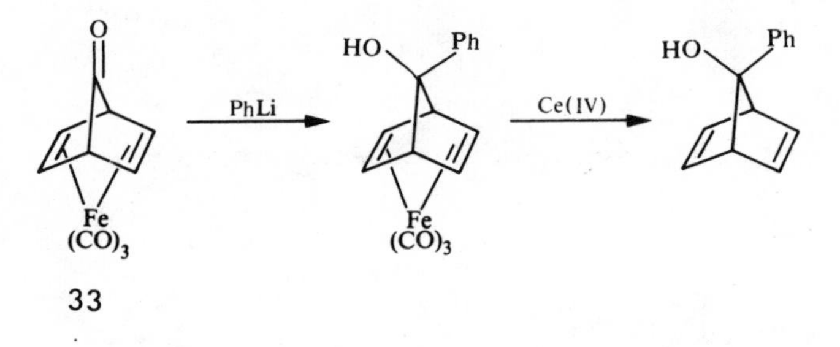

Strained olefins and acetylenes

Bistriphenylphosphine platinum diphenylacetylene 34 has been shown to have a structure in which the two acetylenic phenyl groups are displaced from the line of the C-C triple bond by 40°.[37] This bending is believed to be due to the effects of backbonding from the Pt to the antibonding acetylene π^* orbitals.[38,39]

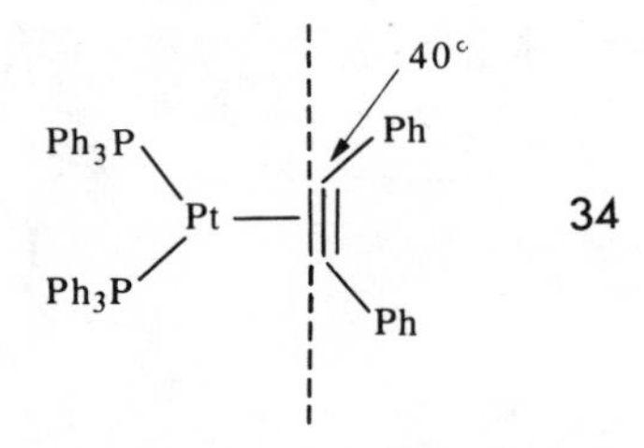

It is not surprising therefore that stable $(Ph_3P)_2$Pt(acetylene) complexes of medium ring acetylenes are formed where the strain of the medium ring acetylene is relieved on coordination to the Pt. The smallest unsubstituted cycloalkyne isolable is cyclo-octyne, however stable

$(Ph_3P)_2Pt$(acetylene) complexes of cyclohexyne and cycloheptyne can be formed.[40]

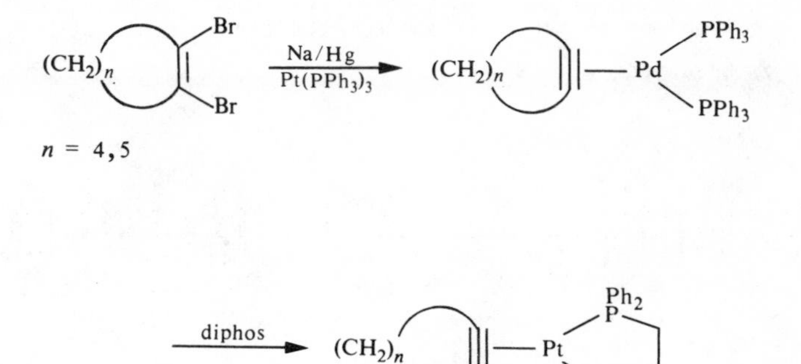

An interesting feature of the chemistry of the cyclohexyne platinium diphos complex 35 is its ease of protonation by weak acids.

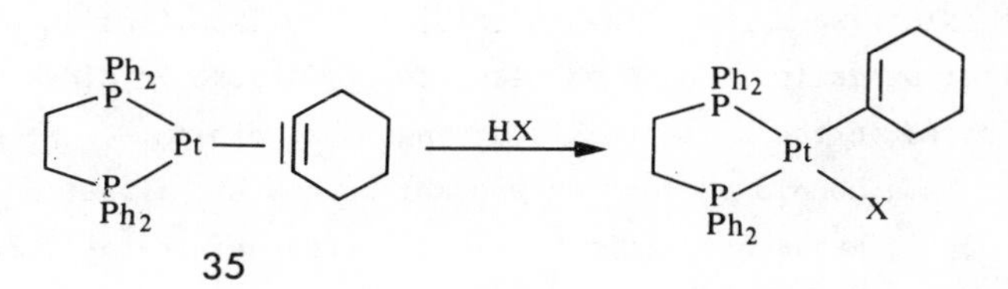

$HX = H_2O, MeOH, RCOCH_3, MeNO_2, RCH_2CN, RCONH_2, PhSH$, etc.

A stable complex of benzyne has also been reported.[41]

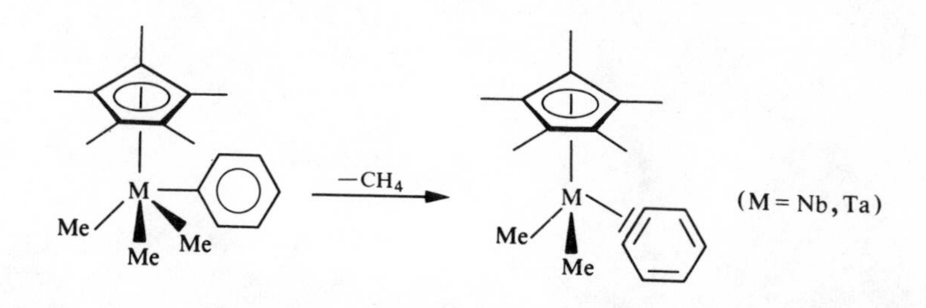

Coordination of strained olefins to Pt also leads to the isolation of stable complexes. For example, reaction of a solution of $\Delta^{1,4}$-bicyclo-[2.2.0]hexene with $(Ph_3P)_2Pt(C_2H_4)$ gave the stable complex 36.[42] Formation of this complex provided a means of purification of the olefin and a convenient method of storage. The olefin could be regenerated by treatment with CS_2.

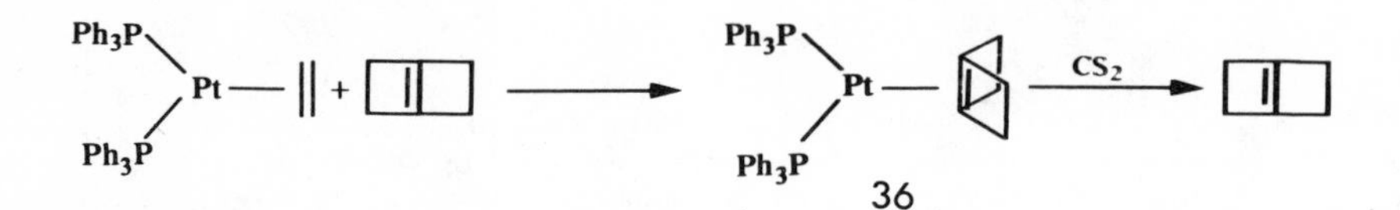

The stability of 36 can again be attributed to relief of strain caused by back donation from the metal.

Dihydropyridines

It is well established that dihydropyridines are important intermediates in biological oxidation-reduction reactions. Dihydropyridines have also been implicated in the biosynthesis of indole alkaloids.[43] Metal carbonyl derivatives have been used for the protection and stabilisation of 1,2-dihydropyridines since it is possible to regenerate the 1,2-dihydro-pyridines under mild conditions.[43-45]

Treatment of the 1,2-dihydropyridine derivative 37 with $Cr(CO)_3(CH_3CN)_3$ gave a separable mixture of the 1,2- and 1,6-dihydropyridine complexes 38 and 39. Decomplexation with pyridine produced the pure isomer 40.

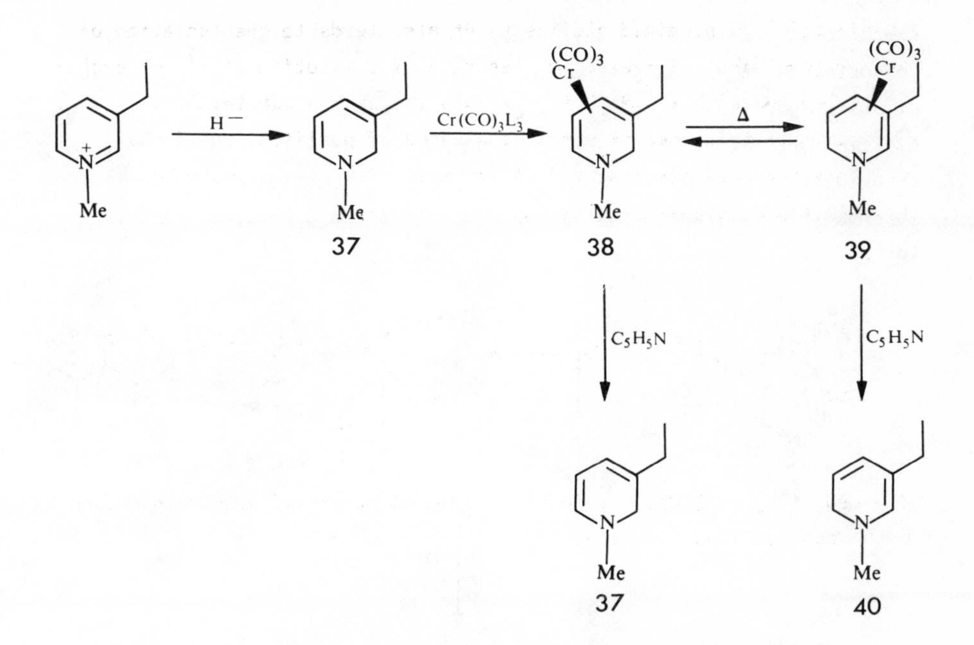

It is also possible to prepare complex 42 in which the $Cr(CO)_3$ preferentially complexes the dihydropyridine rather than the indole moiety.[44]

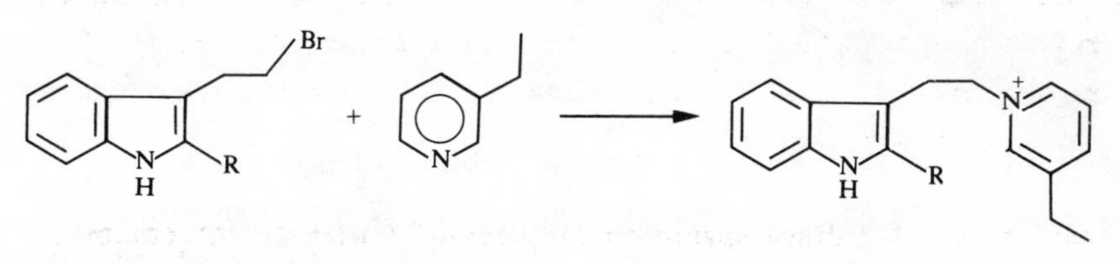

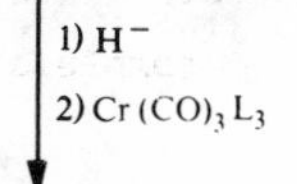

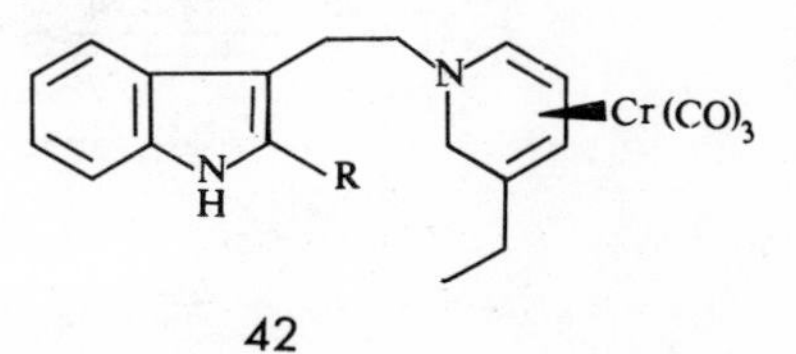

Coordination of the dihydropyridines to $Cr(CO)_3$ renders them susceptible to nucleophilic attack (see chap. 4) as demonstrated in the preparation of 3,5-disubstituted tetrahydropyridines 43 and substituted 1,6-dihydropyridine $Cr(CO)_3$ complexes 44.[46] Treatment of 38 with LiCHMeCN followed by oxidation generated the new complex 44. Protonation of the initial intermediate followed by oxidative decomplexation generated the tetrahydropyridine 43.

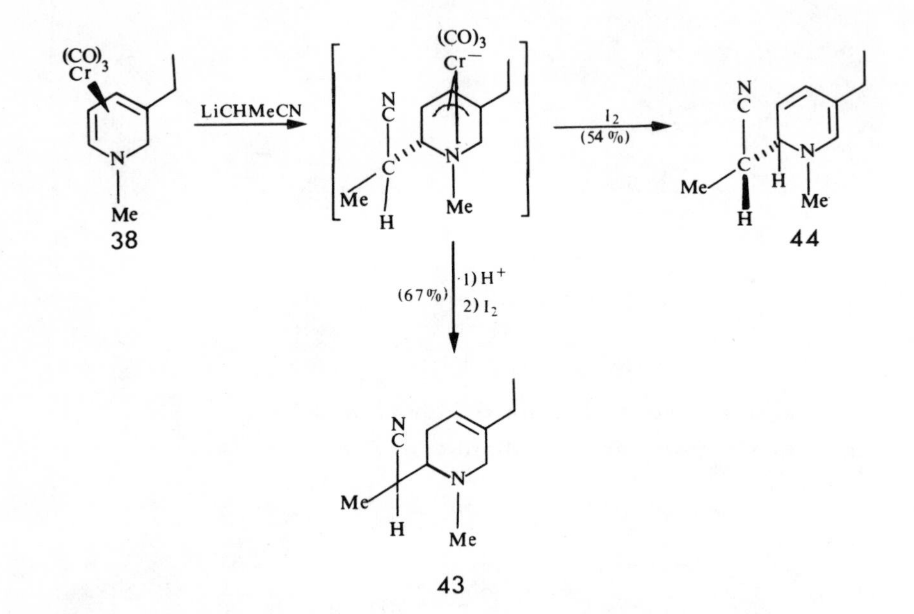

Treatment of either 1,2- or 1,4-dihydro-N-carbomethoxypyridine with $Fe_2(CO)_9$ in benzene at room temperature gave the 1,2-dihydropyridine iron tricarbonyl derivative 45 (~ 40%). The 1,2-dihydropyridine can be liberated from 45 using Me_3NO.[45]

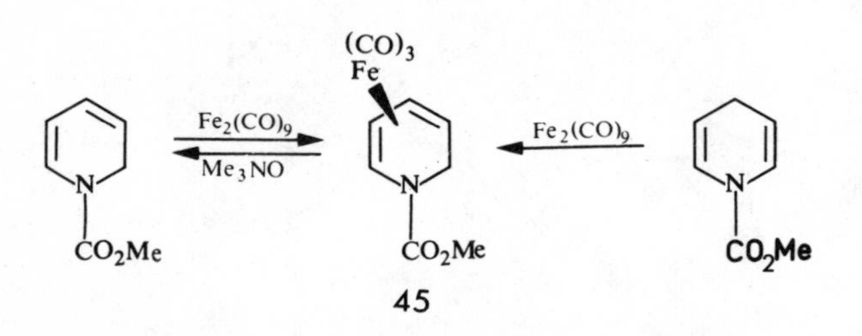

3.2.2 Stabilisation of thermodynamically disfavoured tautomers

Vinyl alcohol is approximately 12.9-14.6 Kcal mol^{-1} less stable than its tautomer, acetaldehyde, and thus it has not been possible to prove its existence as a free molecule. However it has been possible to prepare transition metal complexes of vinyl alcohol for example <u>via</u> trimethylsilyl vinyl ether[47-49] or α-chloroacetaldehyde.[50]

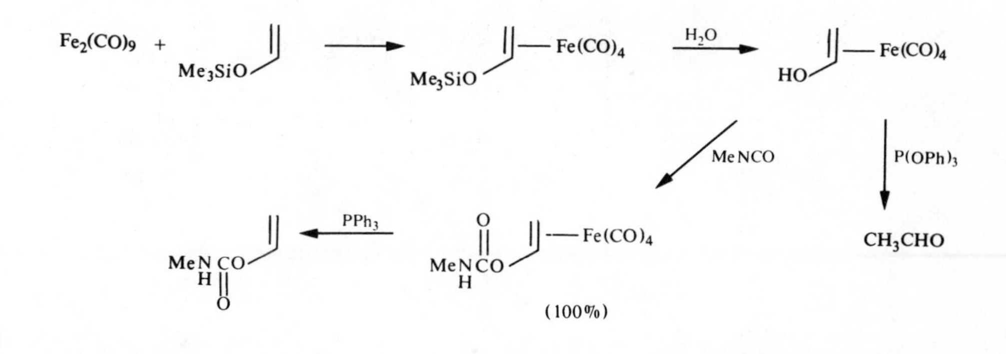

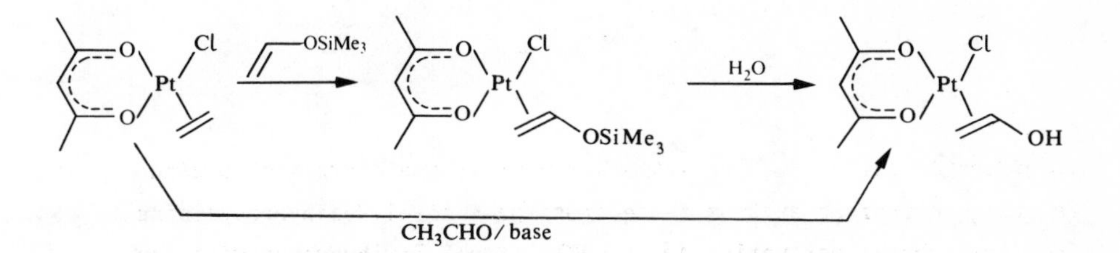

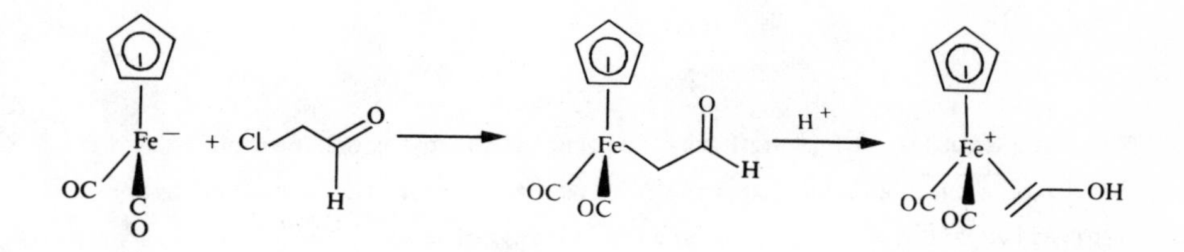

Butadiene-1-ol and -2-ol may be stabilised over their respective keto tautomers by coordination to $Fe(CO)_3$.[51,52]

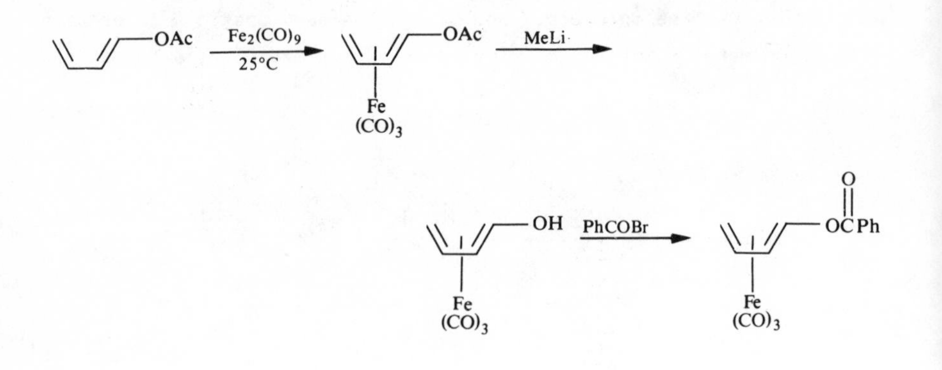

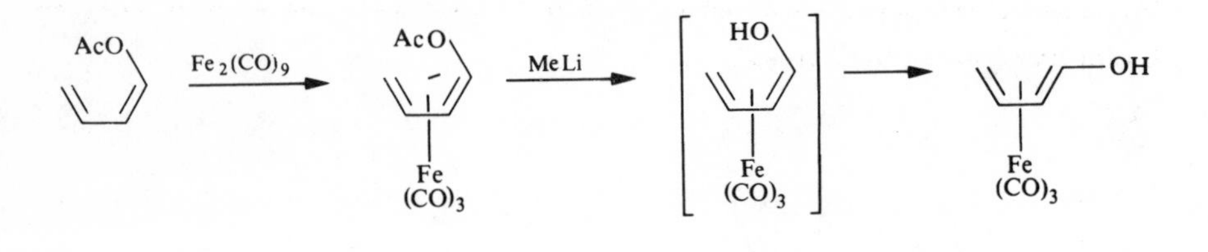

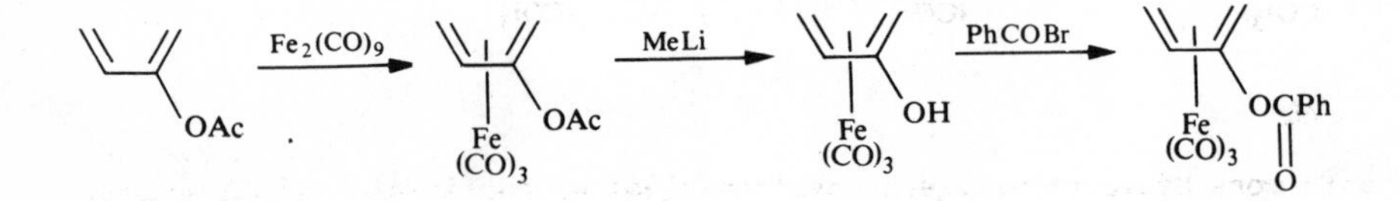

The keto tautomer of phenol may be stabilised by coordination to $Fe(CO)_3$.[53] The ketone complex 46 gives a 2,4-dinitrophenylhydrazone derivative and is reduced by sodium borohydride.

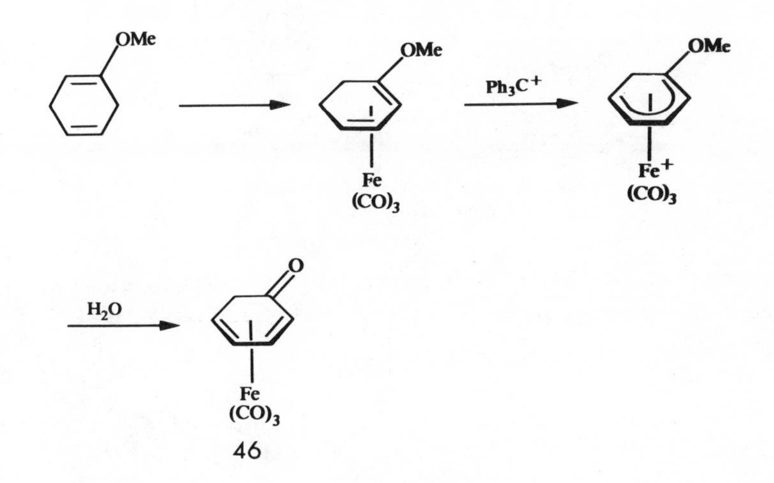

Tricarbonyl iron cyclohexadienone 46 can be used as a phenylating reagent for primary aromatic amines.[54]

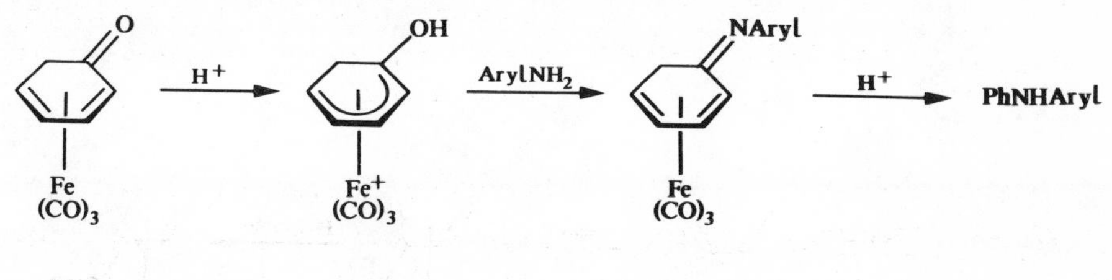

46

With more basic amines, e.g. cyclohexylamine, enolisation of 46 occurs. However conversion of 46 to the ethoxy cyclohexadienyl iron tricarbonyl cation 47 gives a reagent effective for the phenylation of amines under mild and essentially neutral conditions.

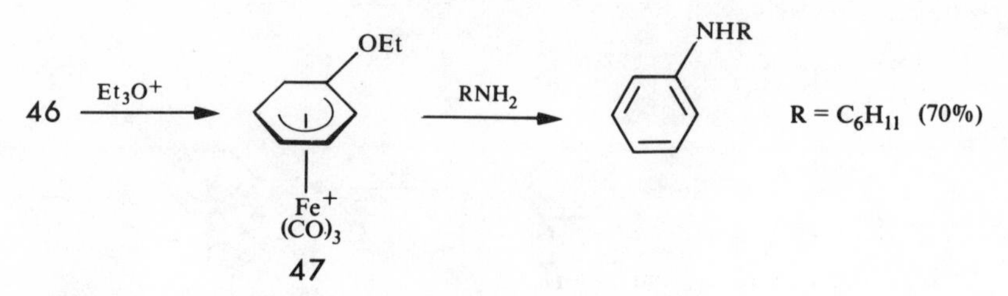

3.2.3 Trapping of reaction intermediates

Trapping of tachysterol$_2$ as its iron tricarbonyl complexes by slow addition of $Fe(CO)_5$ during the photolysis of ergosterol is the best method available for preparing tachysterol$_2$.[55]

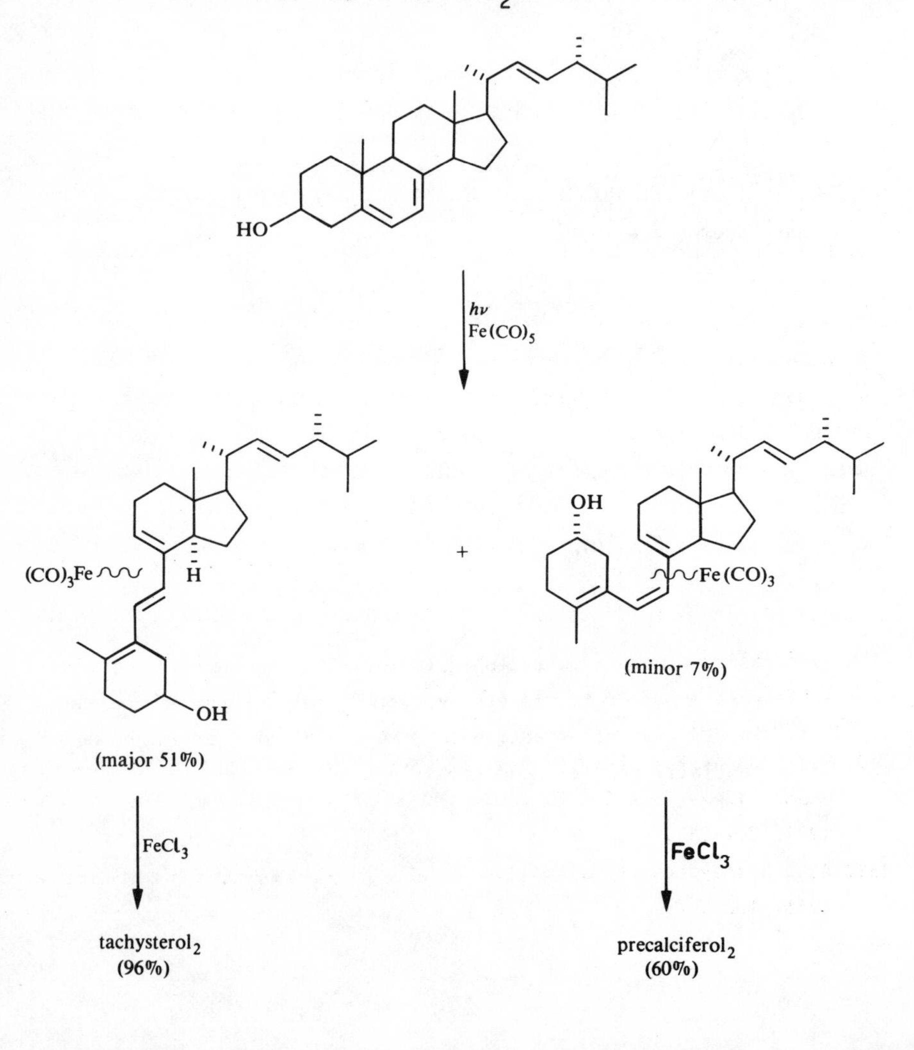

3.3 REFERENCES

1. A. Rosan, M. Rosenblum and J. Tancrede, *J. Amer. Chem. Soc.*, 1973, *95*, 3062.
2. P.F. Boyle and K.M. Nicholas, *J. Org. Chem.*, 1975, *40*, 2682.
3. K.M. Nicholas, *J. Amer. Chem. Soc.*, 1975, *97*, 3254.
4. M.L.H. Green and G. Wilkinson, *J. Chem. Soc.*, 1958, 4314.
5. A. Sanders and W.P. Giering, *J. Amer. Chem. Soc.*, 1975, *97*, 919.
6. D. Seyferth and A.T. Wehman, *J. Amer. Chem. Soc.*, 1970, *92*, 5520.
7. H. Greenfield, H.W. Sternberg, R.A. Friedel, J.H. Wotiz, R. Markby and I. Wender, *J. Amer. Chem. Soc.*, 1956, *78*, 120.
8. K.M. Nicholas and R. Pettit, *Tet. Letters*, 1971, 3475.
9. J.E. O'Boyle and K.M. Nicholas, *Tet. Letters*, 1980, 1595; K.M. Nicholas, M.Mulvaney and M. Bayer, *J. Amer. Chem. Soc.*, 1980, *102*, 2508.
10. C. Descoins and D. Samain, *Tet. Letters*, 1976, 745.
11. C.H. Mauldin, E.R. Biehl and P.C. Reeves, *Tet. Letters*, 1972, 2955.
12. D.V. Banthorpe, H. Fitton and J. Lewis, *J.C.S. Perkin I*, 1973, 2051.
13. A. Salzer and W. von Philipsborn, *J. Organometal.Chem.*, 1979, *170*, 63.
14. B.F.G. Johnson, J. Lewis and D.G. Parker, *J. Organometal.Chem.* 1977, *141*, 319; B.F.G. Johnson, J. Lewis, D.G. Parker, P.R. Raithby and G.M. Sheldrick, *J. Organometal.Chem.*, 1978, *150*, 115; B.F.G. Johnson, J. Lewis and G.R. Stephenson, *Tet. Letters*, 1980, 1995.
15. D.H.R. Barton, A.A.L. Gunatilaka, T. Nakanishi, H. Patin, D.A. Widdowson and B.R. Worth, *J. Chem. Soc. Perkin I*, 1976, 821, G. Evans, B.F.G. Johnson and J. Lewis, *J. Organometal. Chem.* 1975, *102*, 507.
16. G.A. Taylor, *J. Chem. Soc. Perkin I*, 1979, 1716.
17. H. Alper and S. Amaratunga, *Tet. Letters*, 1980, 1589.
18. H. Olsen and J.P. Snyder, *J. Amer. Chem. Soc.*, 1978, *100*, 285.
19. A.J. Birch and A.J. Pearson, *Chem. Comm.*,1976, 601.
20. M. Franck-Neumann and D. Martina, *Tet. Letters*, 1975, 1759.
21. A.J. Birch and H. Fitton, *Aust. J. Chem.*, 1969, *22*, 971.
22. K. Weiss and E.O. Fischer, *Chem. Ber.*, 1976, *109*, 1868.
23. T. Bally and S. Masamune, Tetrahedron Report, No. 74, *Tetrahedron*, 1980, *36*, 343.

24. M.D. Rausch and A.V. Grossi, *Chem. Comm.*, 1978, 401.
25. M. Rosenblum and C. Gatsonis, *J. Amer. Chem. Soc.*, 1967, *89*, 5074.
26. S.A. Gardner and M.D. Rausch, *J. Organometal. Chem.* 1973, *56*, 365.
27. M. Rosenblum and B. North, *J. Amer. Chem. Soc.*, 1968, *90*, 1060.
28. M. Rosenblum, B. North, D. Wells and W.P. Giering, *J. Amer. Chem. Soc.*, 1972, *94*, 1239.
29. J. Agar, F. Kaplan and B.W. Roberts, *J. Org. Chem.*, 1974, *39*, 3451.
30. J.D. Fitzpatrick, L. Watts, G.F. Emerson and R. Pettit, *J. Amer. Chem. Soc.*, 1965, *87*, 3255.
31. G.F. Emerson, L. Watts and R. Pettit, *J. Amer. Chem. Soc.*, 1965, *87*, 131.
32. A. Sanders and W.P. Giering, *J. Amer. Chem. Soc.*, 1974, *96*, 5247.
33. L. Watts, J.D. Fitzpatrick and R. Pettit, *J. Amer. Chem. Soc.*, 1965, *87*, 3253.
34. E.K.G. Schmidt, *Chem. Ber.*, 1975, *108*, 1609.
35. J. Rebek Jr. and F. Gaviña, *J. Amer. Chem. Soc.*, 1975, *97*, 3453.
36. J.M. Landesberg and J. Sieczkowski, *J. Amer. Chem. Soc.*, 1971, *93*, 972.
37. J.O. Glanville, J.M. Stewart and S.O. Grim, *J. Organometal. Chem.*, 1967, *7*, P9.
38. R. Mason, *Nature*, 1968, *217*, 543.
39. A.C. Blizzard and D.P. Santry, *J. Amer. Chem. Soc.*, 1968, *90*, 5749.
40. M.A. Bennett and T. Yoshida, *J. Amer. Chem. Soc.*, 1978, *100*, 1750.
41. S.J. McLain, R.R. Schrock, P.R. Sharp, M.R. Churchill and W.J. Youngs, *J. Amer. Chem. Soc.*, 1979, *101*, 263.
42. M.E. Jason, J.A. McGinnety and K.B. Wiberg, *J. Amer. Chem. Soc.*, 1974, *96*, 6531.
43. J.P. Kutney, *Heterocycles*, 1977, *7*, 593 and references therein.
44. J.P. Kutney, R.A. Badger, W.R. Cullen, R. Greenhouse, M. Noda, V.E. Ridaura-Sanz, Y.H. So, A. Zanarotti and B.R. Worth, *Can. J. Chem.*, 1979, *57*, 300.
45. H. Alper, *J. Organometal. Chem.*, 1975, *96*, 95.
46. J.P. Kutney, M. Noda and B.R. Worth, *Heterocycles*, 1979, *12*, 1269.
47. H. Thyret, *Angew. Chem. Int. Edn.*, 1972, *11*, 520 and references therein.
48. M. Tsutsui, M. Ori and J. Francis, *J. Amer. Chem. Soc.*, 1972, *94*, 1414.

49. J. HIllis and M. Tsutsui, *J. Amer. Chem. Soc.*, 1973, *95*, 7907.
50. J.K.P. Ariyaratne and M.L.H. Green, *J. Chem. Soc.*, 1964, 1.
51. C.H. DePuy, R.N. Greene and T.E. Schroer, *Chem. Comm.*, 1968, 1225.
52. C.H. DePuy, C.R. Jablonski, *Tet. Letters*, 1969, 3989.
53. A.J. Birch, P.E. Cross, J. Lewis, D.A. White and S.B. Wild, *J. Chem. Soc. A.*, 1968, 332.
54. A.J. Birch and I.D. Jenkins, *Tet. Letters*, 1975, 119.
55. A.G.M. Barrett, D.H.R. Barton and G. Johnson, *J. Chem. Soc. Perkin I*, 1978, 1014.

CHAPTER 4

ORGANOMETALLICS AS ELECTROPHILES

Unsaturated hydrocarbons, e.g. ethylene, butadiene or benzene do not normally undergo nucleophilic addition or substitution reactions. However when these molecules are coordinated to electron withdrawing transition metal centres they are attacked by a wide range of nucleophiles such as H^-, R^-, CN^-, MeO^-, R_3N etc.

Nucleophilic addition to 18 electron organotransition metal cations has been extensively studied. The products are generally stable neutral 18 electron complexes. The utility of this reaction for synthesis has been greatly improved by the introduction of a series of rules that allow the prediction of the regioselectivity of nucleophilic addition to 18-electron transition metal cations.[1]

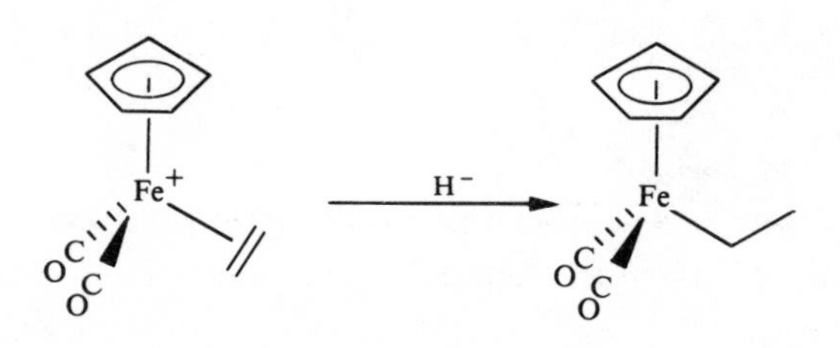

Nucleophilic addition to neutral complexes e.g. (arene)$Cr(CO)_3$ or (olefin)$Fe(CO)_4$, also occurs readily although the anionic primary products are not generally isolable. Nucleophilic addition to olefins may also be catalysed by transition metal species.

The less electron rich a complex the faster nucleophilic addition occurs. This is illustrated by the exchange of halogen for methoxide in the complexes $(C_6H_5X)[M]$ (X = F, Cl) where the rate increases along the series $[M] = (CO)_3Cr < (CO)_3Mo << CpFe^+ < (CO)_3Mn^+$.[2]

4.1 STOICHEIOMETRIC NUCLEOPHILIC ADDITIONS TO ORGANOTRANSITION METAL CATIONS

4.1.1 General rules governing the regioselectivity

The greater reactivity towards nucleophilic attack of unsaturated hydrocarbons coordinated to transition metal cations can be attributed broadly to metal - ligand bonding effects which result in a net withdrawal of electron density from the unsaturated hydrocarbon ligand to the positively charged metal centre. Coordination to the metal cation is therefore akin to the introduction of electron withdrawing substituents onto the hydrocarbon chain or to species such as bromonium ions.

$[M]^+$ —|| ←Y^- RO–C(=O)–CH=CH$_2$ ←Y^- Y^- → bromonium ion (+Br)

X-ray crystallographic and spectroscopic studies have shown that nucleophilic attack invariably occurs on the *exo*-face of the ligand, i.e. on the side of the ligand away from the metal.[3,4] These nucleophilic addition reactions may be considered as involving S_N2 displacement of the metal-carbon bond by the nucleophile with the carbon atom being attacked undergoing inversion of configuration.

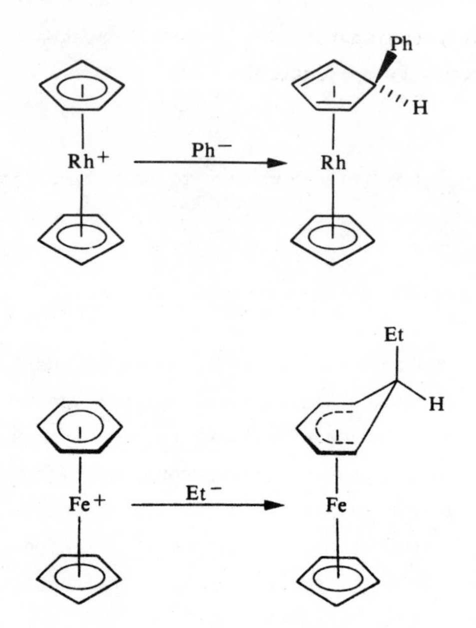

Organotransition metal cations often contain several unsaturated hydrocarbon ligands. However nucleophilic addition reactions are generally very regiospecific occurring only on one of the ligands. A series of rules have been proposed which if applied sequentially allow the prediction of the most favourable position of nucleophilic attack on 18 electron organotransition metal cations for reactions that are kinetically rather than thermodynamically controlled.[1]

Unsaturated hydrocarbon ligands may be classified as *even* or *odd* according to the parity of the ligand hapto number (η). Furthermore ligands are described as *closed* if they are cyclically conjugated and *open* if they are not. It is possible to define all unsaturated hydrocarbon ligands in terms of *even* or *odd* and *open* or *closed*.

Even $\eta = 2, 4, 6....$

Odd $\eta = 3, 5, 7....$

Closed cyclically conjugated

Open not cyclically conjugated

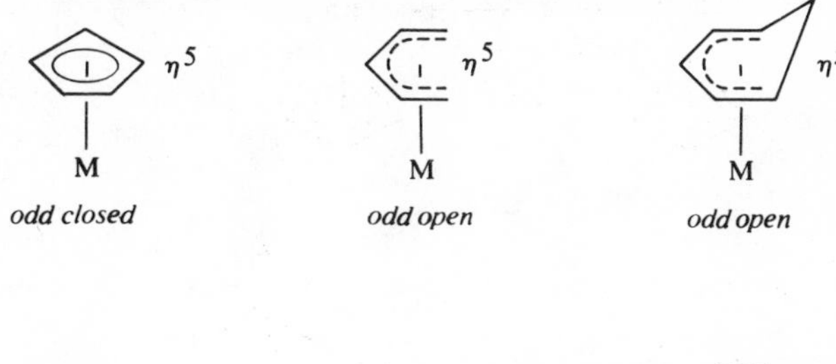

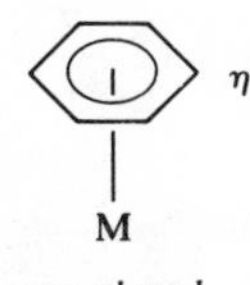

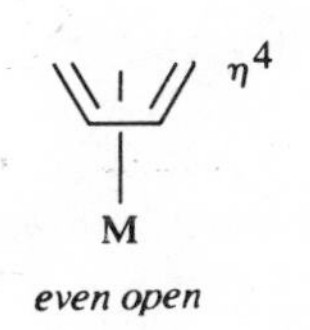

η^4

M

even open

Rule 1 Nucleophilic attack occurs preferentially at *even* coordinated polyenes which have no unpaired electrons in their h.o.m.o.'s. (Although cyclobutadiene is an *even* polyene it has unpaired electrons in its h.o.m.o. and therefore according to Rule 1 nucleophilic addition to other *even* polyenes is preferred. It is however attacked in preference to *odd* polyenyl ligands.)

Rule 2 Nucleophilic addition to *open* coordinated polyenes is preferred to addition to *closed* polyenes.

Rule 3 For *even open* polyenes nucleophilic attack at the terminal carbon atom is always preferred. For *odd open* polyenyls attack at a terminal carbon atom occurs only if $[M^+]$ is a strong electron withdrawing group.

Rules 1 and 2 allow the prediction of which ligand will be attacked. Rule 3 is concerned with regiospecifity on the particular ligand determined by the first two rules. Rules 1 and 2 may be simplified as follows:

Rule 1 : *Even* before *odd*

Rule 2 : *Open* before *closed*

The following examples illustrate applications of the rules.[4,5]

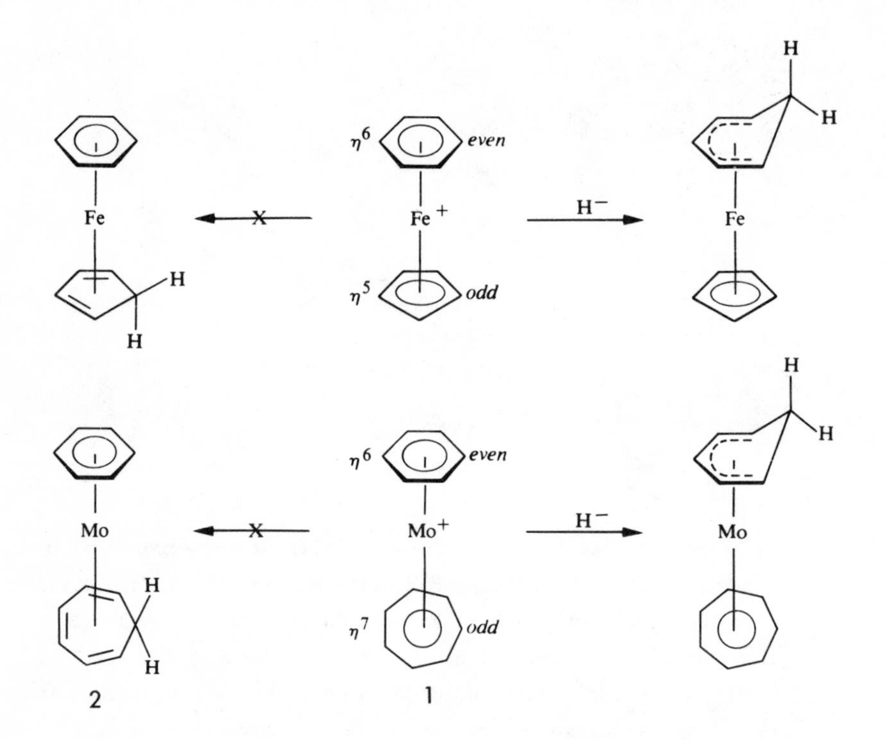

The regioselectivity of the above reactions follows from the application of Rule 1. In the latter example compound 2 is not formed even though it is isoelectronic with the well known $(C_6H_6)_2Mo$ complex and would have been the intuitively predicted product if one had regarded the η^7-cycloheptatrienyl ligand in 1 as approximating to the free aromatic tropylium cation.

It follows from Rule 2 that the *open* ligand in the cation 3 is attacked in preference to the *closed* cyclopentadienyl.[6]

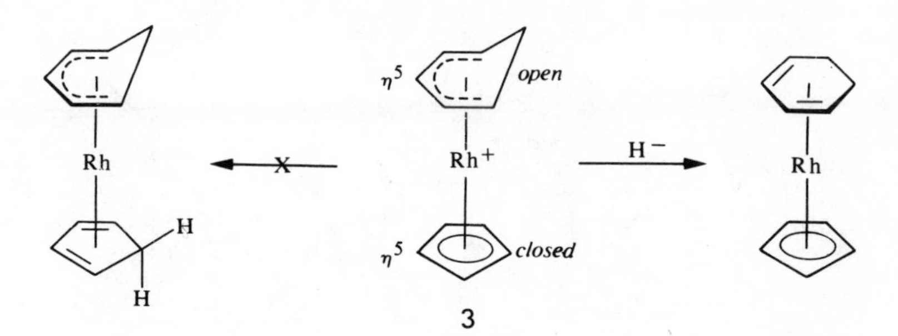

Cation 4 demonstrates the need to apply the rules sequentially. Rule 1 (*even* before *odd*) eliminates the possibility of attack on the allyl ligand. Rule 2 (*open* before *closed*) suggests nucleophilic attack on the butadiene and attack at the terminal position of this ligand follows from Rule 3.[7]

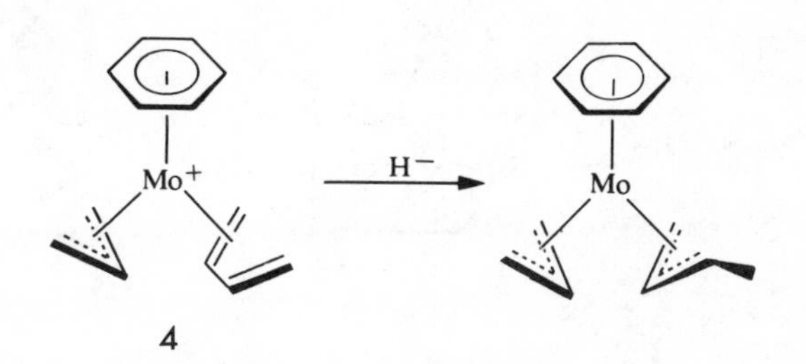

The position of attack on a number of cations is given below.[3-8]

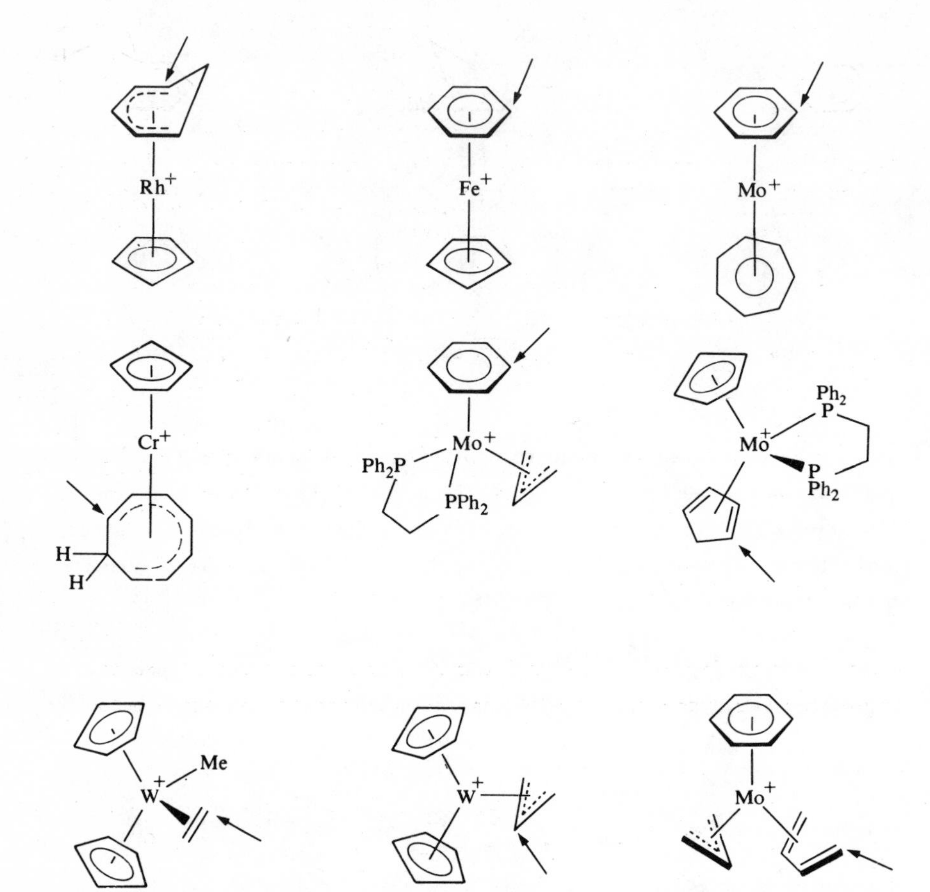

The rules provide a ready explanation for the different modes of attack on the dications 5 and 6 which result in addition to both rings for 5 and the same ring twice for 6.[9]

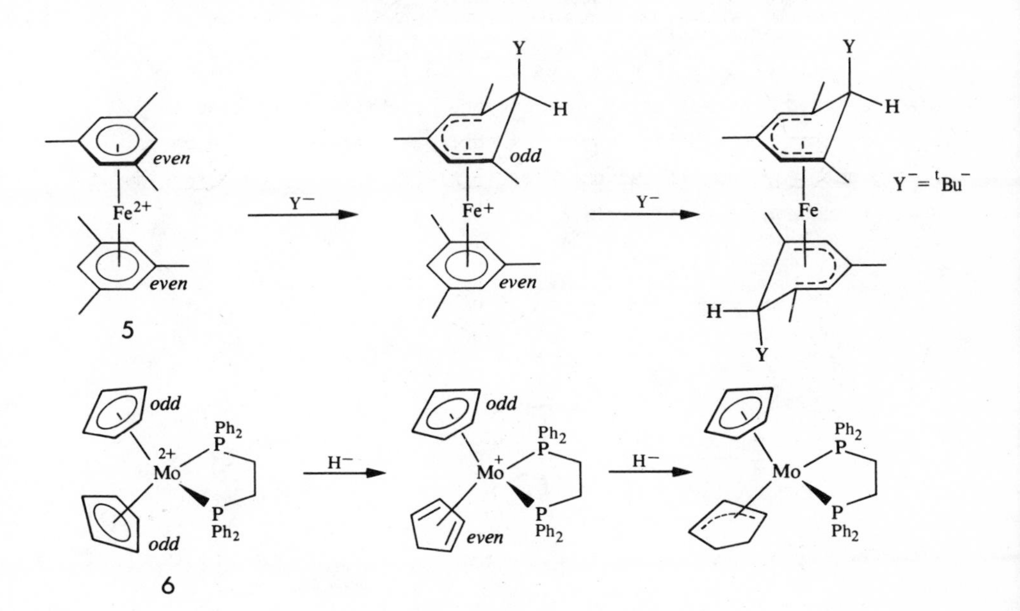

Another way of expressing Rules 1 and 2 is that the order of reactivity of unsaturated hydrocarbons coordinated to cations is as shown below.

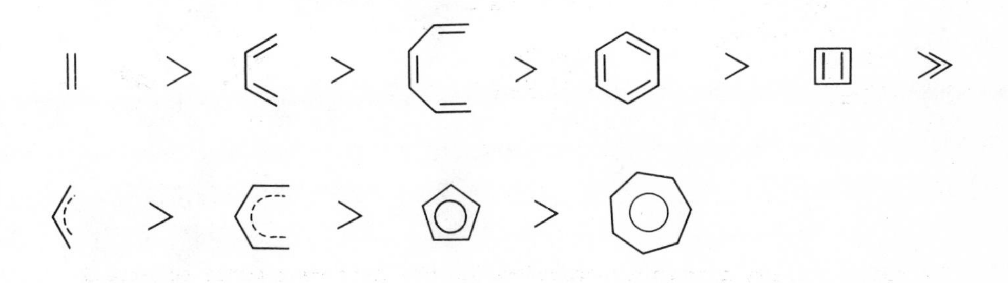

It can be seen from this series that C_5H_5 is relatively unreactive towards nucleophilic addition and therefore acts as a useful ligand in designing complexes for nucleophilic addition reactions to *even* ligands and allyl and pentadienyl ligands.

Cations that contain unsaturated hydrocarbon ligands and at least one CO ligand may undergo nucleophilic attack either on the hydrocarbon ligand or on the carbon atom of a CO ligand. Both types of reaction have been observed. Attack at CO is, however, generally restricted to heteroatomic nucleophiles where the choice is between CO and an η^5-ligand. For example cation 7 is initially attacked by methoxide at CO. However subsequent equilibration leads to the *exo* product 8 .[10]

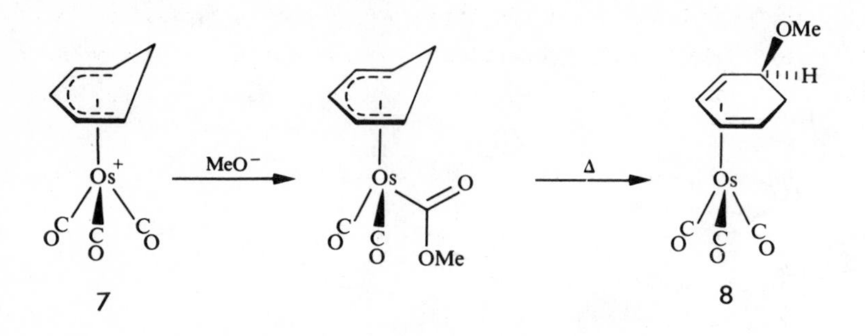

For cations containing two or more unsaturated hydrocarbon ligands and at least one CO ligand the rules described above may again be used to predict the kinetically favoured products of nucleophilic attack.[11]

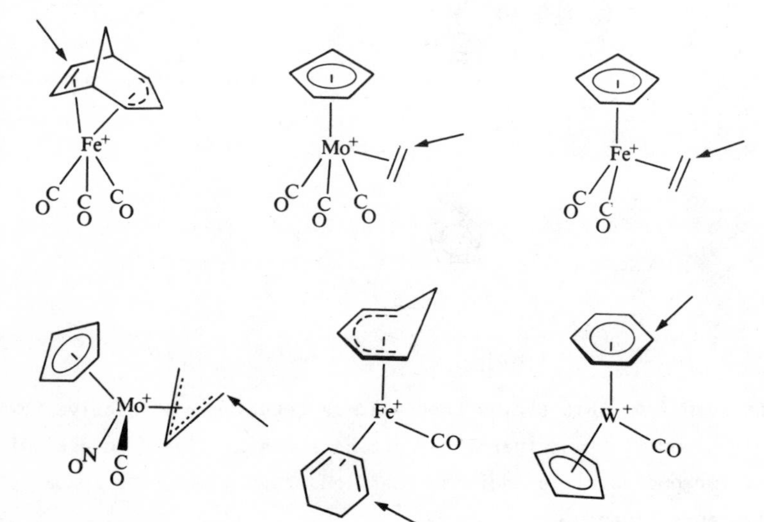

The charge transfer process that leads to a net withdrawal of electron density from the ligand to the positively charged metal centre can be described in terms of the forward and back donation components implicit in the Chatt-Dewar-Duncanson bonding model (section 1.2). For a positively charged metal-olefin complex the dominant charge transfer process will arise from donation of electron density from the ligand orbitals to the metal. Back donation effects will be relatively less important because of the positive charge on the ML_n^+ fragment. Electron donation from the h.o.m.o. of the ligand to an orbital of appropriate symmetry on the metal will be particularly influential in determining the distribution of electron density on the coordinated polyene. The fact that the h.o.m.o. of an *even* polyene is doubly occupied whereas the h.o.m.o. of an *odd* polyenyl is singly occupied leads to an important general difference between these two types of ligand. In the former case all the electron density in the bonding molecular orbital originates from the polyene h.o.m.o. whereas in the latter electron density is contributed by both the polyene and the metal. This means that the charge on an *even* polyene will vary from 0 to +2 and that for an *odd* polyenyl from -1 to +1. Also in the same cation an *even* polyene will have approximately one unit of positive charge more than an *odd* polyenyl.

The rates of nucleophilic addition reactions to 18-electron cationic polyene complexes are likely to be charge rather than orbital controlled, especially if the nucleophile is small and highly charged. Therefore the regioselectivity of such reactions is probably determined by the positive charges on particular carbon atoms; hence Rule 1 - nucleophilic attack occurs preferentially at *even* coordinated polyenes.

When the hydrocarbon ligand being attacked is *open* then it is necessary to apply Rule 3, since unlike the simple *closed* systems the carbon atoms are nonequivalent. For the allyl ligand there are two possible positions of attack i.e. at the terminal carbon atoms or at the central carbon atom. Examples of both modes of addition have been observed.[8,11]

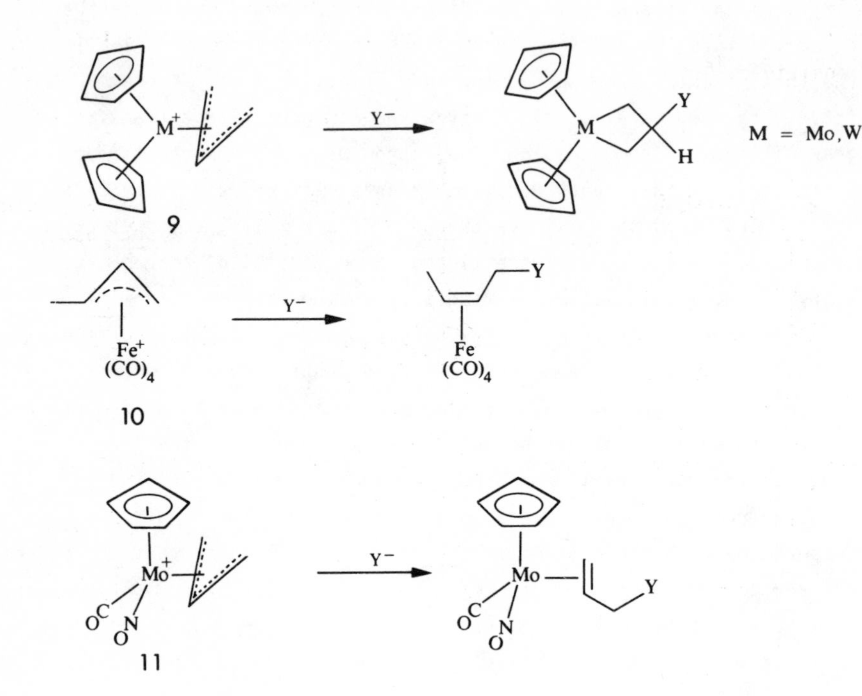

It is possible to correlate the position of attack with the electron richness of the metal. When the metal is electron rich as in 9 the allyl ligand behaves like an allyl anion and the nucleophile attacks the carbon atom with the least electron density namely the central carbon atom. When, however, the metal is electron poor, by virtue of electron withdrawing ligands such as CO or NO as in 10 and 11 the allyl ligand behaves like an allyl cation and the nucleophile attacks one of the terminal carbon atoms, since these are now the carbon atoms with the least electron density.

For *even open* ligands electron density will always tend to be least at the terminal carbon atoms and as expected attack always occurs there.

Further evidence supporting the view that the nucleophile attacks the carbon atom with the least electron density is provided by the examples below which have electron withdrawing and donating groups on the polyene. Nucleophilic addition to the cations 12 and 13 occurs preferentially onto the ring bearing the electron withdrawing carboxyl group and in the β-positions.[12]

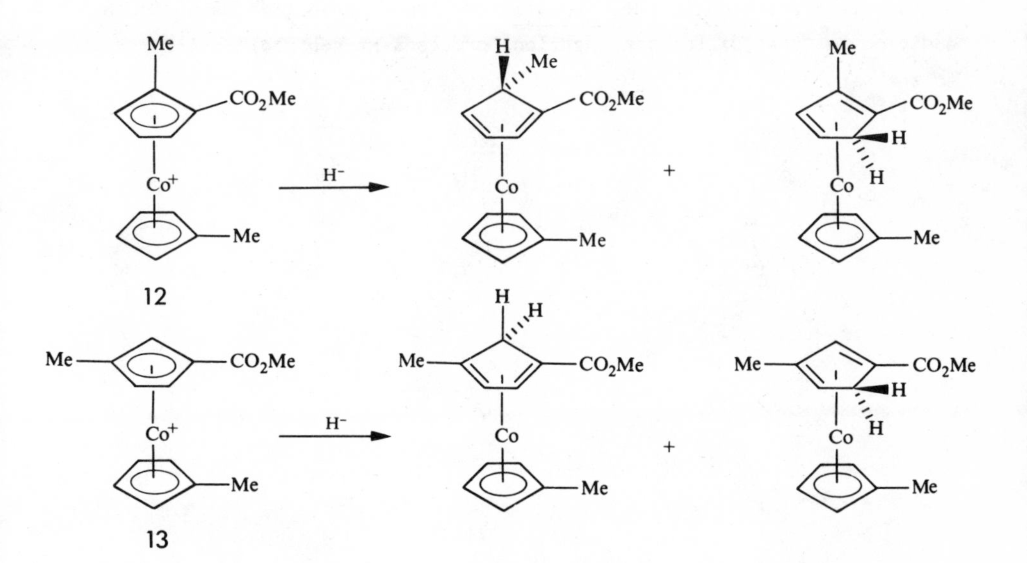

For the cation 14 when X is an electron withdrawing group (e.g. $-CO_2Me$) nucleophilic attack occurs at the 2-position but when X is an electron releasing substituent (e.g. -OMe) attack occurs at the 3-position.[12a]

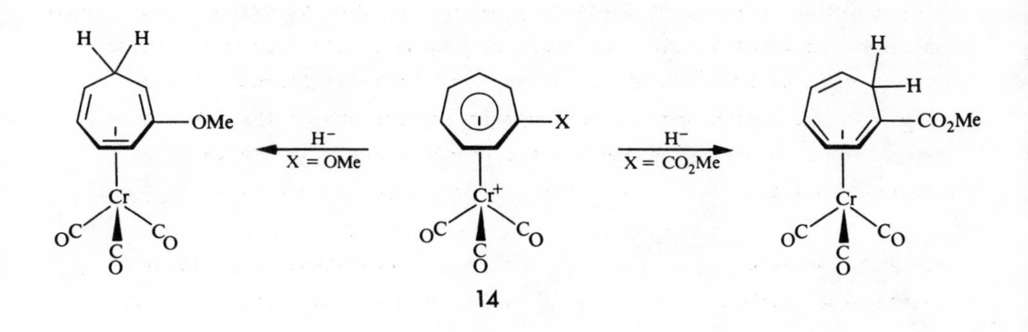

4.1.2 *Nucleophilic addition to cationic η^2-olefin complexes*

The cations $CpFe(CO)_2(olefin)^+$ undergo nucleophilic addition to the coordinated olefin with a variety of nucleophiles.[13] In common with nucleophilic addition to other unsaturated hydrocarbon ligands nucleophilic addition to $CpFe(CO)_2(olefin)^+$ cations proceeds stereospecifically *trans* to the metal.[1,13,14]

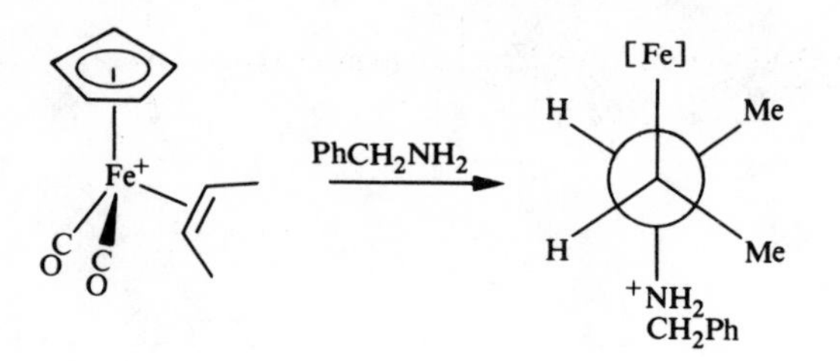

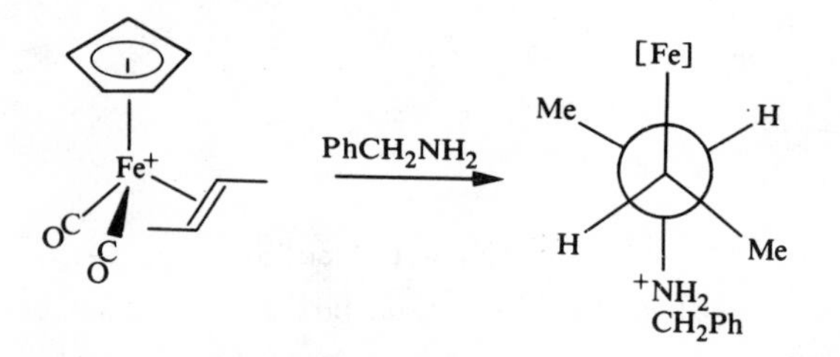

$CpFe(CO)_2(olefin)^+$ cations react readily with a large number of stabilised carbanions such as lithium enolates etc.

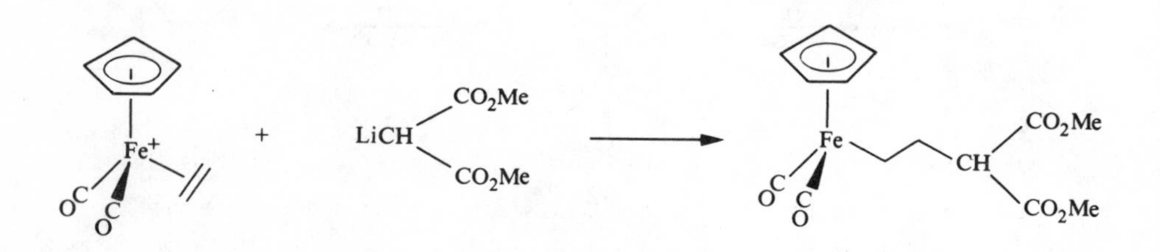

Nucleophilic addition to monosubstituted-olefin complexes is not very regiospecific for simple alkyl substituted alkenes. For the styrene complex, however, only addition to the carbon bearing the phenyl group is observed. This is presumably because the phenyl group stabilises the carbonium ion resonance structure 15 .

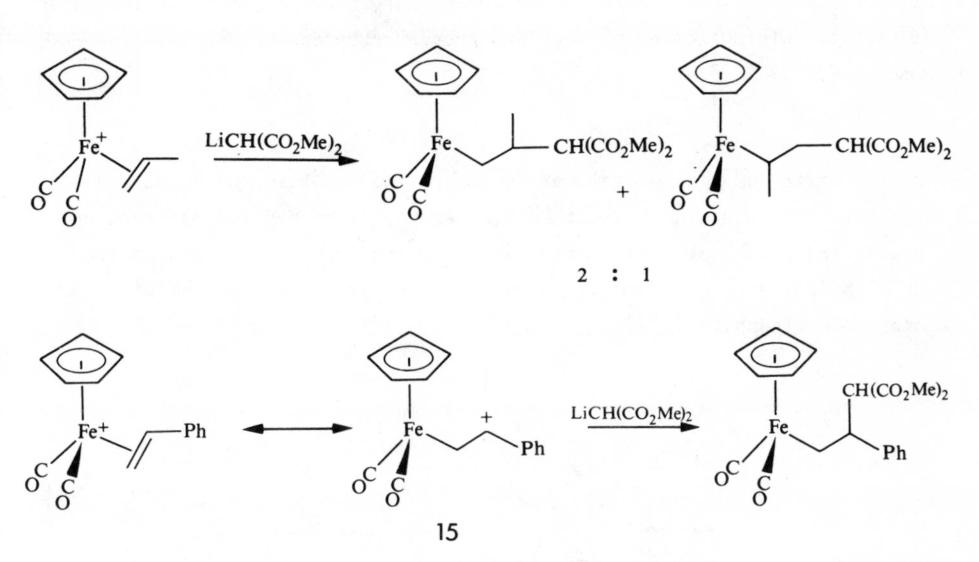

Nucleophilic additions are highly regiospecific when there are electron withdrawing substituents (CHO,C(O)R, CO_2R) on the olefin.[15]

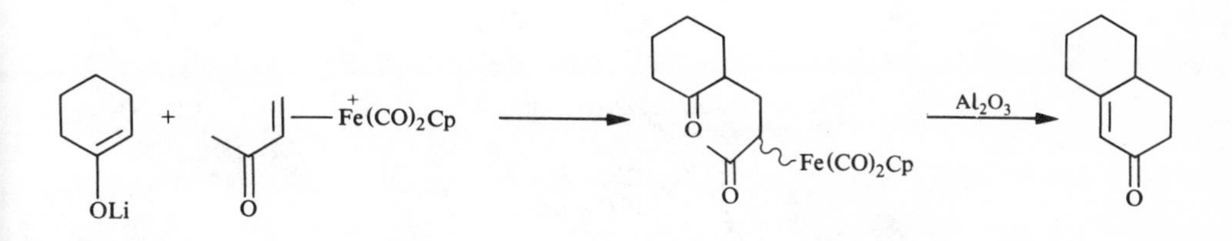

$CpFe(CO)_2(olefin)^+$ cations also readily undergo nucleophilic addition with silyl enol ethers as the source of enolate anions. This allows Michael additions to be performed under mild conditions with regiospecific generation of enolate anions.[15] It is probable that the enolate is generated by removal of the trimethyl-silyl protecting group by fluoride provided by the counterion (BF_4^- or PF_6^-) of the organometallic cations.

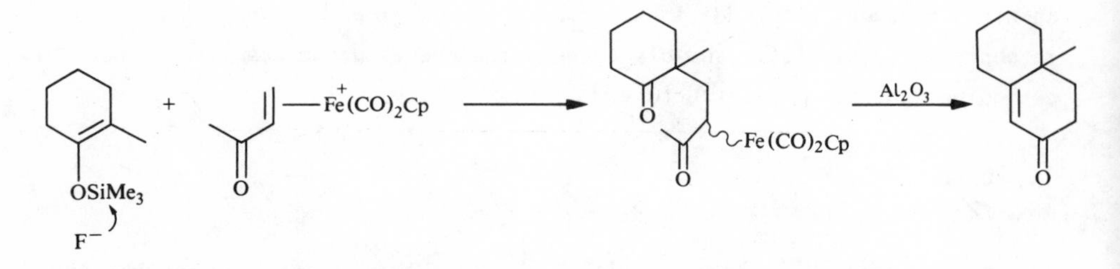

Nucleophilic addition of malonate anion to the cyclopentene complex 16 generates the addition product 17 from which *trans*-hydride abstraction yields the new η^2-olefinic cation 18. The olefin can be liberated from 18 with iodide to give overall allylic substitution, or reacted with another nucleophile.[13,15]

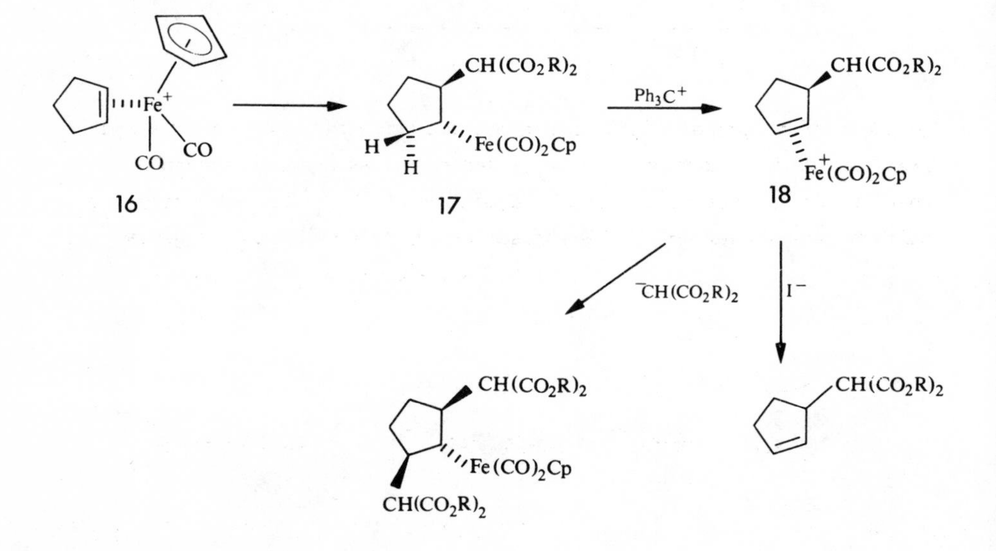

$CpFe(CO)_2(olefin)^+$ cations also undergo nucleophilic addition reactions with enamines.[13,15]

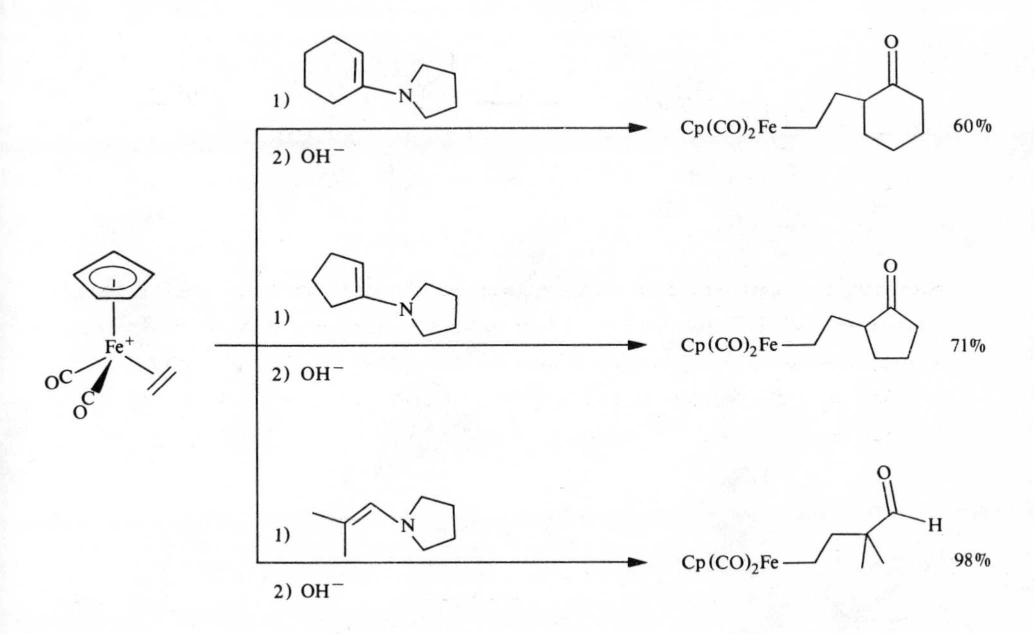

The addition of unstabilised carbanions such as RLi or RMgX (R = alkyl) to $CpFe(CO)_2(olefin)^+$ cations generally leads to reductive formation of $[CpFe(CO)_2]_2$ and not to nucleophilic addition. The problem of alkyl addition can be overcome in some cases by the use of $LiCuR_2$ reagents.[13,15]

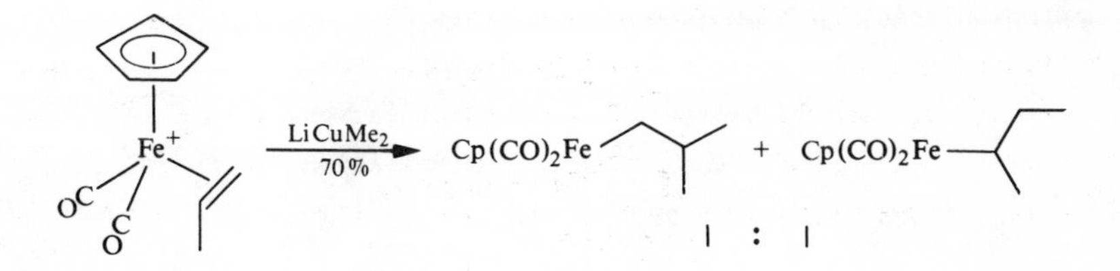

Some heteroatomic nucleophiles such as alkoxides, amines, mercaptides, phosphines and phosphites may also be added efficiently to $CpFe(CO)_2$ $(olefin)^+$ cations.[16,17]

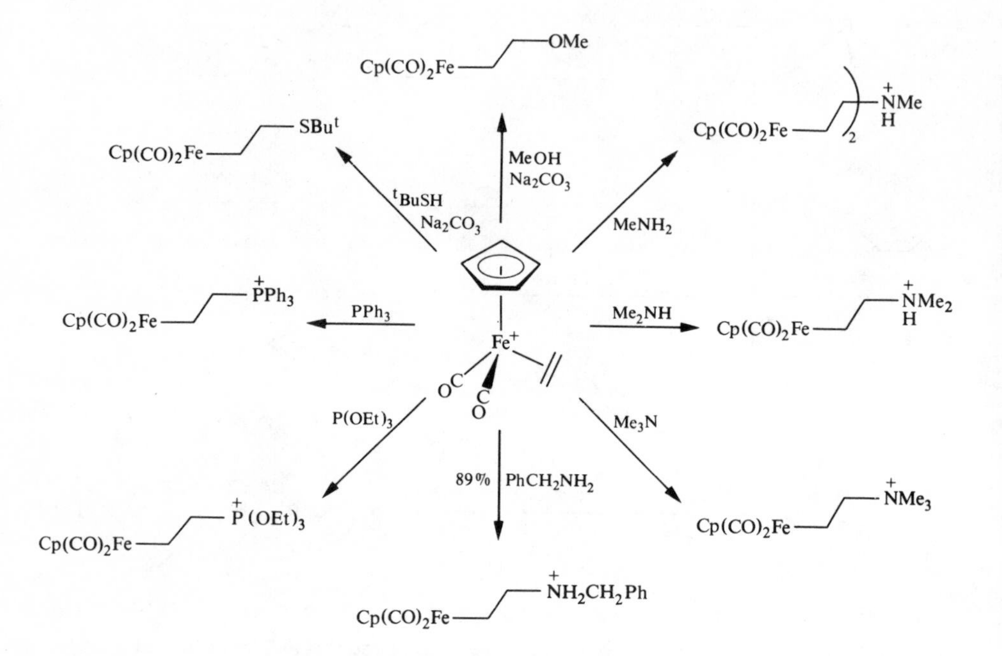

For cations that contain an acidic allylic proton, basic nucleophiles such as amines convert the η^2-olefinic cations into η^1-allyl complexes by proton abstraction. Nucleophiles such as CN^-, NCO^-, N_3^-, or I^- normally lead to decomplexation of the olefin. For olefinic complexes where *trans* addition is hindered then decomplexation of the olefin occurs with most nucleophiles. The general reactions of $CpFe(CO)_2$ $(olefin)^+$ cations are summarised below.

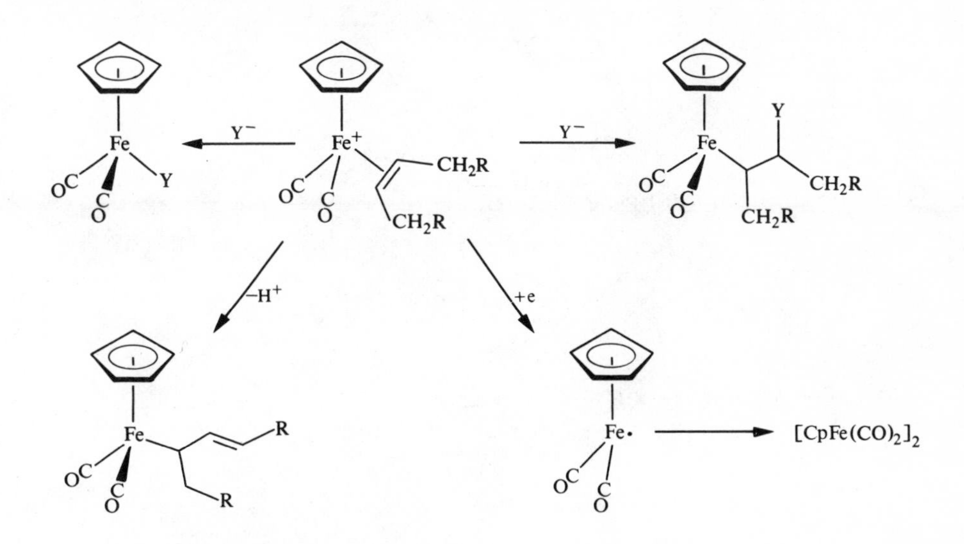

The addition of amines to complexed olefins followed by oxidative carbonylation of the adducts provides a β-lactam synthesis.[18,19]

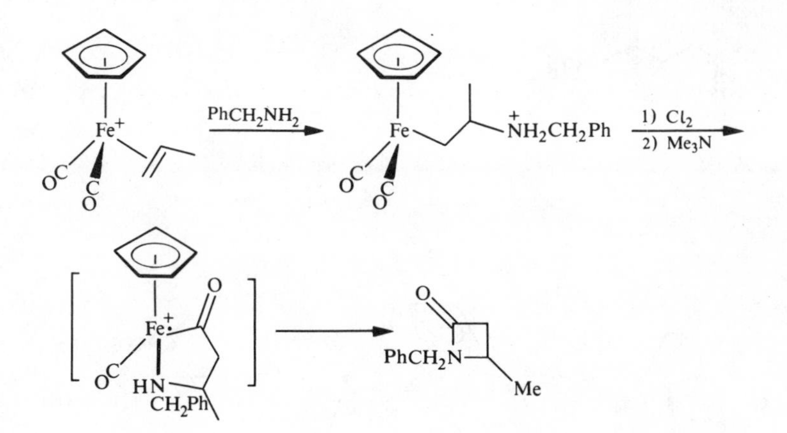

Bicyclic β-lactams may also be prepared.[19]

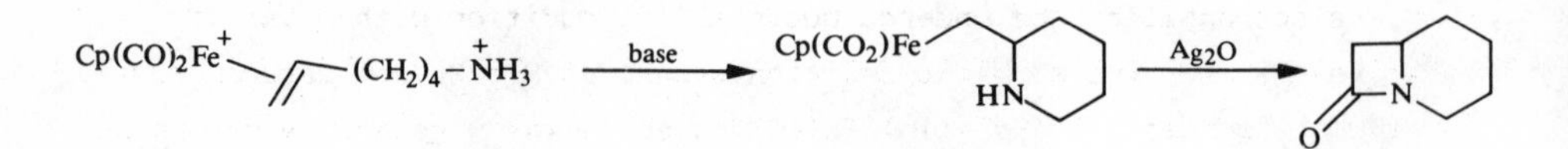

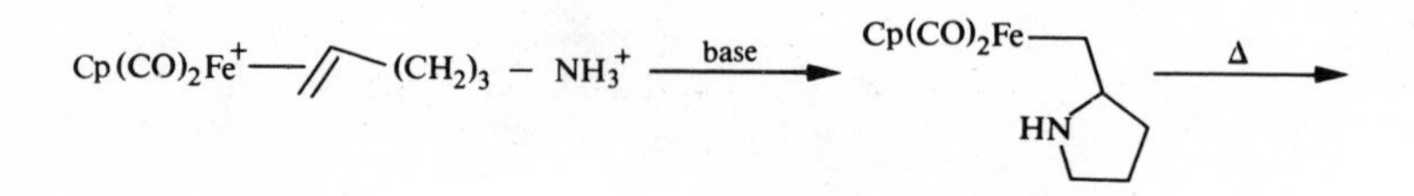

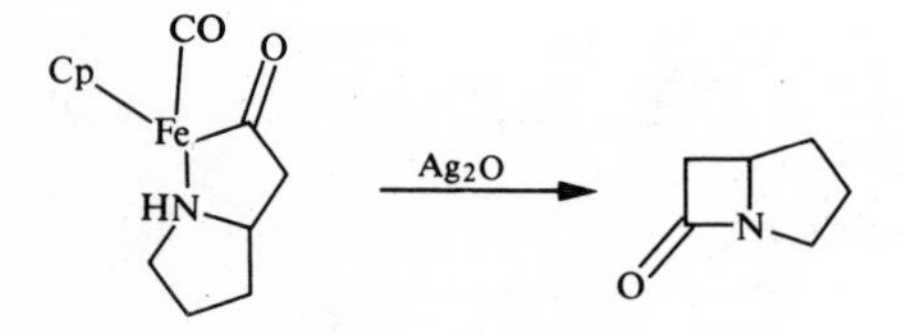

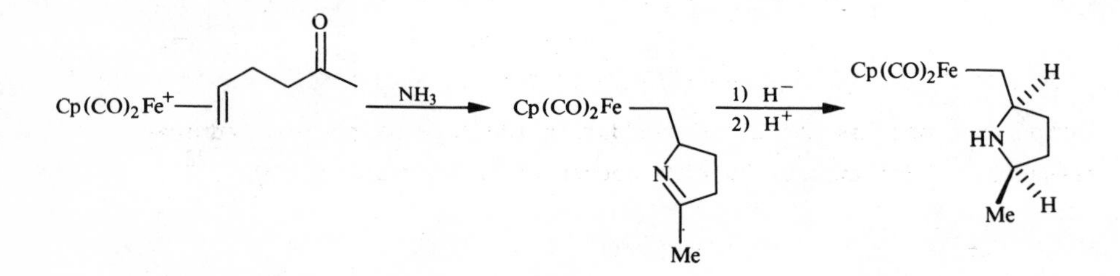

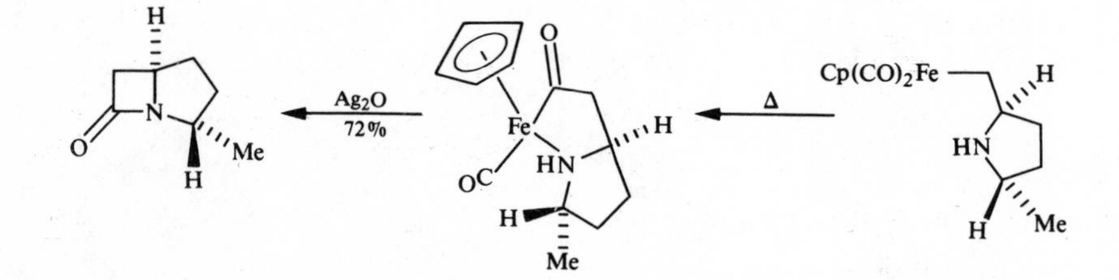

4.1.3 Nucleophilic addition to cationic η^3-allyl complexes

The cations $(\eta^3\text{-allyl})Fe(CO)_4^+$, readily obtainable by protonation of $(\eta^4\text{-diene})Fe(CO)_3$ complexes in the presence of CO (section 2.4.a) are highly electrophilic and undergo nucleophilic addition with a variety of nucleophiles. Nucleophilic addition occurs at a terminal carbon atom to generate unstable $(\eta^2\text{-olefin})\ Fe(CO)_4$ species which generally decompose with release of the olefin. For example stabilised carbanions, PPh_3 and pyridine successfully add to cation 19 .[20]

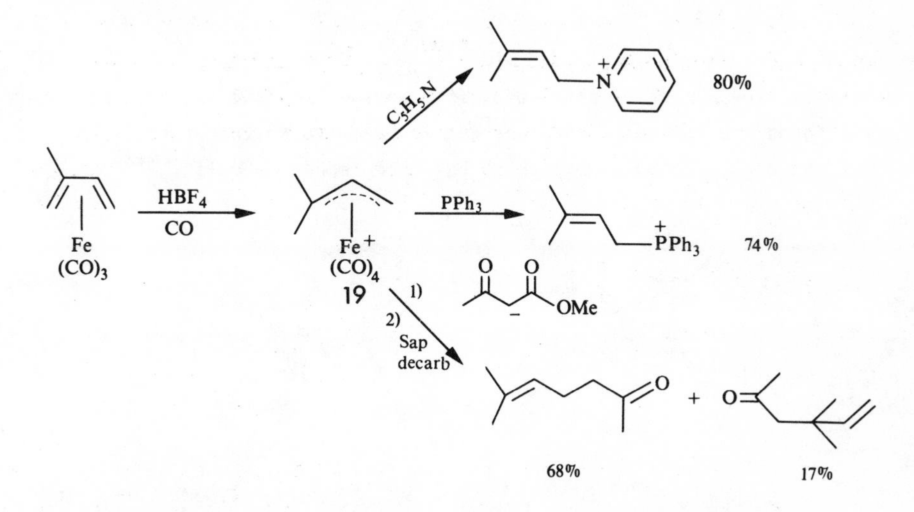

Carbon nucleophiles may also be added in the form of dialkyl cadmium reagents.[21] Addition of hydride occurs on treatment with $NaBH_4$.[22]

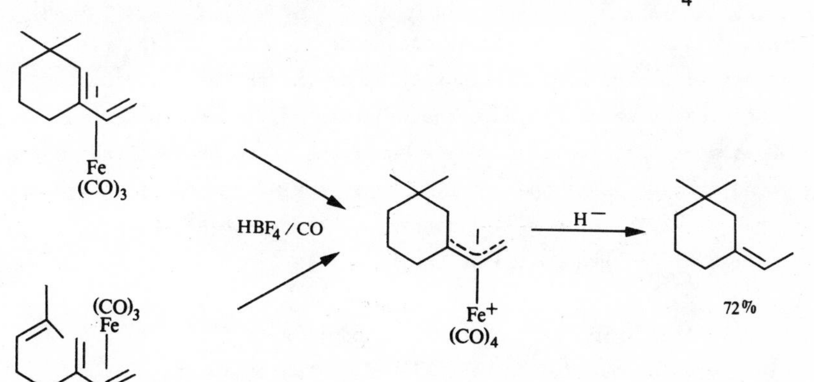

Nucleophilic attack on the cation CpMo(CO)NO(η^3-allyl)$^+$ occurs readily with a number of nucleophiles (e.g. MeO^-, Cp^-, AcO^-, BH_4^-).[23]

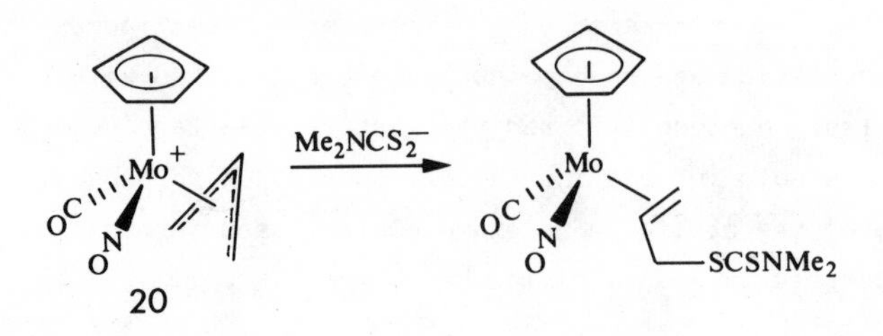

Cations of type 20 are of particular interest because the two ends of the allyl ligand are nonequivalent (one end is *trans* to NO the other is *trans* to CO) and also *cis* and *endo* isomers have been shown to exist.[24,25]

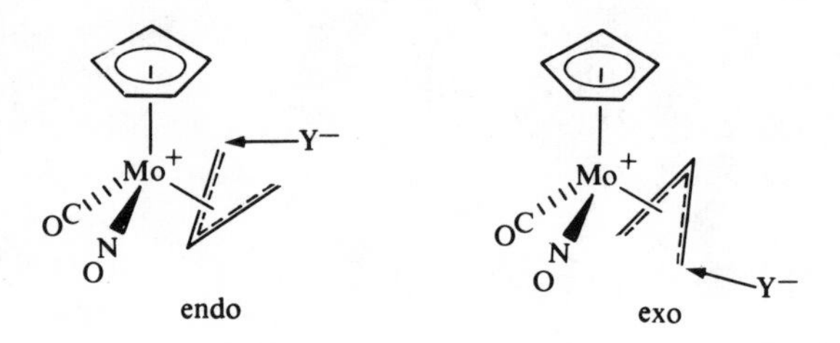

As with other nucleophilic addition reactions the nucleophile attacks the uncoordinated face of the ligand. It has been shown that the nucleophile attacks the allyl ligand of the *endo* isomer *trans* to NO whereas attack on the *exo* isomer occurs trans to CO.[25] This is in agreement with theoretical calculations of the charge distributions in the two isomers which indicate that in the *endo* isomer the carbon *trans* to NO is more positive than the carbon trans to CO whereas the reverse is found for the *exo* isomer.[24] Thus once again, for nucleophilic attack on organometallic cations it appears that the regioselectivity is controlled by charge considerations.

Nucleophilic addition of stabilised carbanions to $[(\eta^3\text{-allyl})PdCl]_2$ complexes can normally only be achieved if two equivalents of a 2 electron phosphorus ligand (R_3P, $(RO)_3P$, $(R_2N)_3P$) are present.[25a]

This is probably due to the need to convert the neutral $[(\eta^3 allyl)PdCl]_2$ complexes to the corresponding $[(\eta^3 allyl)PdL_2]^+$ cations before the reaction can take place. Stabilised carbanions attack the allyl ligand from the uncoordinated side. Nucleophiles such as RLi, R_2Mg, R_2Zn, R_2CuLi etc. irreversibly attack the Pd atom and products of addition to the allyl ligand are not observed.[25b]

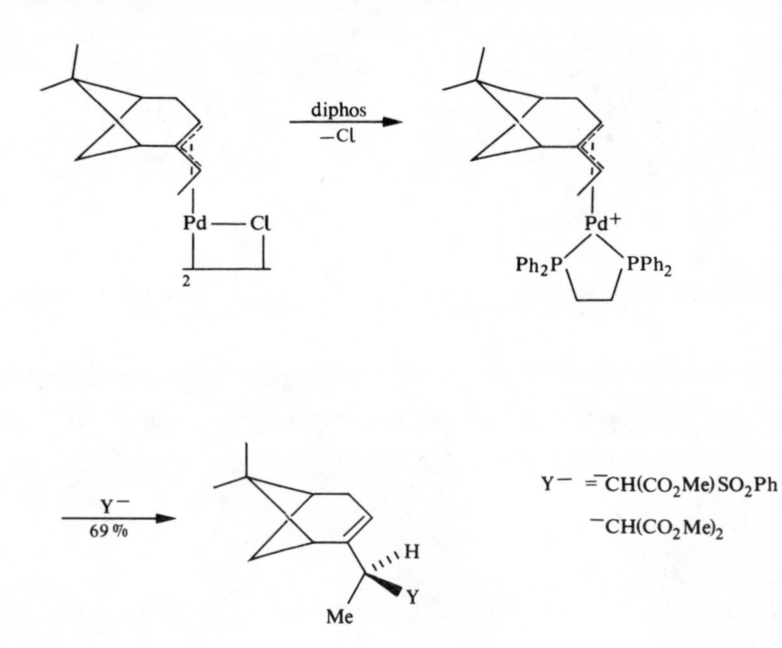

Alkylation occurs at the least sterically hindered end of the η^3-allyl ligand although in certain cases the regioselectivity can be altered by changing the ligands present.

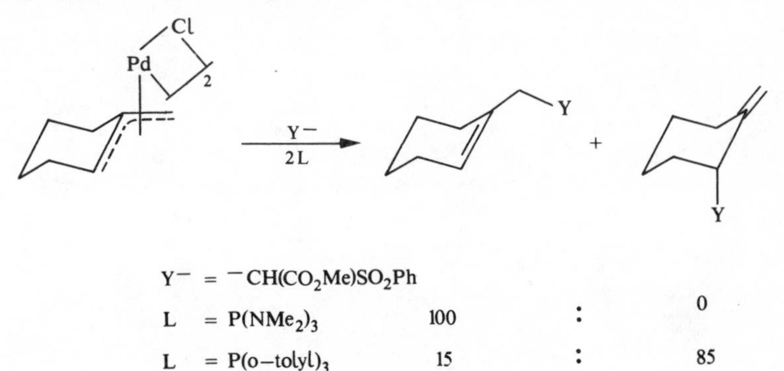

Y−	= ⁻CH(CO$_2$Me)SO$_2$Ph			
L	= P(NMe$_2$)$_3$	100	:	0
L	= P(o−tolyl)$_3$	15	:	85

The regioselectivity of the above reaction depends on the steric requirements of the approaching nucleophile and to what extent the relative stabilities of the initially formed [(η^2-olefin)PdL_2] complexes are reflected in the transition states. The stereochemistry of the products is the result of *exo* addition to the (η^3-allyl)Pd complexes.

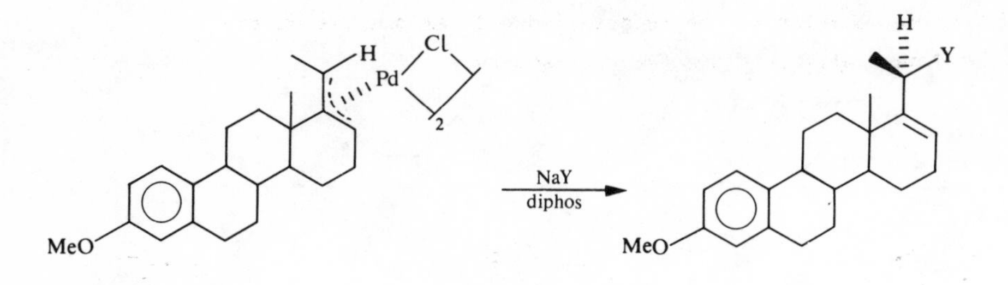

$Y^- = {}^-CH(CO_2Me)_2$ 81%; ${}^-CH(CO_2Me)(SO_2Ph)$ 82%

The use of optically active phosphines has allowed the formation of optically active products in up to 63% optical yield.[25a]

Cl
Pd
2
$NaCH(CO_2Me)(SO_2Ph)$
(+)−DIOP
*
CO_2Me
63% optical yield

4.1.4 Nucleophilic addition to cationic η[4]*-diene complexes*

The readily prepared $CpMo(CO)_2(\eta^4\text{-diene})^+$ cations have been shown to exist as an equilibrating mixture of *exo* and *endo* isomers.[26] Nucleophilic addition to the diene occurs onto a terminal carbon atom with attack onto the face away from the metal as expected. The product of hydride addition to the related indenyl cations yields the thermodynamically less stable *anti*-isomers.

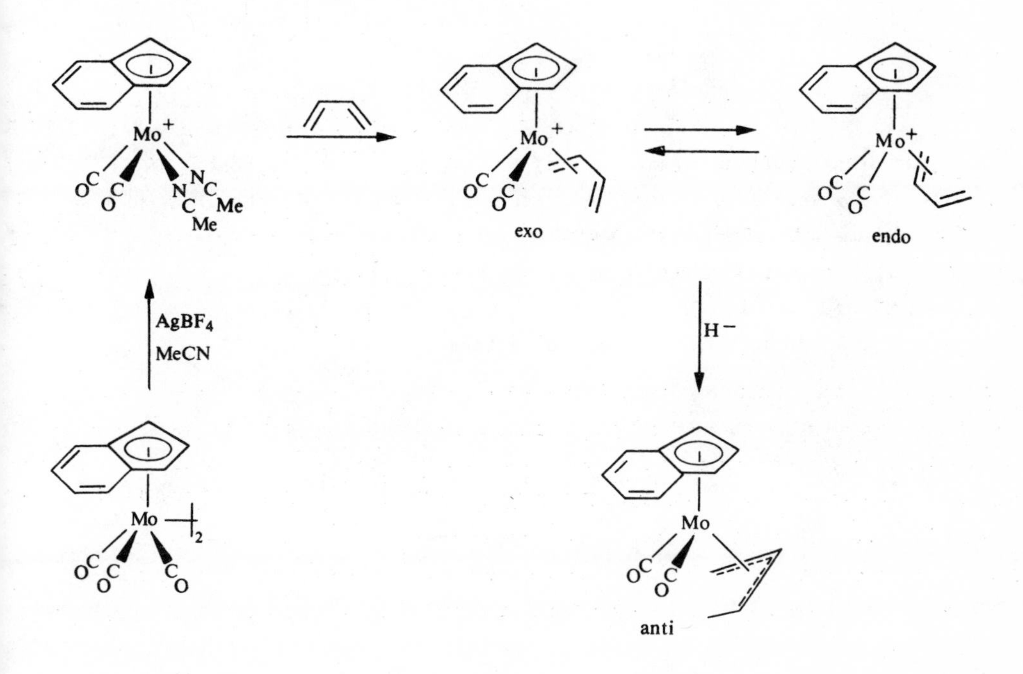

Nucleophilic attack by hydride or the enamine 1-morpholinocyclopent-1-ene on the cationic complexes of isoprene, 2,3-dimethyl butadiene, 1,3-cyclohexadiene and *trans*-penta-1,3-diene affords the corresponding *anti*-η^3-allyl complexes.[26,27]

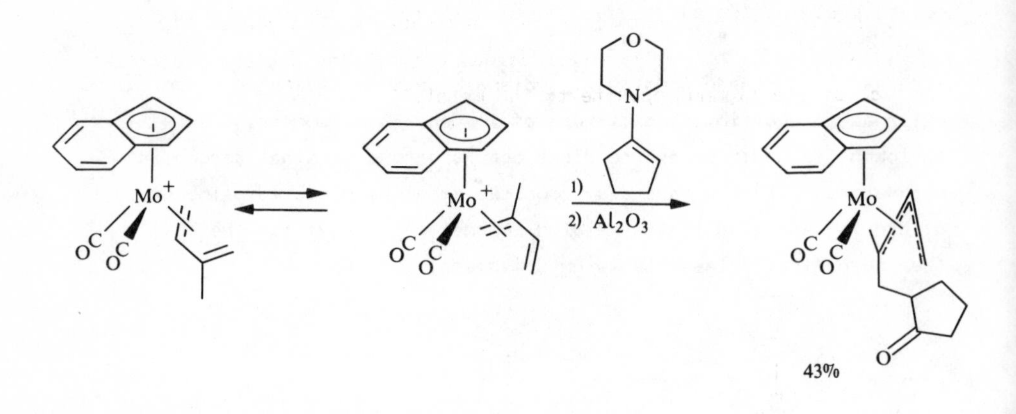

4.1.5 Nucleophilic addition to cationic η^5-dienyl complexes

Of the dienyl cations known those of $Fe(CO)_3$ have been the most extensively studied although complexes with virtually all the transition metals have been prepared.

(Pentadienyl)$Fe(CO)_3^+$ BF_4^- reacts with a variety of nucleophiles (e.g. H^-, R_2Cd, H_2O, ROH, amines) on a terminal carbon atom.[28]

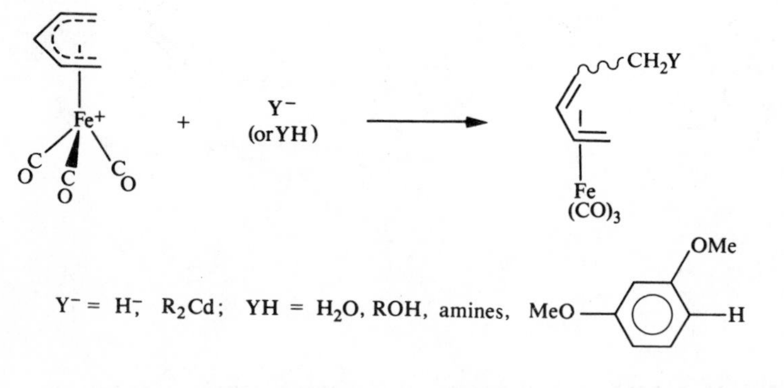

$(Cyclohexadienyl)Fe(CO)_3^+$ cation 21 is susceptible to nucleophilic attack exclusively on the terminal carbon atoms. The nucleophile attacks the face of the ligand opposite to the metal.[29-33]

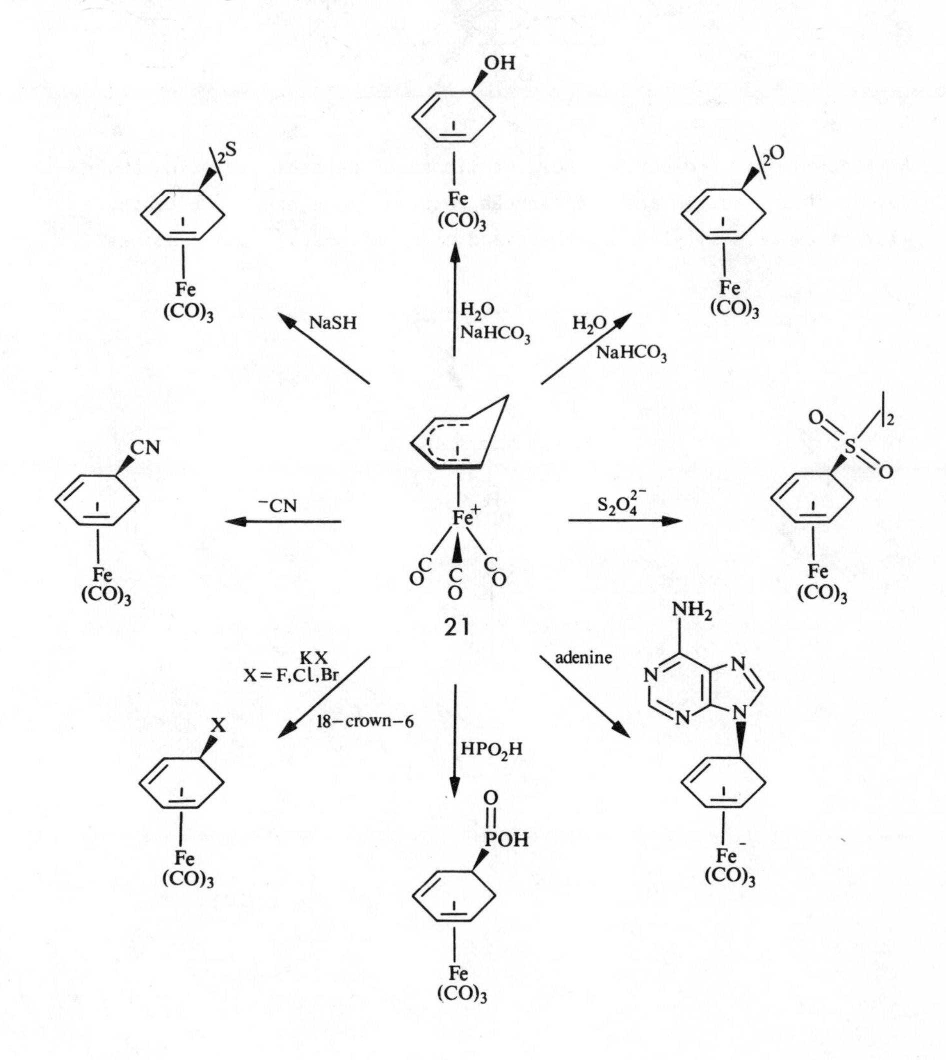

The cyclohexadienyl cation 21 is reduced efficiently by $NaBH_4$[34] and, rather surprisingly, also by Bu_3SnH.[29] Dissolving metal reductions lead to dimerisation.[29]

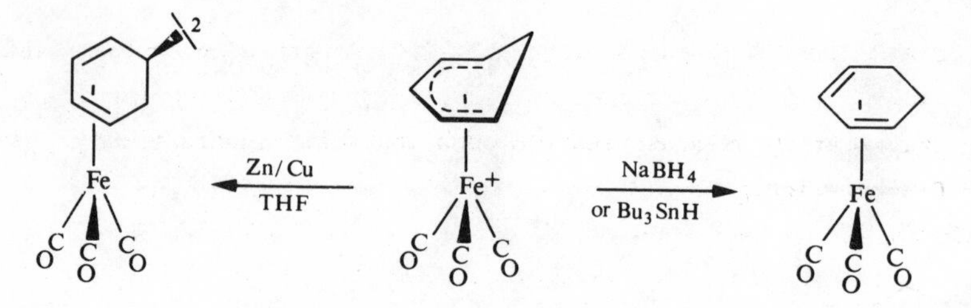

Alkylation of $(\eta^5\text{-dienyl})Fe(CO)_3^+$ cations can be achieved using organoboron, zinc, cadmium and copper reagents,[35] enolates,[33,34] trimethylvinyl silane,[36] silyl enol ethers and allyl silanes,[37] and enamines.[33,38]

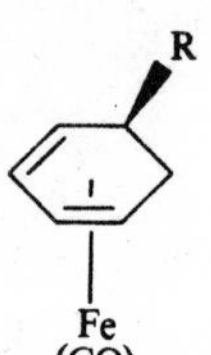

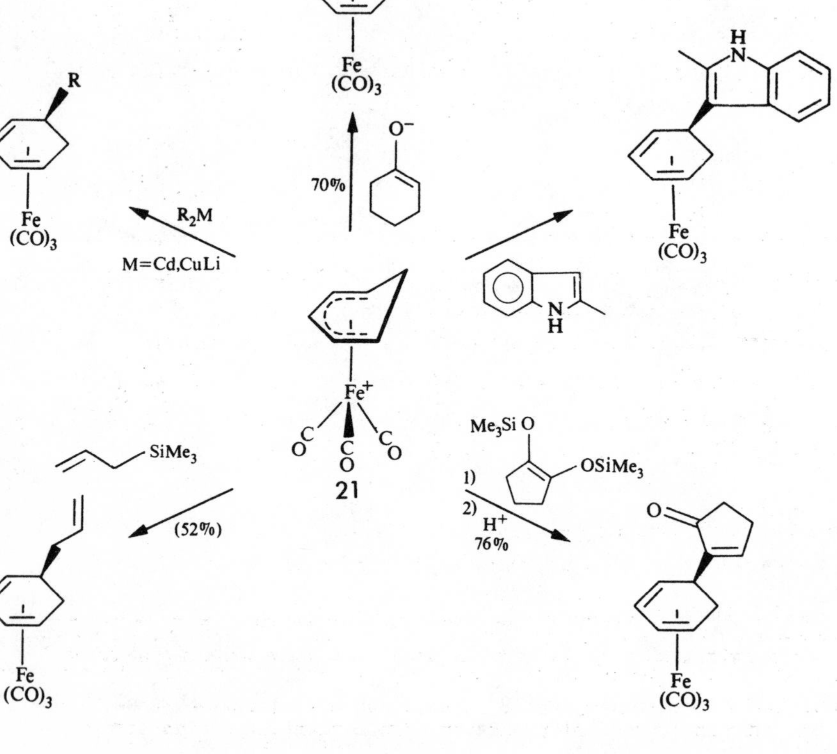

Aromatic amines may undergo N- or C- alkylation depending on the reaction conditions.[39] Electron withdrawing substituents (4-NO_2) on the arene favour N-alkylation whereas electron donating substituents (3-OMe, 3-NR_2) favour C-alkylation.

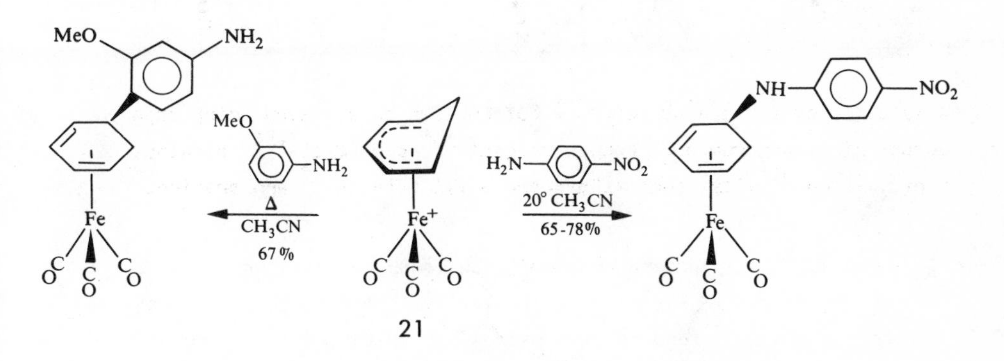

Oxidative decomplexation of the diene from the complex formed from the cation 21 and 3,4-dimethyl aniline gave 2,3-dimethyl carbazole.[39]

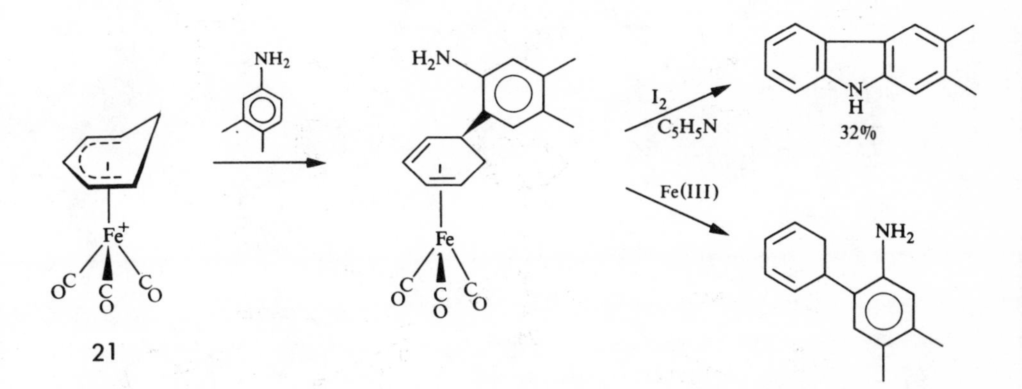

Cation 21 is a strong enough electrophile to react even with weak nucleophiles such as amides, thiourea and potassium phthalamide.[39]

The introduction of alkyl groups onto a terminal carbon atom of cation 21 increases the positive charge on that carbon by stabilisation of the resonance form 22 but also introduces steric constraints to attack at that position.

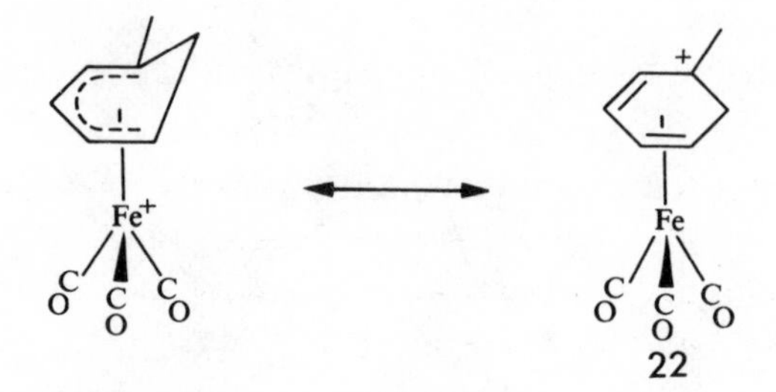

Thus the sterically undemanding nucleophile OH^- attacks predominantly C-1 of cations 23 and 24, whereas the sterically more demanding nucleophiles morpholine and BH_4^- tend to give predominantly C-5 addition products. Borohydride, trimethyl-phosphite and sodium hydrogen sulphide attack cation 25 exclusively at C-5.[29,33]

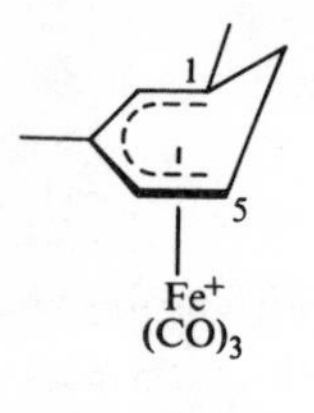

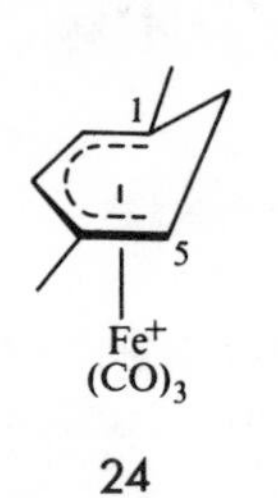

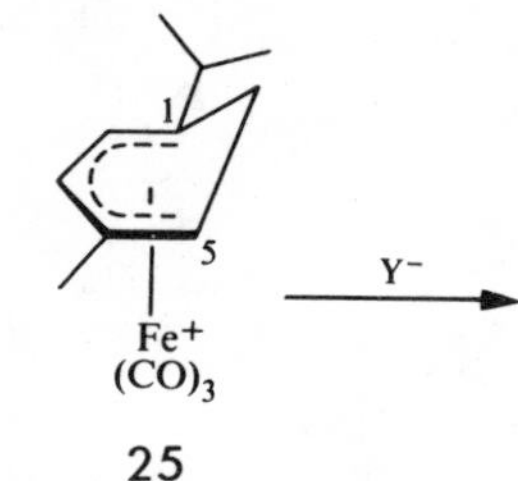

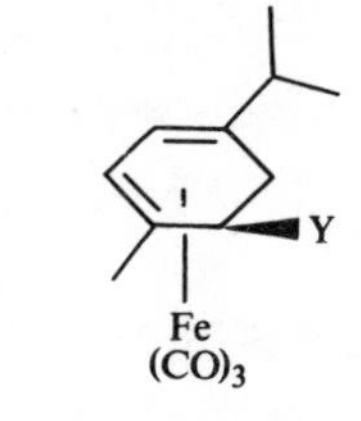

23 24 25

% Attack at C-1 (C-5)

Cation	Hydroxide	Morpholine	Borohydride
23	90 (10)	10 (90)	15 (85)
24	100 (0)	20 (80)	25 (75)
25	--	--	0 (100)

Cations 26 and 27 react regioselectively on the terminal carbon furthest from the alkyl substituents.[36,37] The two reactions shown below are probably initiated by attack of F^- from the PF_6^- counterion on the $SiMe_3$ groups.

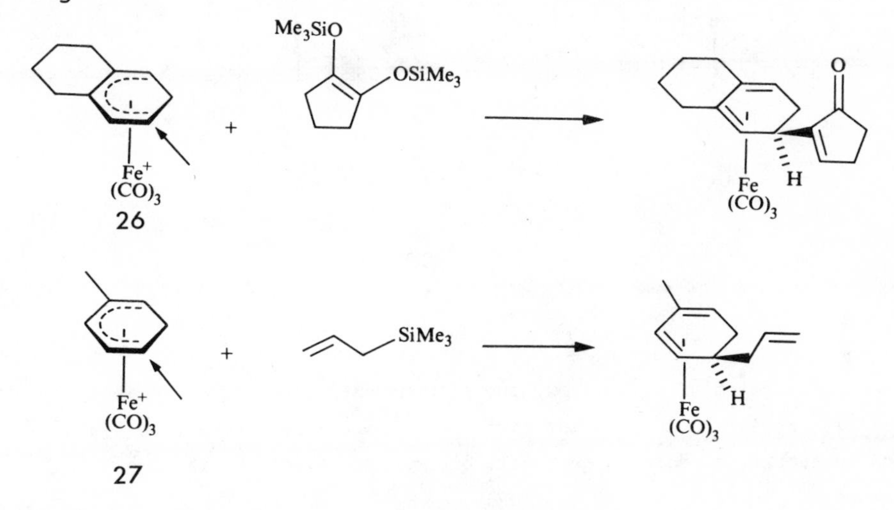

R_2CuLi and R_2Cd reagents add to the cation 26 to give a mixture of C-1 and C-5 addition, C-5 addition being predominant.[35]

Proton abstraction instead of nucleophilic addition can become important in highly substituted (cyclohexadienyl)$Fe(CO)_3^+$ cations.[29,40]

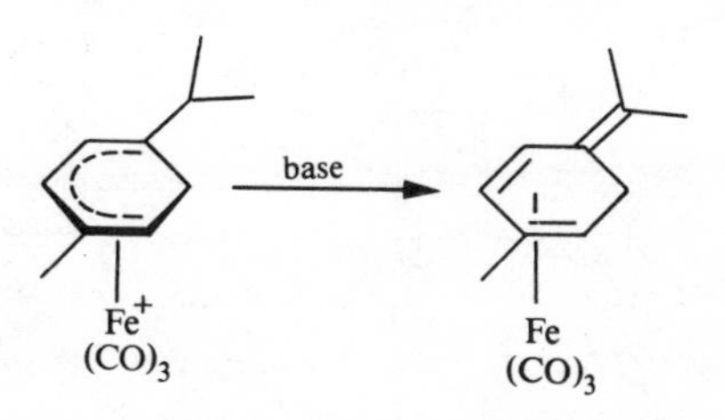

base = $NaHCO_3$, $S_2O_4^{2-}$, amines, enamines

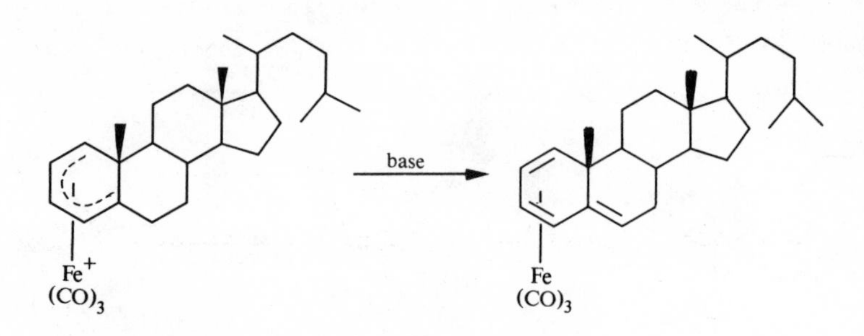

base = H_2O, MeO^-, CN^-, Morpholine

The introduction of a 2-methoxy substituent onto the (cyclohexadienyl)-$Fe(CO)_3^+$ cation lowers the positive charge on C-1 by resonance effects and nucleophilic addition generally occurs exclusively on the C-5 carbon.[36,41]

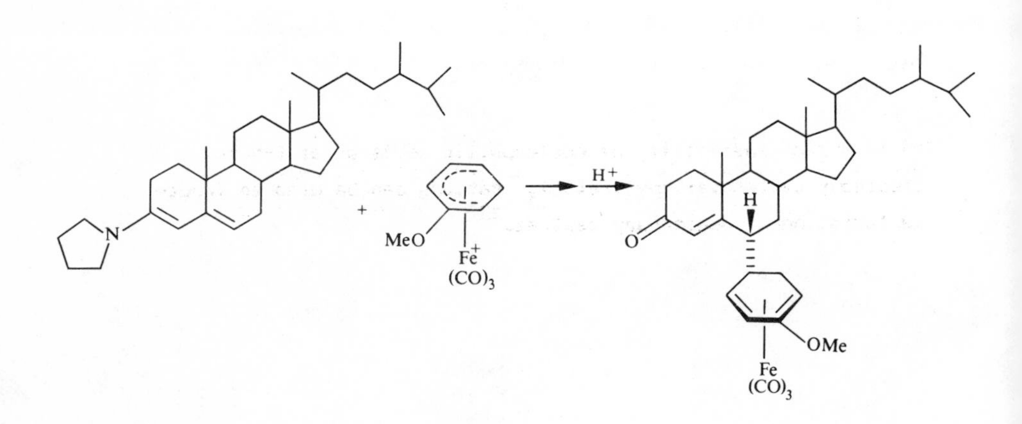

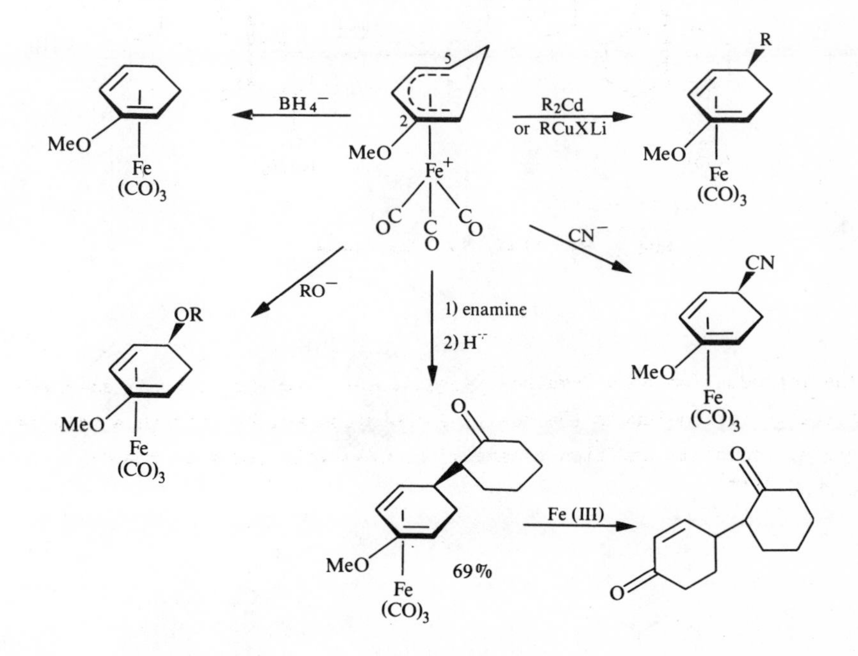

The high regiospecificity of nucleophilic addition at C-5 to (2-methoxy-cyclohexadienyl)Fe(CO)$_3^+$ cations can be used to induce the formation of quaternary centres.[42]

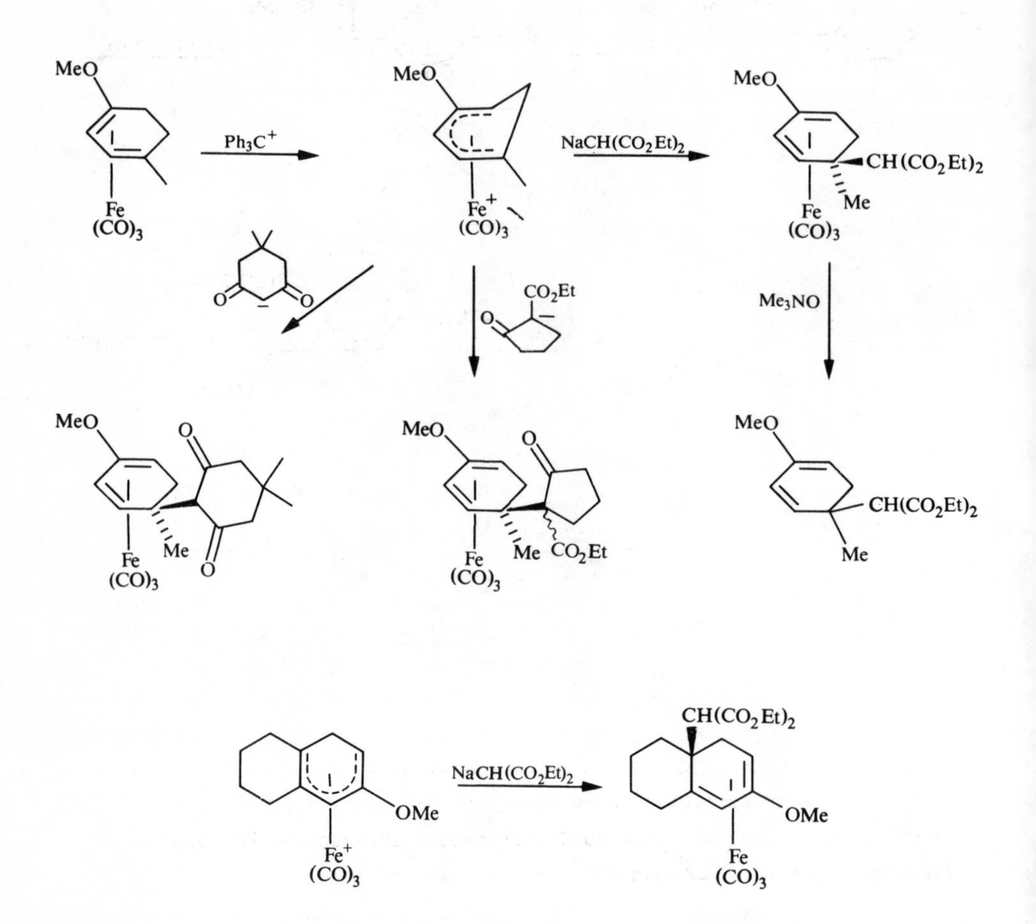

This reaction has been successfully employed in the synthesis of the spiro[4,5]compound 28.[43] This spiro system is present in a number of naturally occurring sesquiterpenes.

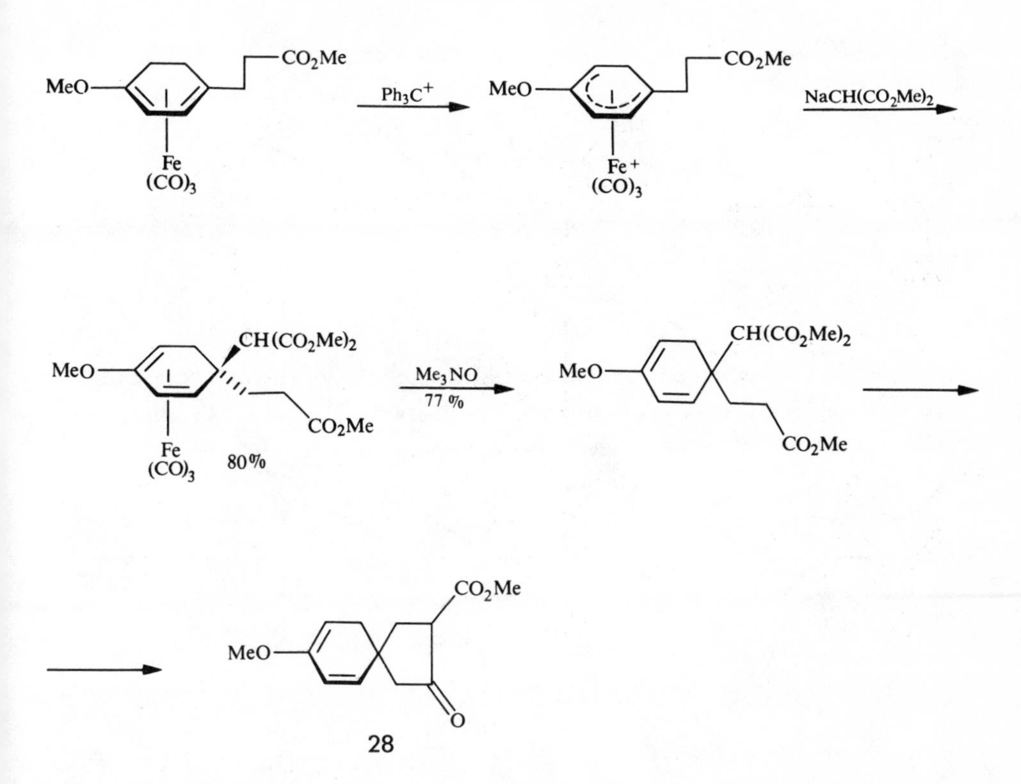

The selective attack at C-5 of (2-methoxy-cyclohexadienyl)$Fe(CO)_3^+$ cations is not observed in molecules where attack at C-5 is very sterically hindered. Thus, only addition to C-1 or proton abstraction has been observed with cations 29[44] and 30[45].

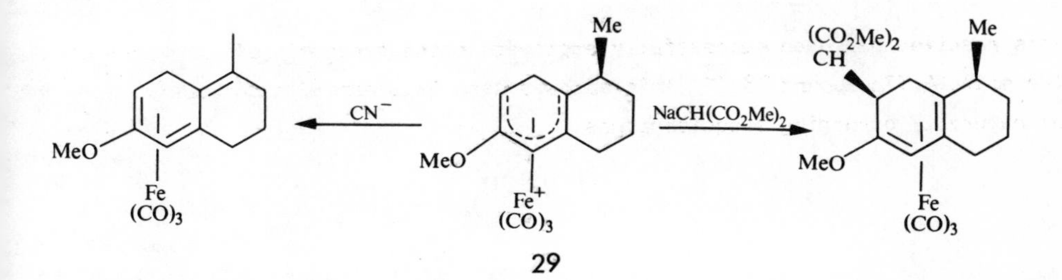

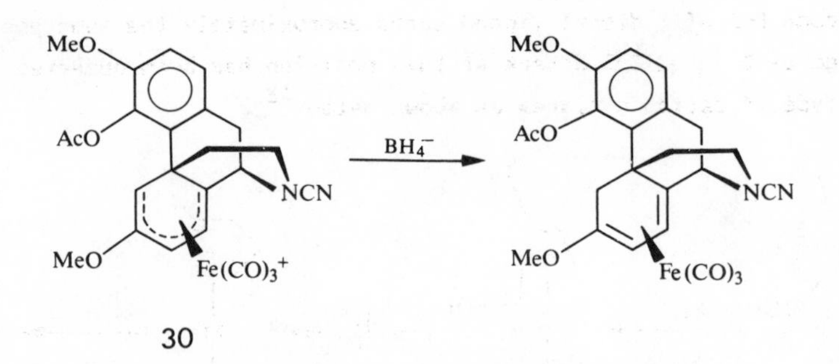

Nucleophilic addition of hydride to (cycloheptadienyl)$Fe(CO)_3^+$ cation 31 is less regioselective than onto the cyclohexadienyl cation 21. Products from attack at C-1 and C-2 have been obtained.[46]

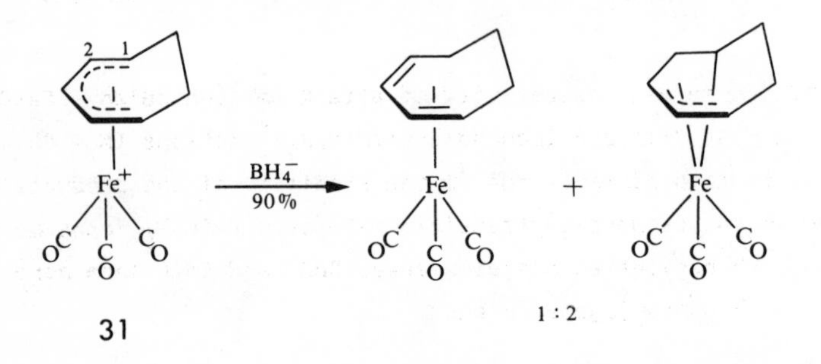

However cation 31 reacts regioselectively at C-1 with a variety of heteroatomic nucleophiles and dialkyl copper reagents and this provides a useful synthesis of 5-substituted cycloheptadienes.[47]

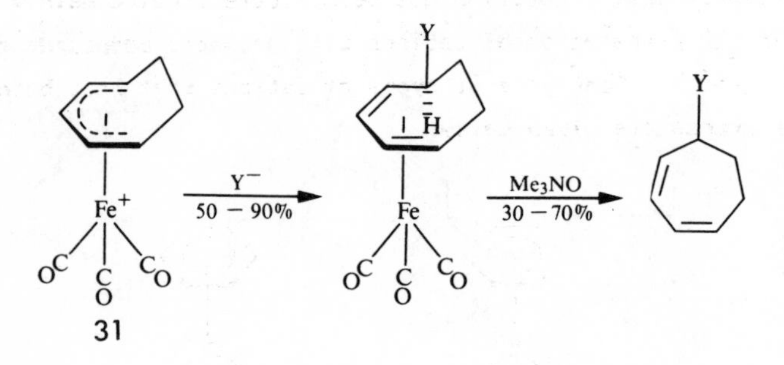

$Y^- = HO^-, RO^-, PhS^-, R^- (R_2CuLi)$,

$YH = R_2NH, RNH_2$

Although C-3 of a dienyl ligand bears approximately the same positive charge as C-1 and C-5 attack at this position has been observed for only one type of cationic system as shown below.[48]

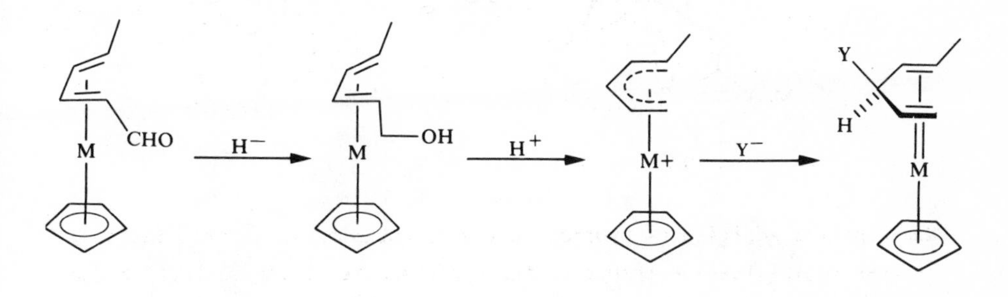

M = Rh, Ir $\qquad\qquad$ Y^- = MeO^-, H^-

The difference in regioselectivity of attack on (pentadienyl)$Fe(CO)_3^+$ cations (only at C-1) and (pentadienyl)$M(C_5H_5)^+$ cations (M = Rh, Ir) (only at C-3) is presumably due to the stability of the products being reflected in the respective transition states. $Fe(CO)_3$ forms more stable complexes with conjugated dienes whereas CpRh and CpIr form more stable complexes with nonconjugated dienes.

4.1.6 *Nucleophilic addition to cationic η^6-arene complexes*

This potentially useful reaction has been little studied mainly due to problems in the preparation of cations with aromatic compounds containing functional groups. Some general types of cations that have been studied for simple arenes are given below.

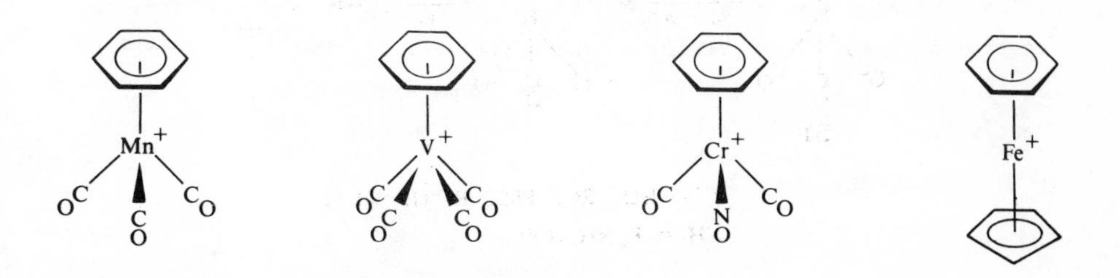

Nucleophilic addition to the reactive (arene)$Mn(CO)_3^+$ cations has been shown to occur with H^- (AlH_4^- and BH_4^-) MeLi and PhLi. CN^- and PPh_3 addition to the complexed arene is readily reversible.

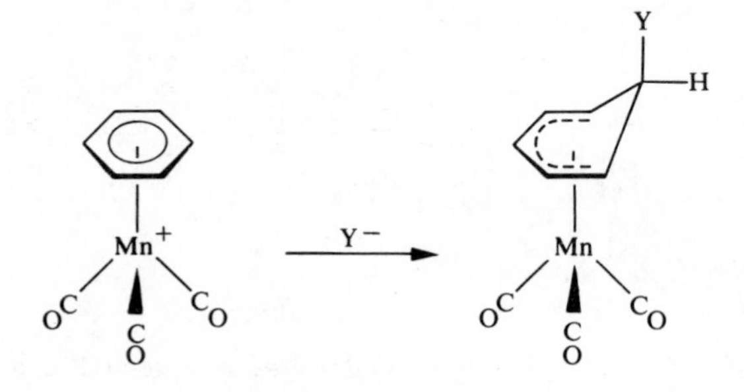

Nucleophilic aromatic substitution of haloarenes occurs easily with a variety of nucleophiles (MeO^-, PhO^-, PhS^-, N_3^-, RNH_2, R_2NH) that can reversibly add to the arene ligand.[49,50]

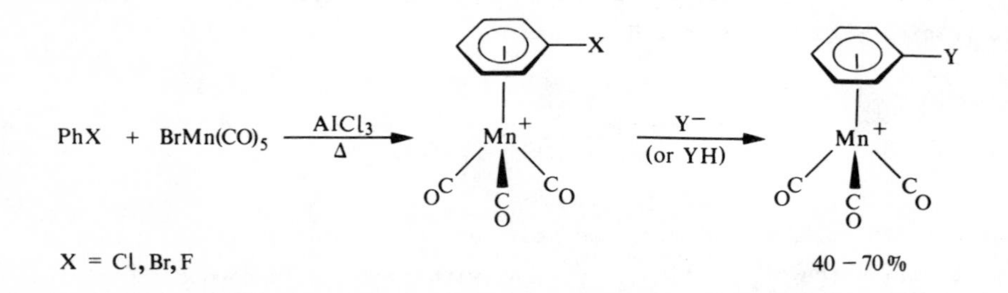

Nucleophiles that cannot add reversibly to the arene ring (H^-, R^-, Ph^-) give addition rather than substitution products; i.e. initial attack onto hetero-substituted arenes is not onto the carbon bearing the heteroatom. Electron attracting substituents such as Cl activate the ortho position to attack whereas electron donating substituents such as MeO or NMe_2 deactivate the ortho position.[50]

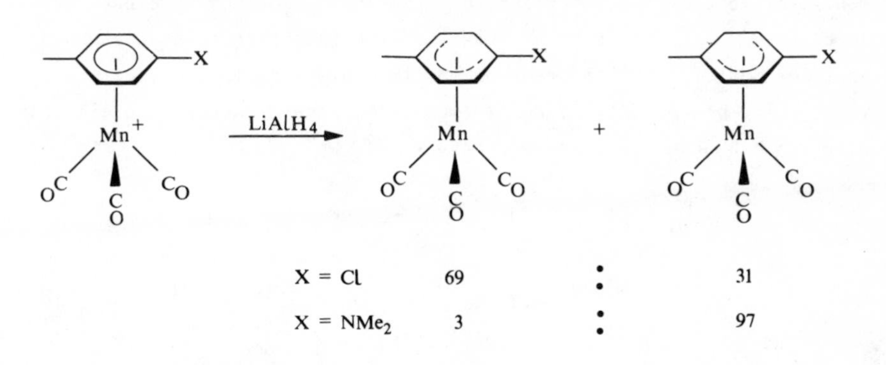

Hydride adds to the (benzene)$V(CO)_4^+$ cation to give (cyclohexadienyl)$V(CO)_4$.[51]

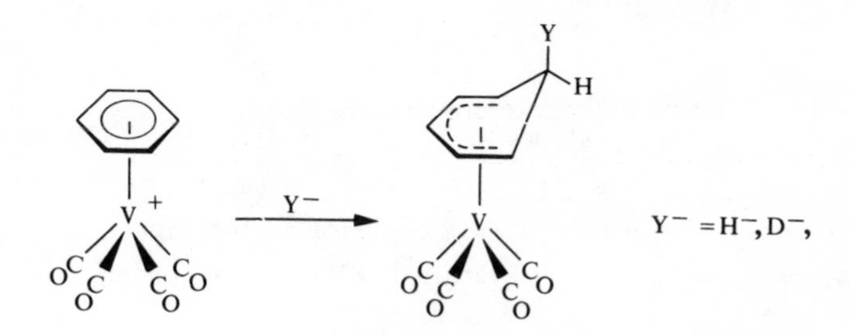

Treatment of (arene)$Cr(CO)_3$ complexes with $NOPF_6$ yields the cations (arene)$Cr(CO)_2(NO)^+$ which readily undergo nucleophilic addition reactions.[52]

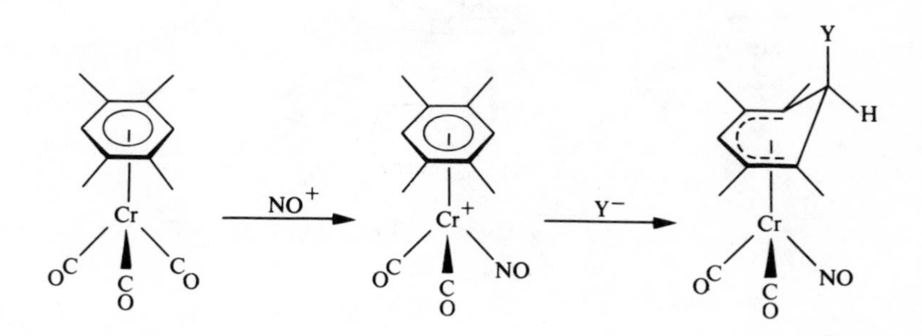

Nucleophilic addition reactions to (arene)$FeCp^+$ cations have been studied in detail.[53] As predicted by Rule 1 of the selection rules discussed earlier, addition to the arene occurs in preference to the Cp. Electron withdrawing groups (Cl, CO_2Me) favour ortho attack whereas the electron donating MeO group favours meta attack.

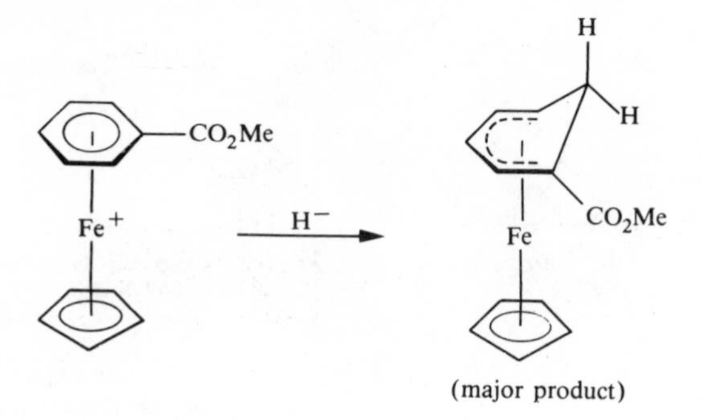

Coordination of aryl halides to $FeCp^+$ activates them to nucleophilic substitution reactions with methoxide.[54] Substitution rather than addition occurs with MeO^- presumably because addition to the ortho, meta or para positions is reversible whereas the substitution reaction is not.

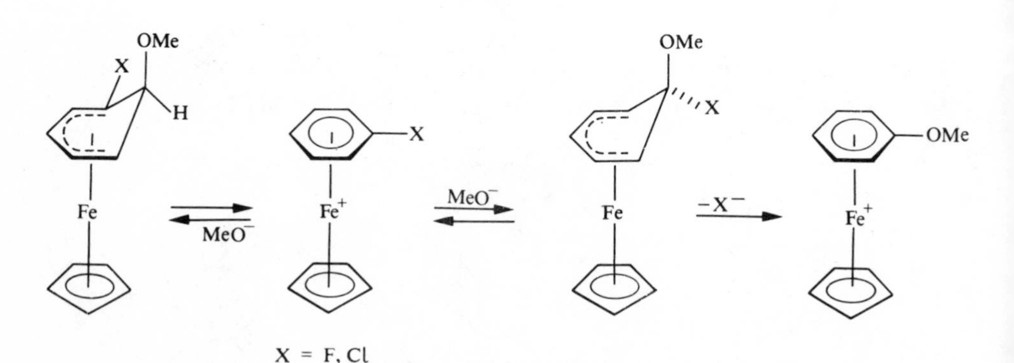

4.1.7 Cyclopropanation reactions

The cationic species 32 acts as a methylene transfer agent converting olefins to cyclopropanes in refluxing dioxane with good yields. The reaction is stereospecific.[55]

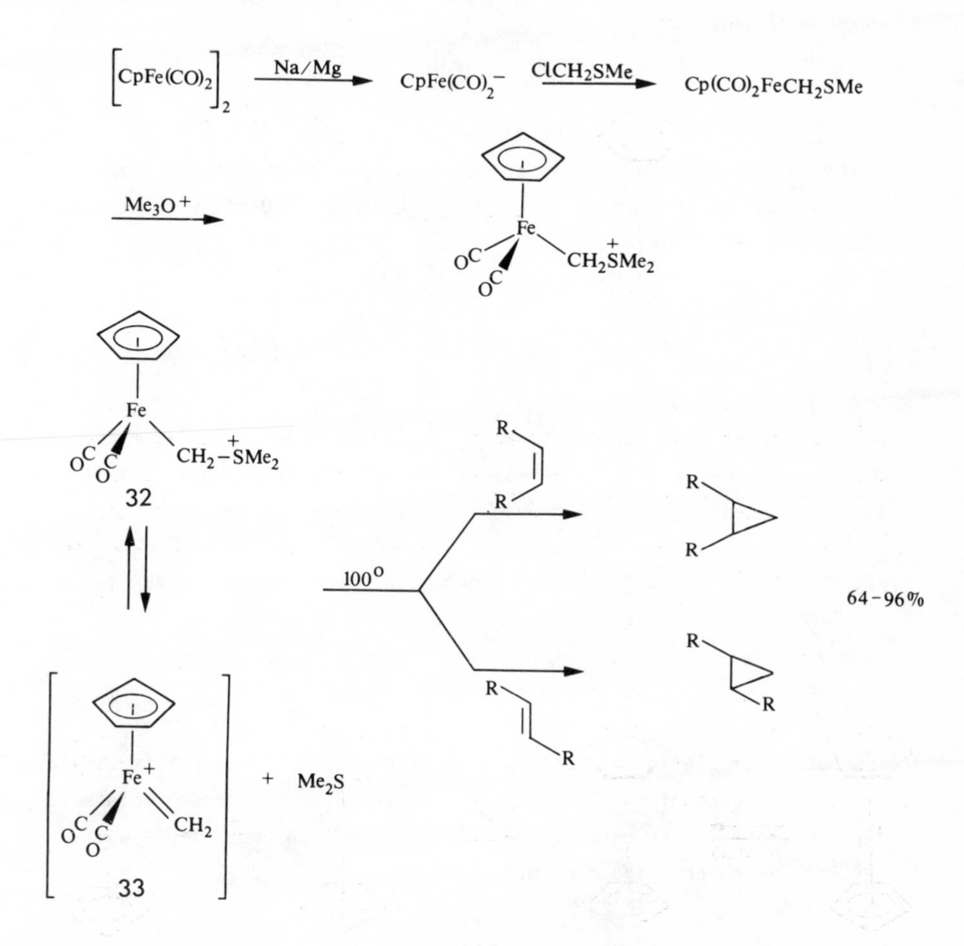

It is not clear whether it is the cation 32 itself or the carbene cation 33 formed by loss of Me_2S that is the methylene transfer agent. Carbene cations have been suggested as intermediates in the acid catalysed cyclopropanation reactions of olefins by $CpFe(CO)(PPh_3)CH_2OR$.[56]

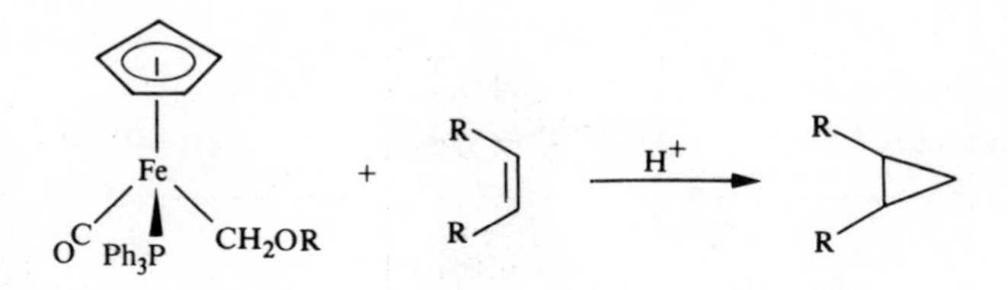

The high temperatures needed to effect methylene transfer with complex 32 can be avoided by using the analogous cation 34. Olefins react with cation 34 at 40° to form cyclopropanes.[57]

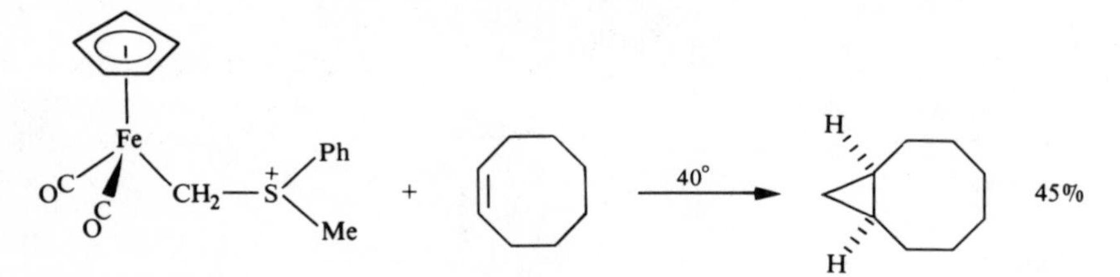

4.2. STOICHEIOMETRIC NUCLEOPHILIC ADDITION AND SUBSTITUTION REACTIONS INVOLVING NEUTRAL ORGANOTRANSITION METAL COMPLEXES

4.2.1 Nucleophilic addition to neutral η^2*-olefin complexes*

Coordination of olefins to $Fe(CO)_4$ activates them to nucleophilic attack by stabilised carbanions.[58] The intermediate anion cannot be isolated. However protonation or carbonylative alkylation (see chap 9) allow the isolation of the corresponding organic products in good yields.

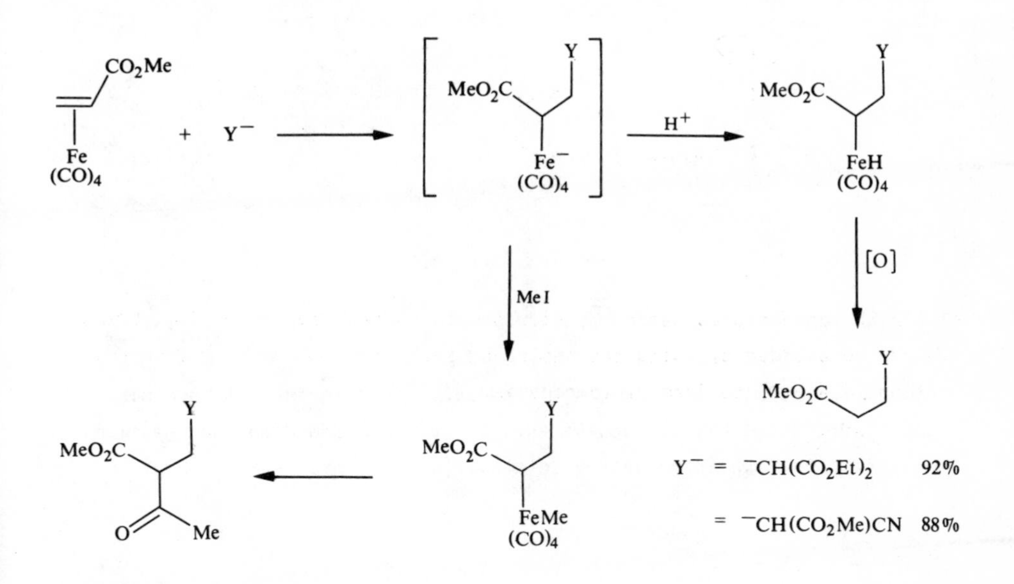

Y^- = $^-CH(CO_2Et)_2$ 85%

The $Fe(CO)_4$ complex of methyl α-chloroacrylate reacts with two equivalents of stabilised carbanions as shown below.[59]

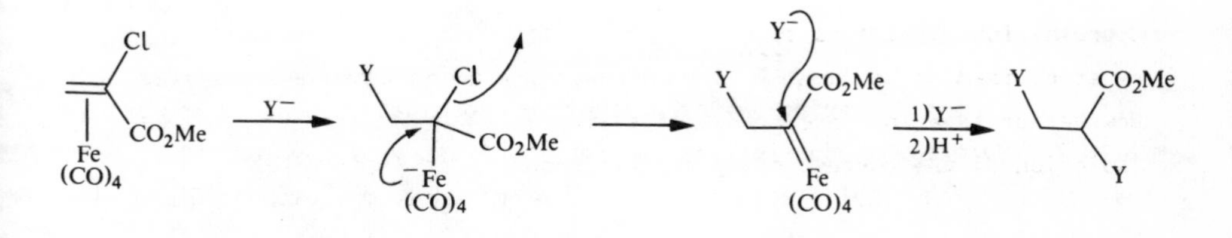

Nucleophilic addition occurs very readily to (olefin)$PdCl_2$ complexes. The intermediate η^1-alkyl complexes are generally too unstable to be isolated and must be reacted further *in situ*. Nevertheless an intermediate η^1-alkyl complex has been isolated in the following example[60]

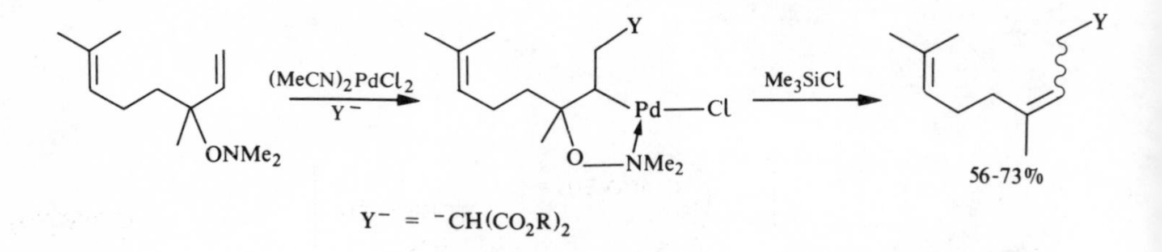

Olefins may be alkylated by stabilised carbanions in the presence of $(CH_3CN)_2PdCl_2$ only if two equivalents of Et_3N are added to the reaction mixture. The Et_3N is necessary to stabilise the intermediate complexes.[61]

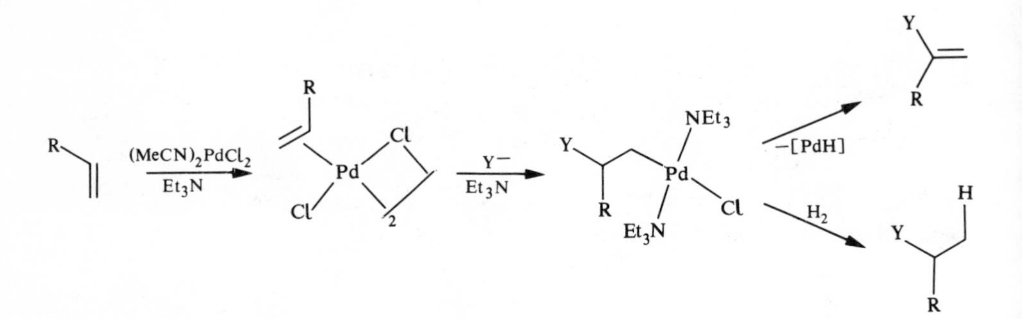

The requirement of having amines present to coordinate to the Pd can be used to advantage in controlling the stereochemistry of complex formation. The allylic amine 35, readily prepared from cyclopentadiene, forms a $PdCl_2$ complex with the amine and Pd *cis*. Nucleophilic addition to this complex followed by base catalysed β-elimination gives the *trans* compound 36 in high yield. Subsequent reaction of 36 with Li_2PdCl_4 yields a new (olefin)$PdCl_2$ complex 37 again with the amine group directing *syn* coordination to the olefin. Nucleophilic addition of an alkoxide to 37 gives the η^1-alkyl derivative 38 which can then be treated with n-pentylvinylketone to yield 39 in 44% overall yield from cyclopentadiene.

Compound 39 is readily converted to the lactone diol 40 from which a variety of prostaglandins have been synthesised in high yield. The above sequence allows the stereocontrolled formation of four centres starting from a simple amine directing group.[62]

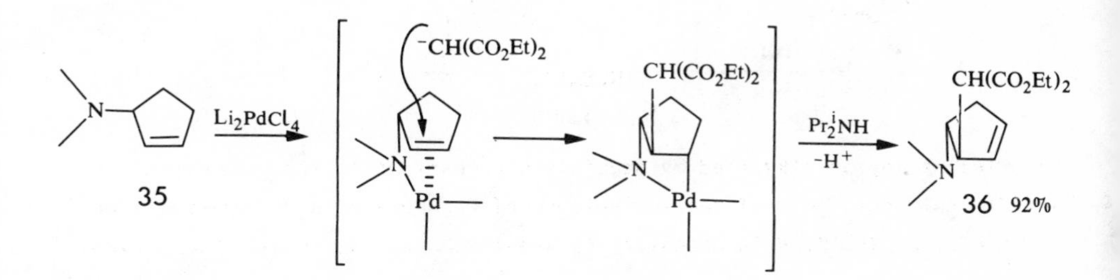

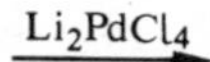

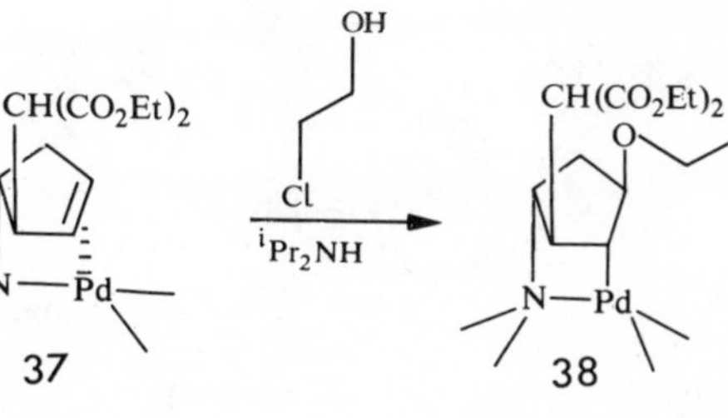

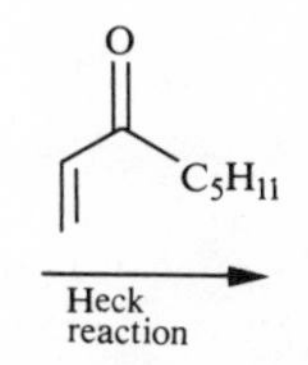

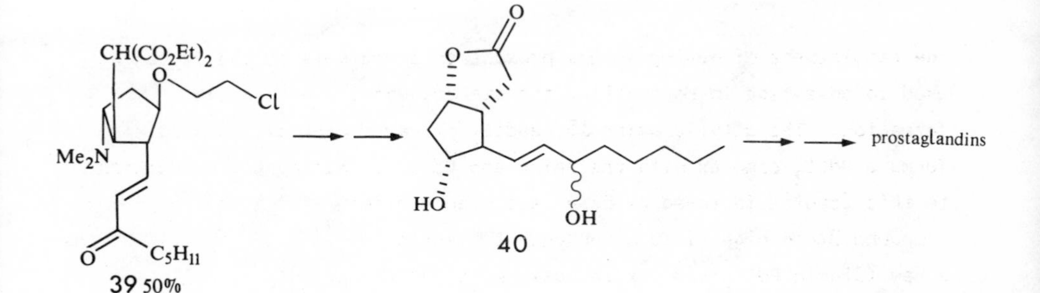

The reaction of olefins with amines in the presence of $PdCl_2$ allows stereospecific formation of amines, aminoacetates, diamines and aziridines.[63]

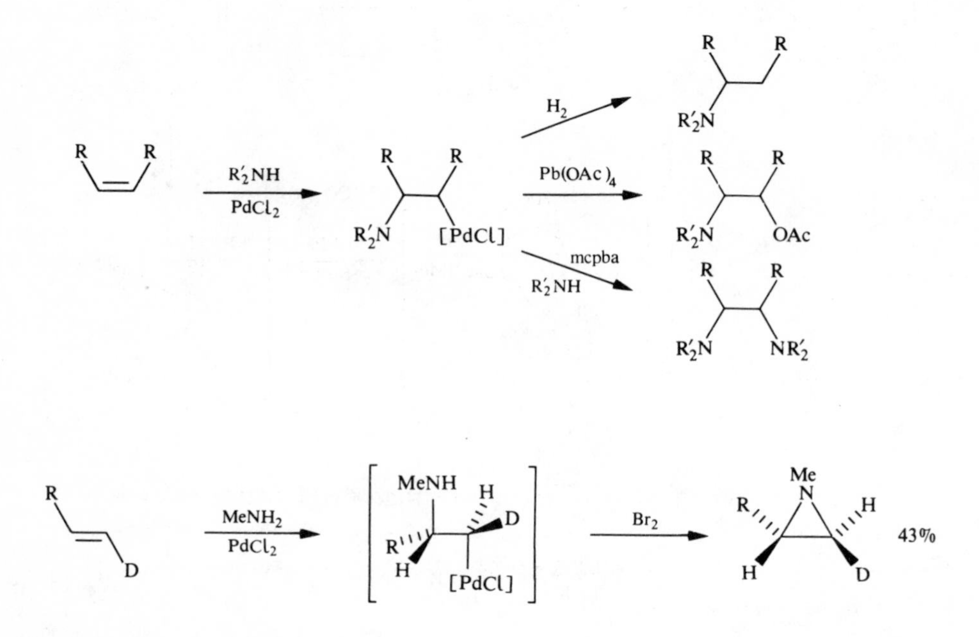

Intramolecular olefin amination leads to the formation of indoles and dihydroquinolines.[64]

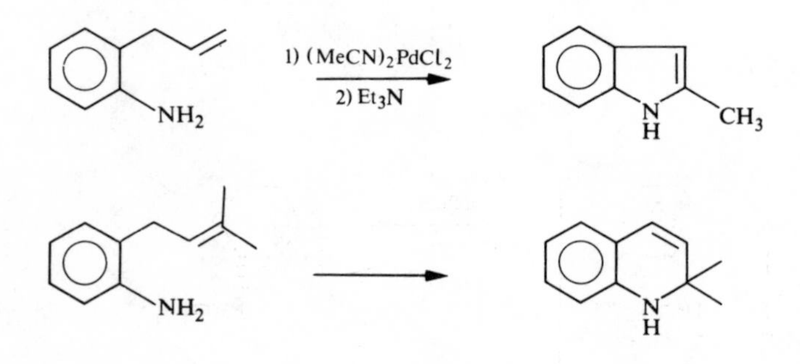

1,3-Dienes may undergo stereospecific *cis*-1.4-diamination in the presence of $(PhCN)_2PdCl_2$.[65]

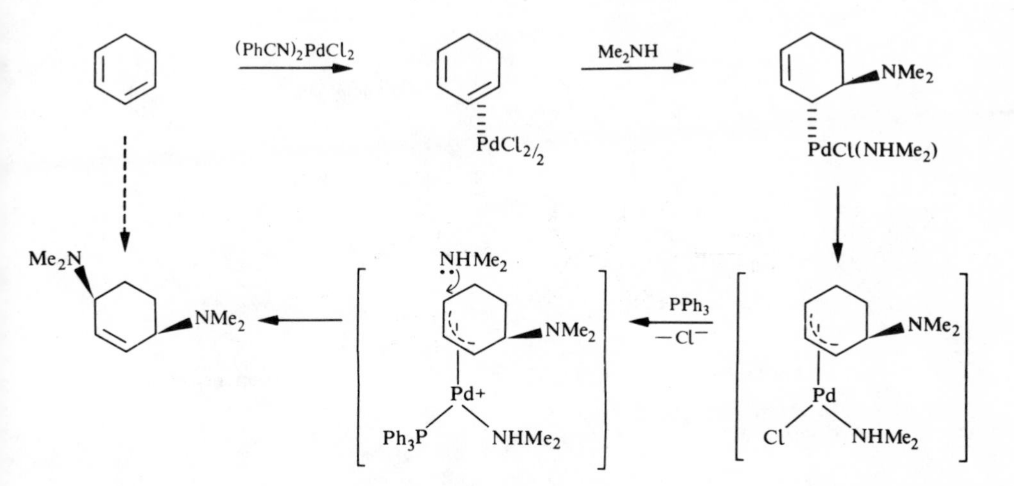

$PdCl_2$ complexes of enol ethers react with nucleophiles regiospecifically on the carbon atom bearing the oxygen substituent via transfer of the nucleophile from the Pd atom. This reaction has been employed in the synthesis of sugar derivatives.[66]

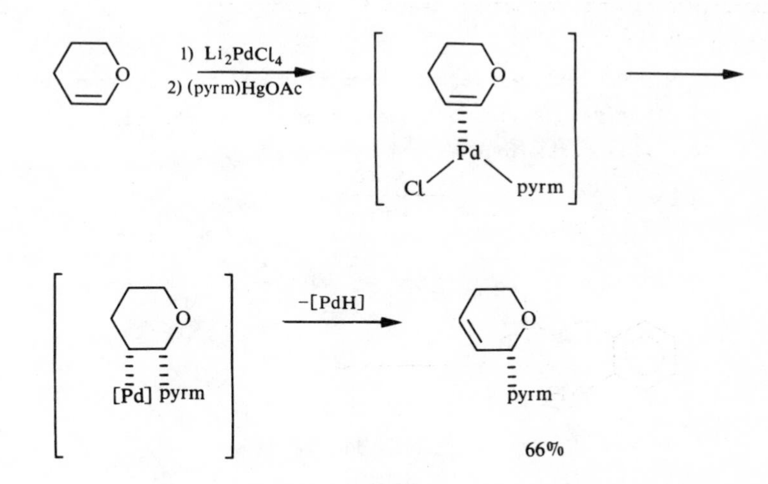

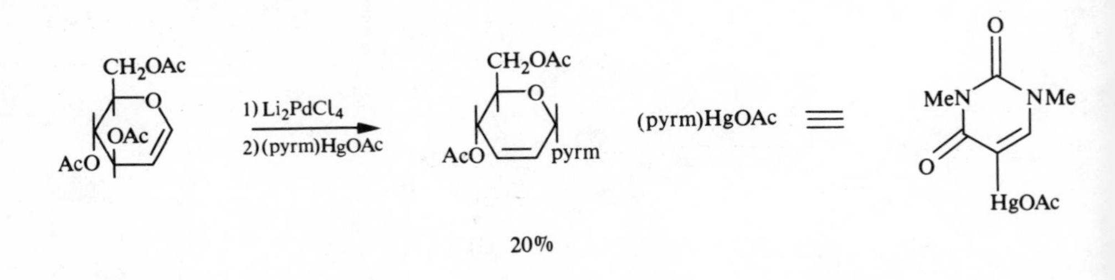

4.2.2. *Nucleophilic addition to neutral η²-carbene complexes*

η^2-Carbene complexes are electrophilic on carbon. They can be represented by the following resonance forms.

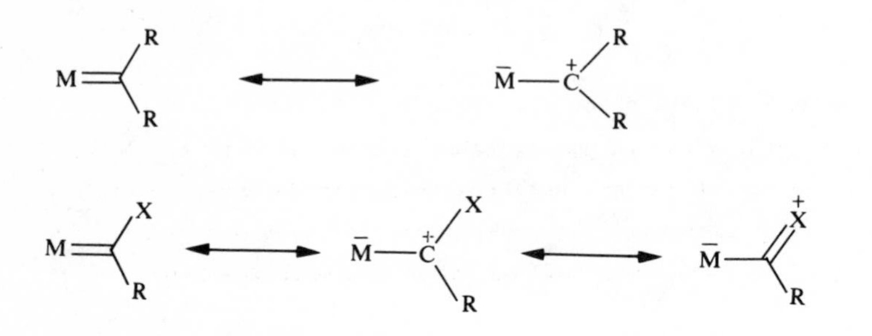

The alkoxy group of alkoxy carbene complexes may be exchanged for a variety of other groups via nucleophilic attack on the carbene by amines, thiols, selenols, carbanions etc.[67]

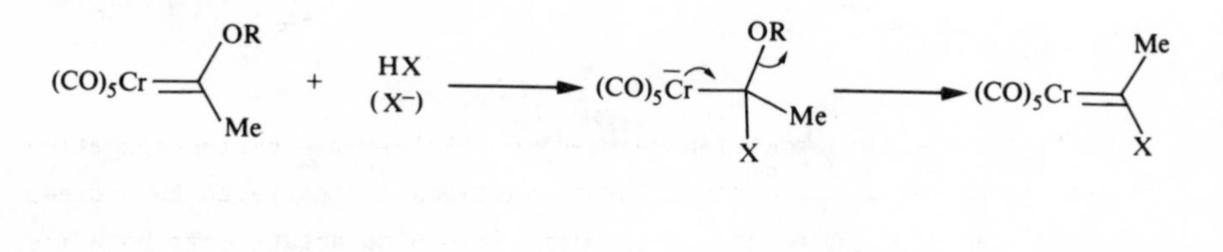

XH = RNH_2, RSH, RSeH X$^-$ = Ph$^-$, Me$^-$

Addition of vinyl lithium to the alkoxy carbene 41 followed by treatment with acid results in the formation of a vinyl ether.[68a]

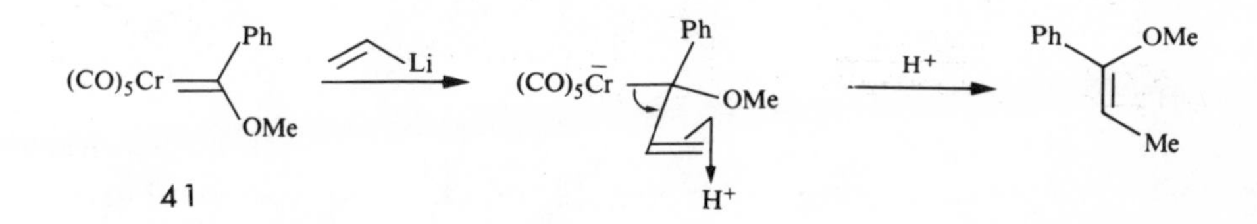

Vinyl ethers may also be formed by the nucleophilic addition of Wittig reagents to carbene complexes.[68b]

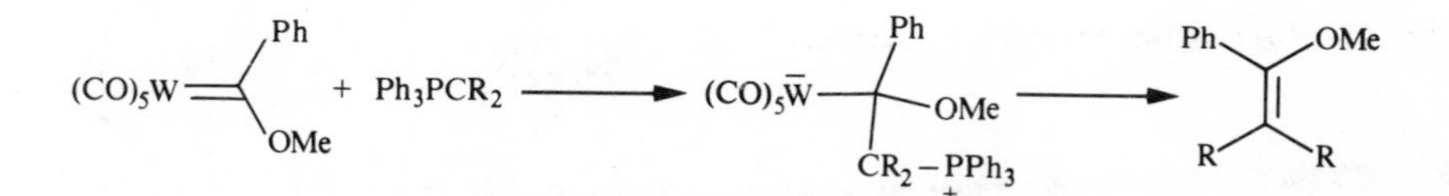

Organotransition metal carbene complexes react with olefins at elevated temperatures to form cyclopropanes.[67] There is a considerable amount of evidence indicating that free carbenes are not involved but rather the reaction proceeds by electrophilic attack of the metal carbene complex on the olefin.[69]

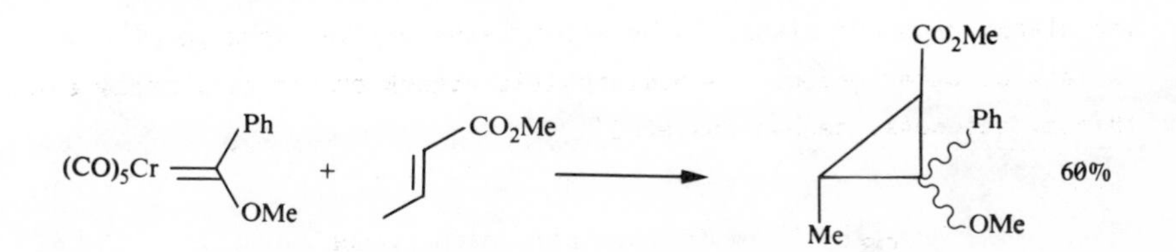

Transition metal complexes can be used to catalyse the cyclopropanation of olefins by diazo compounds. These reactions are believed to proceed via metal carbene complexes. Mechanisms involving attack onto both free and complexed olefins have been suggested.[70] The reaction can be used for many different types of functionalised olefins.

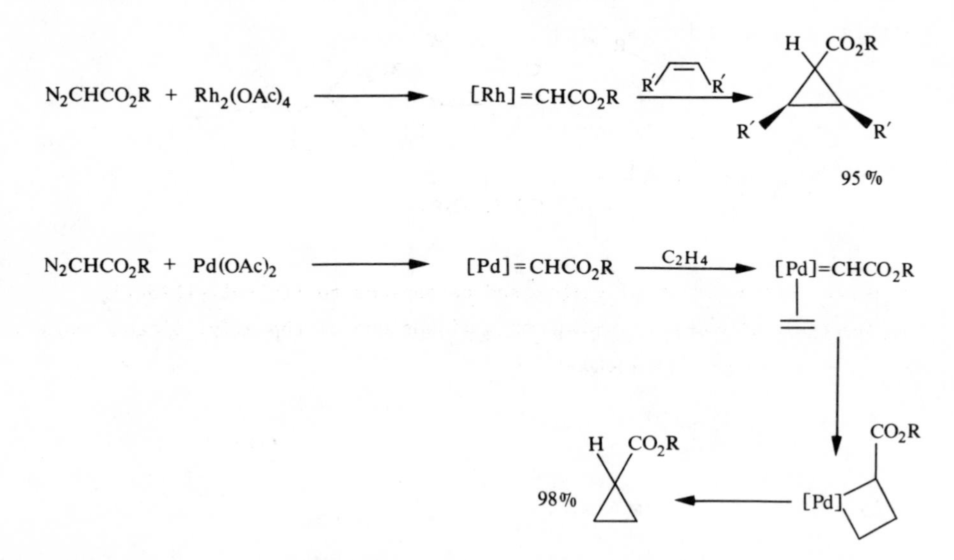

Metal carbene complexes have also been suggested for the cyclopropanation of α,β-unsaturated nitriles and esters by diazo compounds in the presence of $Mo(CO)_6$.[71]

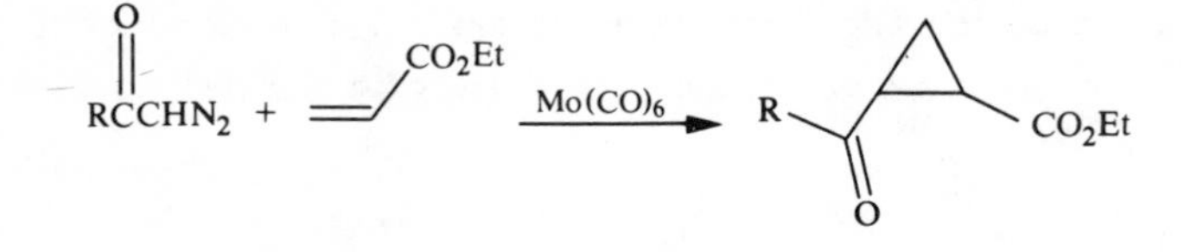

4.2.3 Nucleophilic addition to neutral η^3-allyl complexes

Nucleophilic addition of stabilised carbanions to $(\eta^3\text{-allyl})Fe(CO)_2NO$ complexes produces the corresponding olefinic products in high yields (85–95%). Nucleophilic attack generally occurs preferentially at the least sterically hindered end of the allyl ligand.[72]

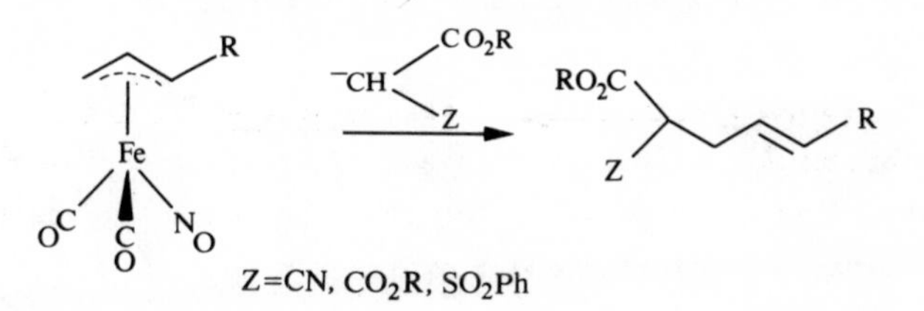

Nucleophilic addition of stabilised carbanions to $[(\eta^3\text{-allyl})PdCl]_2$ complexes proceeds efficiently only if one end of the allyl ligand bears an electron withdrawing group.[73]

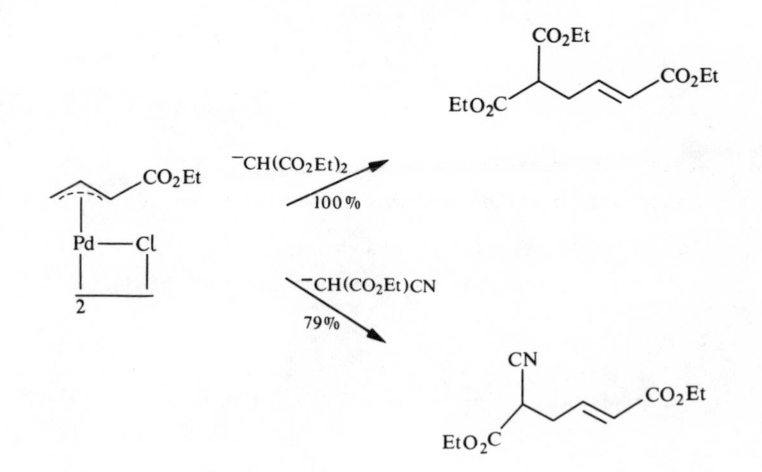

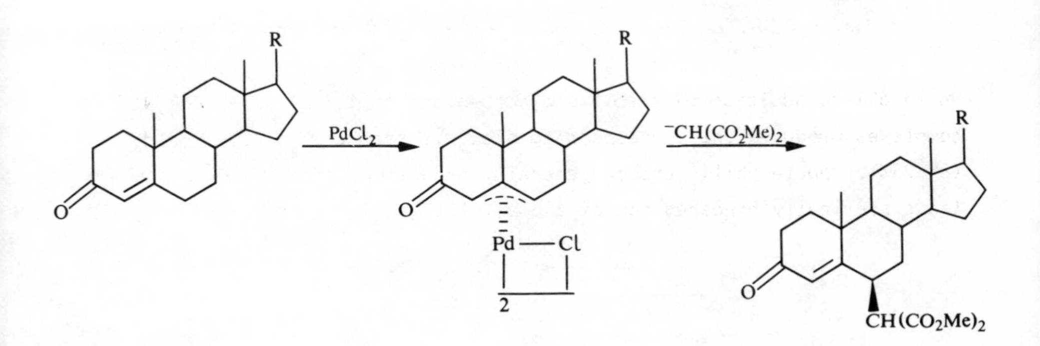

4.2.4 *Nucleophilic addition to neutral η^6-arene complexes*

The complexation of aromatic rings to $Cr(CO)_3$ activates them to nucleophilic attack.[74] Oxidative cleavage of the intermediate anion produces substituted arenes while prior treatment with acid before oxidation results in the formation of cyclohexadienes.[75]

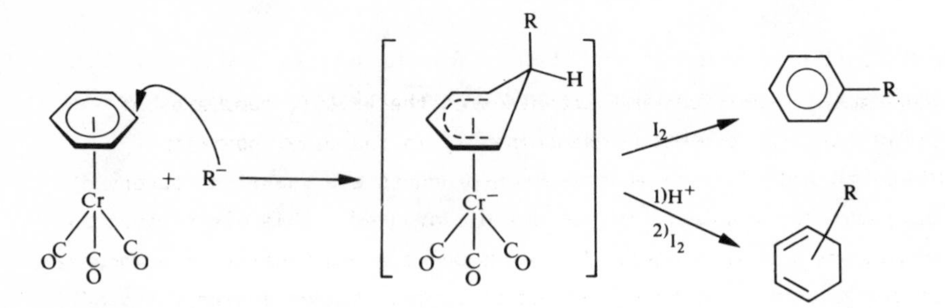

Phenylation of a variety of stabilised carbanions has been achieved. Unfortunately ketone enolates fail to give addition products.[75]

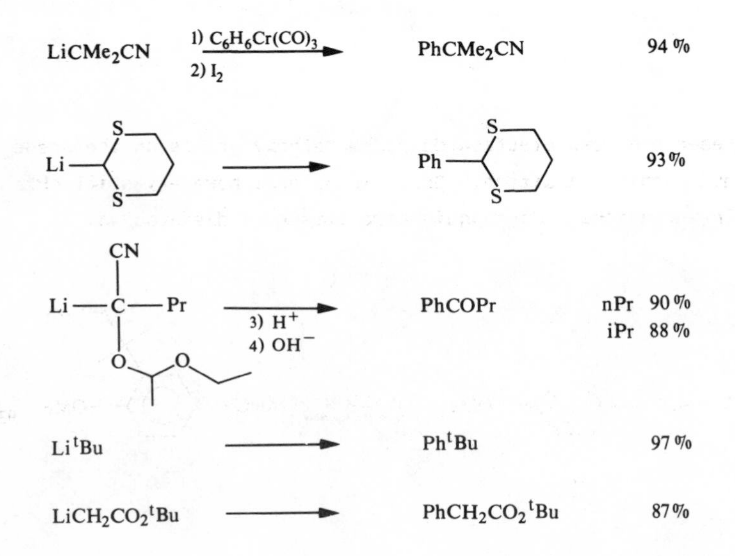

$LiCHMeCO_2{}^tBu$ →	$PhCHMeCO_2{}^tBu$	88 %
$LiCMe_2CO_2{}^tBu$ →	$PhCMe_2CO_2{}^tBu$	75 %
$KCMe_2CO_2{}^tBu$ →	$PhCMe_2CO_2{}^tBu$	88 %

Nucleophilic additions of carbanions to the $Cr(CO)_3$ complexes of anisole and toluene occur predominantly in the meta position.[76,77] Minor amounts of ortho substitution products are sometimes observed and para attack appears to be very disfavoured. This discrimination against para attack may be due to it being energetically disadvantageous to have a substituent on the central carbon atom of a dienyl ligand.

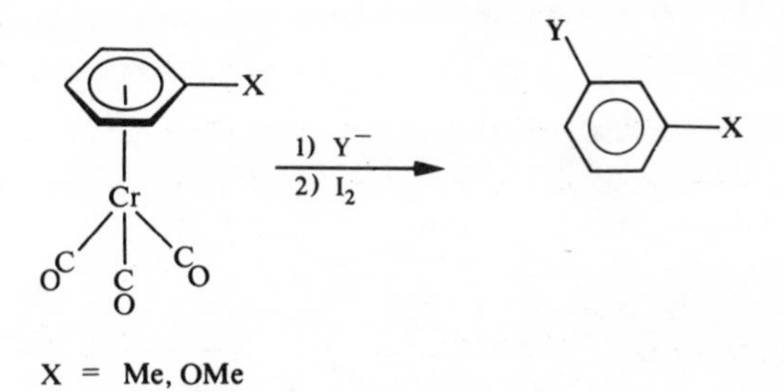

The presence of two electron donating methoxy groups on the arene slows down nucleophilic addition. Good yields are, however, still obtained with certain carbanions. Once again para attack is disfavoured.

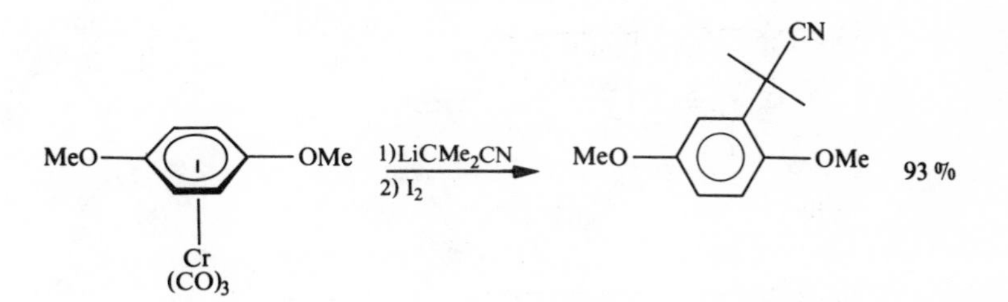

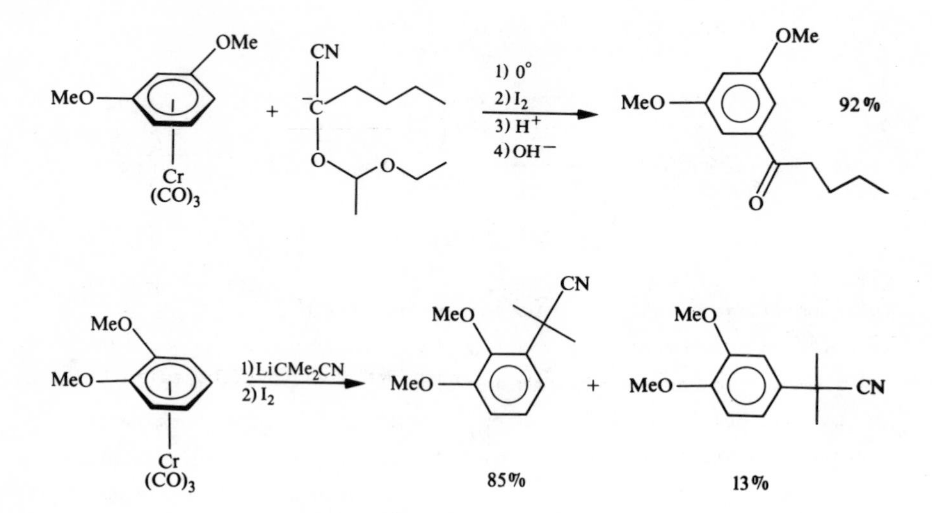

Addition of $LiCMe_2CN$ to (anisole)$Cr(CO)_3$ followed by treatment with acid and work up allowed the isolation of the cyclohexadiene 42 which could be converted into the corresponding α,β or β, γ-unsaturated ketones.

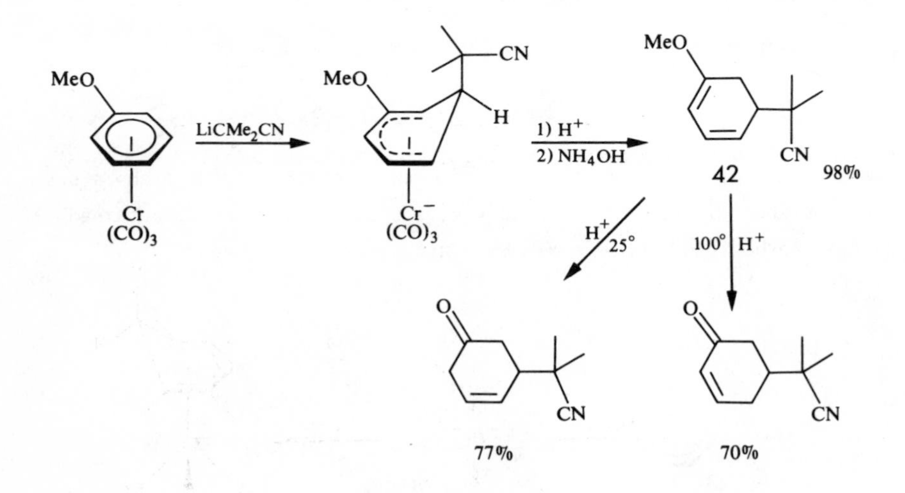

Treatment of the complex 43 with base and work up as above provides a useful entry into spiro-compounds such as 44.[77]

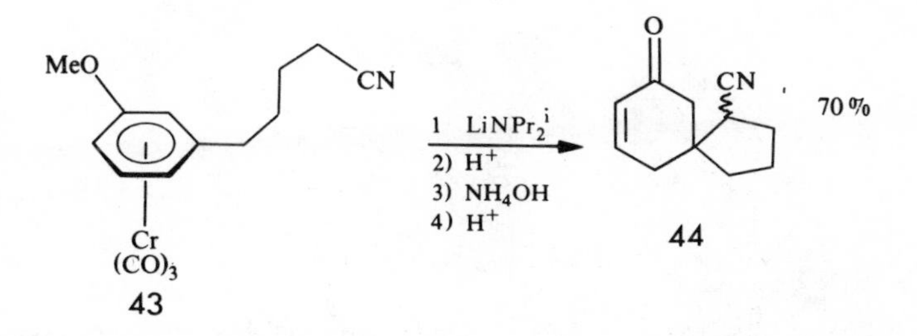

The $Cr(CO)_3$ complex of N-methyl indole undergoes nucleophilic attack in the 7-position.[78]

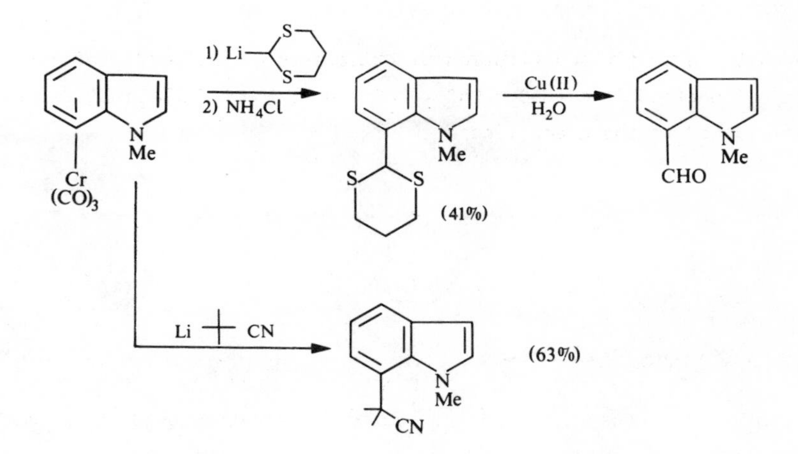

(Haloarene) $Cr(CO)_3$ complexes undergo nucleophilic displacement reactions of halide with a variety of nucleophiles.[79]

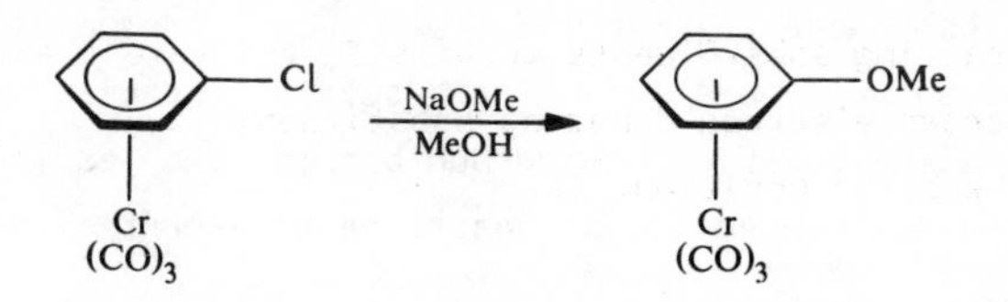

The phenylation of some carbanions that do not possess a second acidic hydrogen may also be achieved.[80]

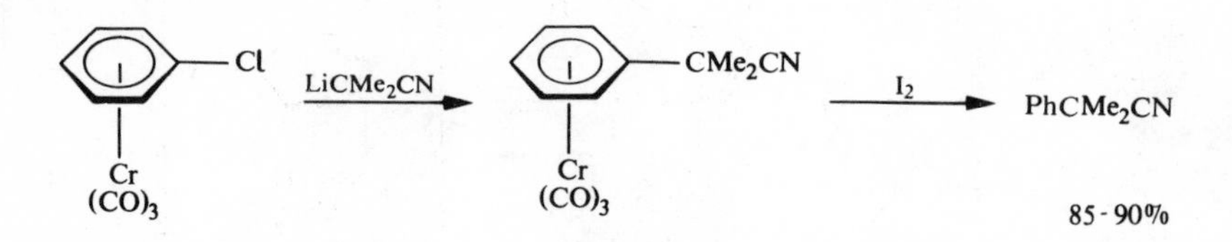

The reactivity of the (haloarene)$Cr(CO)_3$ complexes towards substitution is in the order F > Cl > Br > I. The mechanism of this substitution reaction is not completely clear.[74] One possibility is certainly the classical nucleophilic aromatic substitution mechanism.

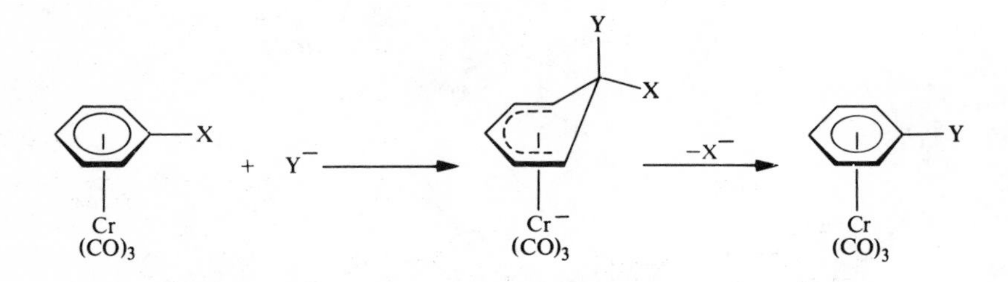

It has been shown that nucleophiles add to carbon atoms other than the one bearing the halogen. Other pathways that have been suggested are reversible addition of the nucleophile and intermolecular transfer of the nucleophile.[75]

An alternative mechanism involving hydrogen migration via the metal, as shown below, has, however, not been ruled out. Analogies for this type of hydrogen shift are known[81] and such a mechanism occurring concurrently with attack on the carbon bearing the leaving group would explain why partial racemisation occurs on treatment of complex 45 with nucleophiles.[82]

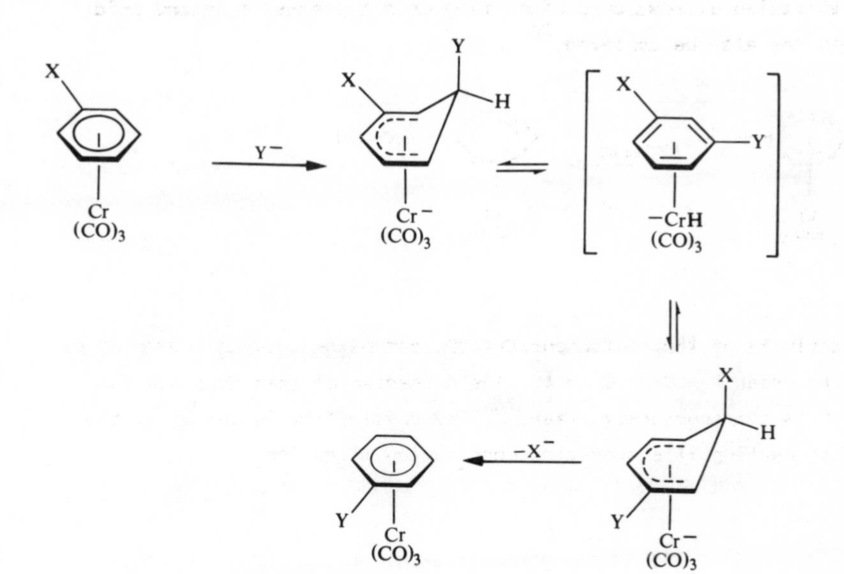

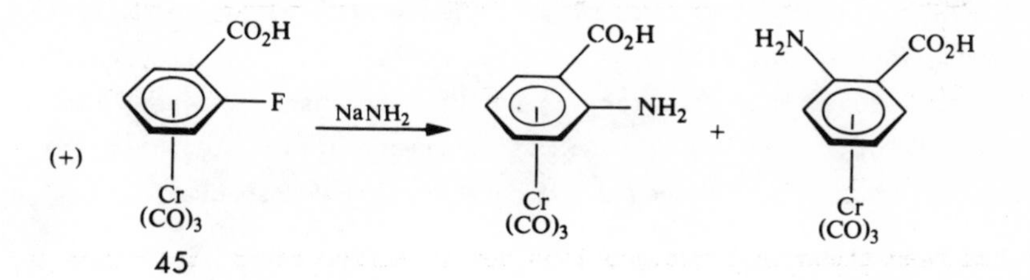

Presumably carbanions bearing a second acidic hydrogen do not undergo the substitution reaction because the chromium anion in the initially formed anion removes this acidic hydrogen to give stable Cr-H species.

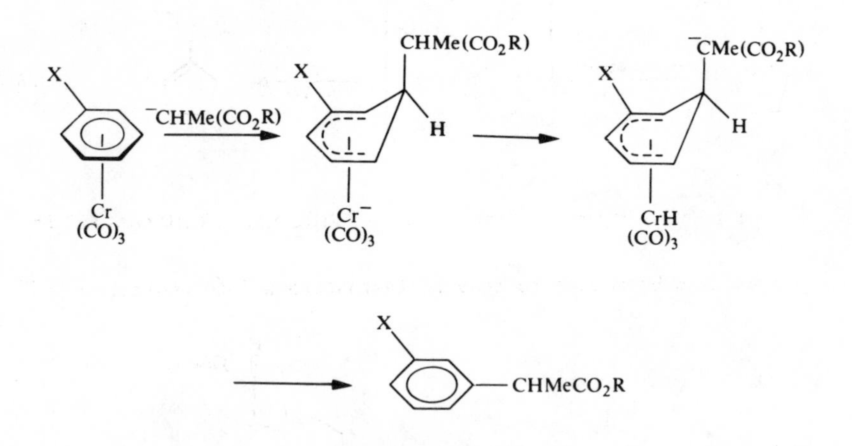

4.3 CATALYTIC NUCLEOPHILIC ADDITION AND SUBSTITUTION REACTIONS

Transition metal catalysed nucleophilic addition and substitution reactions are potentially much more useful for synthesis than their stoicheiometric counterparts. An extensive chemistry of Pd catalysts has been developed for these reactions. Pd^0 complexes react with a variety of allylic compounds (chlorides, alcohols, acetates, ethers etc.) to produce, in the presence of suitable 2-electron ligands L (phosphines or amines), $[(\eta^3\text{-allyl})PdL_2]^+$ cations. These cations react with nucleophiles to generate a new allylic compound and regenerate a Pd^0 species.

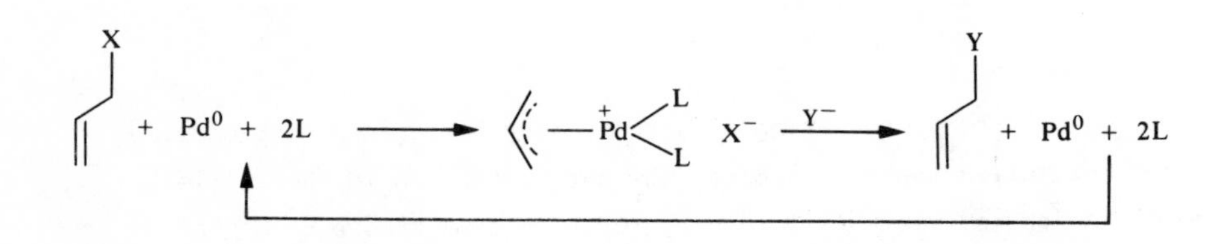

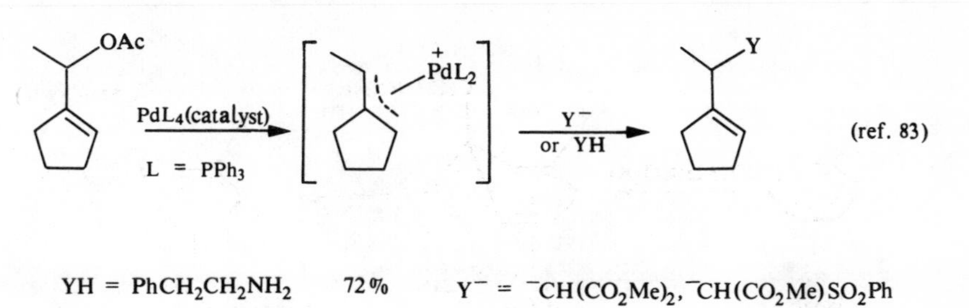

This process has been used to convert testosterone into cholestanone.[84]

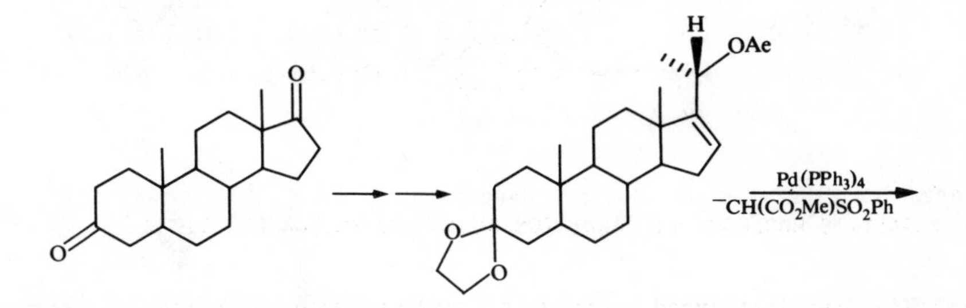

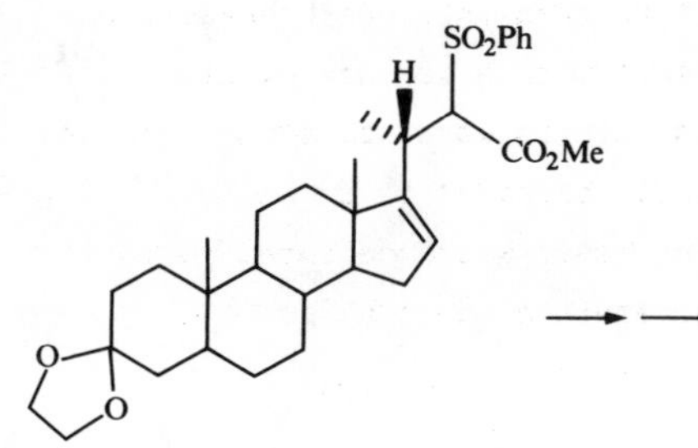

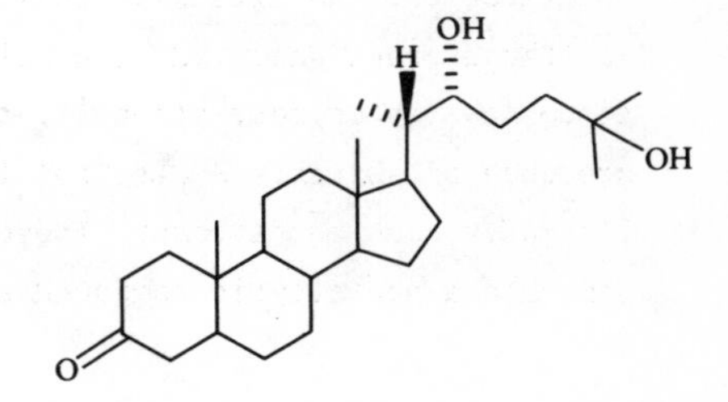

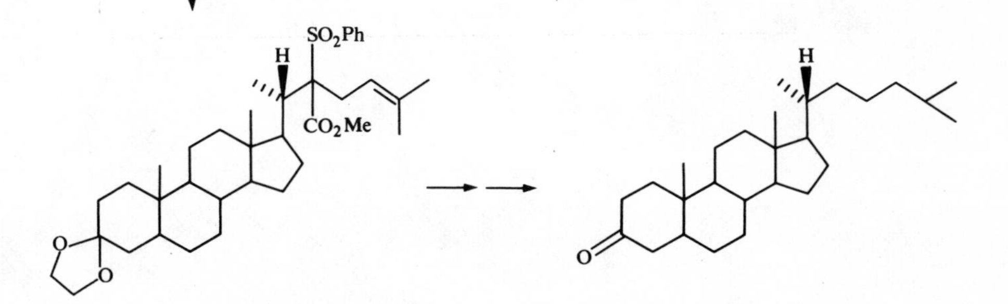

$Pd(PPh_3)_4$ catalyses the allylic amination of an unsaturated sugar acetate with several primary and secondary amines.[85]

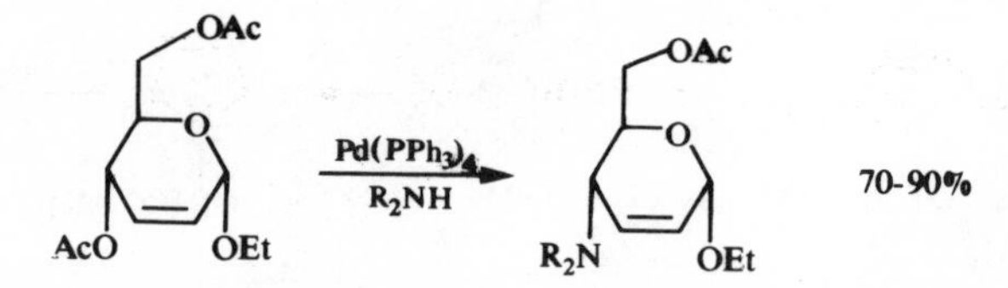

Pd(II) salts (e.g. $Pd(OAc)_2$, $PdCl_2$) may also be used as catalysts for this type of substitution reaction. The Pd(II) must undergo reduction to Pd(0) in the reaction media.

Interesting diallyl substitutions have been observed.[86]

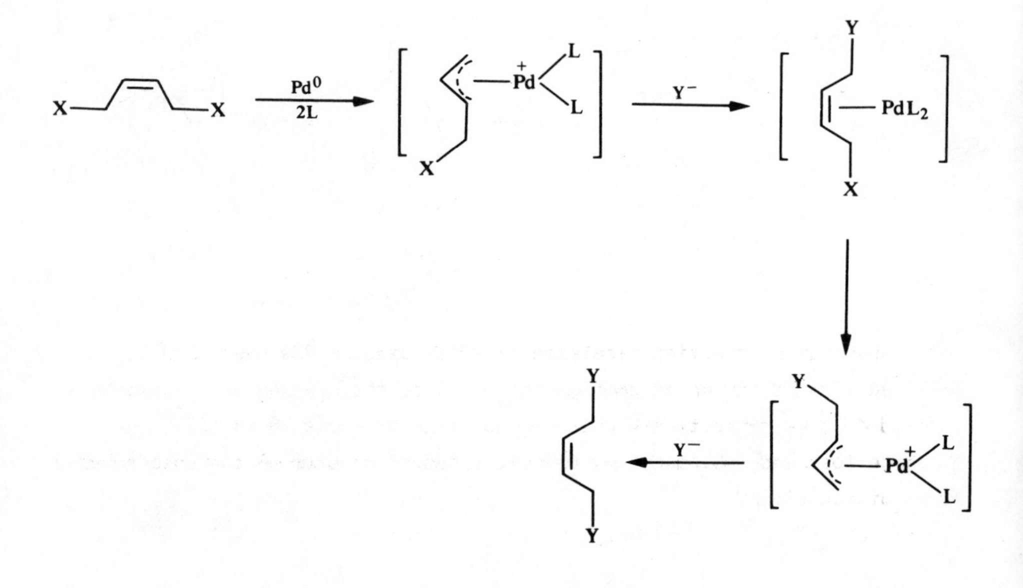

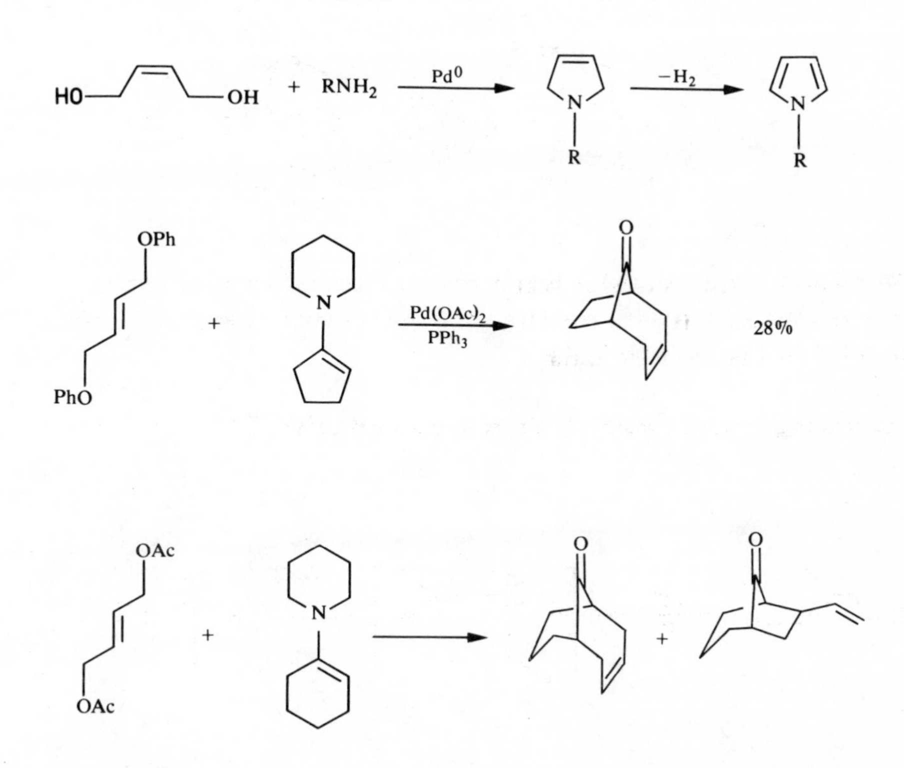

The substitution reaction catalysed by Pd(0) species has been shown to proceed with retention of configuration, i.e. the *cis* and *trans* compounds 46 and 47 give respectively the *cis* and *trans* products 48 and 49. Acetate loss and subsequent nucleophilic addition occur on the uncoordinated face of the ligand.

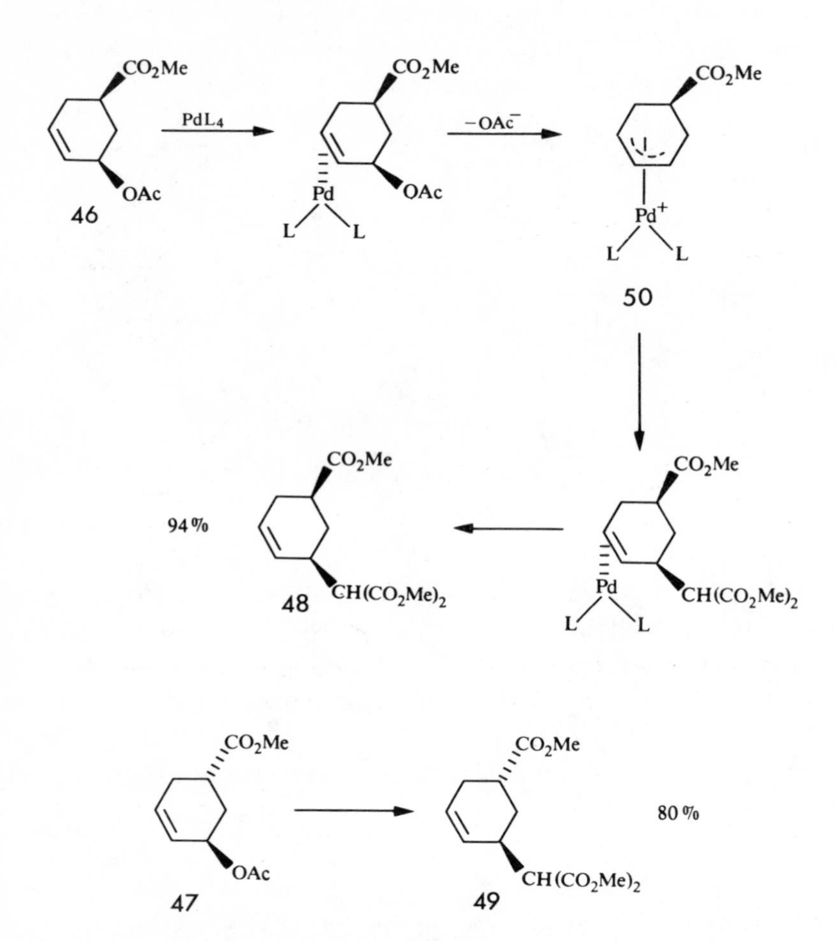

It should be noted that, although the relative configuration is maintained, the above reaction proceeds through the symmetrical η^3-allyl complex 50, which will undergo nucleophilic attack equally at either end of the allyl moiety. Hence, optically active 46 would give racemic 48. However the use of the optically active diphosphine (+)DIOP allows the conversion of racemic 46 to optically active 51.[87]

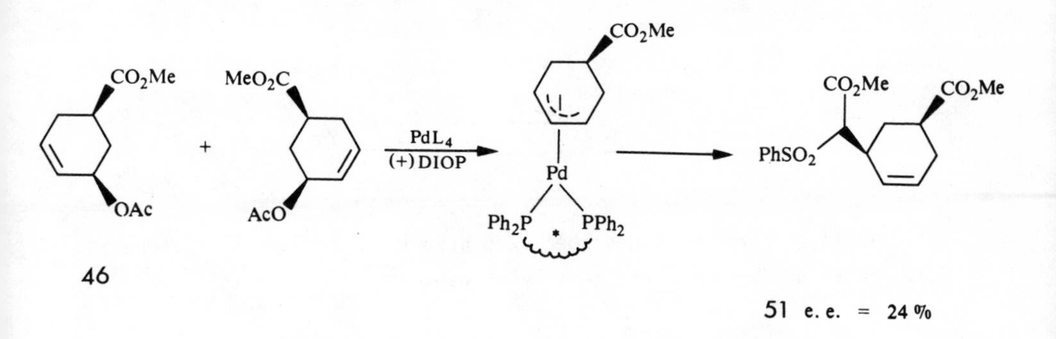

Intramolecular nucleophilic additions lead to cyclisation reactions and permit the synthesis of the ring skeletons of some alkaloids and macrolide antibiotics.[84,88,89]

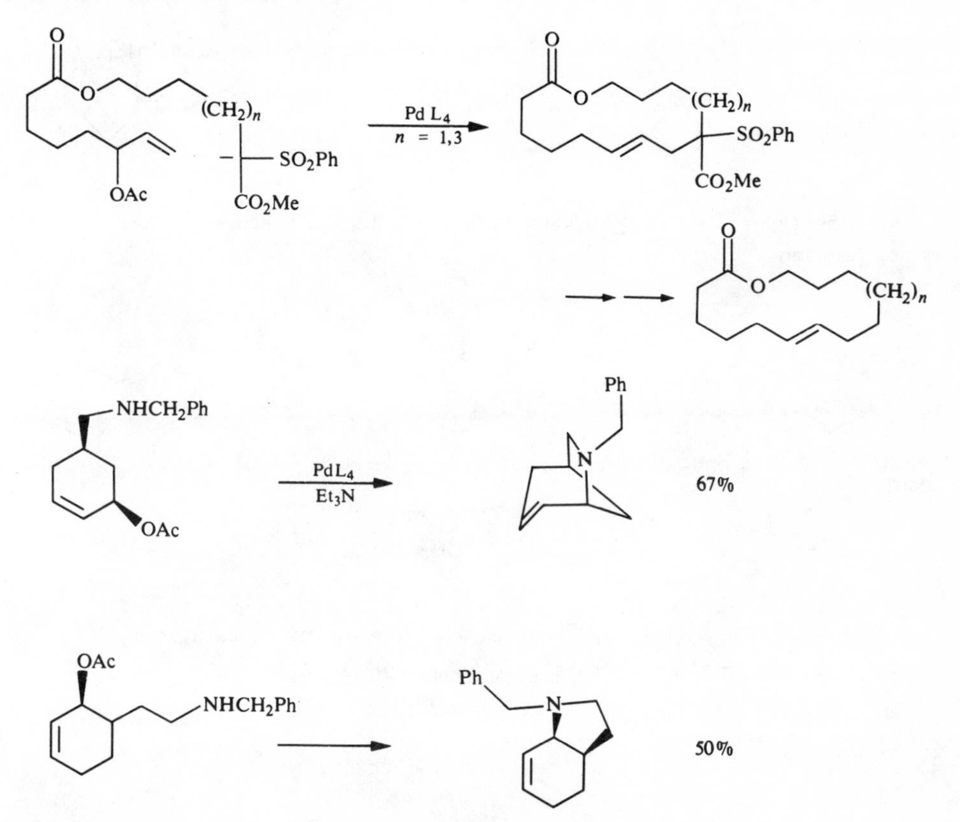

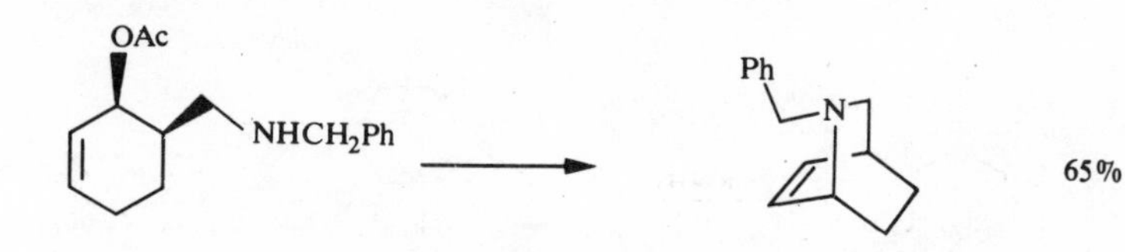

This cyclisation reaction has been used as one of the essential steps in a synthesis of ibogamine and desethylibogamine.[89]

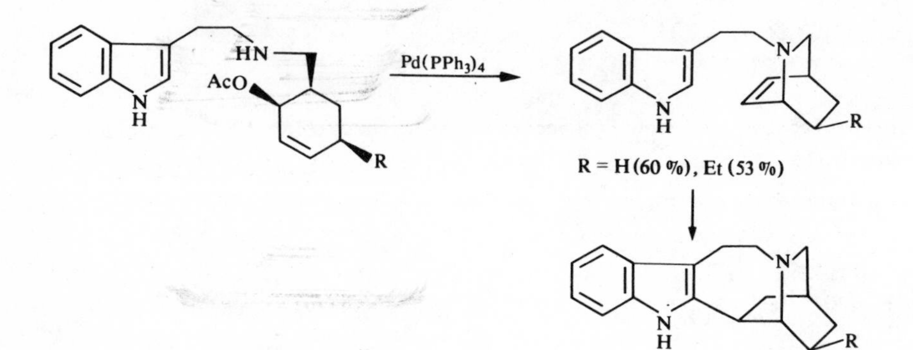

Cyclisation of 52 has been used to prepare 53, a precursor of humulene.[90]

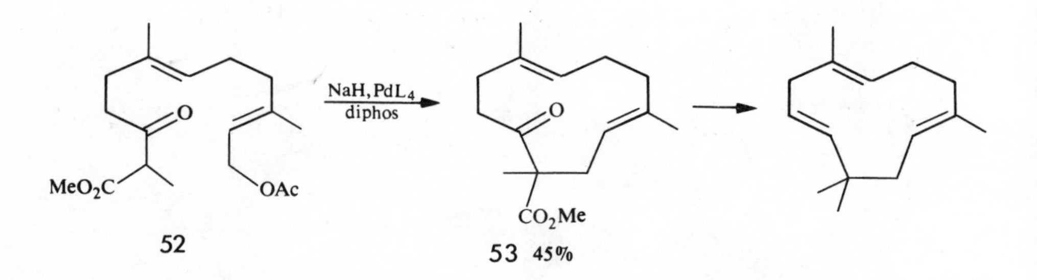

Amino acids can also be synthesised with the aid of Pd catalysis,[91,92] for example, in the synthesis of (±)-gabaculine 54.[92]

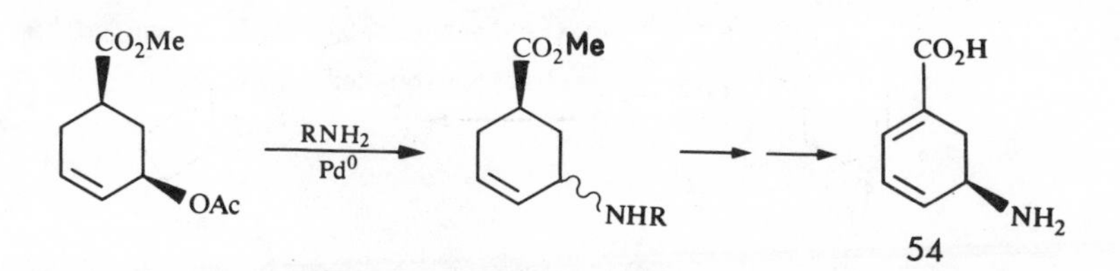

Unlike the Pd catalysed reaction of 46 with the anions from dimethyl malonate or methyl benzenesulphonylacetate the reaction with amines is not stereospecific. Products from attack on both faces of the allyl ligand are observed i.e. normal external attack as well as attack on the metal atom followed by migration to the allyl ligand.[92]

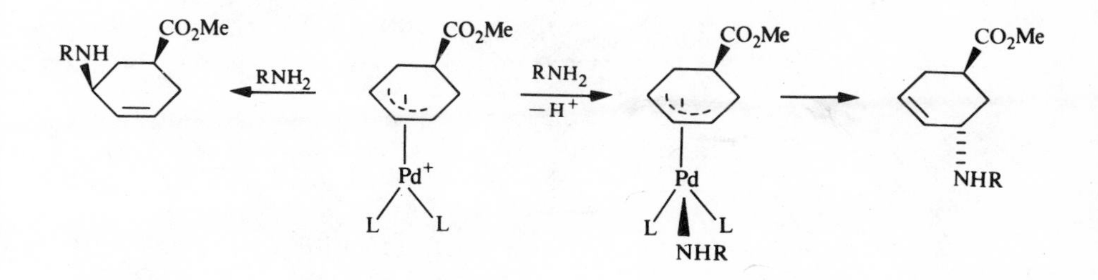

Indeed treatment of 46 with $Pd(PPh_3)_4$ results in isomerisation into a mixture of 46 and 47 by reversible elimination-addition of acetate.

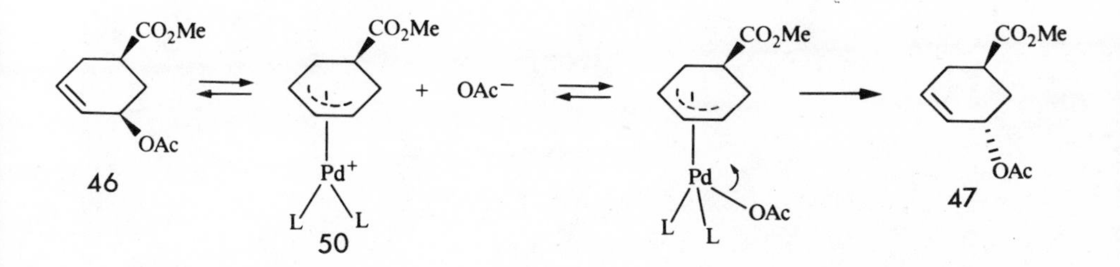

The anions of dimethyl malonate and methyl benzenesulphonyl-acetate substitute 46 stereospecifically because they react rapidly with the intermediate 50. The less reactive anion from bis(benzenesulphonyl) methane gives a mixture of *cis* and *trans* products with 46 presumably because it allows time for the above rearrangement to take place. Such a rearrangement is impossible for the lactone 55 and as expected 55 reacts stereospecifically in the substitution reaction below.[93]

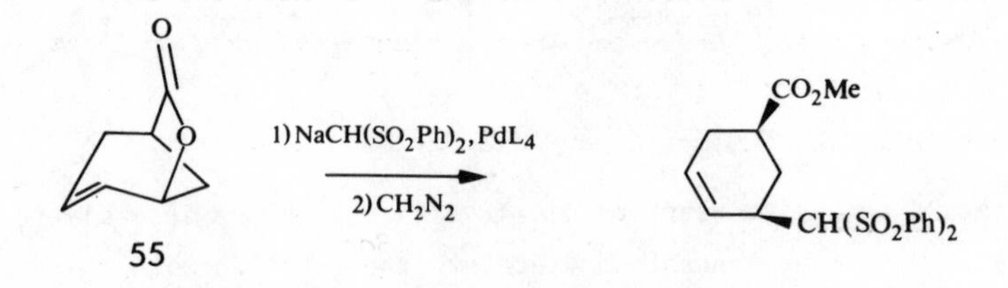

The addition of HCN to olefins is catalysed by $Pd(P(OPh)_3)_4$.[94]

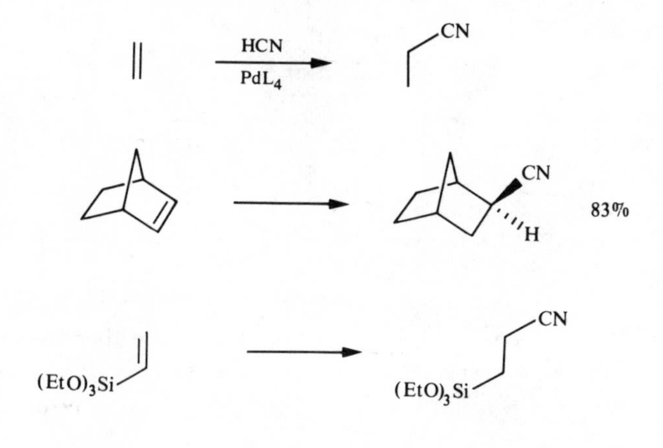

4.4 REFERENCES

1. S.G. Davies, M.L.H. Green and D.M.P. Mingos, Tetrahedron Report No.57, *Tetrahedron*, 1978, *34*, 3047 and references therein. S.G. Davies, M.L.H. Green and D.M.P. Mingos, *Nouveau J. Chimie*, 1977, *1*, 445.
2. A.C. Knipe, S.J. McGuinness and W.E. Watts, *Chem. Comm.*, 1979, 842; *J.C.S. Perkin II*, 1981, 193.
3. M.R. Churchill and R. Mason, *Proc. Roy. Soc. Ser. A.*, 1964, *279*, 191.
4. A.N. Nesmeyanov and N.A. Vol'Kenau, *Izvest. Akad. Nauk. S.S.S.R. Ser. Khim.*, 1975, 1151.
5. E.F. Ashworth, M.L.H. Green and J. Knight, *Proc. Ist. Int. Conf. Chem. Uses of Molybdenum*, 1973, 114.
6. B.F.G. Johnson, J. Lewis and D. Yarrow, *Chem. Comm.*, 1972, 235; *J.C.S. Dalton*, 1972, 2084.
7. M.L.H. Green, L.C. Mitchard and W.E. Silverthorn, *J.C.S. Dalton*, 1973, 1952.
8. J. Müller and H. Menig, *J. Organometal. Chem.*, 1975, *96*, 83; M.L.H. Green, L.C. Mitchard and W.E. Silverthorn, *J.C.S. Dalton*, 1973, 2177; A.R. Dias, C. Ramao and T. Aviles-Perea, personal communication; N.J. Cooper and M.L.H. Green, *Chem. Comm.*, 1974, 761; F.W. Benfield, B.R. Francis, M.L.H. Green, N.I. Luong-Thi, G. Moser, J.S. Poland and D.M. Rose, *J. Less Common Metals*, 1974, *36*, 187.
9. J.F. Helling and D.M. Braitsch, *J. Amer. Chem. Soc.*, 1970, *92*, 7207; T. Aviles, M.L.H. Green, A.R. Dias and C. Romao, *J.C.S. Dalton*, 1979, 1367.
10. E.G. Bryan, A.L. Burrows, B.F.G. Johnson, J. Lewis and G.M. Schiavon, *J. Organometal. Chem.*, 1977, *129*, C19.
11. A. Eisenstadt, *J. Organometal. Chem.*, 1973, *60*, 335; 1976, *113*, 147; W.H. Knoth, *Inorg. Chem.*, 1975, *14*, 1566; A. Rosan, M. Rosenblum and J. Tancrede, *J. Amer. Chem. Soc.*, 1973, *95*, 3062; N.A. Bailey, W.G. Kita, J.A. McCleverty, A.J. Murray, B.E. Mann and N.W. Walker, *Chem. Comm.*, 1974, 592; B.F.G. Johnson, J. Lewis, T.W. Matheson, I.E. Ryder and M.V. Twigg, *Chem. Comm.*, 1974, 269; E.O. Fischer and F.J. Kohl, *Chem. Ber.*, 1965, *98*, 2134.

12. N. El Murr and E. Laviron, *Can. J. Chem.*, 1976, *54*, 3357.

12a. D.W. Clack, M. Monshi and L.A.P. Kane-Maguire, *J. Organometal. Chem.*, 1976, *107*, C40.

13. P. Lennon, A.M. Rosan and M. Rosenblum, *J. Amer. Chem. Soc.*, 1977, *99*, 8426 and ref. therein.

14. K.M. Nicholas and A.M. Rosan, *J. Organometal. Chem.*, 1975, *84*, 351; A. Sanders, C.V. Magatti, and W.P. Giering, *J. Amer. Chem. Soc.*, 1974, *96*, 1610.

15. A. Rosan and M. Rosenblum, *J. Org. Chem.*, 1975, *40*, 3621.

16. L. Busetto, A. Palazzi, R. Ros and U. Belluco, *J. Organometal. Chem.*, 1970, *25*, 207.

17. P. Lennon, M. Madhavarao, A. Rosan and M. Rosenblum, *J. Organometal. Chem.*, 1976, *108*, 93.

18. P.K. Wong, M. Madhavarao, D.F. Marten and M. Rosenblum, *J. Amer. Chem. Soc.*, 1977, *99*, 2823.

19. S.R. Berryhill and M. Rosenblum, *J. Org. Chem.*, 1980, *45*, 1984.

20. T.H. Whitesides, R.W. Arhart and R.W. Slaven, *J. Amer. Chem. Soc.*, 1973, *95*, 5792.

21. A.J. Pearson, *Tet. Letters*, 1975, 3617.

22. A.J. Pearson, *Aust. J. Chem.*, 1976, *29*, 1841.

23. N.A. Bailey, W.G. Kita, J.A. McCleverty, A.J. Murray, B.E. Mann and N.W.J. Walker, *Chem. Comm.*, 1974, 592.

24. B.E.R. Schilling, R. Hoffmann and J.W. Faller, *J. Amer. Chem. Soc.*, 1979, *101*, 592.

25. R.D. Adams, D.F. Chodosh, J.W. Faller and A.M. Rosan, *J. Amer. Chem. Soc.*, 1979, *101*, 2570.

25a. B.M. Trost, *Tetrahedron*, 1977, *33*, 2615; J. Tsuji "Organic Synthesis with Palladium Compounds", Springer-Verlag, Berlin, 1980. (Tetrahedron Report No. 32).

25b. B.M. Trost, L. Weber, P. Strege, T.J. Fullerton and T.J. Dietsche, *J. Amer. Chem. Soc.*, 1978, *100*, 3426.

26. J.W. Faller and A.M. Rosan, *J. Amer. Chem. Soc.*, 1977, *99*, 4858.

27. M. Bottrill and M. Green, *J.C.S. Dalton*, 1977, 2365.

28. R.S. Bayoud, E.R. Biehl and P.C. Reeves, *J. Organometal. Chem.*, *174*, 297.

29. A.J. Birch, I.D. Jenkins and A.J. Liepa, *Tet. Letters*, 1975, 1723.

30. F. Franke and I.D. Jenkins, *Aust. J. Chem.*, 1978, *31*, 595.

31. B.F.G. Johnson, J. Lewis, D.G. Parker, P.R. Raithby and G.M.Sheldrick, *J. Organometal. Chem.*, 1978, *150*, 115.
32. B.F.G. Johnson, K.D. Karlin, J. Lewis and D.G. Parker, *J. Organometal. Chem.*, 1978, *157*, C67.
33. A.J. Birch, K.B. Chamberlain, M.A. Haas and D.J. Thompson, *J.C.S. Perkin I*, 1973, 1882.
34. A.J. Birch, K.B. Chamberlain and D.J. Thompson, *J.C.S. Perkin I*, 1973, 1900.
35. A. Pelter, K.J. Gould and L.A.P. Kane-Maguire, *Chem. Comm.*, 1974, 1029; A.J. Pearson, *Aust. J. Chem.*, 1977, *30*, 345 and references therein.
36. A.J. Birch, A.S. Narula, P. Dahler, G.R. Stephenson and L.F. Kelly, *Tet. Letters*, 1980, 979.
37. L.F. Kelly, A.S. Narula and A.J. Birch, *Tet. Letters*, 1980, 871, 2455.
38. L.A.P. Kane-Maguire and C.A. Mansfield, *Chem. Comm.*, 1973, 540.
39. A.J. Birch, A.J. Liepa and G.R. Stephenson, *Tet. Letters*, 1979, 3565; G.R. John and L.A.P. Kane-Maguire, *J.C.S. Dalton*, 1979, 1196.
40. H. Alper and C.C. Huang, *J. Organometal.Chem.*, 1973, *50*, 213.
41. R.E. Ireland, G.G. Brown, R.H. Stanford and T.C. McKenzie, *J. Org. Chem.*, 1974, *39*, 51; A.J. Birch, P.W. Westerman and A.J. Pearson, *Aust. J. Chem.*, 1976, *29*, 1671.
42. A.J. Pearson, *J.C.S. Perkin I*, 1977, 2069; 1980, 400; A.J. Pearson and P.R. Raithby, *J.C.S. Perkin I*, 1980, 395.
43. A.J. Pearson, *J.C.S. Perkin I*, 1979, 1255.
44. A.J. Pearson, *J.C.S. Perkin I*, 1978, 495.
45. A.J. Birch and H. Fitton, *Aust. J. Chem.*, 1969, *22*, 971.
46. R. Aumann and J. Knecht, *Chem. Ber.*, 1976, *109*, 174.
47. B.Y. Shu, E.R. Biehl and P.C. Reeves, *Synth. Comm.*, 1978, *8*, 523; G.R. John and L.A.P. Kane-Maguire, *J.C.S.Dalton*, 1979, 873; Y. Becker, A. Eisenstadt and Y. Shvo, *J. Organometal.Chem.*, 1978, *155*, 63.
48. P. Powell, *J. Organometal. Chem.*, 1979, *165*, C43.
49. P.L. Pauson and J.A. Segal, *J.C.S. Dalton*, 1975, 1677 and references therein; P.L. Pauson, *Isr. J. Chem.*, 1977, *15*, 258; L.A.P. Kane-Maguire and D.A. Sweigert, *Inorg. Chem.*, 1979, *18*, 700.
50. P.L. Pauson and J.A. Segal, *J.C.S. Dalton*, 1975, 1683.

51. F. Calderazzo, *Inorg. Chem.*, 1966, *5*, 429.

52. D.E. Ball and N.G. Connelly, *J. Organometal. Chem.*, 1973, *55*, C24.

53. J.F. McGreer and W.E. Watts, *J. Organometal. Chem.*, 1976, *110*, 103; D.W. Clack and L.A.P. Kane-Maguire, *J. Organometal. Chem.*, 1979, *174*, 199.

54. A.C. Knipe, S.J. McGuinness and W.E. Watts, *Chem. Comm.*, 1979, 842.

55. S. Brandt and P. Helquist, *J. Amer. Chem. Soc.*, 1979, *101*, 6473.

56. A. Davison, W.C. Krusell and R.C. Michaelson, *J. Organometal. Chem.*, 1974, *72*, C7; T.C. Flood, F.J. DiSanti and D.L. Miles, *Inorg. Chem.*, 1976, *15*, 1910; P.E. Riley, C.E. Capshew, R. Pettit and R.E. Davis, *Inorg. Chem.*, 1978, *17*, 408; M. Brookhart, J.R. Tucker, T.C. Flood and J. Jensen, *J. Amer. Chem. Soc.*, 1980, *102*, 1203.

57. P. Helquist, personal communication.

58. B.W. Roberts and J. Wong, *Chem. Comm.*, 1977, 20.

59. B.W. Roberts, M. Ross and J. Wong, *Chem. Comm.*, 1980, 428.

60. K. Hirai, N. Ishii, H. Suzuki, Y. Moso-Oka and T. Ikawa, *Chem. Letters*, 1979, 1113.

61. T. Hayashi and L.S. Hegedus, *J. Amer. Chem. Soc.*, 1977, *99*, 7093.

62. R.A. Holton, *J. Amer. Chem. Soc.*, 1977, *99*, 8083.

63. B. Akermark, J.E. Bäckvall, K. Siirala-Hanseń, K. Sjöberg, and K. Zetterberg, *Tet. Letters*, 1974, 1363; B. Akermark, J.E. Bäckvall, L.S. Hegedus, K. Zetterberg, K. Siirala-Hanseń, and K. Sjoberg, *J. Organometal. Chem.*, 1974, *72*, 127; J.E. Bäckvall, *Tet. Letters*, 1975, 2225; 1978, 163; *Chem. Comm.*, 1977, 413.

64. L.S. Hegedus, G.F. Allen, E.L. Waterman, *J. Amer. Chem. Soc.*, 1976, *98*, 2674; L.S. Hegedus, G.F. Allen, J.J. Bozell and E.L. Waterman, *J. Amer. Chem. Soc.*, 1978, *100*, 5800.

65. B. Akermark, J.E. Bäckvall, A. Lowenborg and K. Zetterberg, *J. Organometal. Chem.*, 1979, *166*, C33.

66. I. Arai and G.D. Daves, *J. Amer. Chem. Soc.*, 1978, *100*, 287; *J. Org. Chem.*, 1978, *43*, 4110.

67, C.P. Casey "Transition Metal Organometallics in Organic Synthesis" 1976, *1*, 190, Ed. H. Alper, Academic Press, London,

68a. C.P. Casey and W.R. Brunsvold, *J. Organometal. Chem.*, 1974, *77*, 345.

68b. C.P. Casey and T.J. Burkhardt, *J. Amer. Chem. Soc.*, 1972, *94*, 6543.

69. E.O. Fischer and K.H. Dotz, *Chem. Ber.*, 1970, *103*, 1273; K.H. Dotz and E.O. Fischer, *ibid.*, 1972, *105*, 1356.

70. A.J. Anciaux, A.J. Hubert, A.F. Noels, N. Petiniot and P. Teyssié, *J. Org. Chem.*, 1980, *45*, 695.

71. M.P. Doyle and J.G. Davidson, *J. Org. Chem.*, 1980, *45*, 1538.

72. J.L.A. Roustan and F. Houlihan, *Can. J. Chem.*, 1979, *57*, 2790.

73. D.J. Collins, W.R. Jackson and R.N. Timms, *Tet. Letters*, 1976, 495; *Aust. J. Chem.*, 1977, *30*, 2167; W.R. Jackson and J.U.G. Strauss, *Aust. J. Chem.*, 1977, *30*, 553; *Tet. Letters*, 1975, 2591.

74. M.F. Semmelhack, *Ann. N.Y. Acad. Sci.*, 1977, 36; G. Jaouen, "Transition Metal Organometallics in Organic Synthesis". 1978, *II*, 105, Ed. H. Alper, Academic Press, London.

75. M.F. Semmelhack, H.T. Hall, M. Yoshifuji and G. Clark, *J. Amer. Chem. Soc.*, 1975, *97*, 1247; M.F. Semmelhack, H.T. Hall, and M. Yoshifuji, *J. Amer. Chem. Soc.*, 1976, *98*, 6387; M.F. Semmelhack, H.T. Hall, R. Farina, M. Yoshifuji, G. Clark, T. Bargar, K. Hirotsu and J. Clardy, *J. Amer. Chem. Soc.*, 1979, *101*, 3535.

76. M.F. Semmelhack and G. Clark, *J. Amer. Chem. Soc.*, 1977, *99*, 1675.

77. M.F. Semmelhack, J.J. Harrison and Y. Thebtararonth, *J. Org. Chem.* 1979, *44*, 3275.

78. A.P. Kozikowski and K. Isobe, *Chem. Comm.*, 1978, 1076.

79. B. Nicholls and M.C. Whiting, *J. Chem. Soc.*, 1959, 551.

80. M.F. Semmelhack and H.T. Hall, *J. Amer. Chem. Soc.*, 1974, *96*, 7091, 7092.

81. W. Lamanna and M. Brookhart, *J. Amer. Chem. Soc.*, 1980, *102*, 3490; G.A.M. Munro and P.L. Pauson, *Chem. Comm.*, 1976, 134.

82. G. Jaouen and R. Dabard, *J. Organometal. Chem.*, 1970, *21*, P43.

83. B.M. Trost, L. Weber, P. Strege, T.J. Fullerton and T.J. Dietsche *J. Amer. Chem. Soc.*, 1978, *100*, 3426; B.M. Trost, Tetrahedron Report No. 32, *Tetrahedron*, 1977, *33*, 2615.

84. B.M. Trost and J.R. Verhoeven, *J. Amer. Chem. Soc.*, 1976, *98*, 630; 1977, *99*, 3867; 1978, *100*, 3435; B.M. Trost, Y. Matsumura, *J. Org. Chem.* 1977, *42*, 2036; B.M. Trost and T.R. Verhoeven, *J. Amer. Chem. Soc.*, 1980, *102*, 4743.

85. H.H. Baer and Z.S. Hanna, *Carbohydrate Res.*, 1980, *78*, C11.

86. S.I. Murahashi, T. Shimamura and I. Moritani, *Chem. Comm.*, 1974, 931; H. Onoue, I. Moritani and S.I. Murahashi, *Tet. Letters*, 1973, 121.

87. B.M. Trost and P.E. Strege, *J. Amer. Chem. Soc.*, 1977, *99*, 1649.

88. B.M. Trost and T.R. Verhoeven, *J. Amer. Chem. Soc.*, 1977, *99*, 3867; *Tet. Letters*, 1978, 2275; *J. Amer. Chem. Soc.*, 1979, *101*, 1595.

89. B.M. Trost and J.P. Genet, *J. Amer. Chem. Soc.*, 1976, *98*, 8516; B.M. Trost, S.A. Godleski and J.P. Genet, *J. Amer. Chem. Soc.*, 1978, *100*, 3930.

90. Y. Kitagawa, A. Itoh, S. Hashimoto, H. Yamamoto and H. Nozaki, *J.Amer. Chem. Soc.*, 1977, *99*, 3864.

91. J.P. Haudegond, Y. Chauvin and D. Commereuc, *J. Org. Chem.*, 1979, *44*, 3063.

92. B.M. Trost and E. Keinan, *J. Org. Chem.*, 1979, *44*, 3451.

93. B.M. Trost, T.R. Verhoeven, J.M. Fortunak, *Tet. Letters*, 1979, 2301; B.M. Trost and T.R. Verhoeven, *J. Amer. Chem. Soc.*, 1980, *102*, 4730.

94. E.S. Brown and E.A. Rick, *Chem. Comm.*, 1969, 112; E.S. Brown, E.A. Rick and F.D. Mendicino, *J. Organometal. Chem.*, 1972, *38*, 37.

CHAPTER 5

ORGANOMETALLICS AS NUCLEOPHILES

Electrophiles may react with organometallic complexes either on the metal or on one of the ligands. Reactions of metal centres with electrophiles is one of the major methods of synthesis of organotransition metal complexes. These reactions have been discussed in Chapter 2 and therefore this chapter will deal mainly with the second type of reaction, electrophilic attack on one of the ligands.

It should be noted that coordination of unsaturated hydrocarbon ligands to transition metals generally reduces the electron density on the hydrocarbon and thus makes the hydrocarbon less susceptible to attack by electrophiles. This effect allows the use of transition metal species as protecting groups (see section 3.1). Formation of cationic complexes is a particularly good method of protection against electrophilic attack since the ligands in positively charged complexes are rarely, if ever, attacked by electrophiles.

5.1 NEUTRAL COMPLEXES AS NUCLEOPHILES

η^1-Alkyl and aryl ligands are often removed from transition metal complexes by reaction with electrophiles (section 2.1.8). η^1-Alkyl ligands possessing a β-hydrogen can act as hydride donors to Ph_3C^+ to yield cationic olefin complexes.

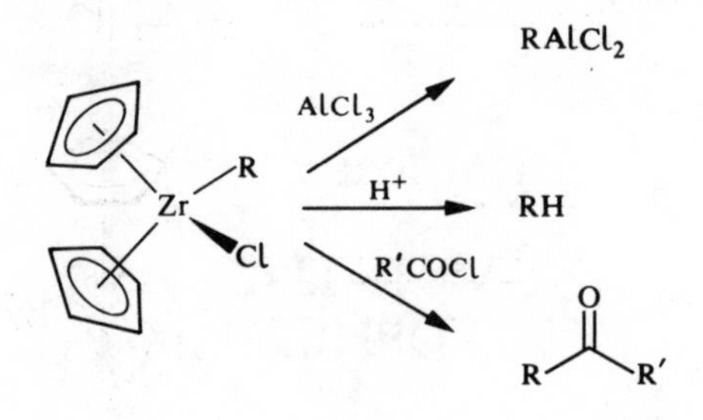

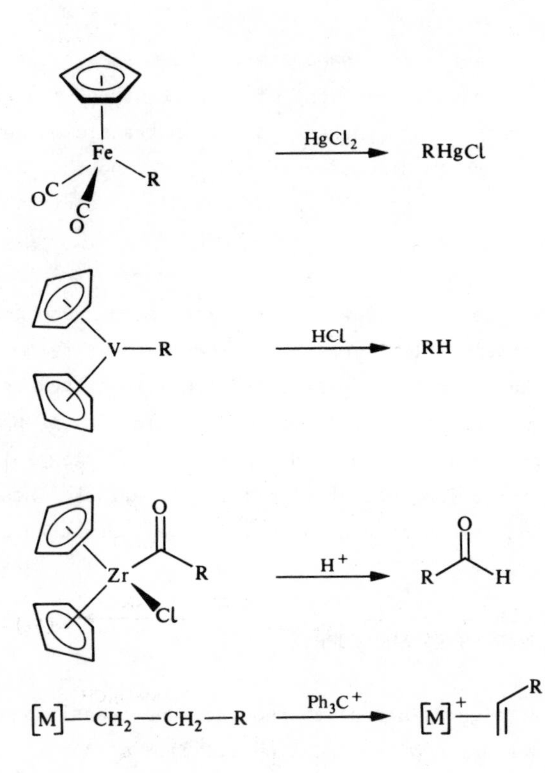

η^1-Allyl ligands are nucleophilic towards strong electrophiles giving cationic η^2-olefin complexes as products.[1]

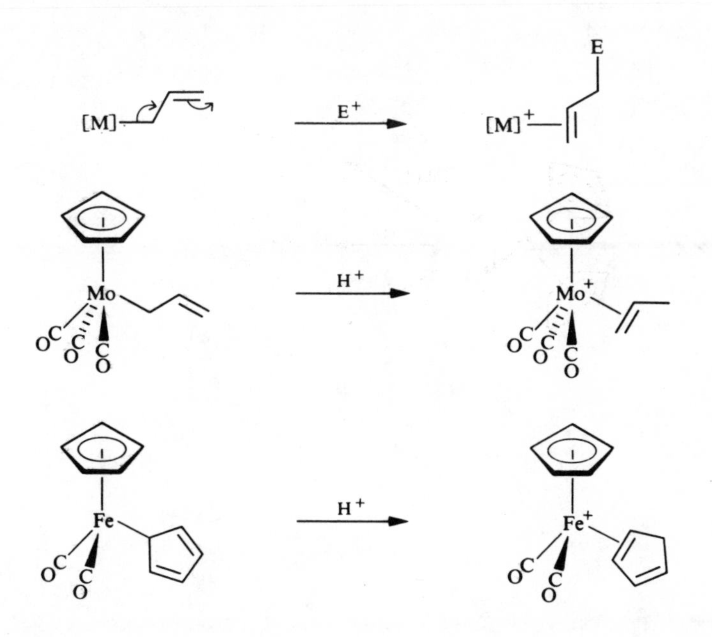

Some examples of the reactions of the readily available complex $CpFe(CO)_2$ (η^1-allyl) 1 are shown below.[2]

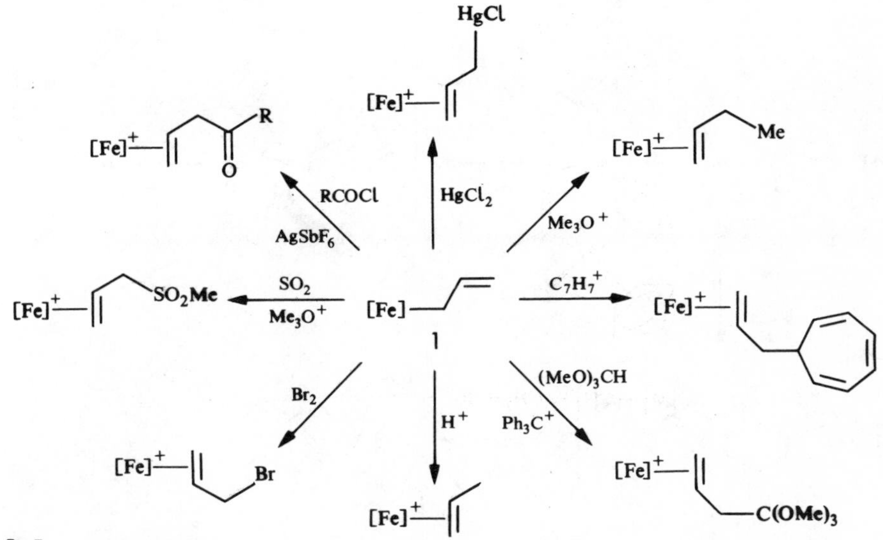

The η^1-allyl complex 1 is also nucleophilic towards $[CpFe(CO)_2\,(olefin)]^+$ cations. One or other of the two organometallic groups can be removed selectively from 3 , the product of the reaction between 1 and the ethylene cation 2.[3]

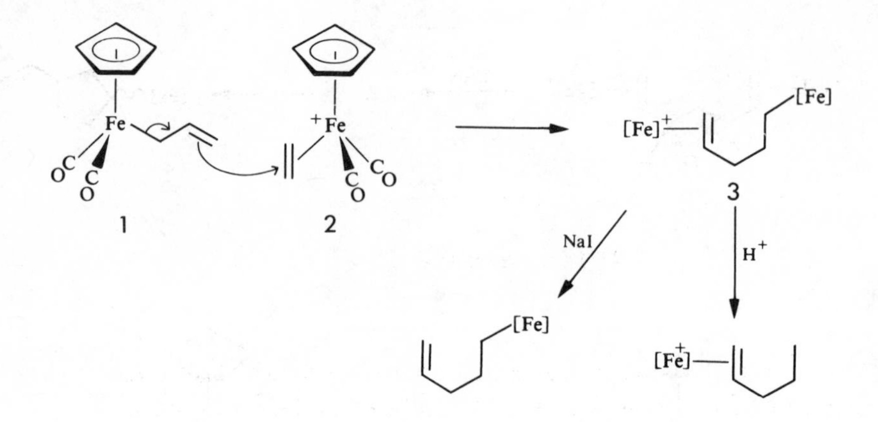

Addition of 1 to the cationic η^2-butadiene complex 4 generates the intermediates 5 and 6 which cyclise as shown below.

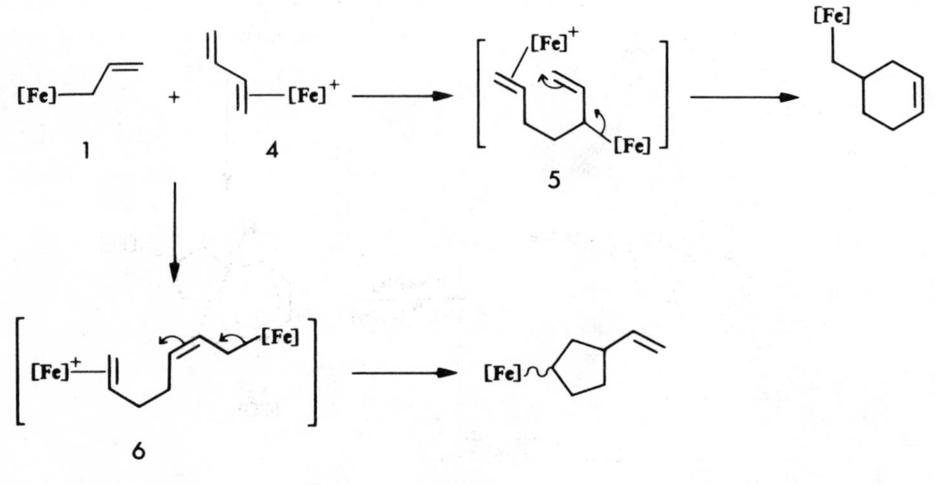

Treatment of the dication 7 with base generates the η^1-allyl complex 6 which is then able to cyclise.

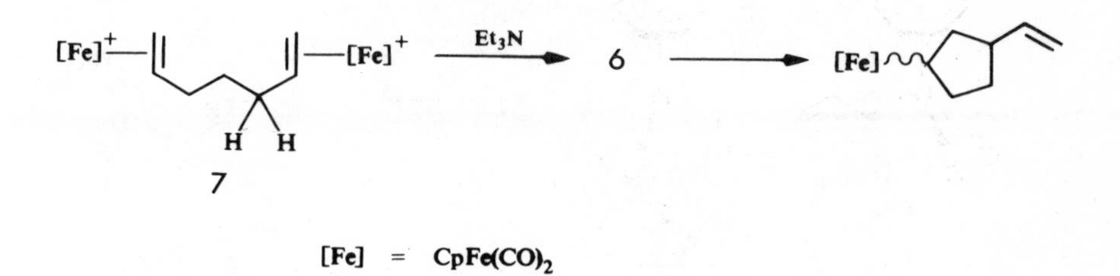

The η^1-allyl complex 1 also acts as a nucleophile towards the dienyl cation 8 and further reactions of the product lead to hydroazulenes.[4]

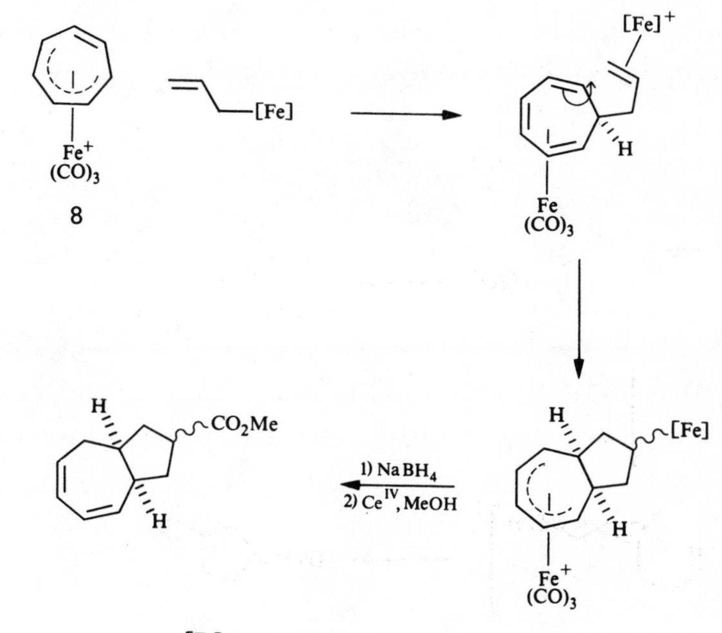

Substituted η[1]-allyl ligands may also act as nucleophiles.[3,4]

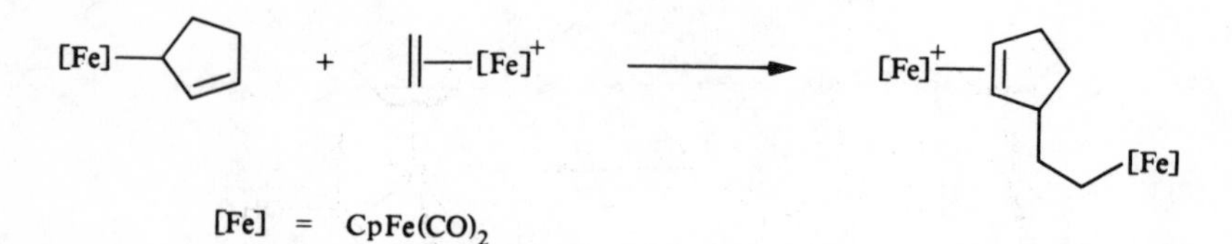

$CpFe(CO)_2$ (η[1]-allyl) complexes react with very electrophilic olefins (i.e. olefins bearing at least two ethyl carboxylate groups) and dimethyl acetylene dicarboxylate to give cyclised products.[4]

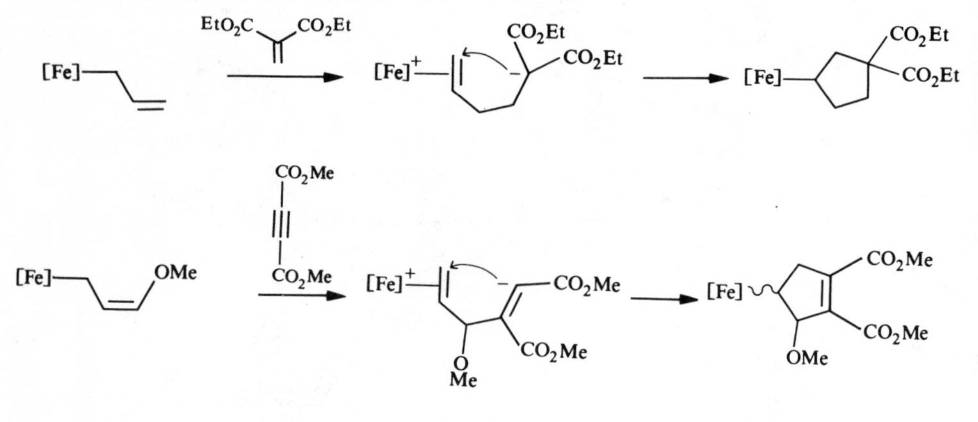

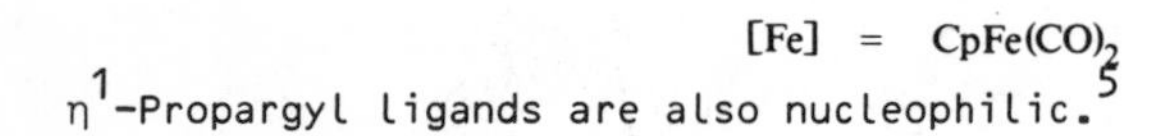

η[1]-Propargyl ligands are also nucleophilic.[5]

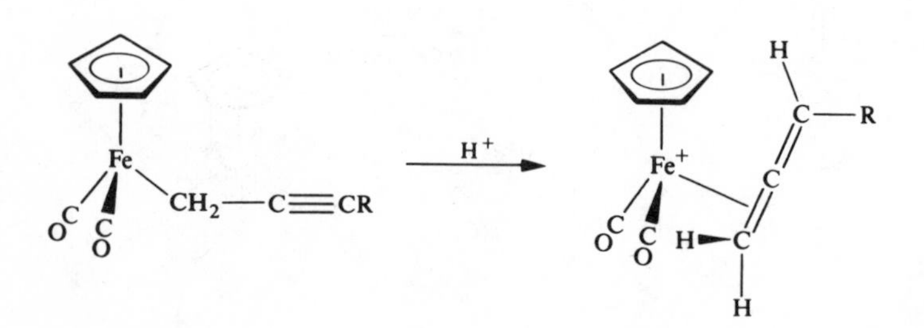

Interesting hetero-substituted analogues of the above reactions occur leading to stabilised enols etc. (see section 3.2.2).[6]

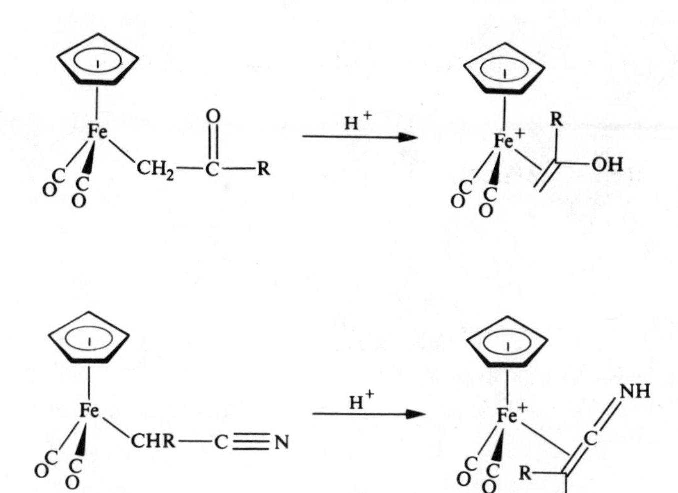

η^2-Carbene complexes are formed by the reaction of η^1-alkyne ligands with electrophiles.[7,8]

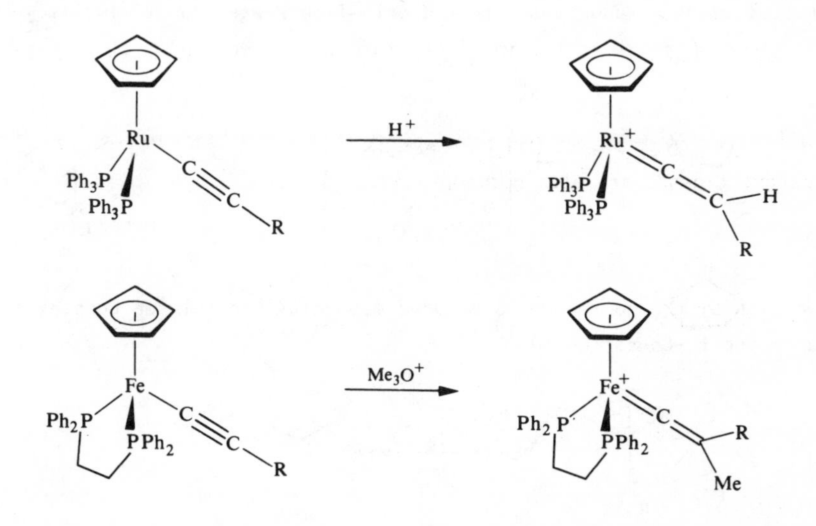

The cyanide ligand in $CpW(CO)_3CN$ is nucleophilic on nitrogen and reacts with methyl iodide to yield the corresponding isocyanide cation.[9]

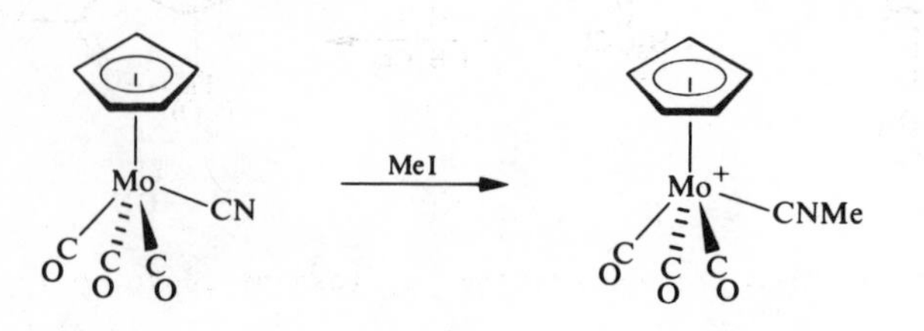

$(Olefin)Fe(CO)_4$ complexes are protonated by hydrogen halides to lead to overall reduction of the olefin.[10]

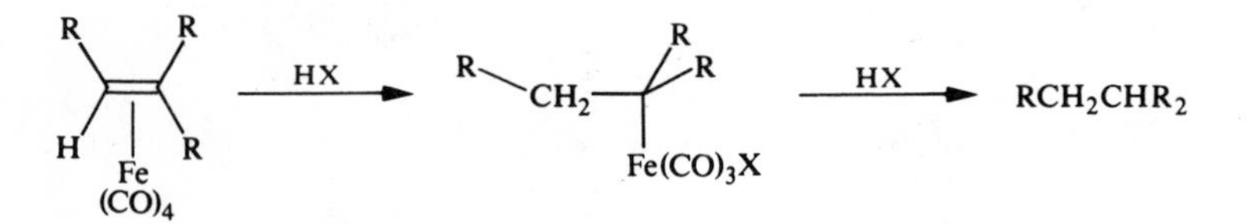

Neutral η^4-diene complexes may be protonated to yield η^3-allyl cations. Generally, stable products are formed only if a 2-electron ligand is present to coordinate to the initially formed 16-electron η^3-allyl cation.[11]

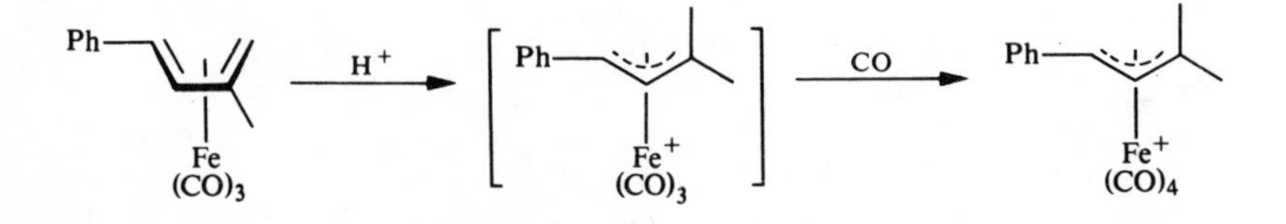

Protonation of the non-coordinated double bond of the fulvene complex produces the η^5-Cp cation 10.[12]

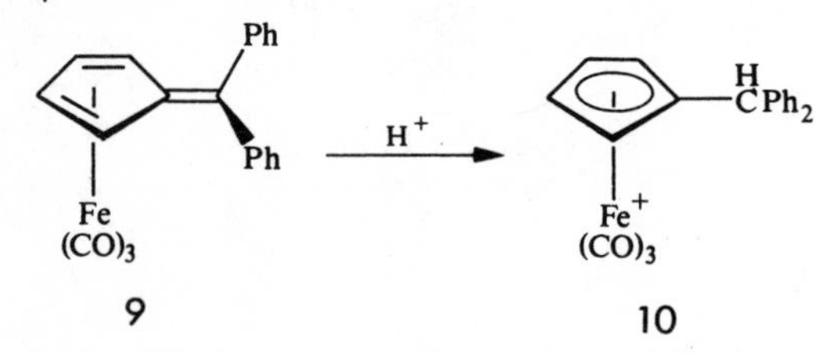

(η^4-Diene)Fe(CO)$_3$ complexes react with a number of other electrophiles, addition occurring to a terminal position.[13]

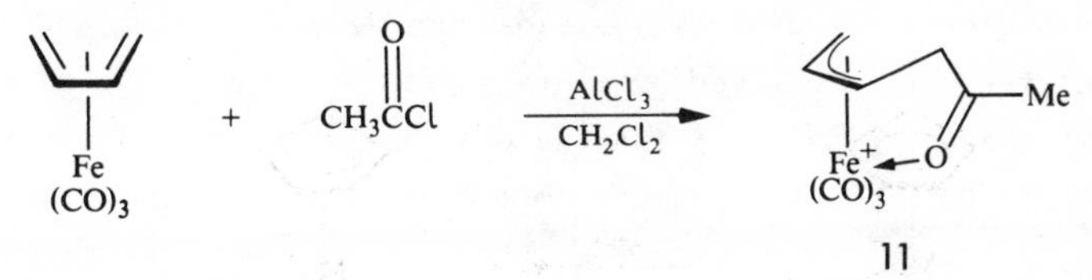

The above reaction often proceeds further by loss of a proton from the intermediate 11.[14] Since dienes are easily decomplexed from their Fe(CO)$_3$ complexes this method can be used for the conversion of dienes into dienones.

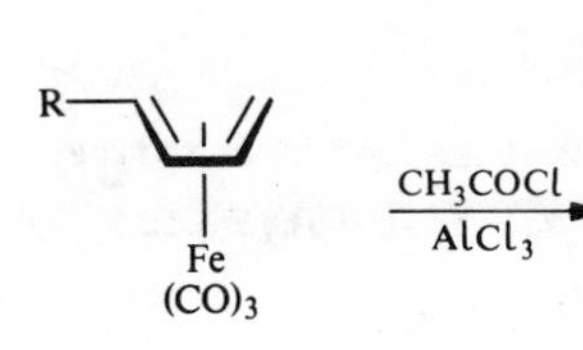

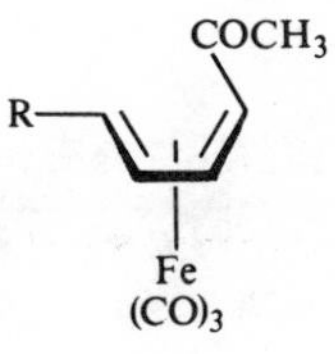

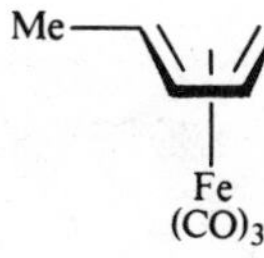

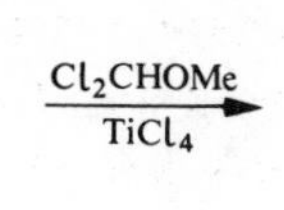

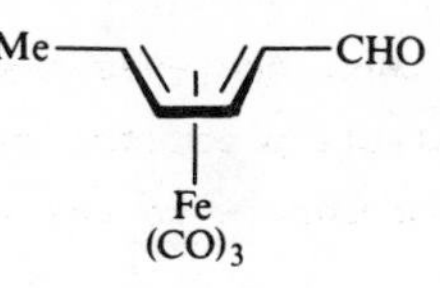

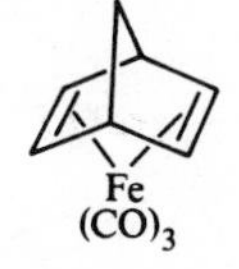

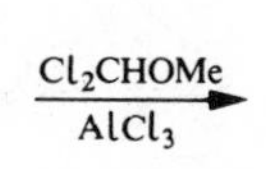

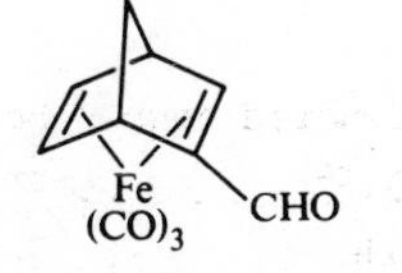

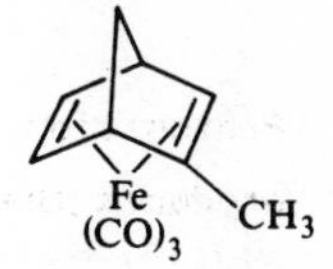

Electrophilic addition to unsaturated hydrocarbon ligands in organometallic complexes tends to be much less stereospecific than nucleophilic addition since both direct attack from the side away from the metal and initial attack on the metal followed by migration to the *endo* side of a hydrocarbon ligand are possible. For example, acetylation of (1,3-cyclohexadiene)$Fe(CO)_3$ gives a mixture of the 5-*exo* and 5-*endo* acetyl complexes.[15]

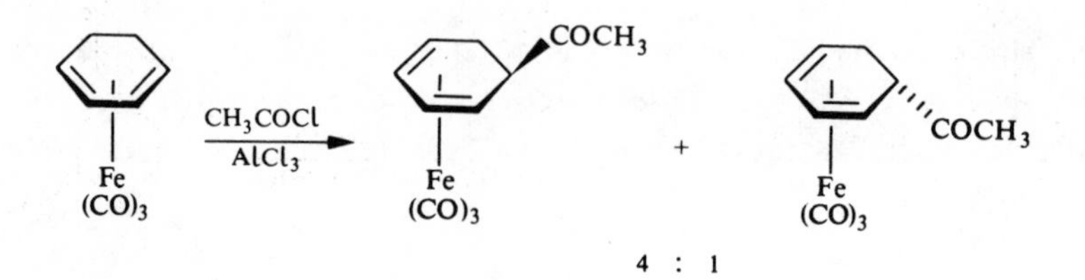

Friedel-Crafts acylation reactions cannot be performed directly on tropone. However, (tropone)$Fe(CO)_3$ undergoes ready acetylation to give the 2-acetyl tropone complex.[16] This reaction forms part of an interesting synthesis of β-thujaplicin 12 and β-dolabrin 13.

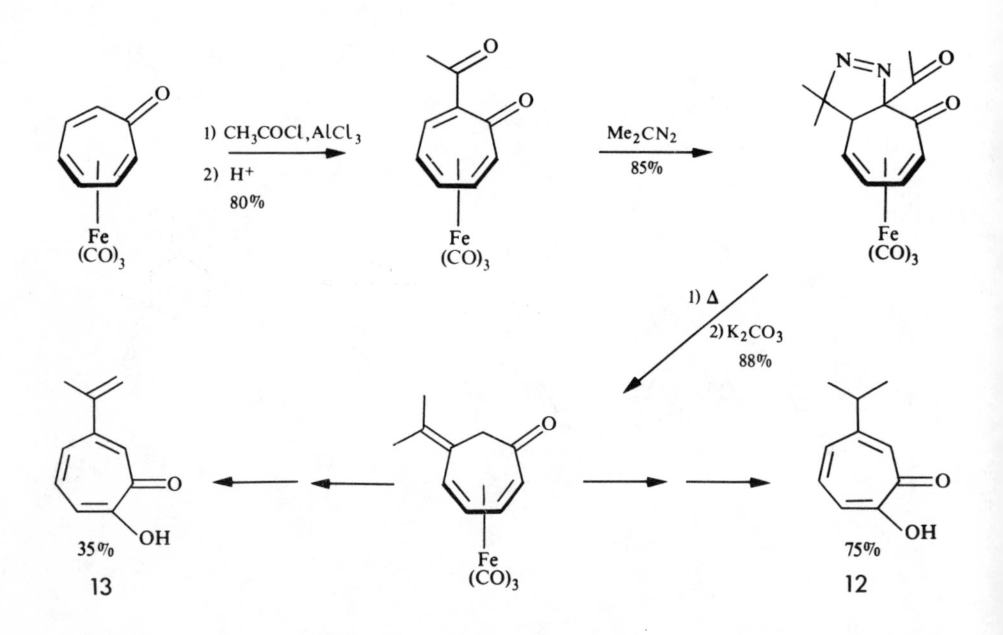

The uncoodinated double bond in (myrcene)Fe$(CO)_3$ is more nucleophilic than the coordinated diene. This allows some interesting cyclisation reactions to proceed.[17]

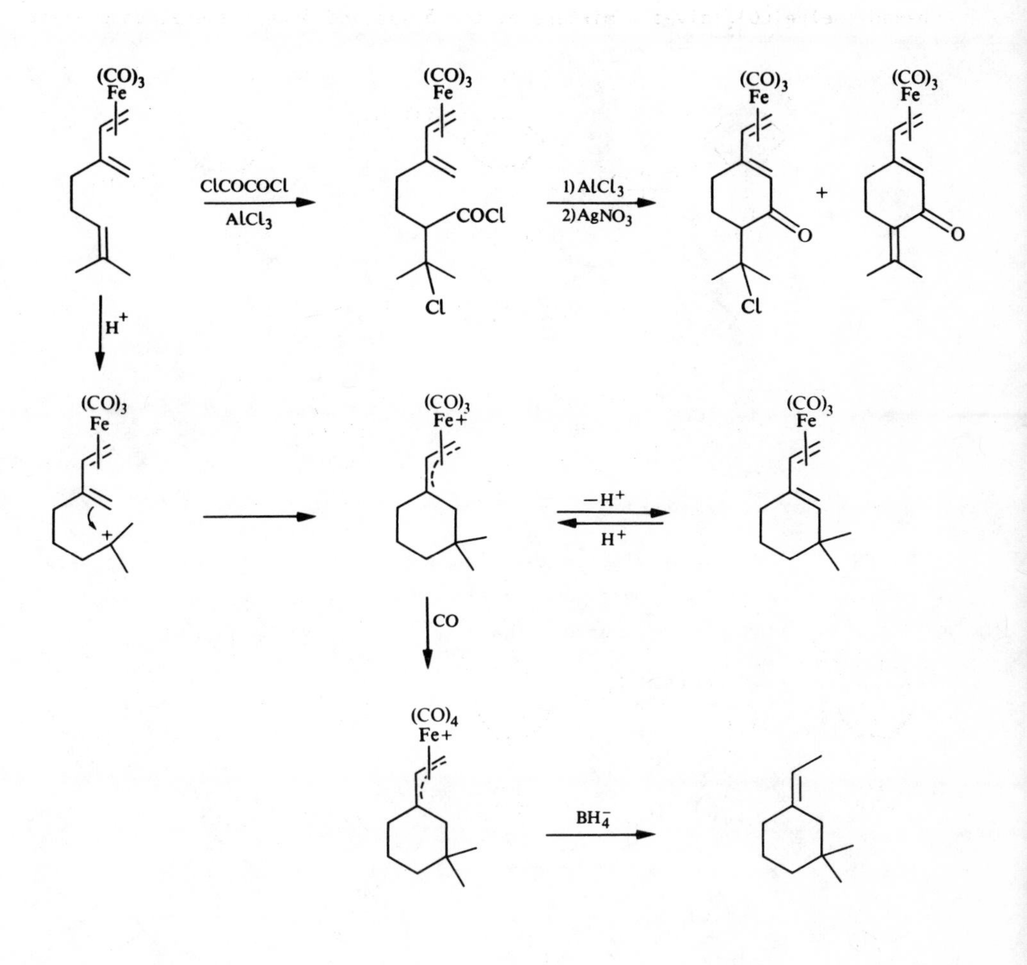

Enone and dienone complexes are nucleophilic on oxygen.[18]

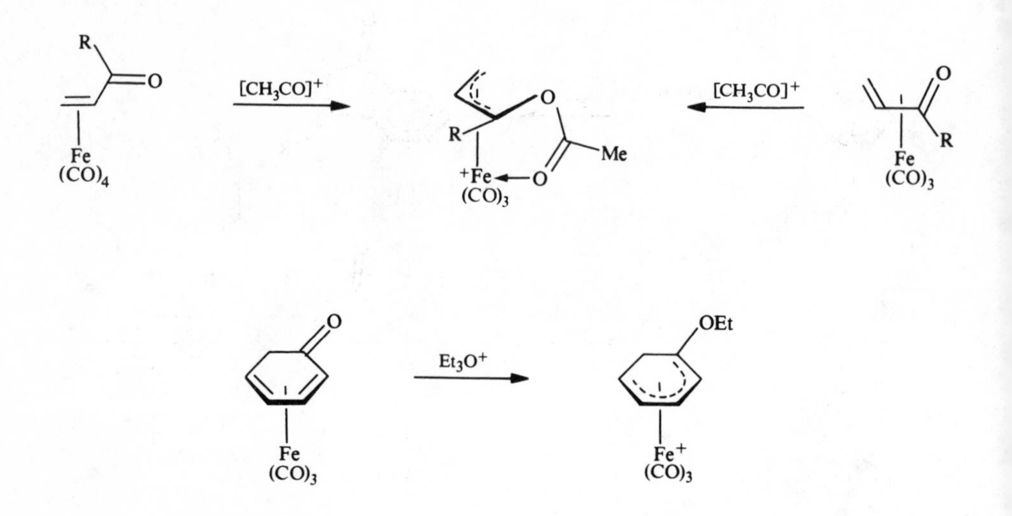

The complex 14 is nucleophilic on the nitrogen atom and reacts with MeI to give the N-methyl pyrrole cationic complex.[19] This is an interesting example of the conversion of a 5 electron ligand into a 6 electron ligand.

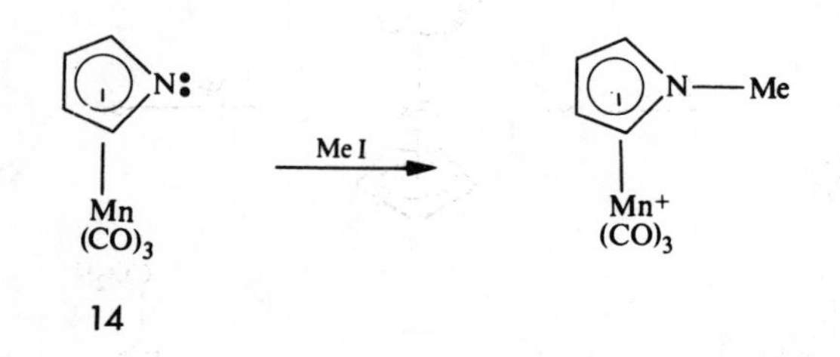

Electrophilic attack on η^6-arene ligands may be illustrated by the acid catalysed cyclisation of the $Cr(CO)_3$ complexes of 3-phenylpropionic acids to the corresponding indanone complexes.[20]

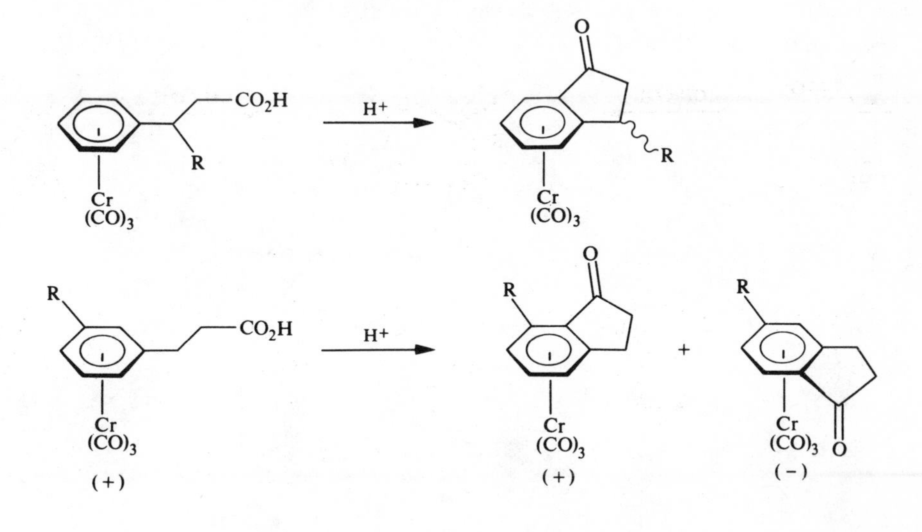

Treatment of (arene)$FeCp^+$ cations containing benzylic protons with base leads to the formation of neutral (cyclohexadienyl)FeCp complexes. These (cyclohexadienyl)FeCp complexes react with electrophiles such as MeI, CO_2, and CS_2 to give substituted (arene)$FeCp^+$ cations.[21]

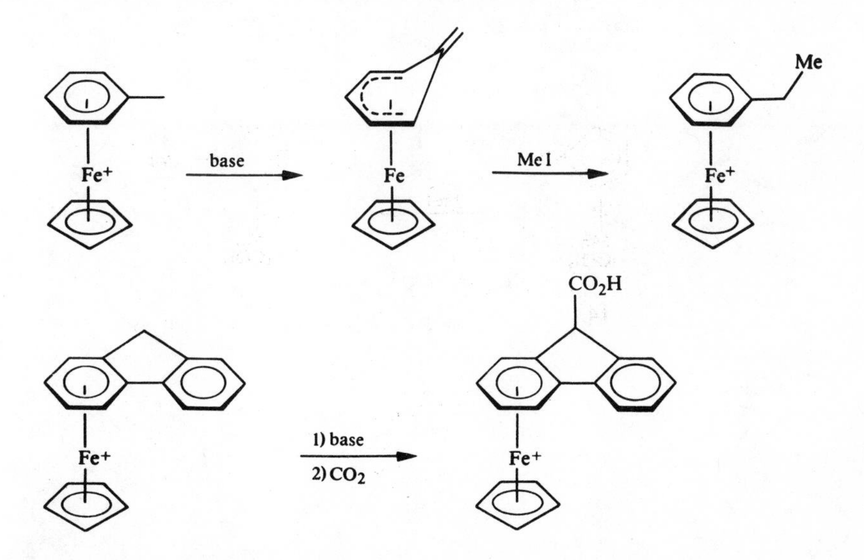

Although many carbene complexes are electrophilic on carbon, there are a few examples which behave as though they were nucleophilic on carbon. Some carbene complexes with 16 electrons have been shown to effect efficient methylene transfer reactions with the carbonyl groups of ketones, aldehydes, esters and amides.[22,24]

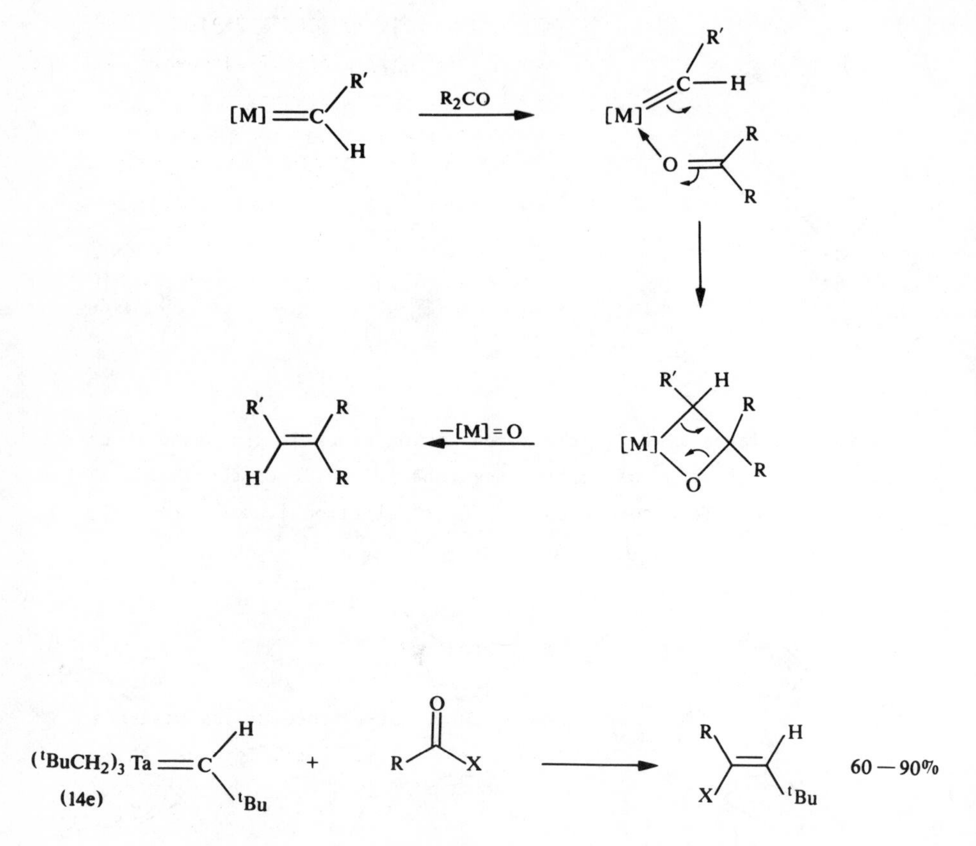

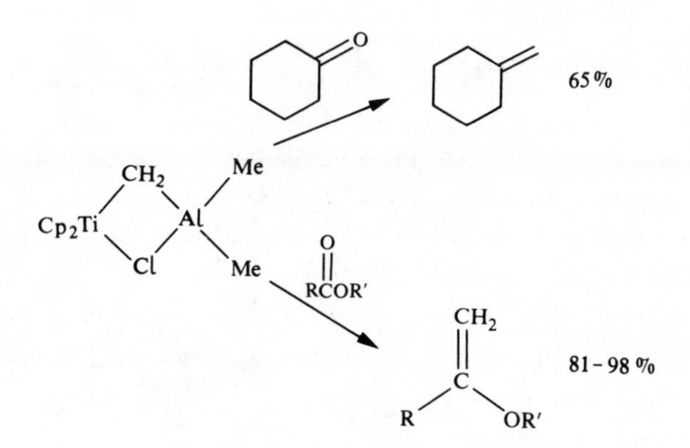

The use of the Ti reagent above provides a useful synthesis of allyl vinyl ethers.[24]

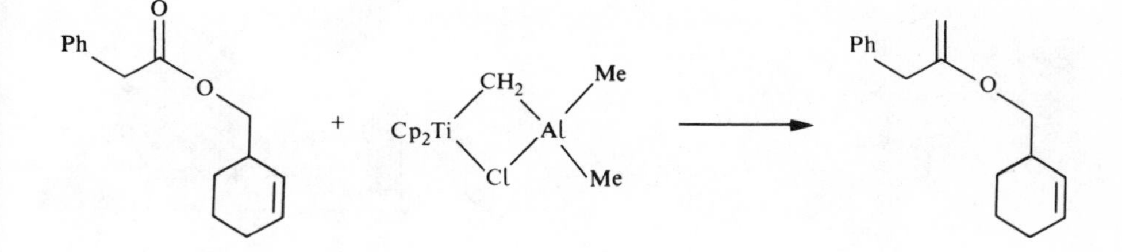

These reactions are obviously related to the Wittig reaction with the advantage that direct methylenation of esters and amides is possible,[22,24]

5.2 ANIONIC COMPLEXES AS NUCLEOPHILES

5.2.1 *Anions derived from carbene ligands*[25]

The α-protons of (alkoxycarbene)$Cr(CO)_5$ complexes are very acidic and treatment of such complexes with BuLi generates the corresponding anions.[26]

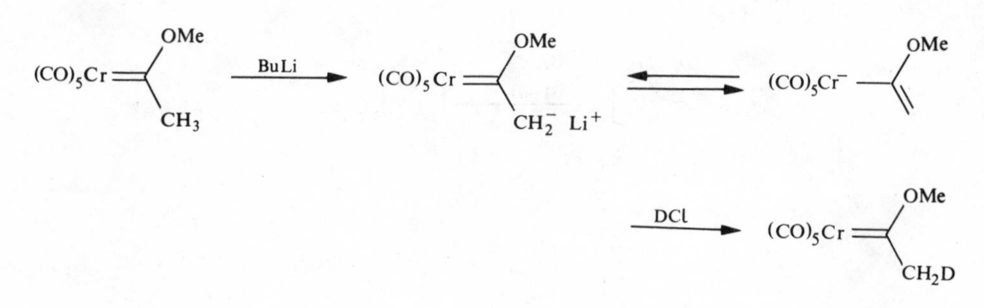

These carbene anions are readily alkylated by carbon electrophiles.[25-32]

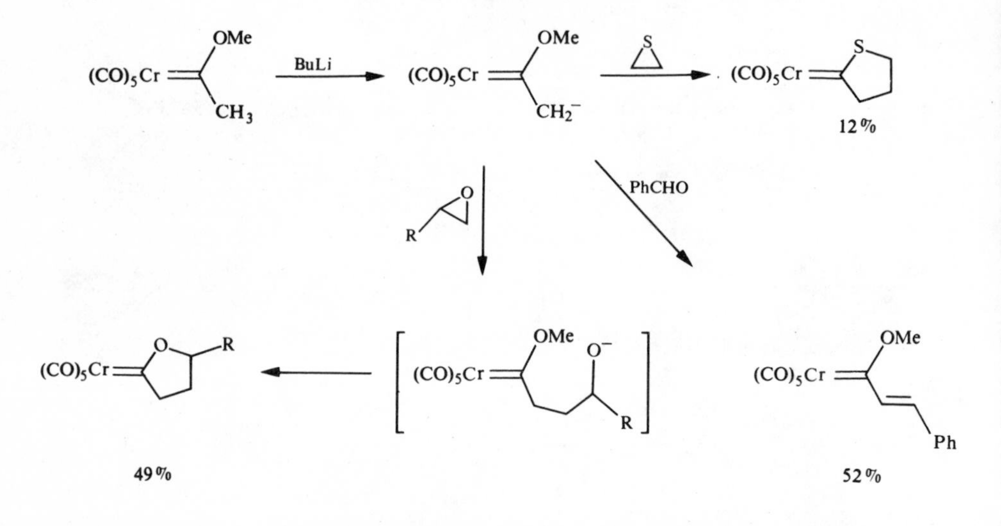

Where the carbene possesses more than one α-hydrogen then addition of 1 equivalent of BuLi followed by 1 equivalent of electrophile generally leads to a mixture of mono, di and non-alkylated compounds.[26,27]

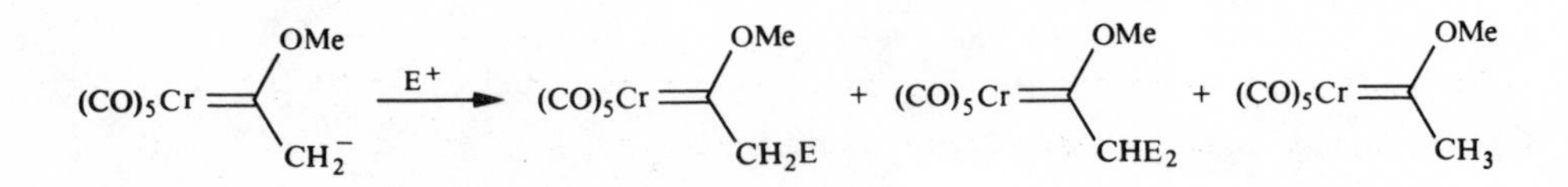

E^+ = $BrCH_2CO_2Me$, $MeSO_3F$, methyl acrylate

Alkylation of the carbene complex 15, containing only one acidic hydrogen, occurs in good yield.[25,28]

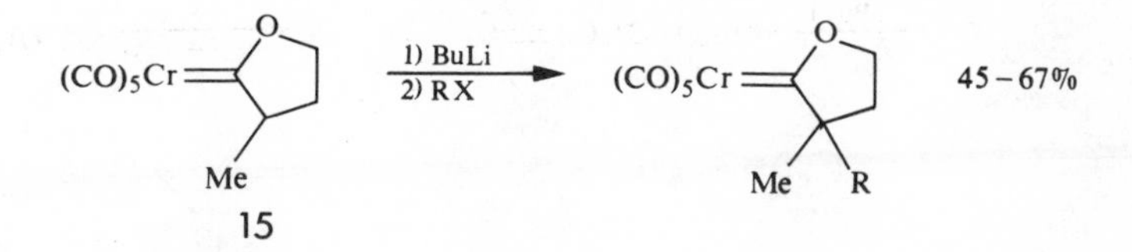

RX = $PhCH_2Br$, $Me_2C=CHCH_2Br$, $CH_2=CHCH_2Br$, CH_3COCl

α-Methylene-γ-butyrolactones can be successfully synthesised via (carbene) $Cr(CO)_5$ complexes.[29]

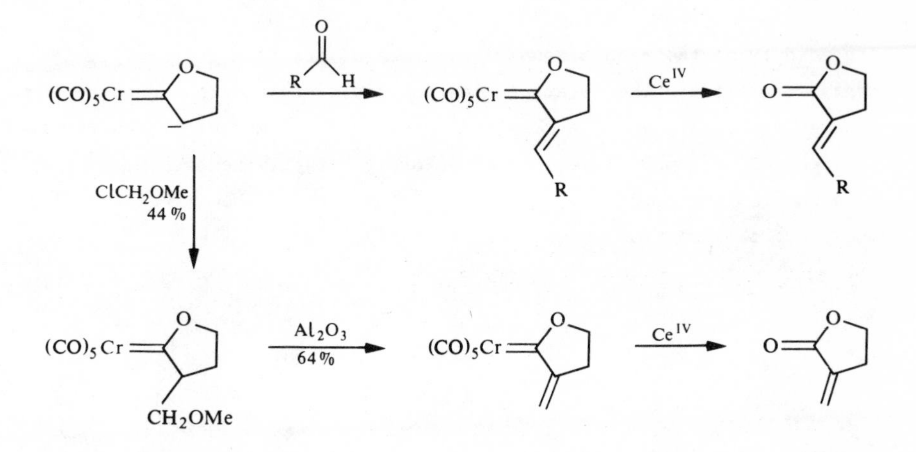

5.2.2 *Anions derived from (η^6-arene)$Cr(CO)_3$ complexes*

The coordination of arene rings to the electron withdrawing $Cr(CO)_3$ group renders the ring susceptible to nucleophilic attack (section 4.2.4). The $Cr(CO)_3$ group also has the effect of making the ring and benzylic protons more acidic than in the free arene. The ring protons are more acidic due to inductive stabilisation of the aryl anion by the $Cr(CO)_3$ group whereas the benzylic protons are more acidic due to resonance stabilisation of the benzyl anion.

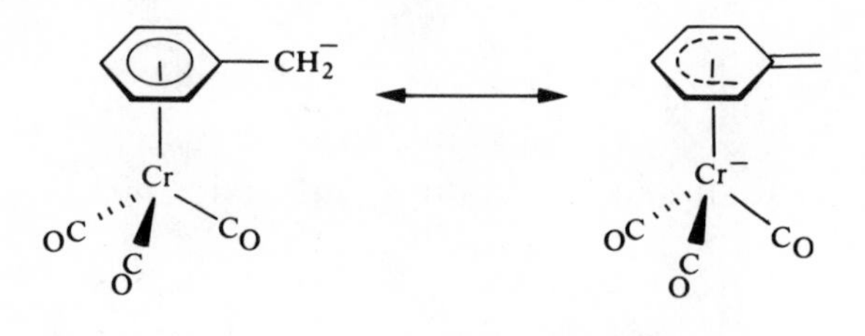

The acidity of the ring protons is increased more than the acidity of the benzylic protons. The relative kinetic acidities seem to depend on the nature of the base and the reaction conditions. A methyl proton rather than a ring proton of free toluene is preferentially removed by strong base whereas the reverse is true for (toluene)$Cr(CO)_3$ on treatment with BuLi, but the reaction is not very selective.[33]

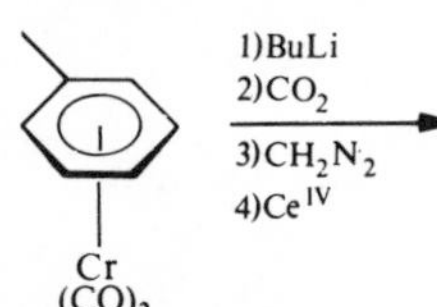

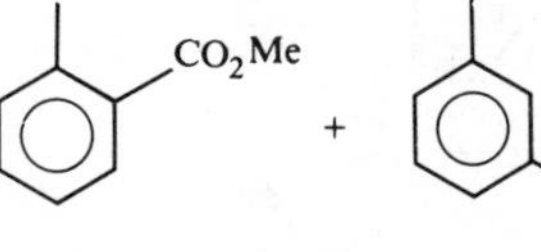

5% 24%

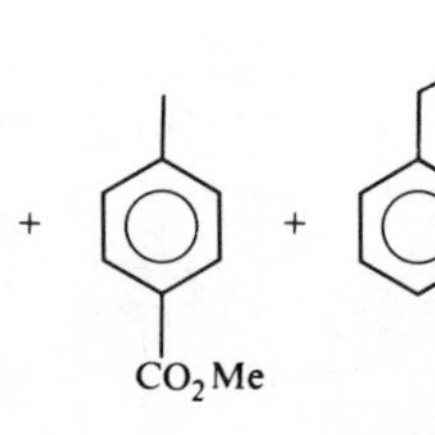

24% 13%

The regioselectivity is greatly improved by the introduction of substituents. For example, metallation is directed to the ortho position by functional groups such as OMe, F and Cl on the arene ring.[33]

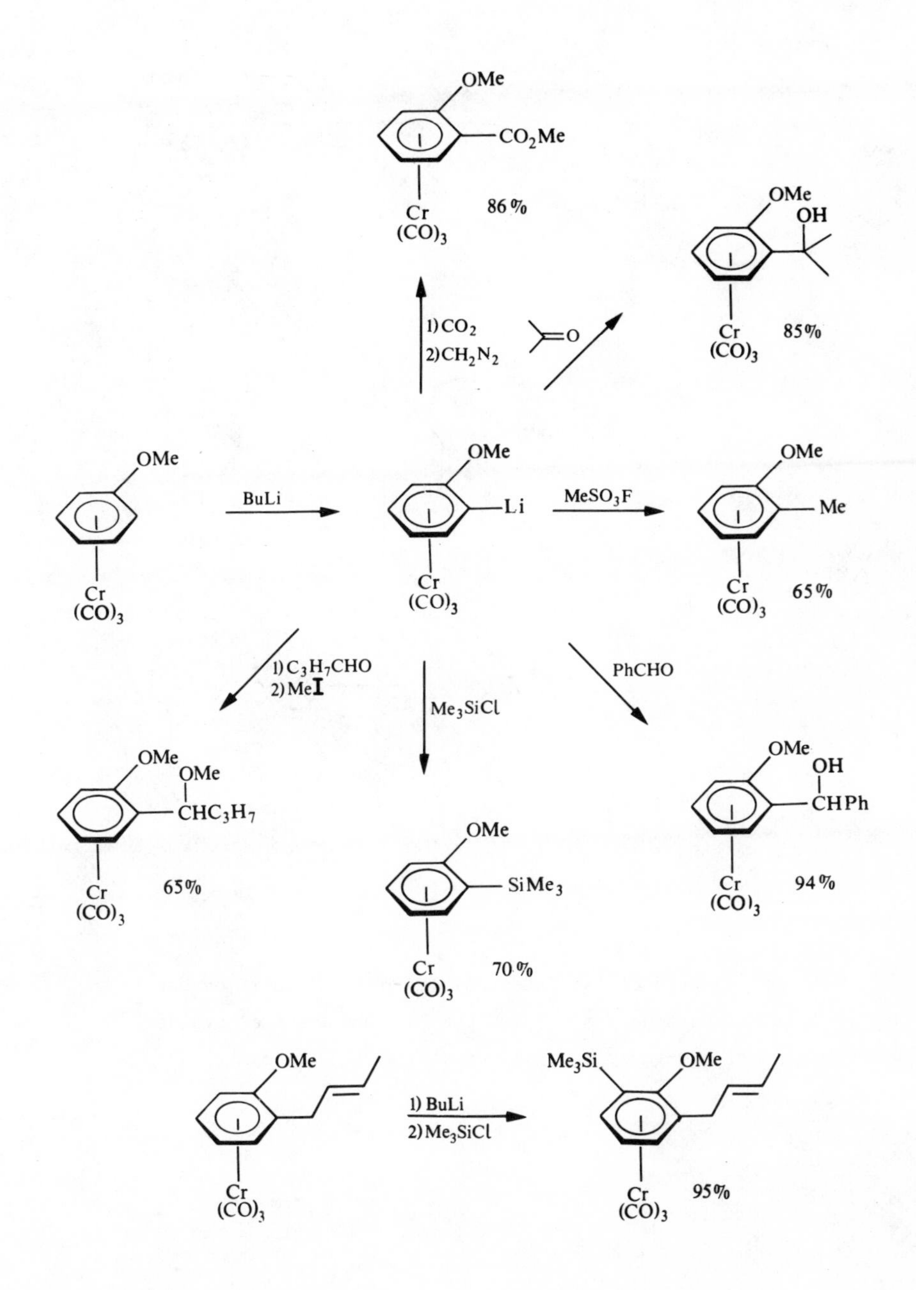

The regiospecific alkylation reaction above can be combined with the nucleophilic addition reaction (see section 4.2.4) to give bicyclic products.[33]

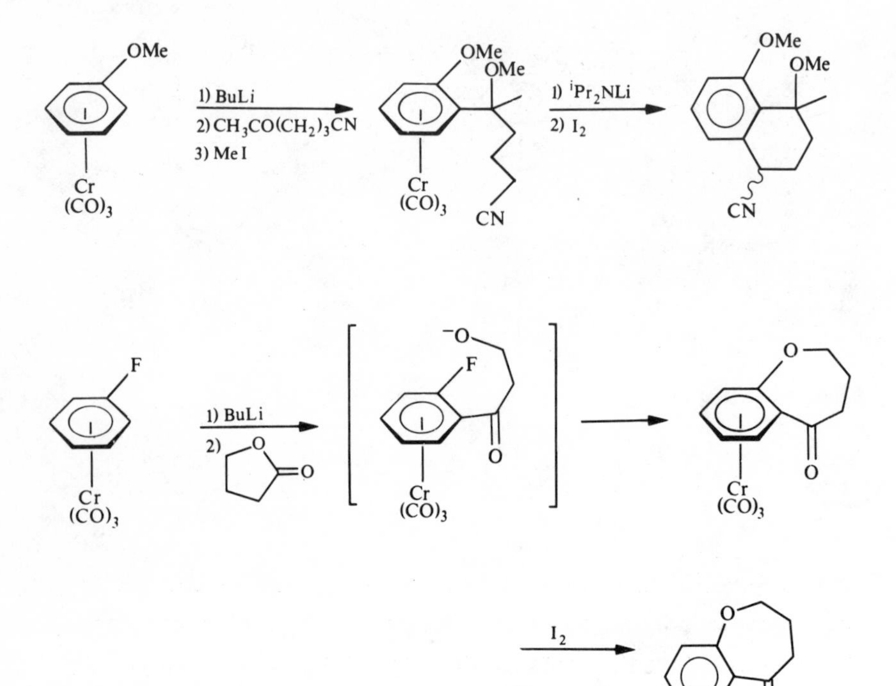

3-Methoxy benzyl alcohol lithiates mainly in the 2-position due to intramolecular coordination between the lithium and the oxygens.[34] Coordination of 3-methoxy benzyl alcohol to $Cr(CO)_3$ results in increased acidity of the ring protons and a change of **regio**selectivity in favour of the 4-position.

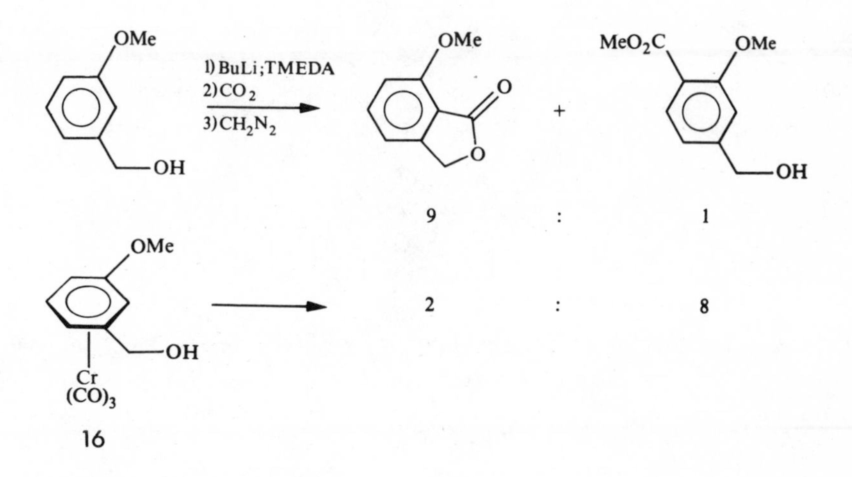

The related tetralin systems **metallate** exclusively in the 6-position presumably because coordination of the oxygen atom at C-1 to a Li atom at C-8 is sterically less favourable than for 16.

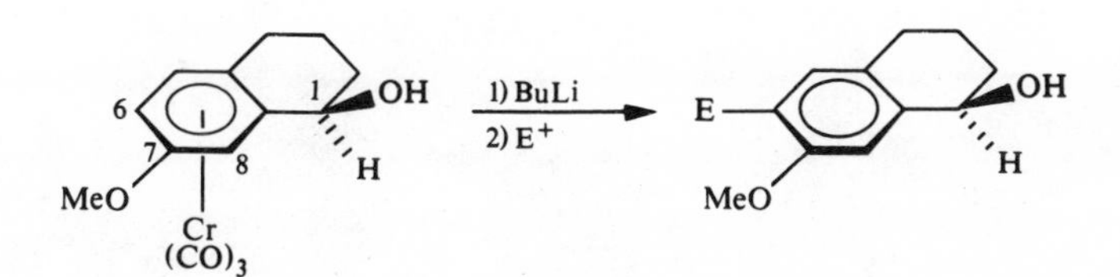

$E^+ = CO_2, Me_3SiCl$, Aryl CHO, Aryl CO_2Me, RCOCl.

The benzamides 17 **metallate** ortho to the amide group with little regioselectivity to give 18 and 19. However, **metallation** of the free arenes gives only 19 and not 18.[35]

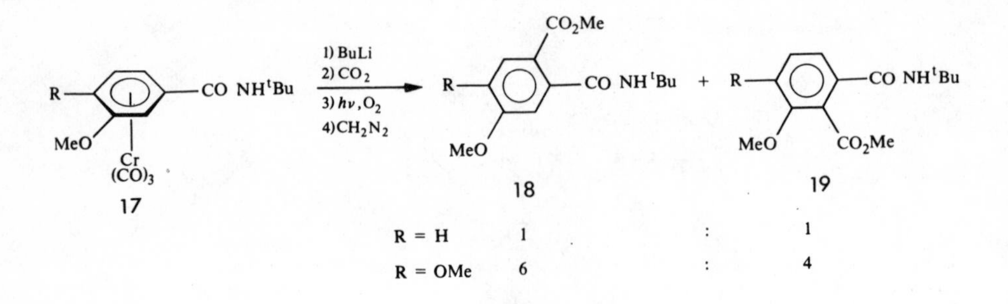

The direct metallation of (benzene)$Cr(CO)_3$ proceeds in low yield unless the conditions are carefully controlled or a transmetallation reaction is employed.[36]

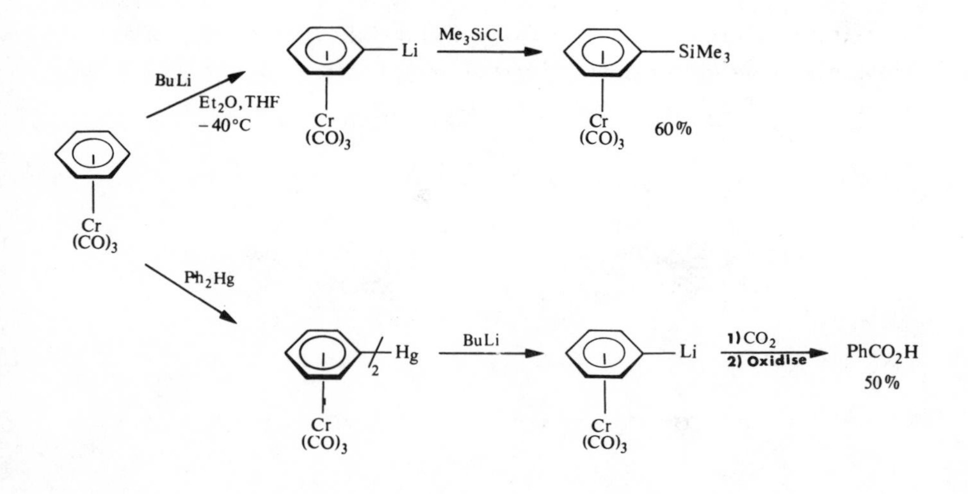

As described above the coordination of toluene to $Cr(CO)_3$ increases the acidity of the ring and benzylic protons. By varying the base used either a ring proton or a benzylic proton can be removed to generate different anions. Thus, whereas BuLi preferentially removes a ring proton (see above) benzylic anions are generated by tBuOK/DMSO or NaH/DMF.[37]

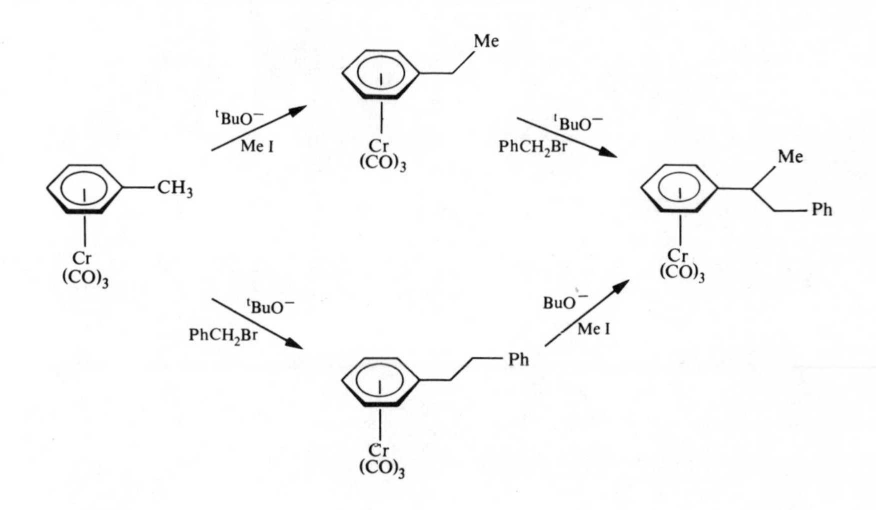

The $Cr(CO)_3$ group is not a powerful enough activating group to allow benzylic alkylation of the tertiary carbon of 20. However substitution of a CO ligand for the more electron withdrawing CS ligand allows the final alkylation to occur.

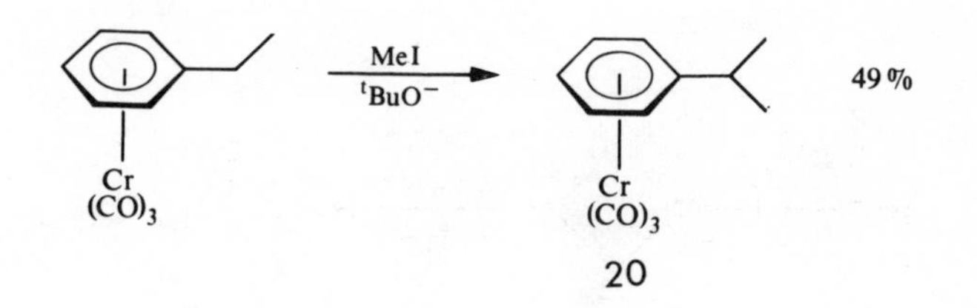

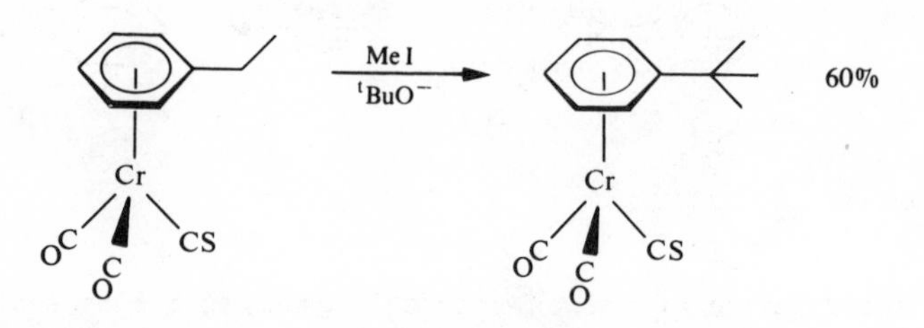

If the benzylic protons are activated by the presence of an ester group as well as a $Cr(CO)_3$ group then benzylic alkylation occurs smoothly even though uncoordinated phenylacetic ester itself is inert to NaH/DMF at 25°.[37,38]

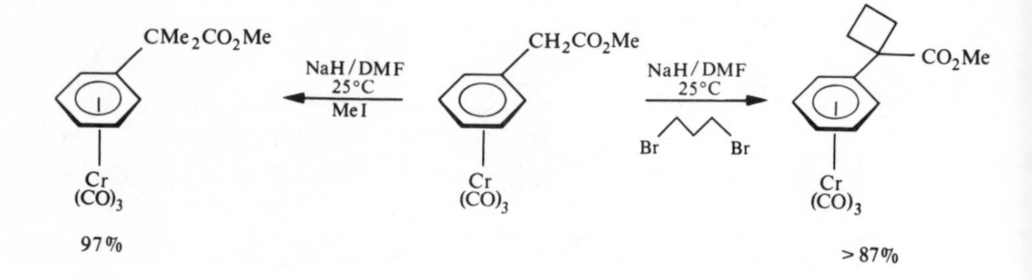

The coordination of $Cr(CO)_3$ to an arene ring does not render the arene or the benzylic protons more acidic than protons α to carbonyl functions. Thus it is the protons α to the ester and ketone functions of 21 and 22 that are abstracted by base.

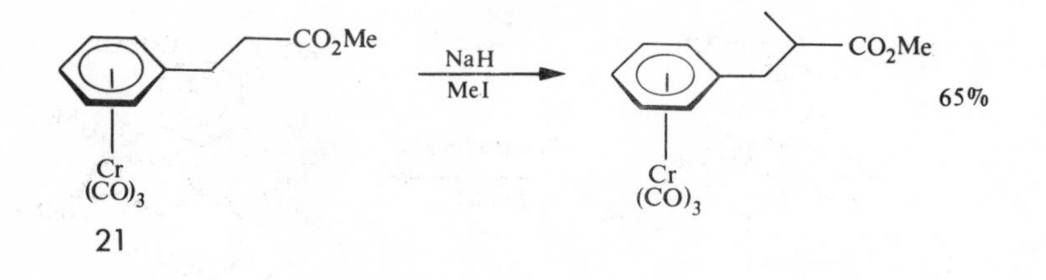

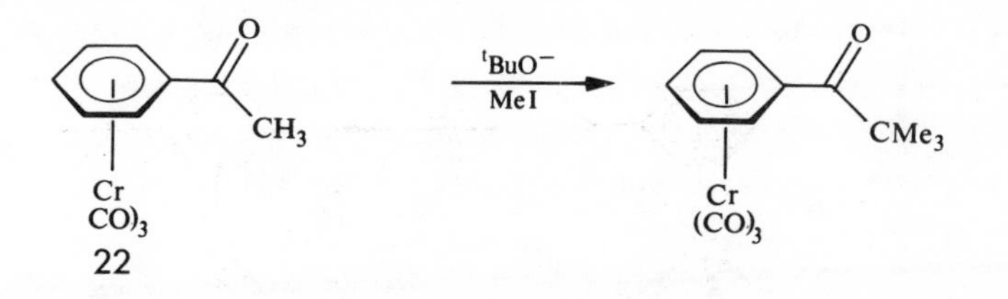

The coordination of the bulky $Cr(CO)_3$ group to one face of the arene ring has important stereochemical consequences on the alkylation of benzylic carbanions. For example, alkylation of either of the two diastereoisomers 23 or 24 gives the same single product 25 where the alkylating agent has approached from the least hindered side of the molecule.[39]

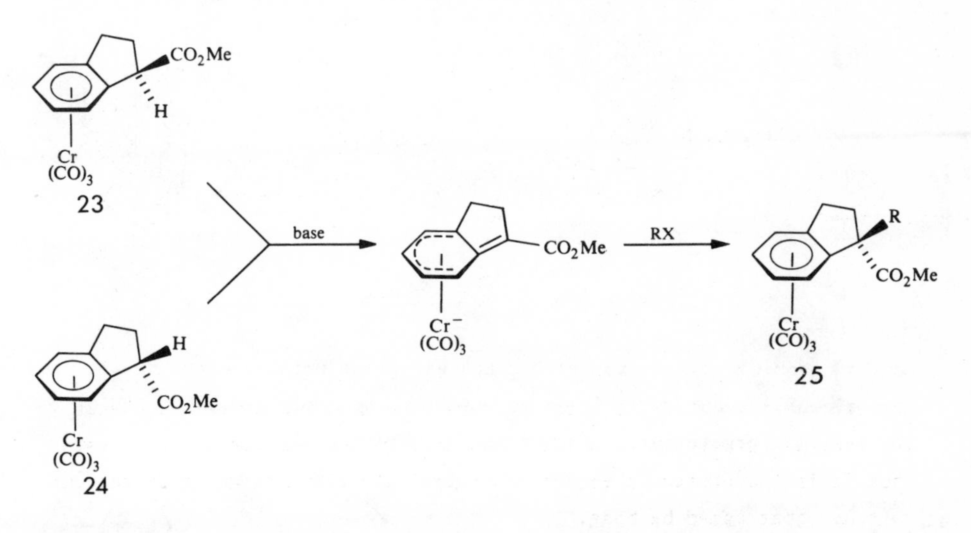

This effect has allowed the introduction of a methyl group stereospecifically into the 10 position of codeine and morphine.[40]

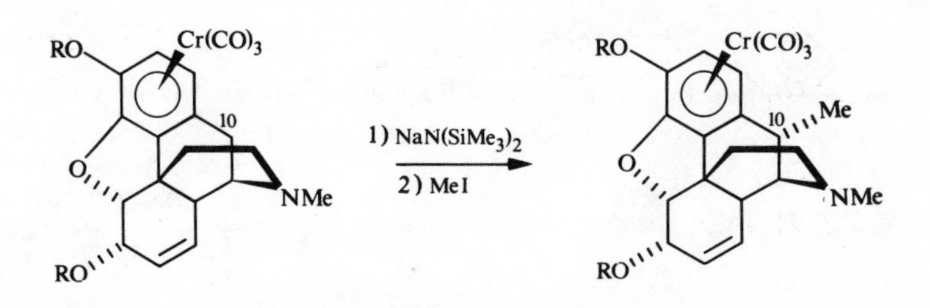

Alkylation of the anions 26 and 27 with $PhCH_2Br$ proved to be completely stereoselective producing only one diastereoisomer, 28 and 29 respectively, in each case. Presumably alkylation is occurring from the uncoordinated face of the molecule and the most stable conformations of the anions 26 and 27 are with the largest group away from the ortho methoxy substituent.[39]

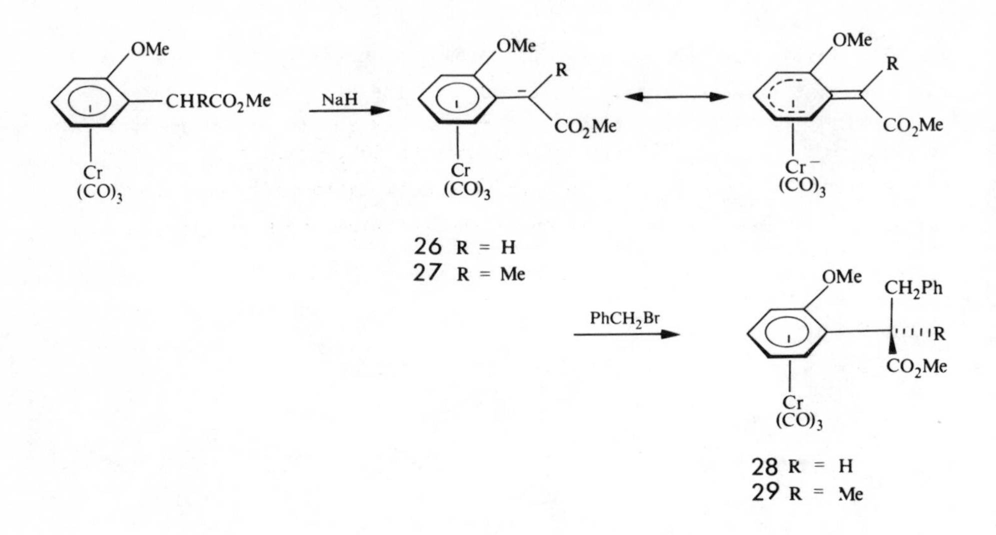

Coordination of styrene to $Cr(CO)_3$ activates the uncoordinated double bond to nucleophilic attack.[41] This potentially extremely useful method of generating benzylic anions has so far been little studied.

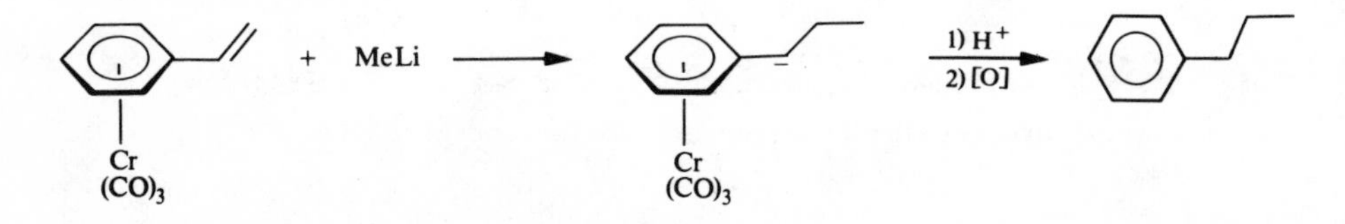

Deprotonation of the fluorene complex 30 at -40° gives the anion 31 which on methylation yields the expected *exo* methyl complex.[42]

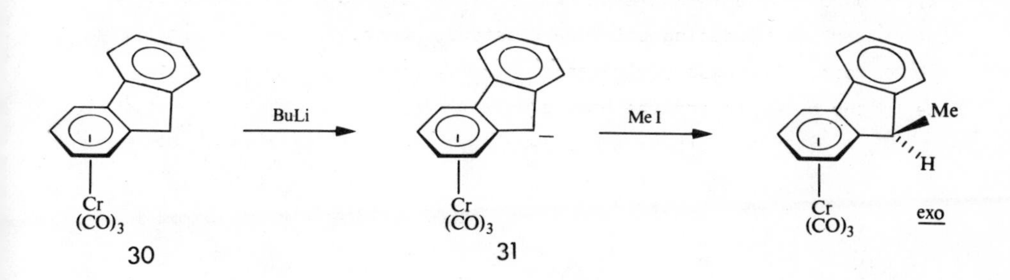

At temperatures above -20° migration of the $Cr(CO)_3$ group from η^6-arene to η^5-cyclopentadienyl occurs to give 32. Methylation of 32 then leads to the *endo* complex via methylation of the chromium and migration.

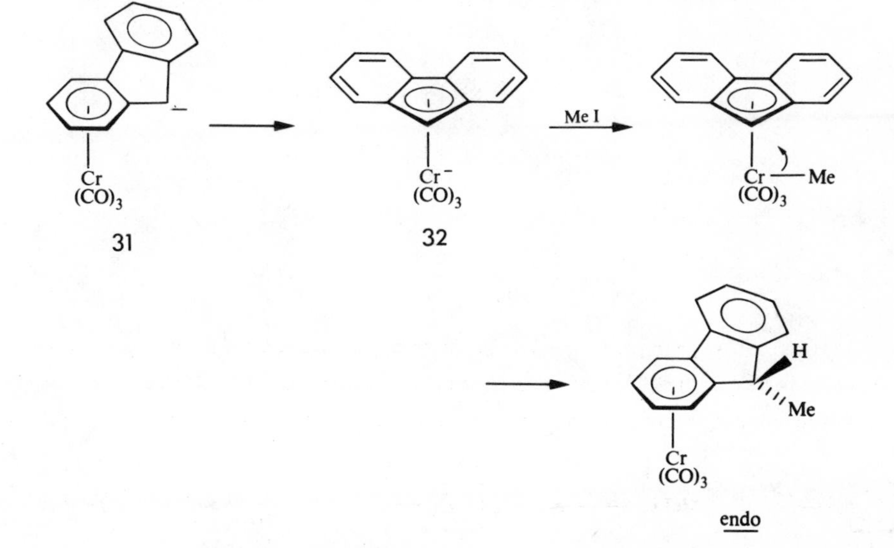

5.2.3 *The anion derived from (η^4-cycloheptatriene)Fe(CO)$_3$.*

Proton abstraction from (η^4-cycloheptatriene)$Fe(CO)_3$ produces the highly nucleophilic anion $(C_7H_7)Fe(CO)_3^-$ which reacts with electrophiles to give *exo* substituted complexes.[43]

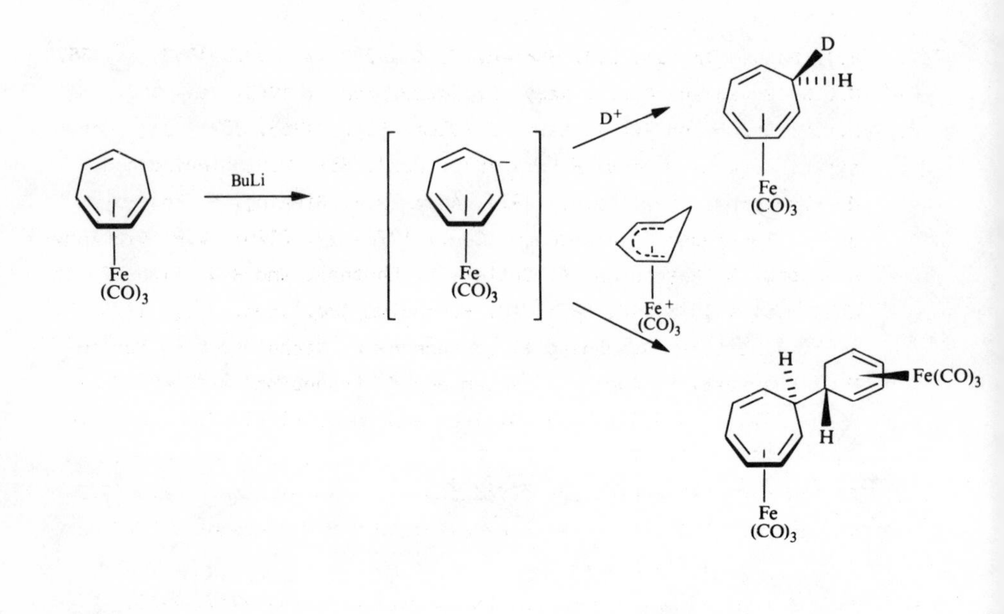

5.3 REFERENCES

1. H.J. Dauben Jr. and L.R. Honnen, *J. Amer. Chem. Soc.*, 1958, *80*, 5570; M.L.H. Green and P.L.I. Nagy, *Z. Naturforsch.*, 1963, *18b*, 162.
2. M.L.H. Green and P.L.I. Nagy, *J. Chem. Soc.*, 1963, 189; J.Y. Merour and P. Cadiot, *Compt. Rend.*, 1970, *C271*, 83; W.P. Giering and M. Rosenblum, *Chem. Comm.*, 1971, 441; W.P. Giering, M. Rosenblum and J. Tancrede, *J. Amer.Chem. Soc.*, 1972, *94*, 7170; W.P. Giering, S. Raghu, M. Rosenblum, A. Cutler, D. Ehntholt and R.W. Fish, *J. Amer. Chem. Soc.*, 1972, *94*, 8251; M. Rosenblum, *Acc. Chem. Res.*, 1974, *7*, 125; A. Cutler, D. Ehntholt, P. Lennon, K. Nicholas, D.F. Marten, M. Madhavarao, S. Raghu, A. Rosan and M. Rosenblum, *J. Amer. Chem. Soc.*, 1975, *97*, 3149; L.J. Dizikes and A. Wojcicki, *J. Organometal. Chem.*, 1977, *137*, 79.
3. A. Rosan, M. Rosenblum and J. Tancrede, *J. Amer. Chem. Soc.*, 1973, *95*, 3062; N. Genco, D. Marten, S. Raghu and M. Rosenblum, *J. Amer. Chem. Soc.*, 1976, *98*, 848.
4. T.S. Abram, R. Baker and C.M. Exon, *Tet. Letters*, 1979, 4103.
5. A. Cutler, S. Raghu and M. Rosenblum, *J. Organometal. Chem.*, 1974, *77*, 381; P.W. Jolly and R. Pettit, *J. Organometal. Chem.*, 1968, *12*, 491.
6. R. Lazzaroni and B.E. Mann, *J. Organometal. Chem.*, 1979, *164*, 79; J.K.P. Ariyaratne and M.L.H. Green, *J. Chem. Soc.*, 1963, 2976; 1964, 1.
7. A. Davison, J.P. Selegue, *J. Amer. Chem. Soc.*, 1980, *102*, 2455; M.I. Bruce and R.C. Wallis, *J. Organometal. Chem.*, 1978, *161*, C1; D.F. Marten, *Chem. Comm.*, 1980, 341.
8. A. Davison and J.P. Selegue, *J. Amer. Chem. Soc.*, 1978, *100*, 7763.
9. C.E. Coffey, *J. Inorg. and Nucl. Chem.*, 1963, *25*, 179.
10. Koerner von Gustorf, K.A.M.J. Jun and G.O. Schenck, *Z. Naturforsch.*, 1963, *18b*, 503, 767.
11. T.H. Whitesides, R.W. Arhart, *Inorg. Chem.*, 1975, *14*, 209; T.H. Whitesides, R.W. Arhart and R.W. Slaven, *J. Amer. Chem. Soc.*, 1973, *95*, 5792.
12. E. Weiss and W. Hübel, *Chem. Ber.*, 1962, *95*, 1186.
13. A.D.U. Hardy and G.A. Sim, *J.C.S. Dalton*, 1972, 2305.
14. R.E. Graf and C.P. Lillya, *J. Organometal. Chem.*, 1976, *122*, 377.

15. B.F.G. Johnson, J. Lewis and D.G. Parker, *J. Organometal. Chem.*, 1977, *141*, 319.
16. M. Franck-Neumann, F. Brion and D. Martina, *Tet. Letters*, 1978, 5033.
17. A.J. Birch and A.J. Pearson, *Chem. Comm.*, 1976, 601; A.J. Pearson, *Aust. J. Chem.*, 1976, *29*, 1841.
18. A.N. Nesmeyanov, N.V. Rybin, N.T. Gubenko, M.I. Rybinskaya and P.V. Petrovskii, *J. Organometal. Chem.*, 1974, *71*, 271; A.J. Birch and I.D. Jenkins, *Tet. Letters*, 1975, 119.
19. K.K. Joshi, P.L. Pauson, A.R. Qazi and W.H. Stubbs, *J. Organometal. Chem.*, 1964, *1*, 471.
20. G. Jaouen and R. Dabard, *Bull. Soc. chim. France*, 1974, 1646.
21. J.F. Helling and W.A. Henrickson, *J. Organometal. Chem.*, 1977, *141*, 99; C.C. Lee, B.R. Steele, K.J. Demchuk and R.G. Sutherland, *Can. J. Chem.*, 1979, *57*, 946; R.G. Sutherland, B.R. Steele, K.J. Demchuk and C.C. Lee, *J. Organometal. Chem.*, 1979, *181*, 411.
22. R.R. Schrock, *J. Amer. Chem. Soc.*, 1976, *98*, 5399.
23. F.N. Tebbe, G.W. Parshall and G.S. Reddy, *J. Amer. Chem. Soc.*, 1978, *100*, 3611.
24. S.H. Pine, R. Zahler, D.A. Evans and R.H. Grubbs, *J. Amer. Chem. Soc.*, 1980, *102*, 3270.
25. C.P. Casey, Transition Metal Organometallics in Organic Synthesis, 1976, *1*, 210, Ed. H. Alper, Academic Press, London.
26. C.P. Casey, R.A. Boggs and R.L. Anderson, *J. Amer. Chem. Soc.*, 1972, *94*, 8947; C.P. Casey and R.L. Anderson, *J. Amer. Chem. Soc.*, 1974, *96*, 1230.
27. C.P. Casey and R.L. Anderson, *J. Organometal. Chem.*, 1974, *73*, C28.
28. C.P. Casey, W.R. Brunsvold and D.M. Scheck, *Inorg. Chem.*, 1977, *16*, 3059.
29. C.P. Casey and W.R. Brunsvold, *J. Organometal. Chem.*, 1975, *102*, 175.
30. C.P. Casey, *Inorg. Synth.*, 1979, *19*, 178.
31. C.P. Casey and W.R. Brunsvold, *J. Organometal. Chem.*, 1976, *118*, 309.
32. C.P. Casey and W.R. Brunsvold, *Inorg. Chem.*, 1977, *16*, 391.
33. M.F. Semmelhack, J. Bisaha and M. Czarny, *J. Amer. Chem. Soc.*, 1979, *101*, 768; R.J. Card and W.S. Trajanovsky, *J. Org. Chem.*, 1980, *45*, 2560.
34. M. Uemura, S. Tokuyama and T. Sakan, *Chem. Letters*, 1975, 1195.
35. M. Uemura, N. Nishikawa and Y. Hayashi, *Tet. Letters*, 1980, 2069.

36. M.D. Rausch and R.E. Gloth, *J. Organometal. Chem.*, 1978, *153*, 59; M.D. Rausch, *Synth. React. Inorg. Met-Org. Chem.*, 1979, *9*, 357.
37. G. Simonneaux and G. Jaouen, *Tetrahedron*, 1979, *35*, 2249.
38. G. Jaouen, A. Meyer and G. Simonneaux, *Chem. Comm.*, 1975, 813.
39. H. des Abbayes and M.A. Boudeville, *J. Org. Chem.*, 1977, *42*, 4104.
40. H.B. Arzeno, D.H.R. Barton, S.G. Davies, X. Lusinchi, B. Meunier and C. Pascard, *Nouveau J. Chimie*, 1980, *4*, 369.
41. G.R. Knox, D.G. Leppard, P.L. Pauson and W.E. Watts, *J. Organometal. Chem.*, 1972, *34*, 347.
42. A.N. Nesmeyanov, N.A. Ustynyuk, L.N. Novikova, T.N. Rybina, Y.A. Ustynyuk, Y.F. Oprunenko and O.I. Trifonova, *J. Organometal. Chem.*, 1980, *184*, 63.
43. H. Maltz and B.A. Kelly, *Chem. Comm.*, 1971, 1390; M. Moll, P. Würstl, H. Behrens and P. Merbach, *Z. Naturforsch*, 1978, *33b*, 1304.

CHAPTER 6

COUPLING AND CYCLISATION REACTIONS

6.1 COUPLING REACTIONS INVOLVING BIS-η^1-COMPLEXES

The generation of two alkyl or aryl groups on a transition metal centre can lead to a coupling reaction by pericyclic elimination involving the formation of a carbon-carbon bond.[1]

$$\text{(Aryl) HgOAc} \xrightarrow{PdCl_2} \text{(Aryl)} - \text{(Aryl)} \qquad 47\% - 95\%$$

Cross coupling reactions may be achieved between aryl or vinyl halides and organo Mg, Li, Zn, Zr, or Al reagents in the presence of Pd (O) or Ni (O) catalysts.[2] The first step in the reaction is pericyclic addition of the aryl or vinyl halide to the metal to give an aryl or vinyl metal complex. Exchange of halide for alkyl then occurs followed by pericyclic elimination of the coupled product.

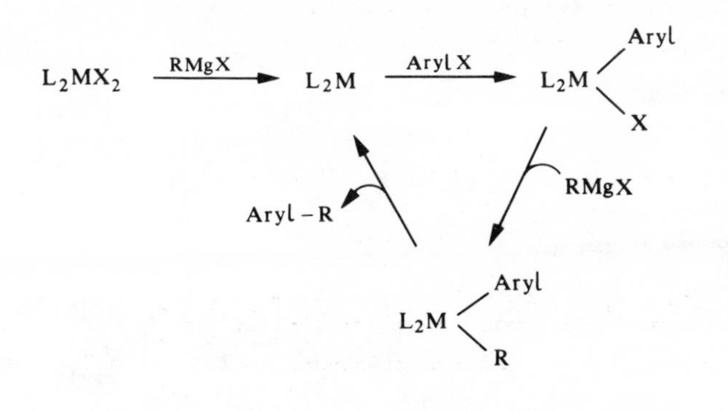

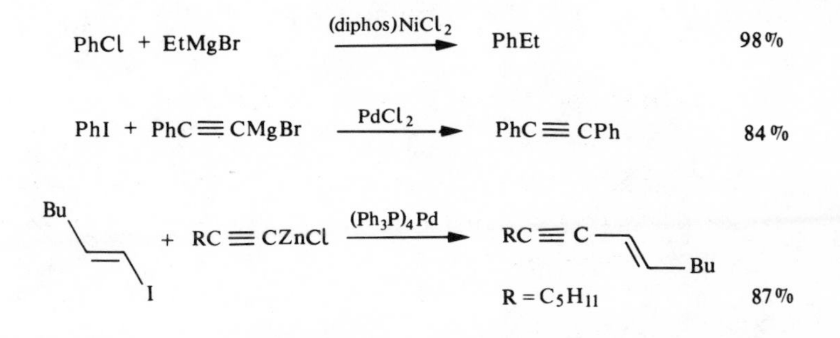

Pericyclic addition of a variety of iodo and bromoarenes to $(Ph_3P)_3RhMe$ leads under mild conditions in dimethylformamide to the corresponding methyl substituted compounds in high yields.[3] For example, iodobenzene is converted into toluene in 99% yield.

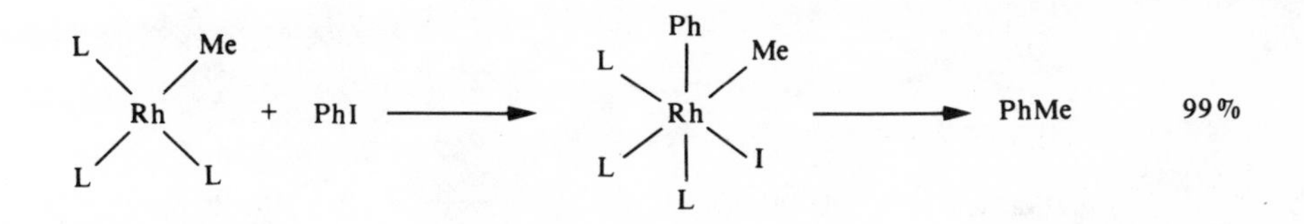

Selective monoalkylation of aromatic dihalides may be achieved with Grignard and organozinc reagents if $Pd(PPh_3)_4$ is present as a catalyst.[4] Monoalkylation of 2,6-dihalopyridines and dihalothiophenes may also be achieved by this method.

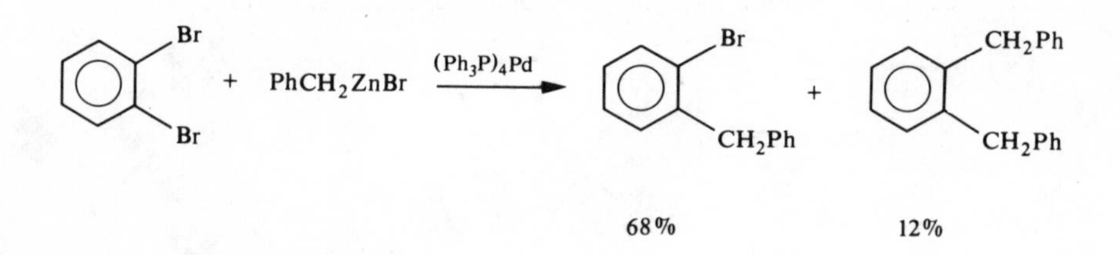

Treatment of the enolate generated from 1 with $(COD)_2Ni$ in THF at 25° produces the cyclised product cephalotaxinone 2 in 30% yield.[5] In this case the zerovalent Ni cyclisation could not compete, however, with a photostimulated $S_{RN}1$ reaction which gave 2 in 94% from 1.

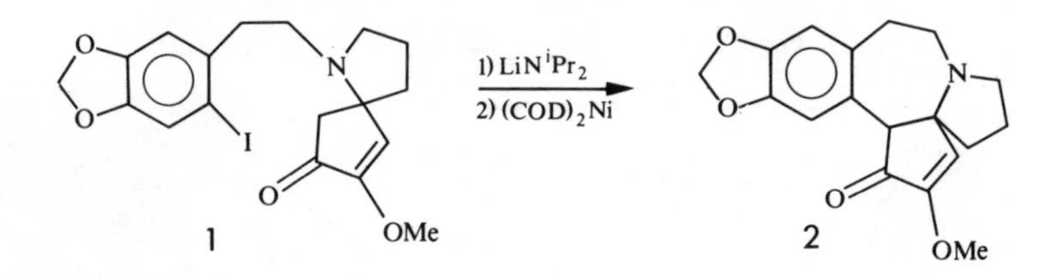

The use of optically active phosphines in the Ni catalysed coupling of Grignard reagents with aryl and vinyl chlorides leads to optically active products.[6] For example, reaction of racemic 2-butylmagnesium chloride with bromobenzene gives 2-phenylbutane with an optical yield of 14.8%. Racemic 2-butyl magnesium bromide and chlorobenzene also gives 2-phenylbutane with an optical yield of 17%.

$$EtCHMeMgBr + (DIO\overset{*}{P})NiCl_2 + PhCl \longrightarrow Ph\overset{*}{C}HMeEt \quad (e.e. = 17\%)$$

Coupling reactions involving vinyl halides are stereospecific.[2,7,8]

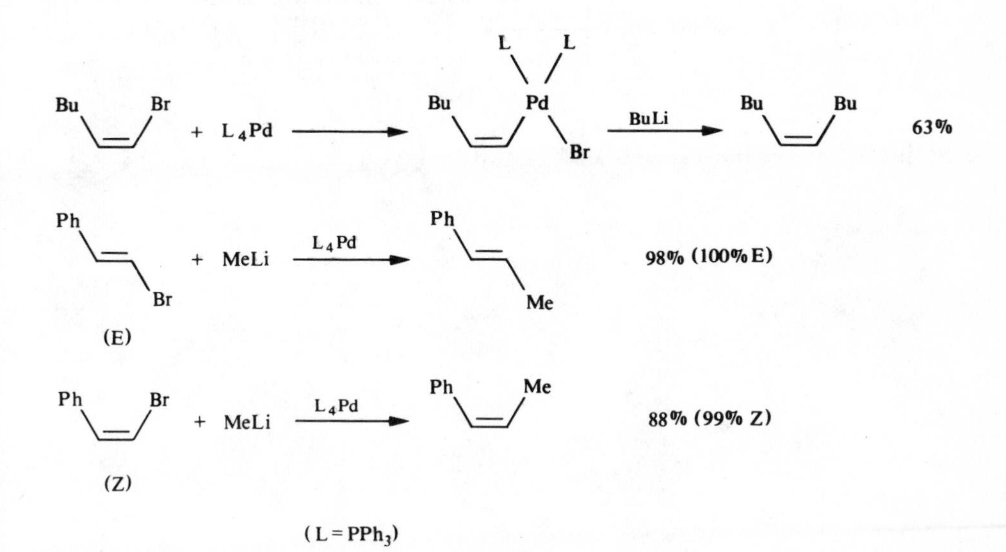

E,E-Farnesol 3 has been synthesised using three consecutive coupling reactions.[9]

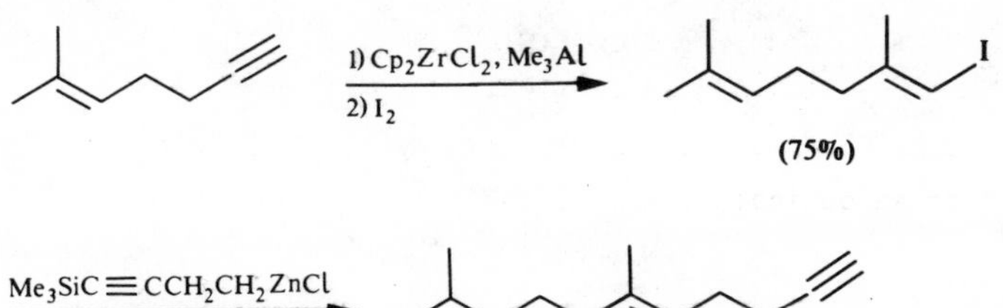

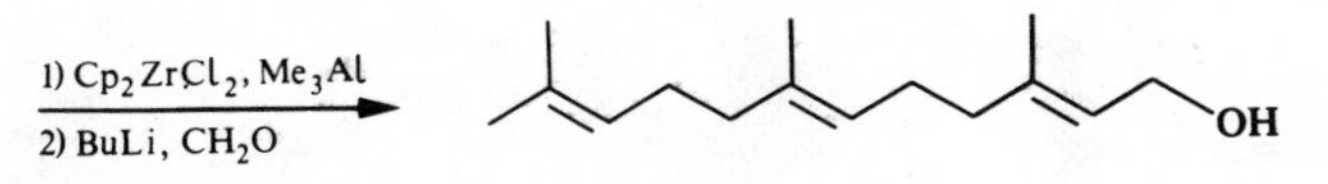

As before, in the presence of chiral ligands optically active products are generated, for example in the synthesis of the **sesquiterpene** α-curcumene **4**.[10]

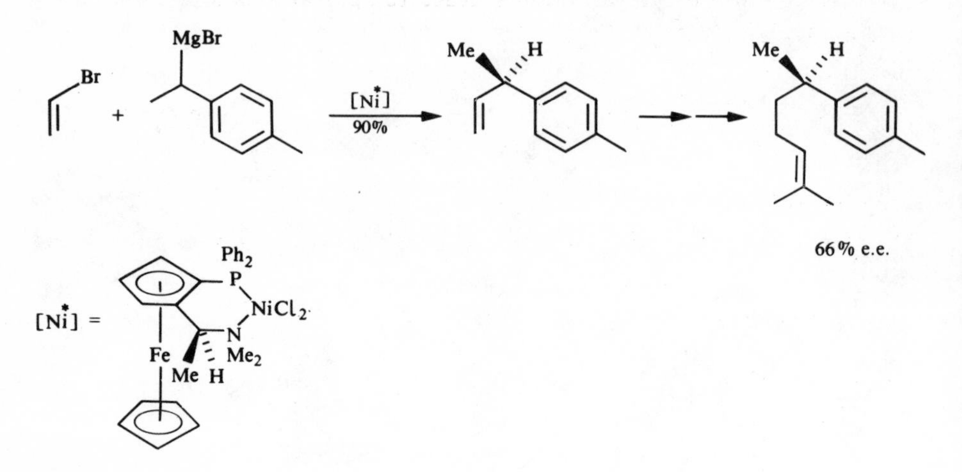

Terminal acetylenes may be coupled to aryl and vinyl halides in the presence of $(Ph_3P)_2PdX_2$ complexes and base.[11]

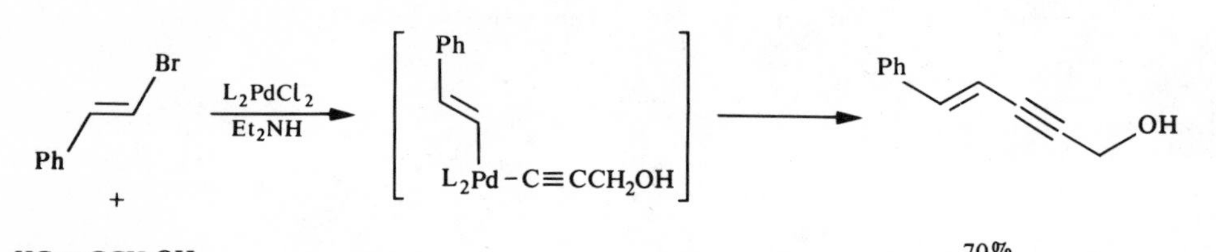

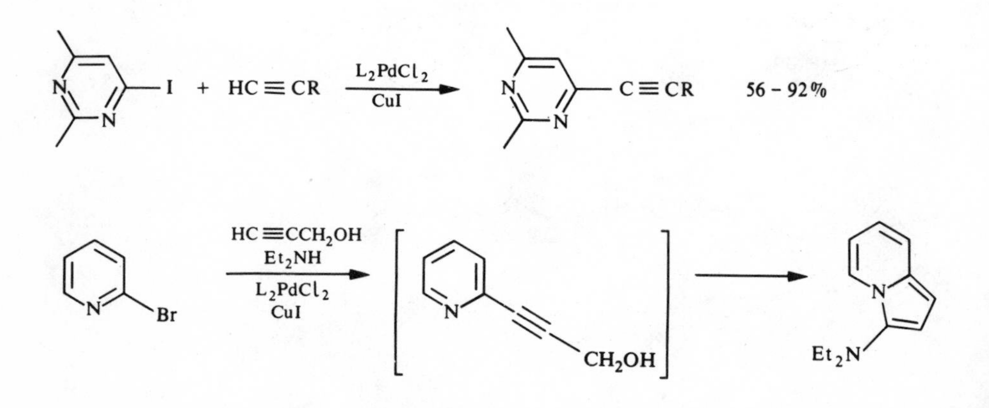

The reaction of 4-bromobut-1-ene with PhMgBr catalysed by L_2NiCl_2 gives 3-phenylbut-1-ene due to isomerisation of the initially formed Ni-alkyl complex. The use of chiral ligands leads to optically active 3-phenylbutene (34% e.e.).[12]

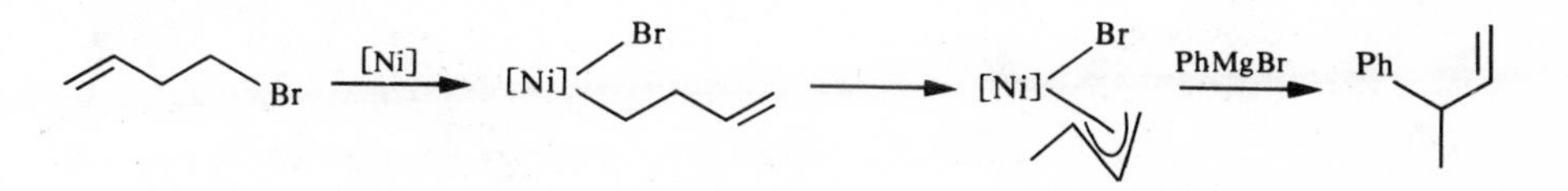

The direct oxidative cyclisation of bis-aryl systems may also be achieved.[13]

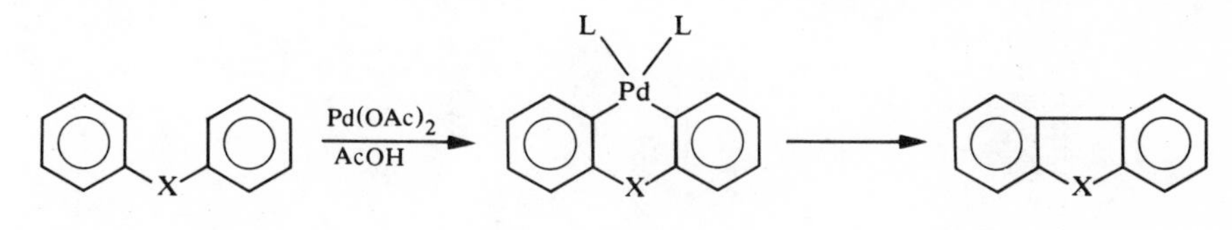

X = O, NH, CO

65 – 90%

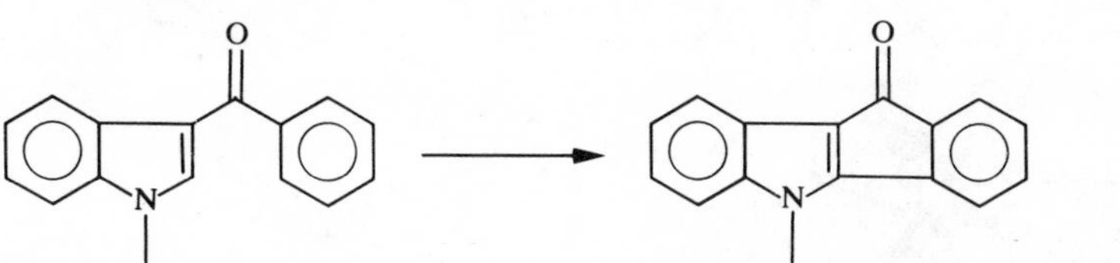

60%

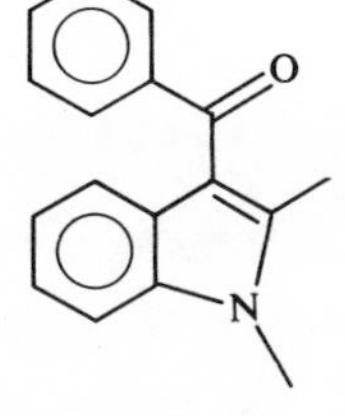

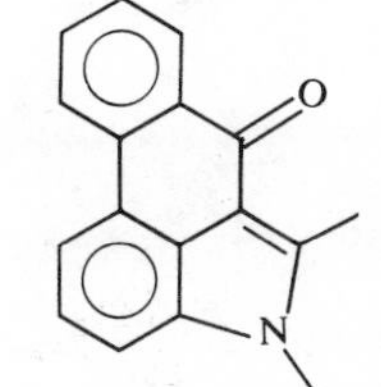

31%

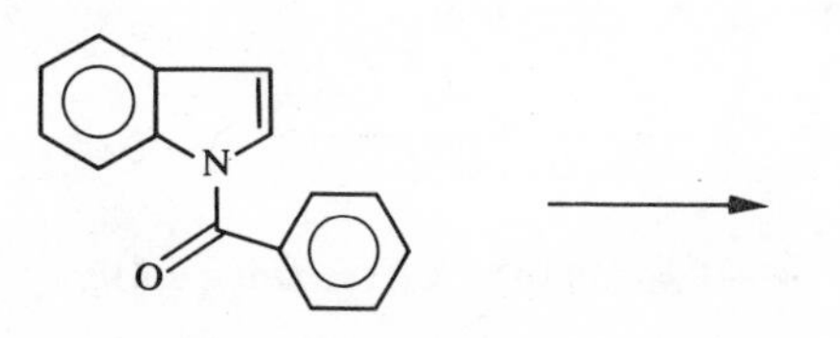

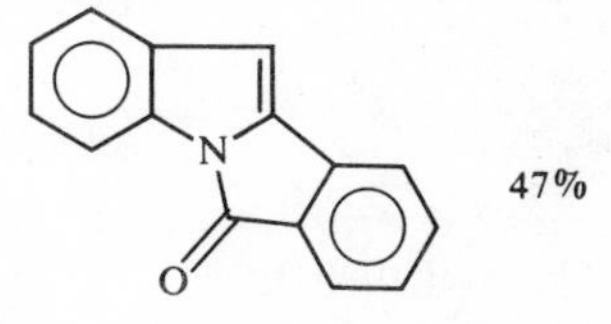

47%

6.2 THE HECK REACTION AND RELATED TRANSFORMATIONS.

6.2.1 Coupling of organic nucleophiles with olefins

The Heck reaction is a general reaction involving the cis addition of η^1-R[Pd]species across a double bond.[14] In cases where this generates a β-hydrogen syn to the [Pd], elimination of [Pd] occurs to give a new olefin. The overall reaction is the substitution of an aryl or vinyl hydrogen. β-elimination of [Pd]H in Pd complexes is so rapid that the initial R[Pd] complex is restricted to R groups that do not have labile β-hydrogens themselves (e.g. vinyl, aryl, benzyl, allyl, methyl).

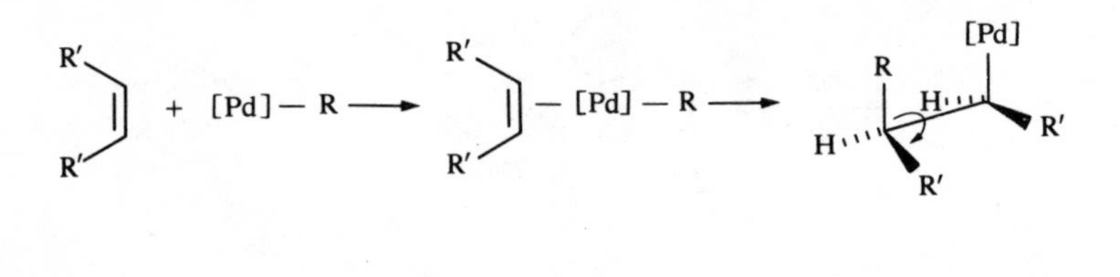

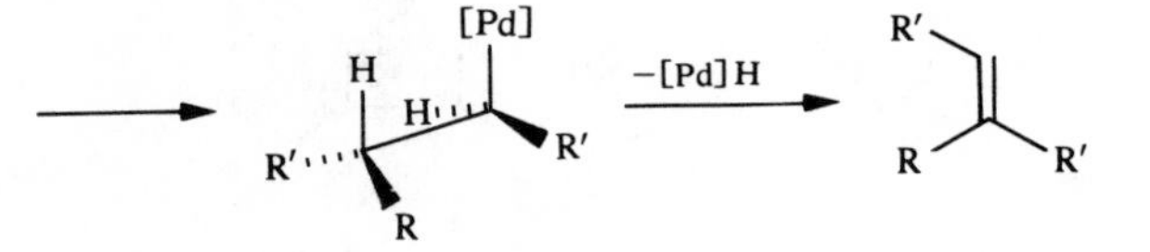

A good method of generating the initial alkyl-Pd complex is via organomercurials.[15]

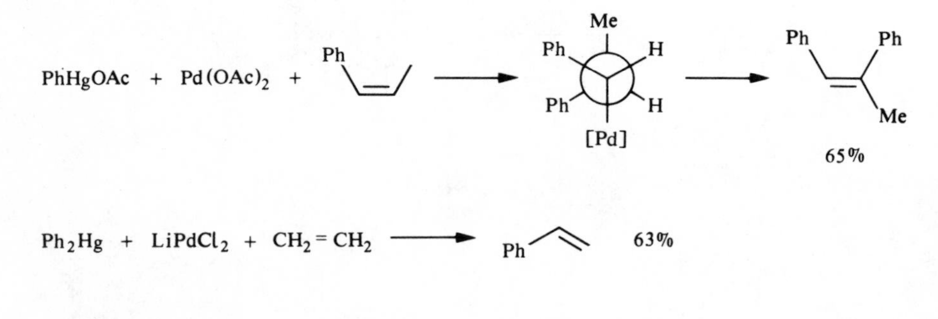

PhHgOAc + $Pd(OAc)_2$ + OAc ⟶ PhCH = $CHCH_2OAc$ 63%

The relative rates of reaction and the regioselectivity depend on both steric and electronic factors.[14,16] The less sterically hindered the olefin, the greater the rate ; the new carbon-carbon bond is formed at the least sterically hindered or most electron deficient carbon.

relative rates:

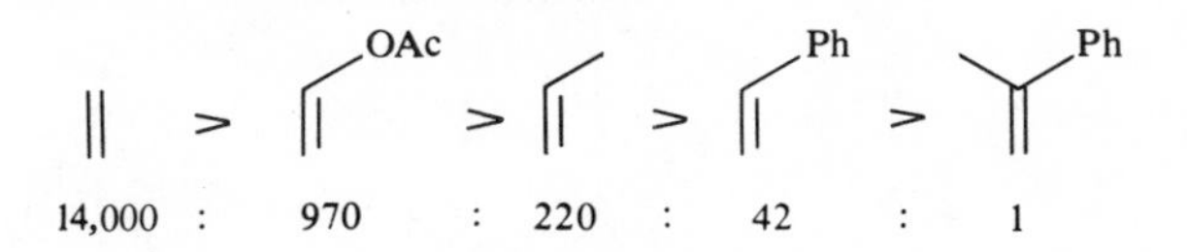

regioselectivity:

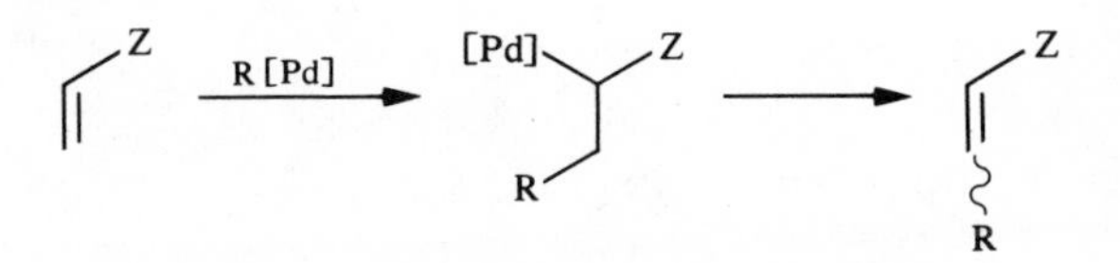

100% Z = CO_2Me, CN, Ph, $CH(OR)_2$

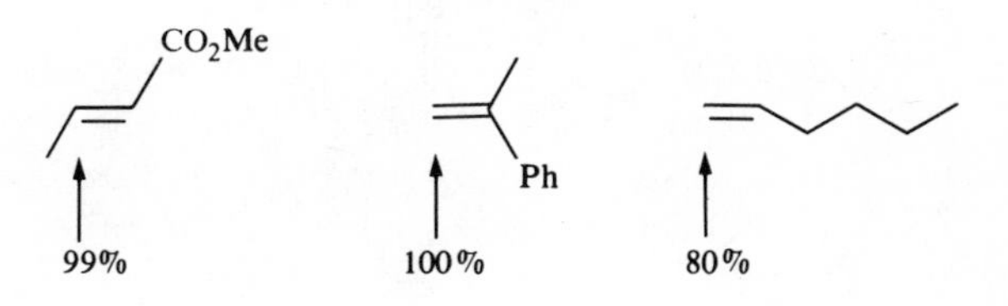

The Heck reaction of aryl mercurials with allyl alcohols as substrates yields 3-aryl ketones as products[17] whereas allyl halides give 3-aryl olefins.[18] Thus while allyl acetates and alcohols undergo the Heck reaction, allyl chlorides undergo overall allylic substitution. This is presumably because chloride is a better leaving group than hydroxide or acetate and therefore in the latter case elimination of $PdCl_2$ is favoured over β-elimination.

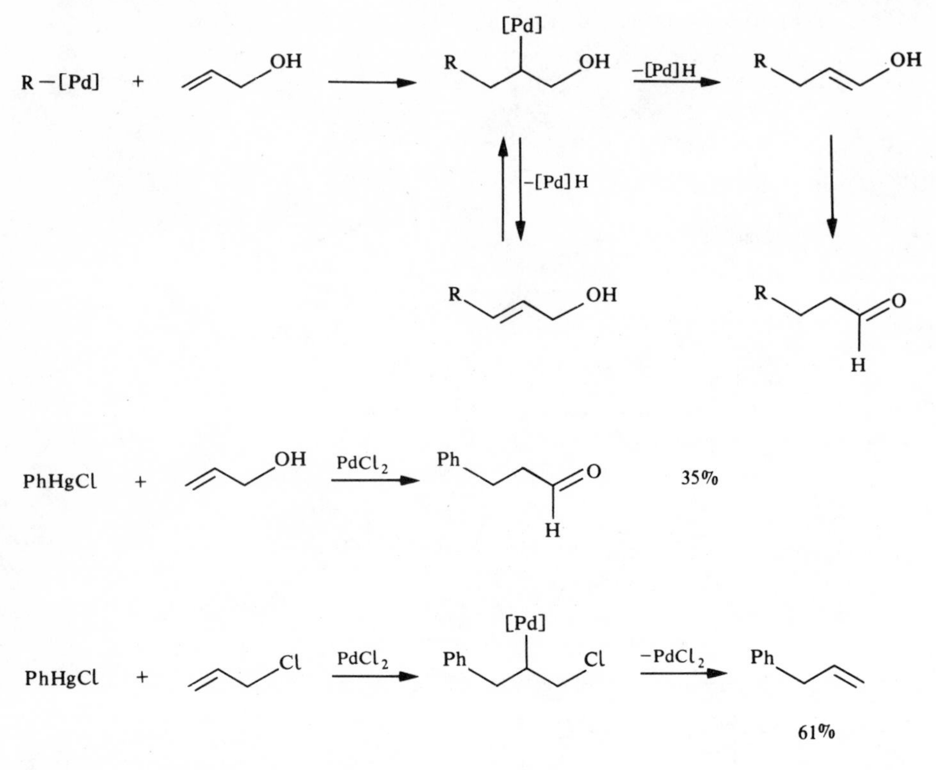

1,4-Dienes result from the regiospecific reaction of (vinyl) HgCl with allyl chlorides in the presence of $PdCl_2$.[19]

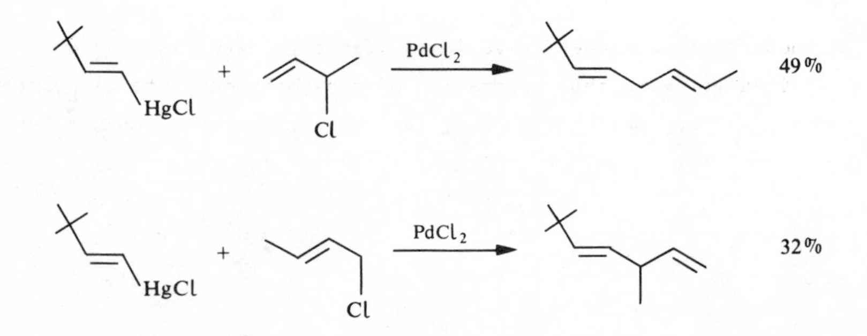

When β-elimination of a *syn*-hydrogen is not possible then new, relatively stable, alkyl-Pd species are formed and may be subjected to further reactions, for example reduction, carbonylation, solvolysis etc.[20]

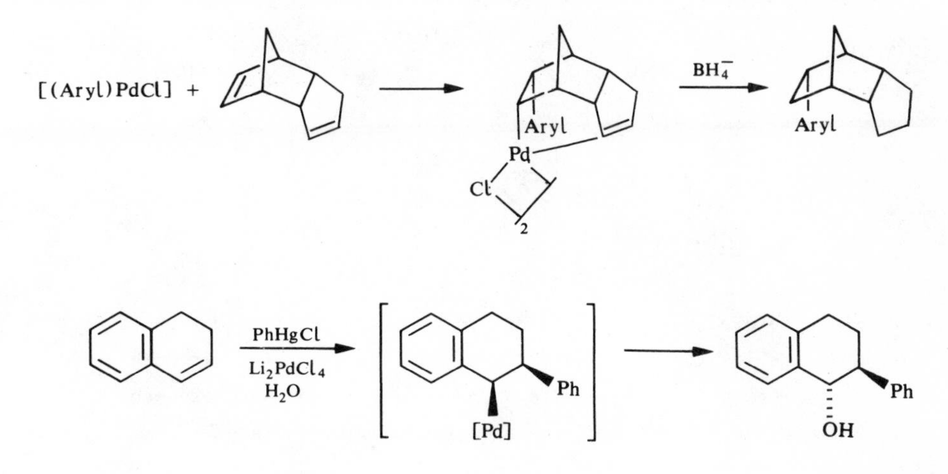

Intramolecular trapping of the alkyl-Pd species by alkoxide can be used for the synthesis of (±)-pterocarpin 5.[21]

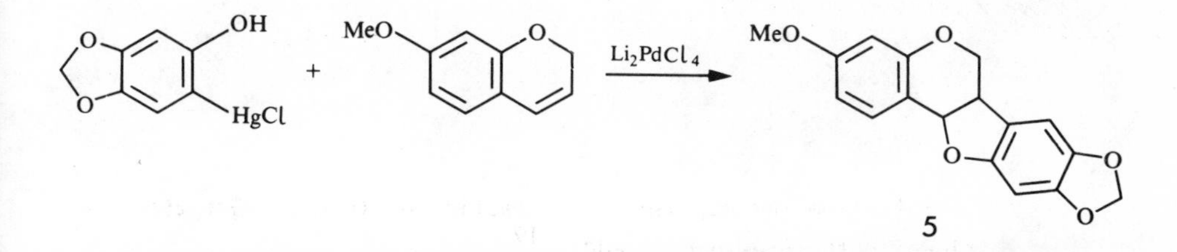

Treatment of norbornene derivatives with vinyl mercurials under similar conditions leads to the preparation of prostaglandin intermediates.[22]

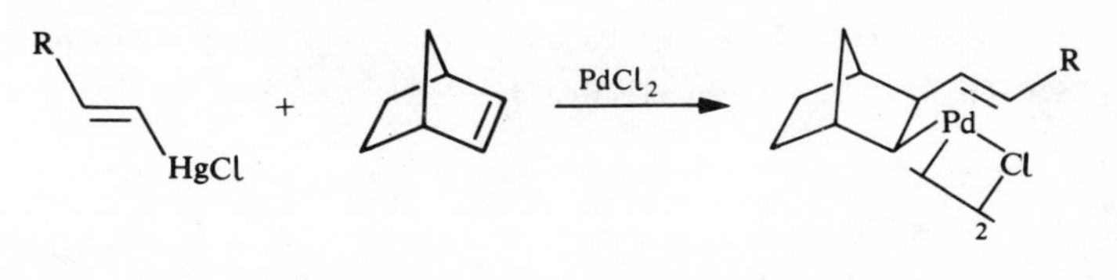

$R = Bu^t$ 89%

The coupling reaction of organomercurials with olefins catalysed by Li_2PdCl_4 can be used to synthesise C-5 substituted pyrimidine nucleosides.[23]

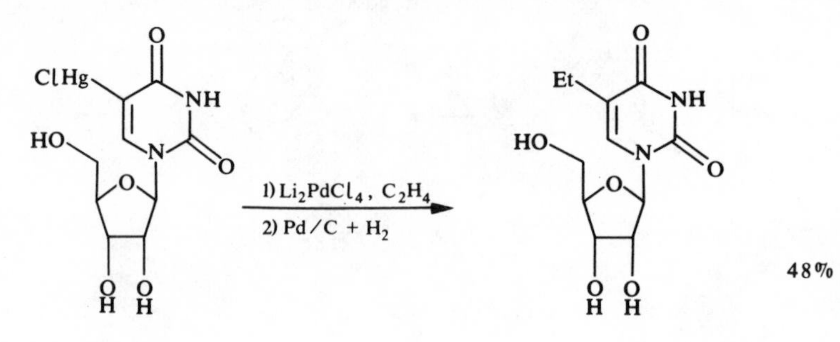

Alkyl-Pd complexes may also be prepared by nucleophilic addition to olefin-Pd species (see section 4.2.a). The coupling of a cyclopentyl-Pd complex , formed in this way, to α,β-unsaturated ketones has been employed for the synthesis of a prostaglandin intermediate.[23a]

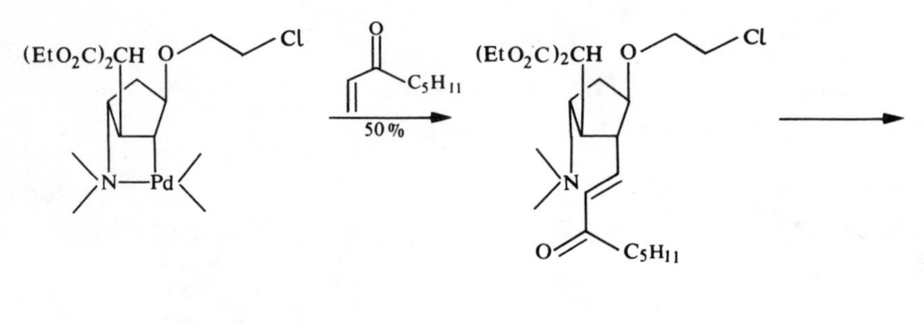

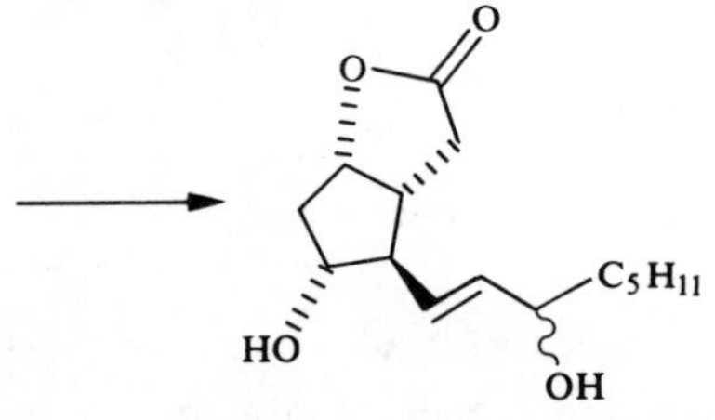

The reaction is not restricted to Pd as a catalyst. $Cp_2ZrCl(vinyl)$ complexes, which are readily available from the addition of Cp_2ZrHCl to acetylenes, couple to α,β-unsaturated ketones in the presence of $Ni(acac)_2$.[24]

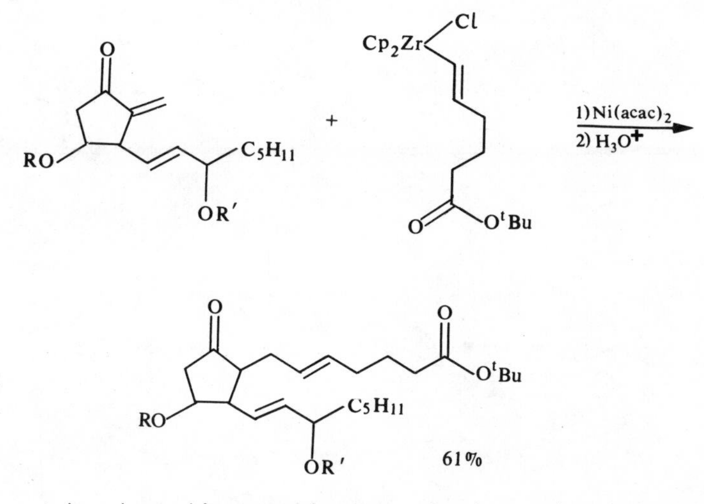

$Pd(OAc)_2$ catalysed coupling reactions may also be used to prepare cyclic α,β-unsaturated ketones by cyclisation of trimethylsilyl enol ethers.[25]

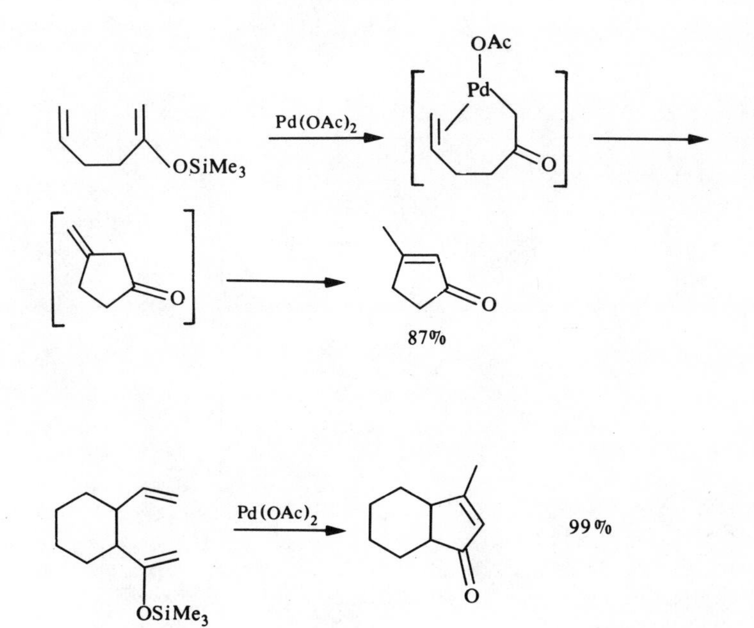

η^1-Allyl ligands do not have any removable β-hydrogens and may be efficiently coupled to olefins.[26]

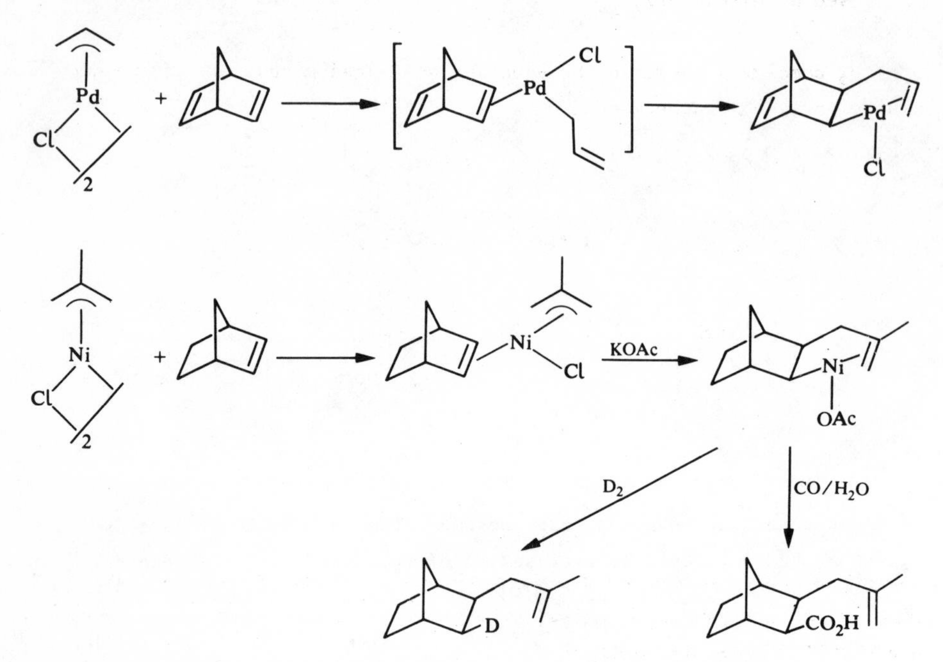

This type of coupling mechanism is believed to form part of the

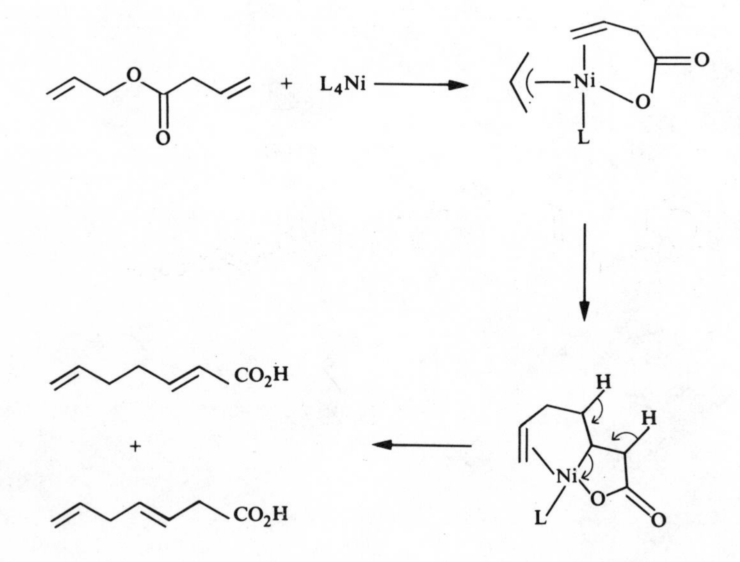

rearrangement reaction of allyl but-3-enoate to heptadienoic acids catalysed by $Ni(P(O^iPr)_3)_4$.[26a]

L_2NiCl_2 complexes catalyse the coupling of Grignard reagents with vinyl ethers and vinyl sulphides.[26b].

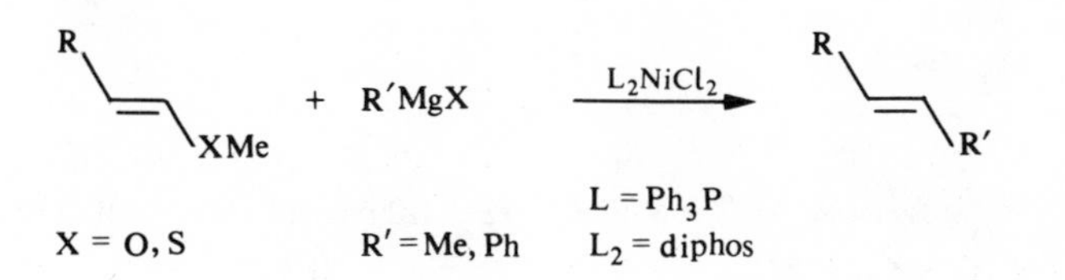

6.2.2 *Coupling of organic halides with olefins*

(Aryl)PdX complexes formed by the oxidative addition of aryl halides to Pd(0) may also undergo the Heck reaction.[27] The Pd(0) catalyst is often generated *in situ* by reduction of PdX_2.

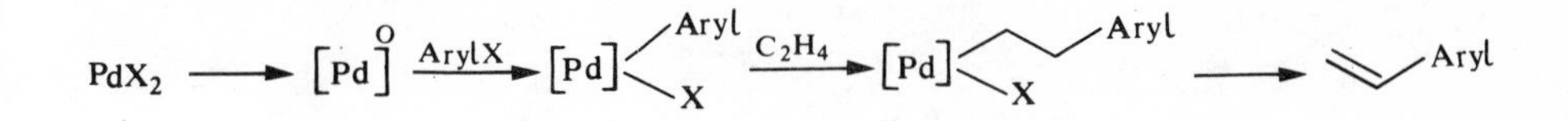

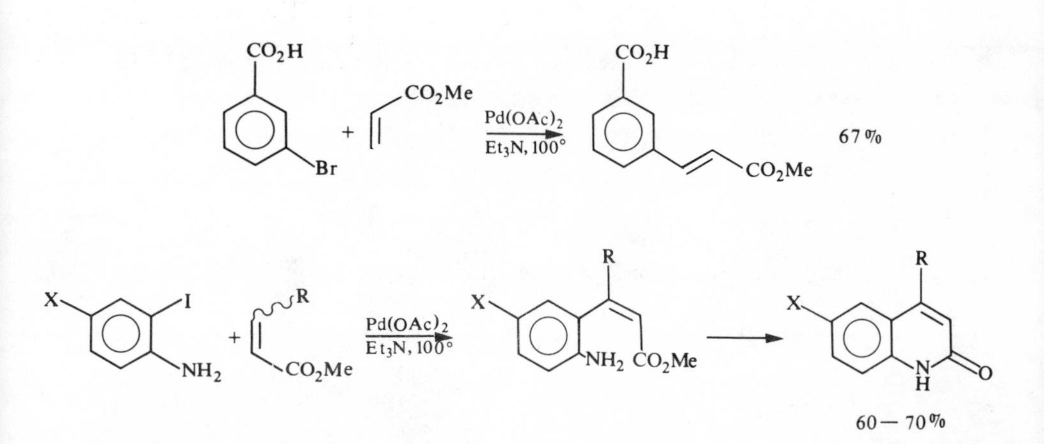

Many interesting cyclisation reactions catalysed by Pd occur by this mechanism to give, for example, indoles and quinolines.[27]

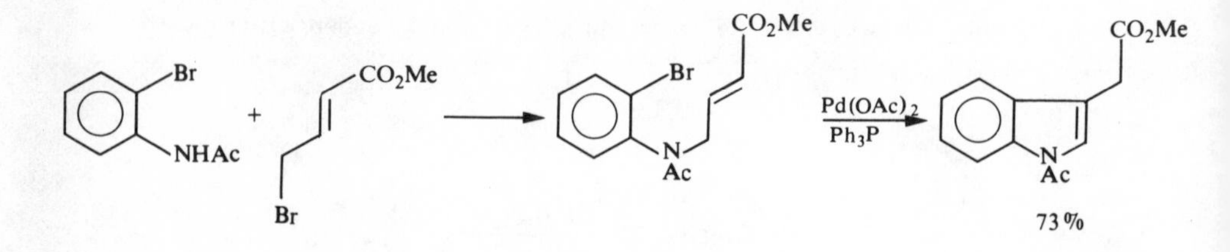

Zero valent Ni complexes also catalyse the intramolecular coupling of aryl halides with olefins.[27,28]

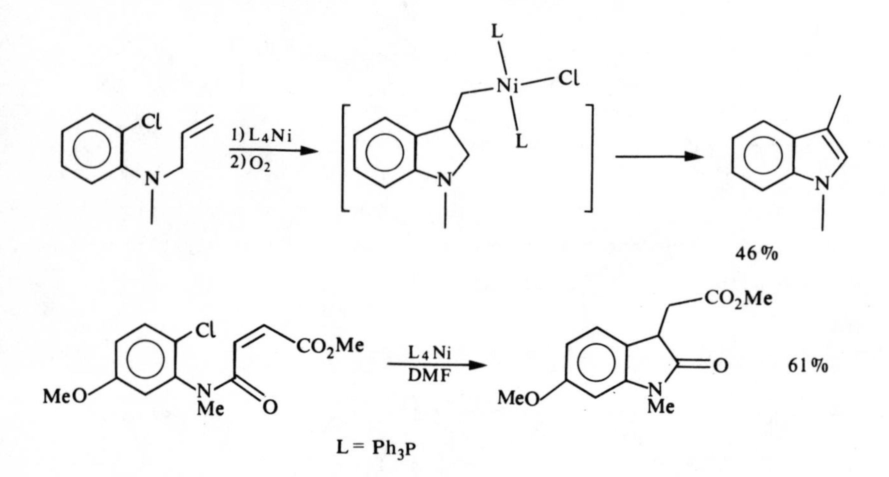

The vinylation of 1,3-dienes can be achieved with vinyl halides and Pd catalysts.[29]

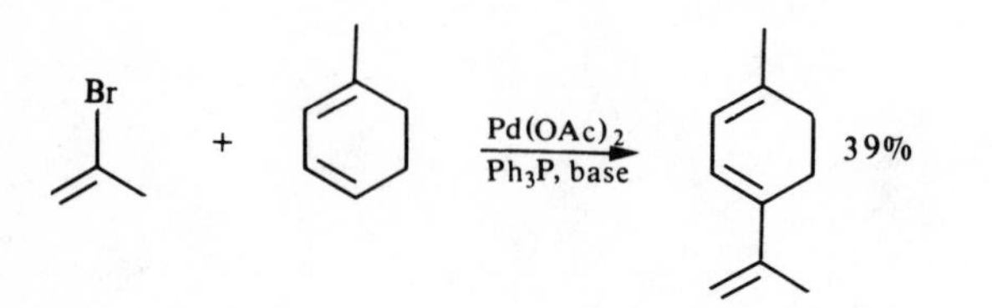

The coupling of aryl halides to allyl alcohols catalysed by Pd in the presence of base leads to a variety of substituted ketones and aldehydes.[30] Tertiary allylic alcohols undergo substitution without rearrangement. 2- and 3-bromo thiophenes also couple to allyl alcohols in the presence of $Pd(OAc)_2$ and this reaction has been used to **prepare a honey bee pheromone,** queen substance 6 .[31]

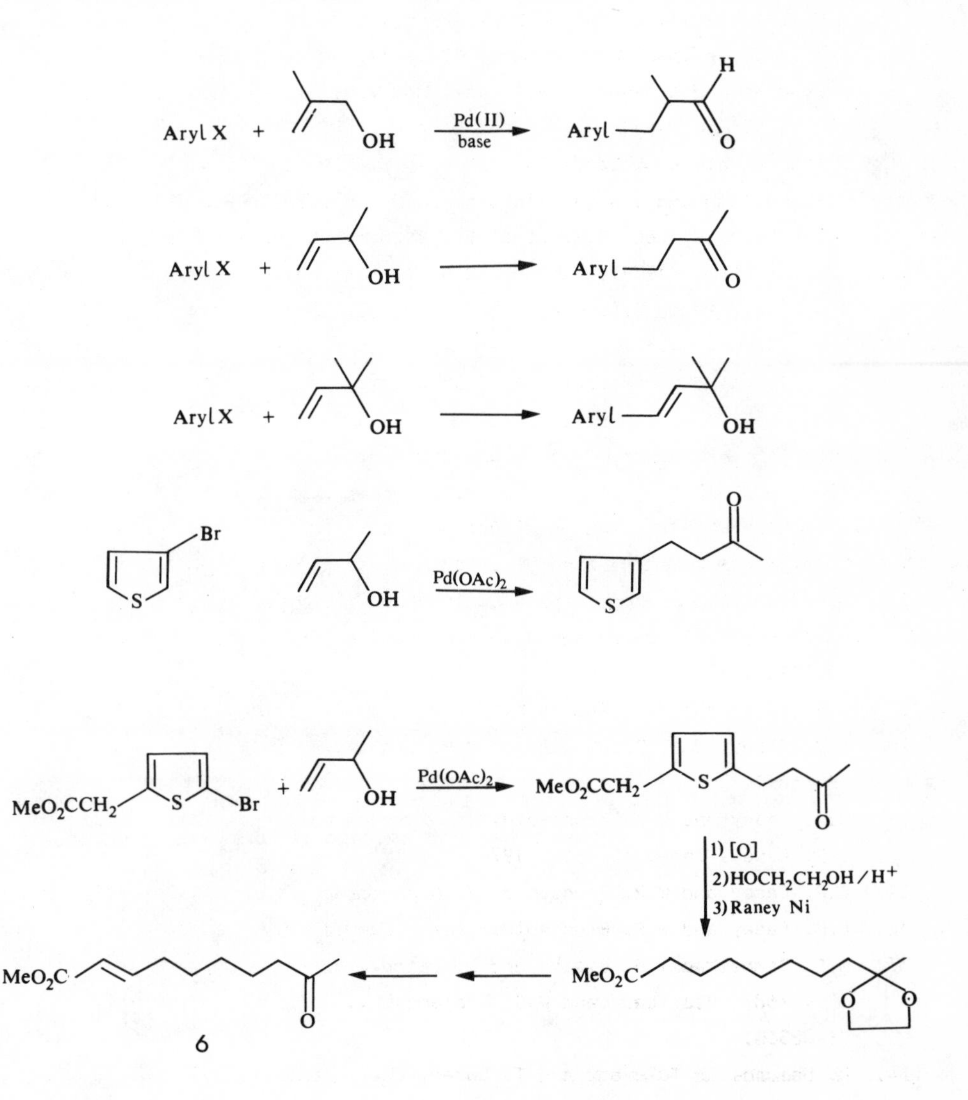

3-Allyl pyridines are prepared by the coupling of 3-bromopyridines with olefins[32] and allyl alcohols[33], for example, in the synthesis of nornicotine.[32]

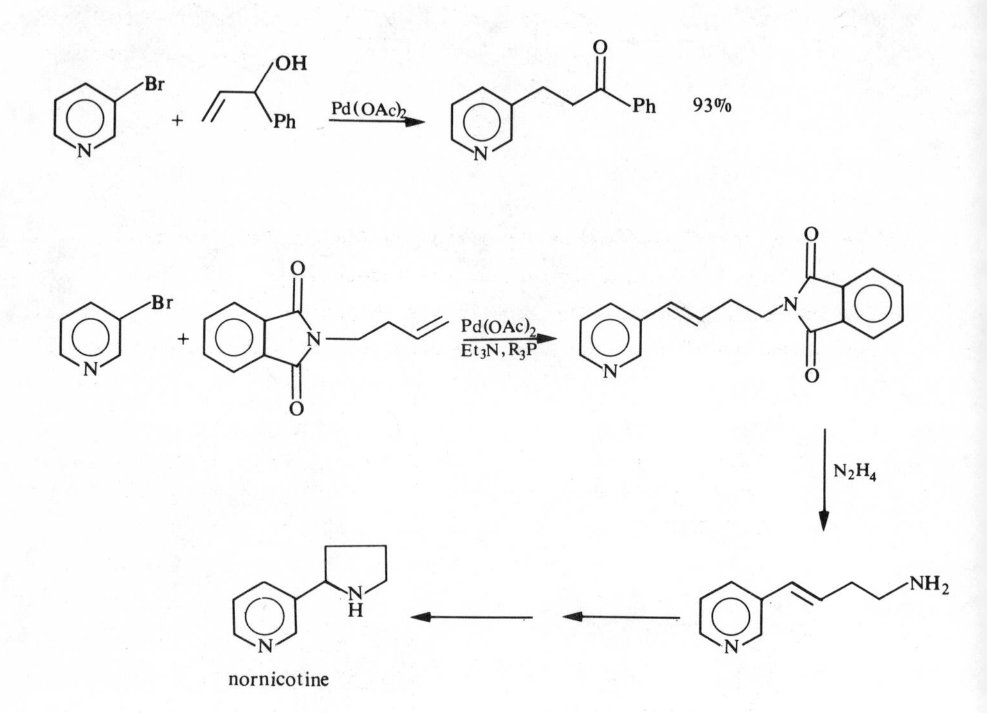

6.2.3 Coupling reactions involving C-H bond activation

(Aryl)Pd complexes also result from the oxidative addition of aryl-H bonds to Pd(II). In this case stoicheiometric amounts of Pd are needed unless Cu(II)/O_2 are present to reoxidise the Pd(0) formed in the reaction.[34]

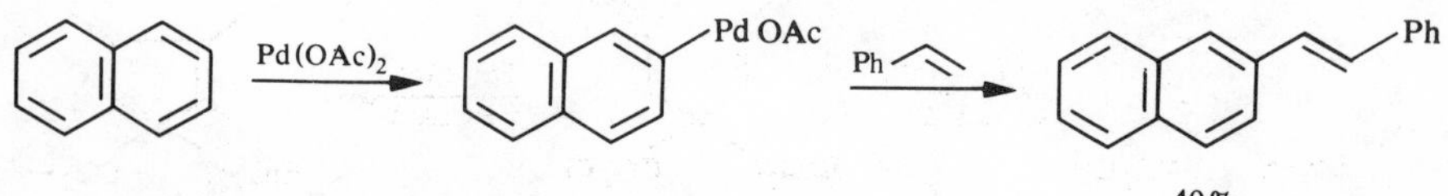

Coordinating functional groups can direct the position of metalation.[35]

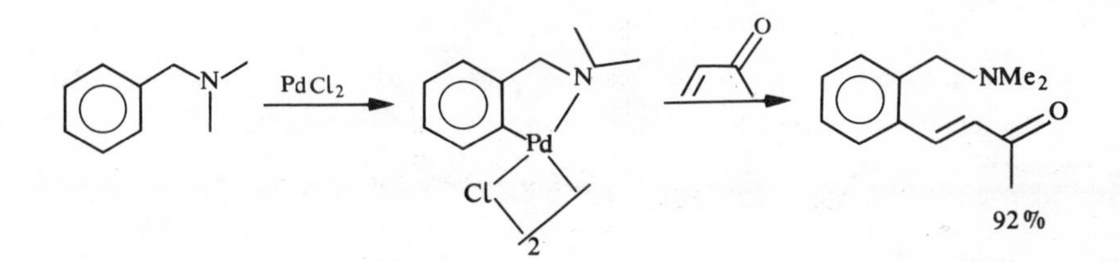

Indoles may be metallated in the 2-position by $(CH_3CN)_2PdCl_2$ in the presence of $AgBF_4$. Thus, treatment of the indole 7 (R = Et) with $(CH_3CN)_2PdCl_2/AgBF_4$ followed by $NaBH_4$ gives ibogamine.[36-38] The related compounds with R = H, O_2C^tBu may also be cyclised (see section 4.3).

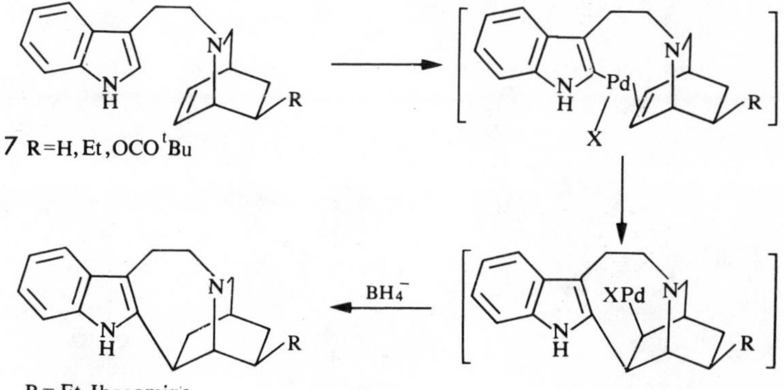

1,6-Dienes which do not bear hydrogen in the 4-position undergo coupling to cyclopentenes in good yields.[39] If a 4-hydrogen is present then isomerisation reactions occur. These reactions presumably occur through formation of an alkyl-[M] complex by addition of [M]H to an olefin followed by an intramolecular Heck reaction, the metal hydride catalyst being formed initially from the metal salt and chloroform.

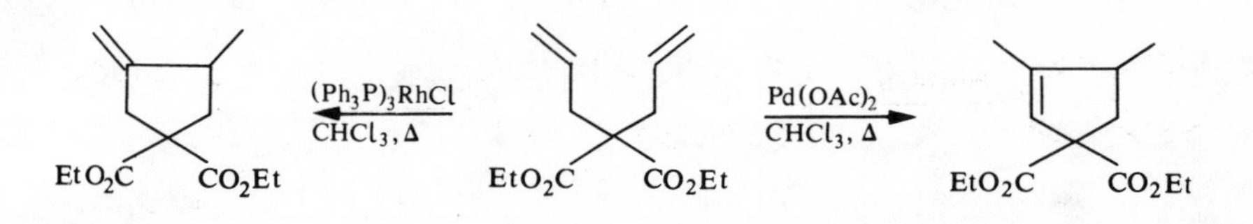

6.3 COUPLING REACTIONS INVOLVING η^3-ALLYL INTERMEDIATES

The generation of organometallic complexes of Ni, Pd and Fe containing an η^3-allyl ligand and another hydrocarbon ligand (alkyl, aryl, allyl, olefin, acetylene etc.) often leads to a coupling reaction with formation of a carbon-carbon bond.

$[(\eta^3\text{-Allyl})NiBr]_2$ complexes, which are readily available from allyl bromides and $Ni(COD)_2$ or $Ni(CO)_4$, react with a variety of alkyl halides to give substituted olefins. The mechanism is believed to proceed via addition of the alkyl halide to $(\eta^3\text{-allyl})NiBr$ to give 8 which then undergoes elimination of the two alkyl groups with carbon-carbon bond formation. Alkyl, aryl and vinyl halides undergo this reaction in high yields.[40]

The tolerance of substituents on the allyl ligand makes this reaction even more useful for synthesis.

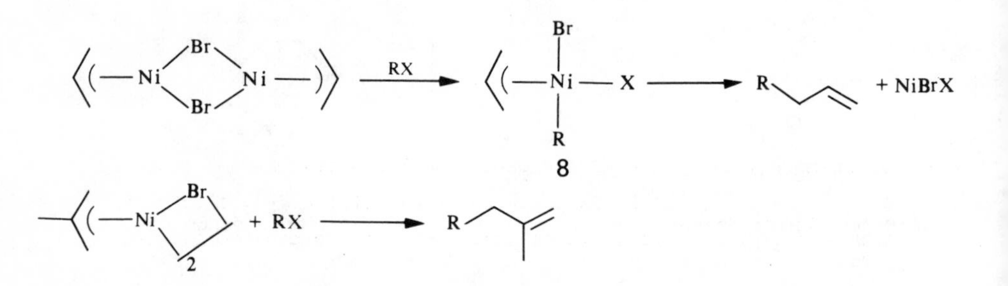

RX = MeI (90%); PhI (98%); vinyl bromide (70%); $PhCH_2Br$ (91%)

Isoprenylation reactions are possible using $[(1,1\text{-dimethylallyl})NiBr]_2$

as for example in the synthesis of α-santalene 9, epi-β-santalene 10,[40] and dictyolene 11.[41]

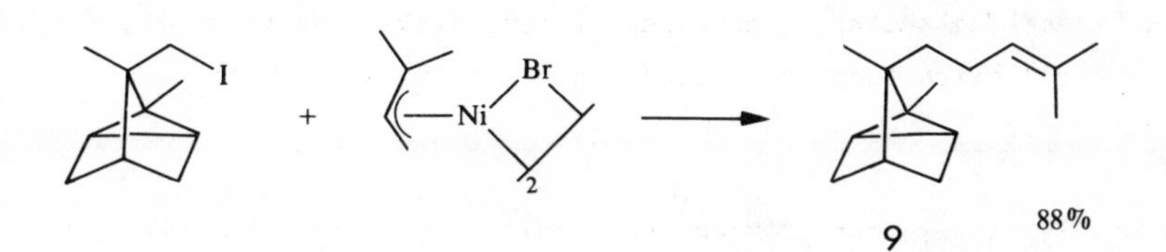

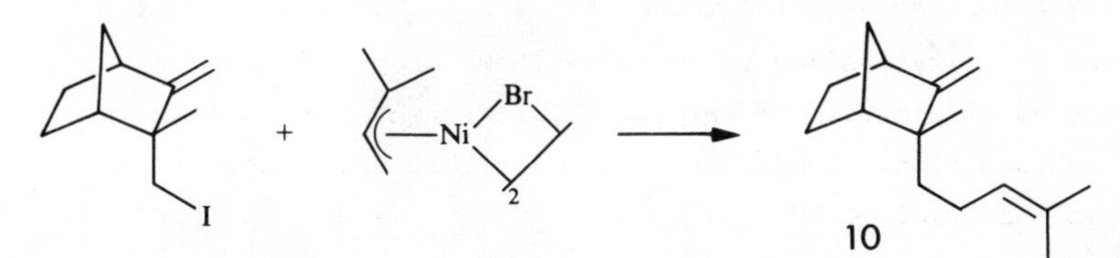

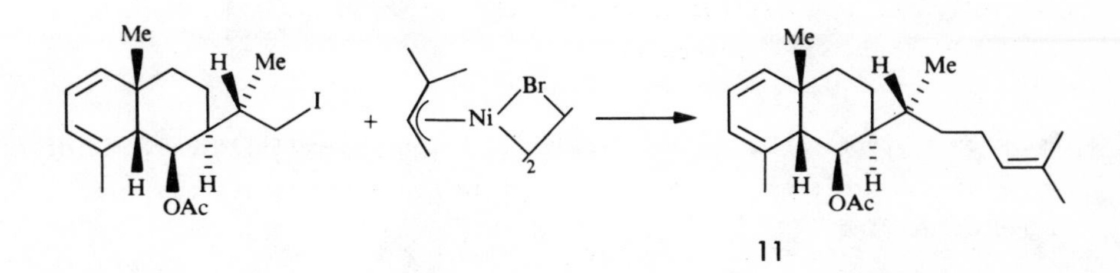

Groups other than alkyl (e.g, 2-CO_2Et, 2-OMe) can also be accommodated on the allyl ligand.[40,42] The use of [(2-methoxyallyl)NiBr]$_2$ allows the introduction of the acetonyl functional group into organic molecules.[42]

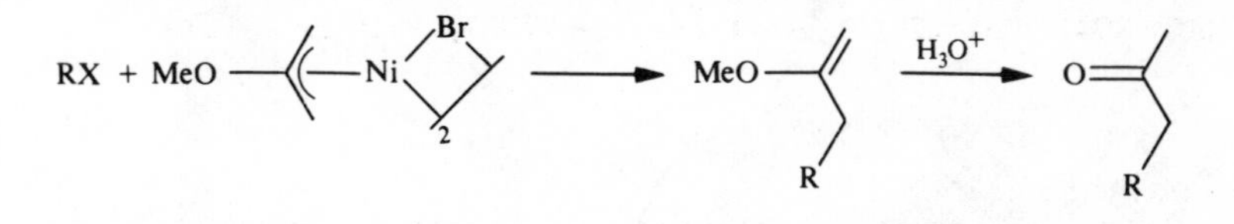

RX = PhI (73%); $PhCH_2Br$ (82%); PhCH = CHBr (cis 84%, trans 87%); etc.

A combination of the above allylation of aryl halides and the activation of olefins to nucleophilic attack by Pd (Chap.4) results in efficient syntheses of a number of heterocyclic compounds starting from suitably substituted aryl halides.[43] For example, 2-bromoanilines lead to the formation of indoles.

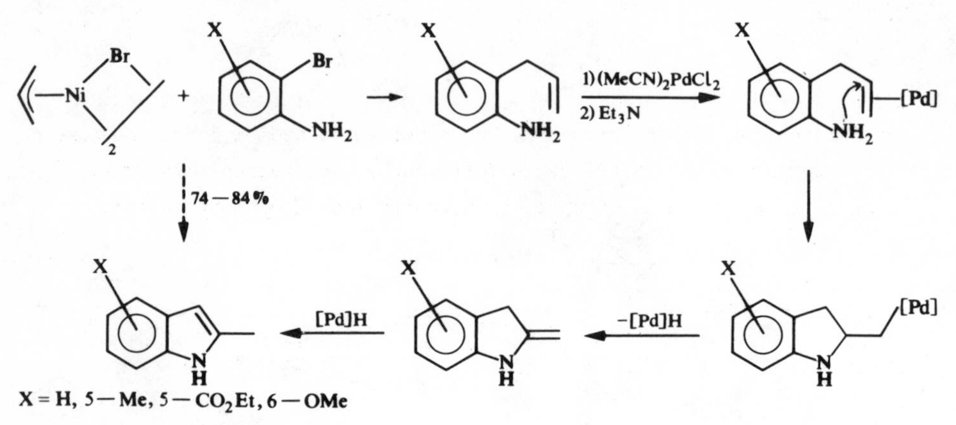

Reaction of 2-bromobenzoic esters with $[(\eta^3\text{-2-methoxyallyl})NiBr]_2$ yields 2-acetonyl benzoic esters which are readily converted to isocoumarins or dihydrocoumarins.[44]

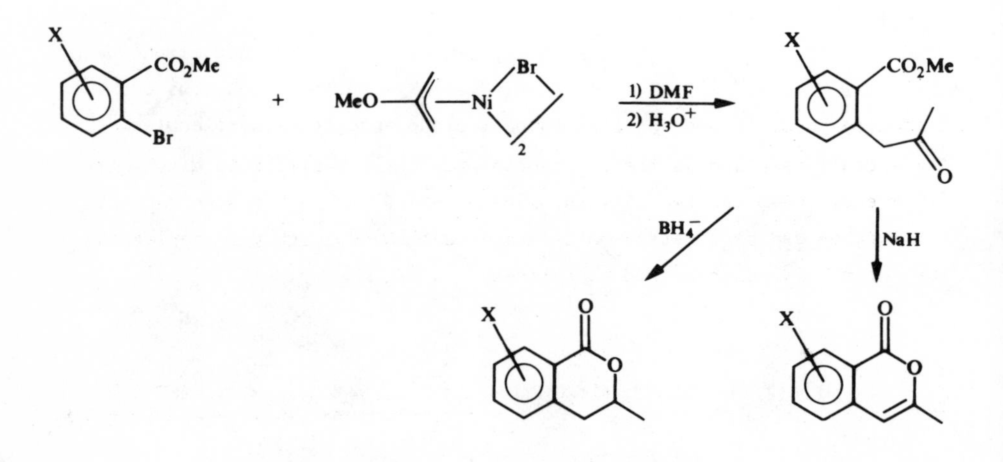

2-Bromopyridines are also reactive towards $[(\eta^3\text{-allyl})NiBr]_2$ complexes.

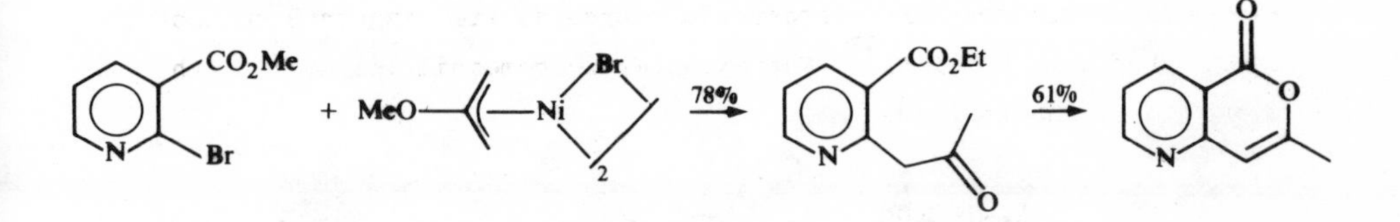

Isocoumarins may also be prepared from substituted sodium 2-bromobenzoates by reaction with allyl Ni reagents followed by cyclisation with $PdCl_2$.[44] In the same way isoquinolones may be prepared from 2-bromobenzamides.

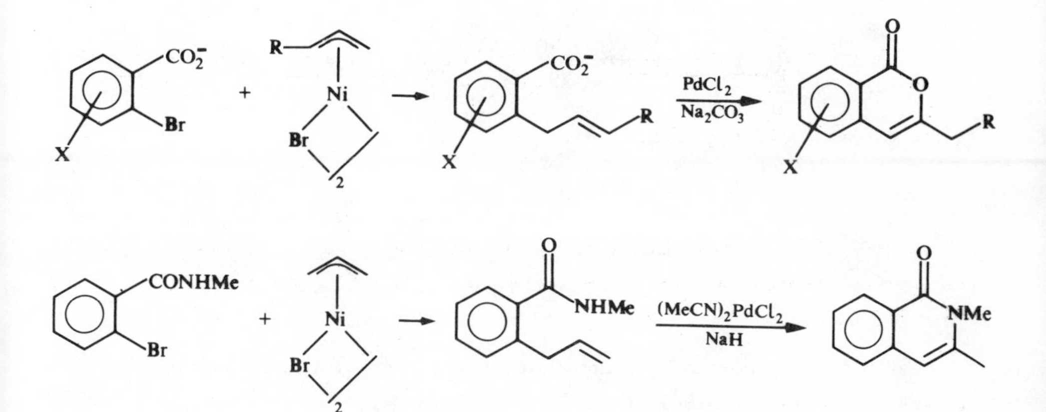

$[(\eta^3\text{-Allyl})NiX]_2$ complexes react with allyl halides to give 1,5-dienes.[45] The overall reaction is the coupling of two allyl bromides to give a 1,5-diene. When the two allyl bromide moieties are in the same molecule cyclisation can be effected on treatment with $Ni(CO)_4$ to give medium and large ring 1,5-dienes[46] and macrolides.[47]

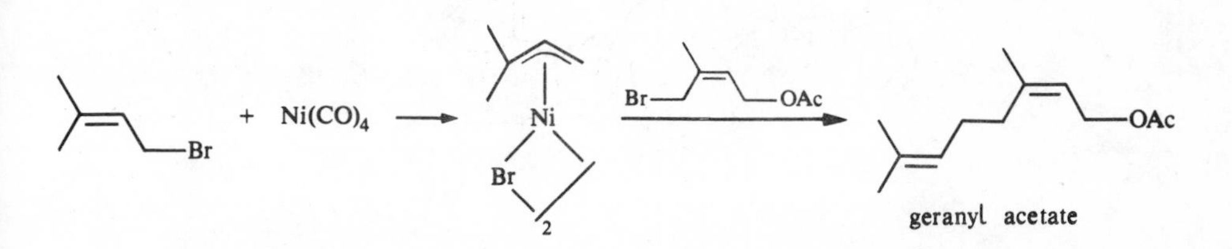

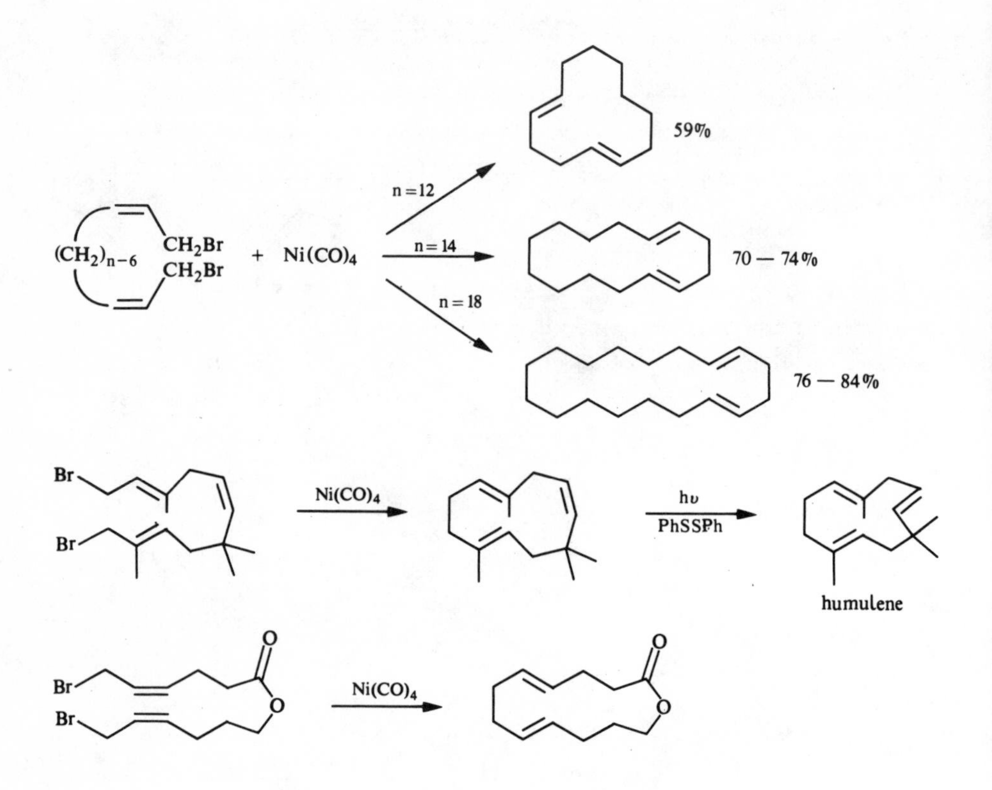

Homoallylic alcohols and 3-hydroxy ketones may be synthesised from the reaction of $[(allyl)NiBr]_2$ complexes and ketones.[40,42]

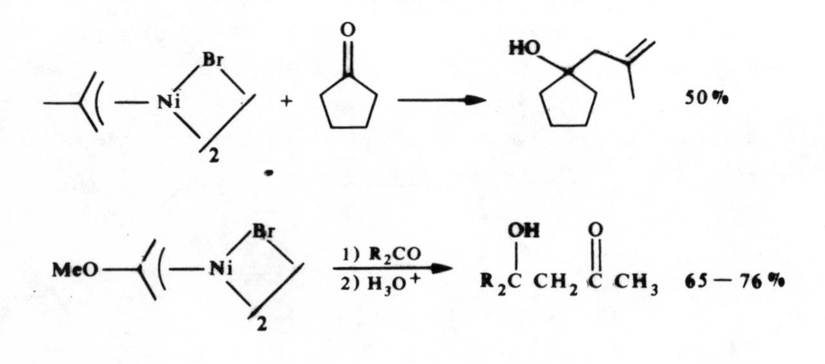

The reaction of [(2-carbethoxyallyl)NiBr]$_2$ with aldehydes and ketones results in the formation of α-methylene-γ-butyrolactones.[48]

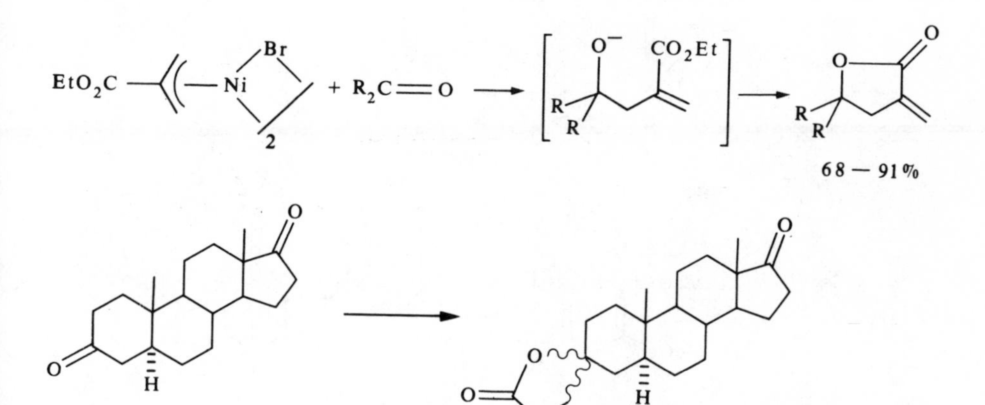

Intramolecular versions of the above reaction are also possible [49] and this allows the synthesis of sesquiterpene α-methylene-γ-lactones such as confertin 12 .[50]

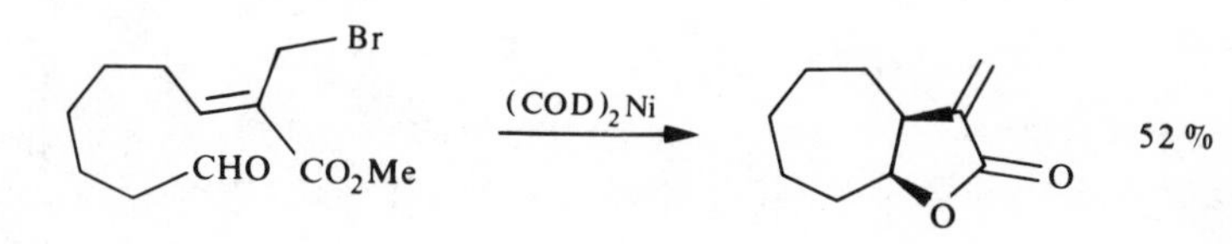

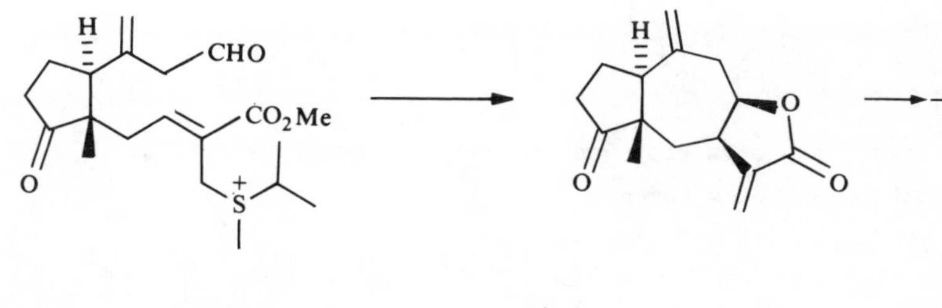

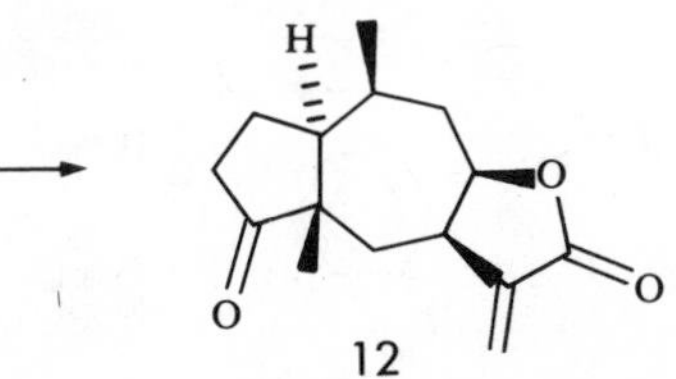

$[(\eta^3\text{-Allyl})NiBr]_2$ complexes can be used to convert quinones to allyl hydroquinones. The synthesis of coenzyme Q_1 **13** is possible by this method.[51]

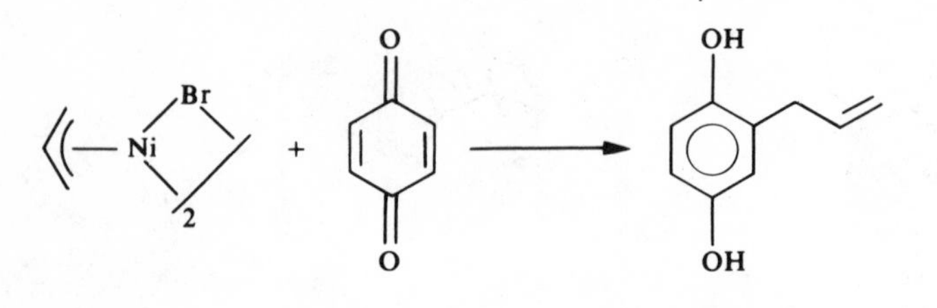

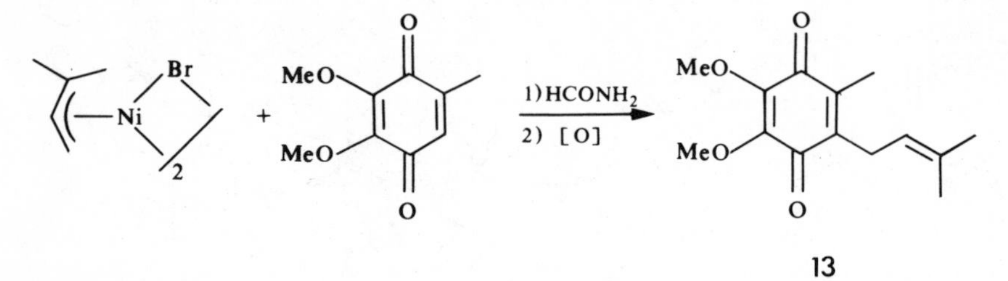

$Fe_2(CO)_9$ reacts with 1,1'-dibromoketones to generate (2-oxyallyl)$Fe(CO)_x$ complexes **14**, which react with olefins that possess substituents capable of stabilising an adjacent positive charge (e.g. vinyl, aryl, NR_2, OR etc.) to give cyclic products.[52]

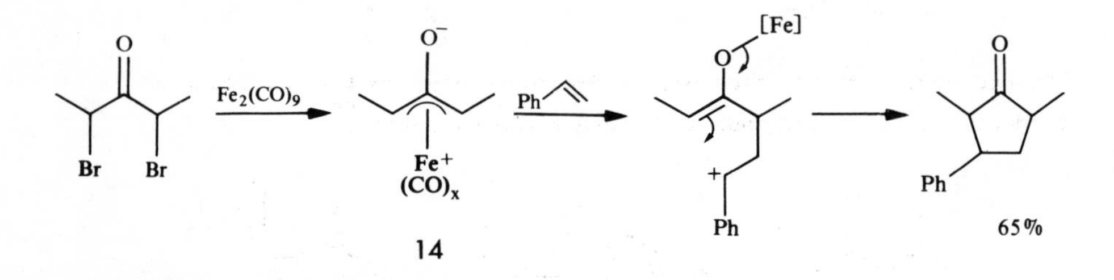

This reaction has been employed in a single step synthesis of (±)α-cuparenone **15** and camphor.

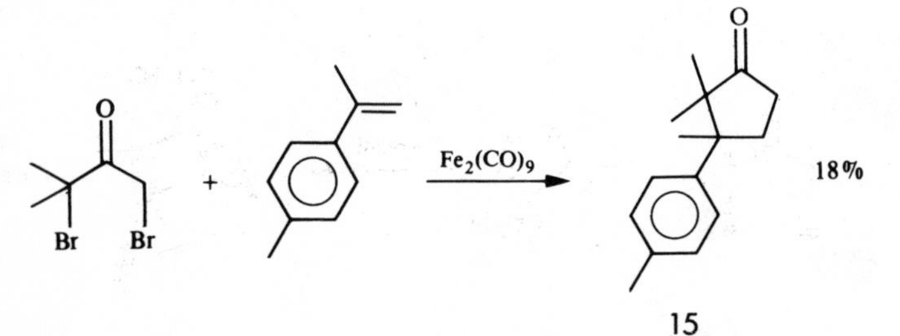

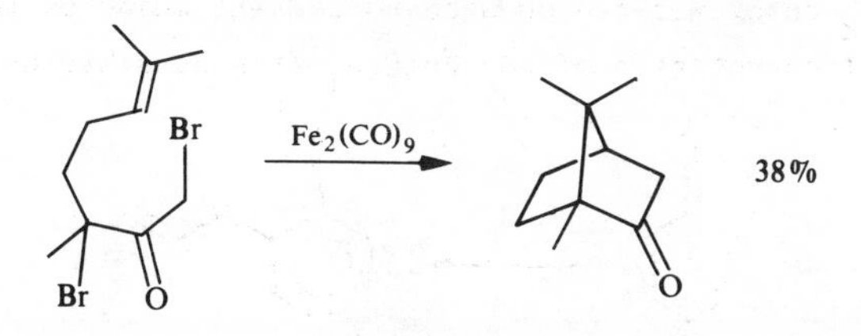

Cyclopentenones are formed with enamines[54] whereas 1,3-dienes give cyclohepten-5-ones.[55]

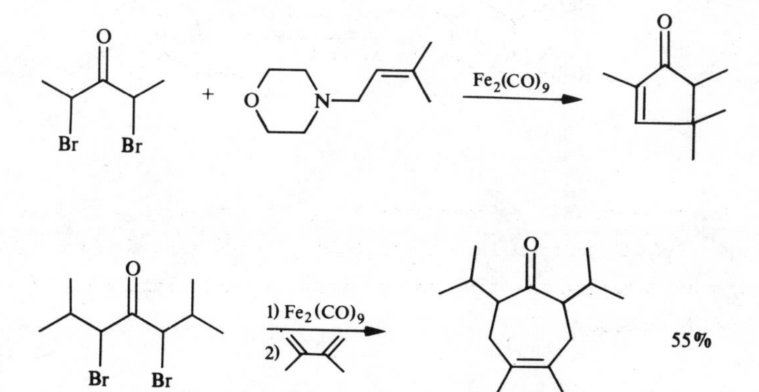

The cycloaddition reaction with 1,3-dienes leads to the formation of bridged bicyclic compounds with furans and pyrroles.[55]

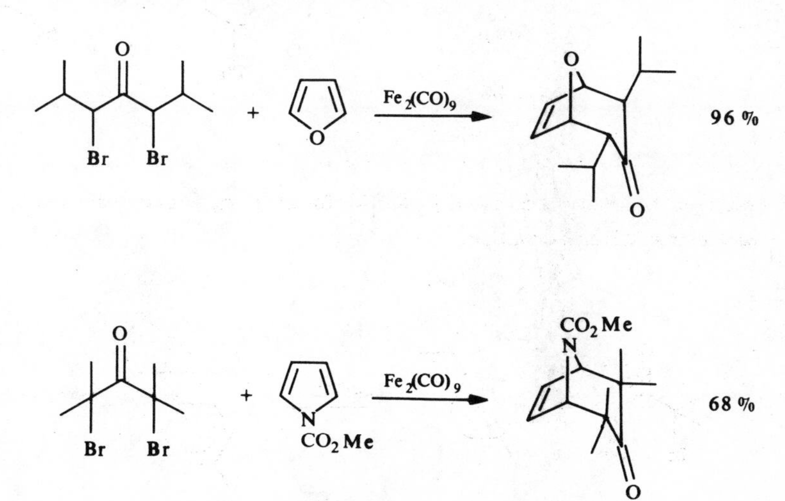

The reaction with furan has allowed the synthesis of several different classes of natural products as shown below for the synthesis of

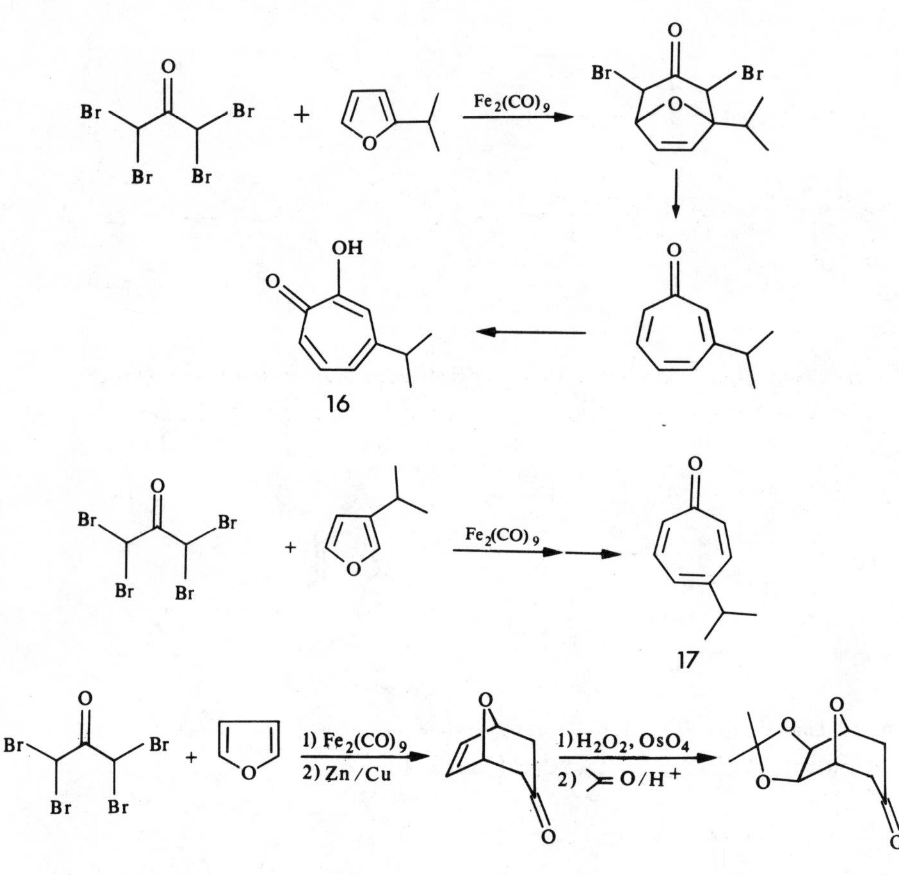

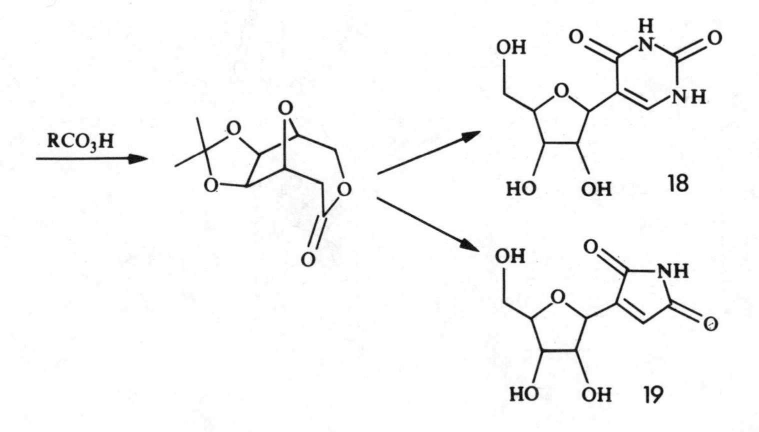

β-thujaplicin 16,[56] nezukone 17,[57] pseudouridin 18,[58] showdomycin 19 [59] and α-thujaplicin 20.[56]

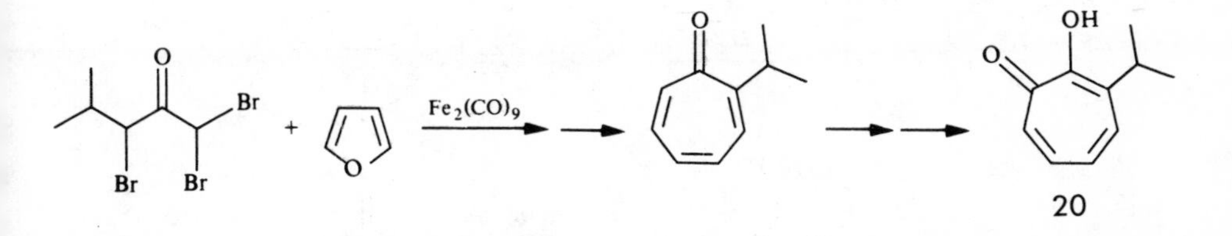

The cycloaddition reaction with pyrroles leads to the synthesis of the tropane alkaloids scopine 21, tropine 22 and ψ-tropine 23.[60]

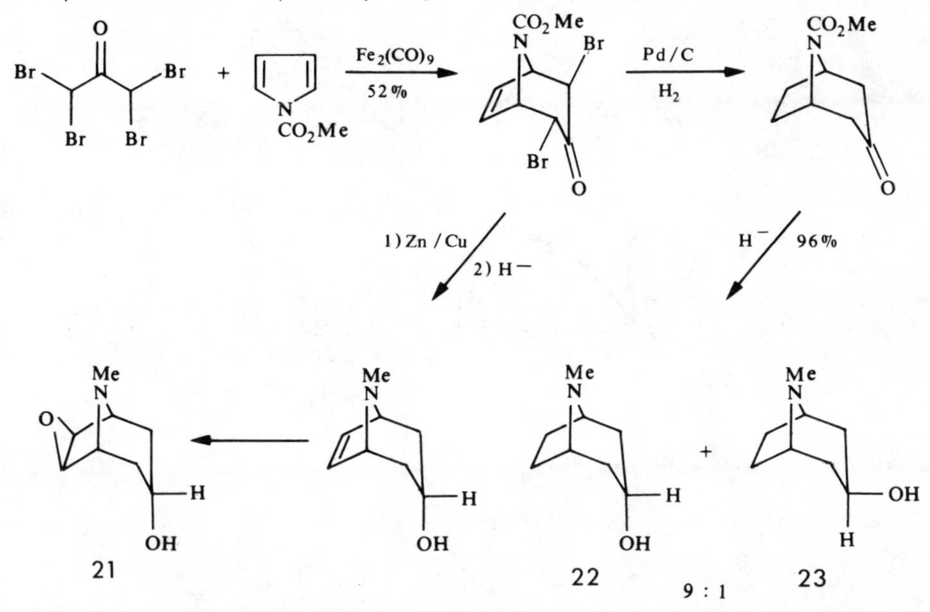

The reaction of 1,1'-dibromoketones with dimethylformamide in the presence of $Fe_2(CO)_9$ leads to the formation of furanones 24, which are useful precursors of the muscaine alkaloids.[61]

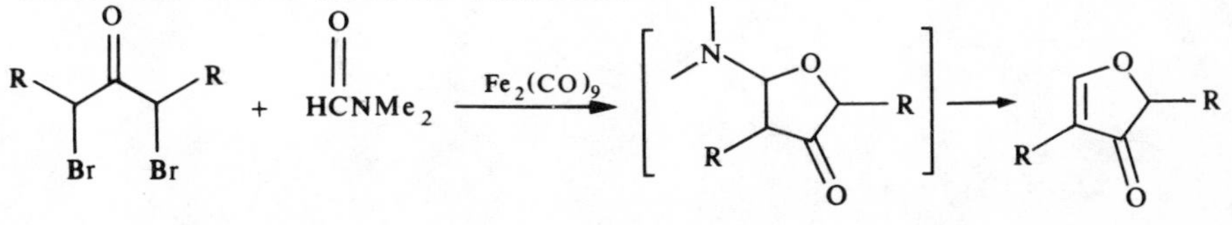

Allyl alcohols react with non-reducing Grignard reagents (i.e. those that do not bear a labile β-hydrogen) in the presence of L_2NiCl_2 (L = phosphine) catalysts to give products that result from substitution of the hydroxyl function (the Felkin reaction).[62]

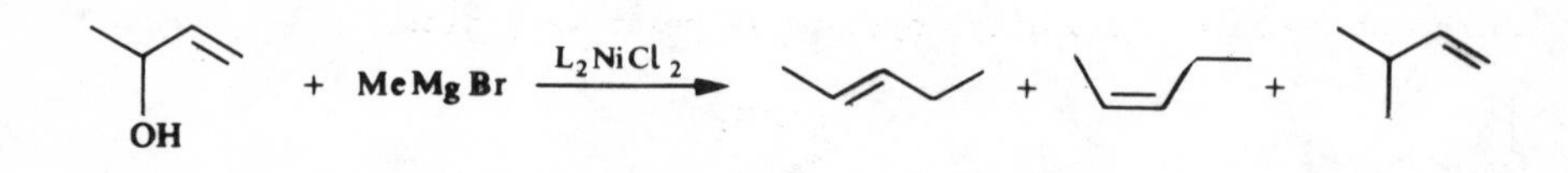

The mechanism of the Felkin reaction is shown below and involves the formation of an (η^3-allyl)Ni species 25. The coupling step occurs via formation of a Ni(alkyl) complex followed by migration of the alkyl group from the metal to the allyl ligand.

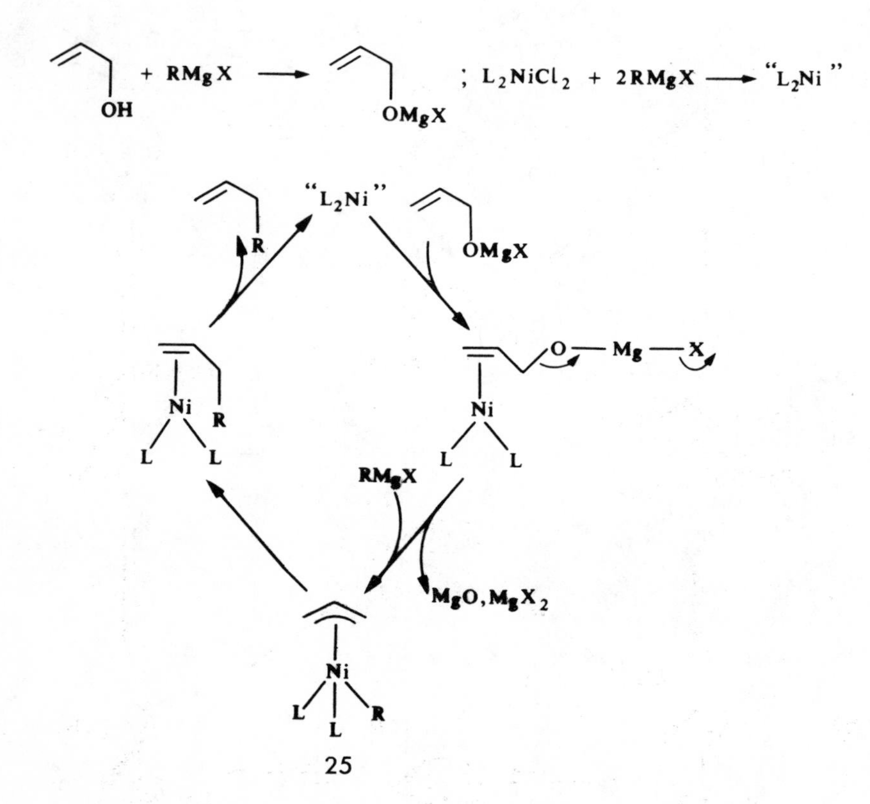

The above mechanism suggests that inversion of configuration occurs in the Felkin reaction. This has been verified by the reaction of (S)(+)-but-1-en-3-ol 26 with PhMgBr in the presence of $(Ph_3P)_2NiCl_2$ which results in the formation of (R)(-)-3-phenylbut-1-ene 27 with an optical yield of 23%.[63] The accompanying racemisation is due to the η^3-η^1-η^3 equilibrium shown, which for monoalkylated allyl systems is relatively slow.

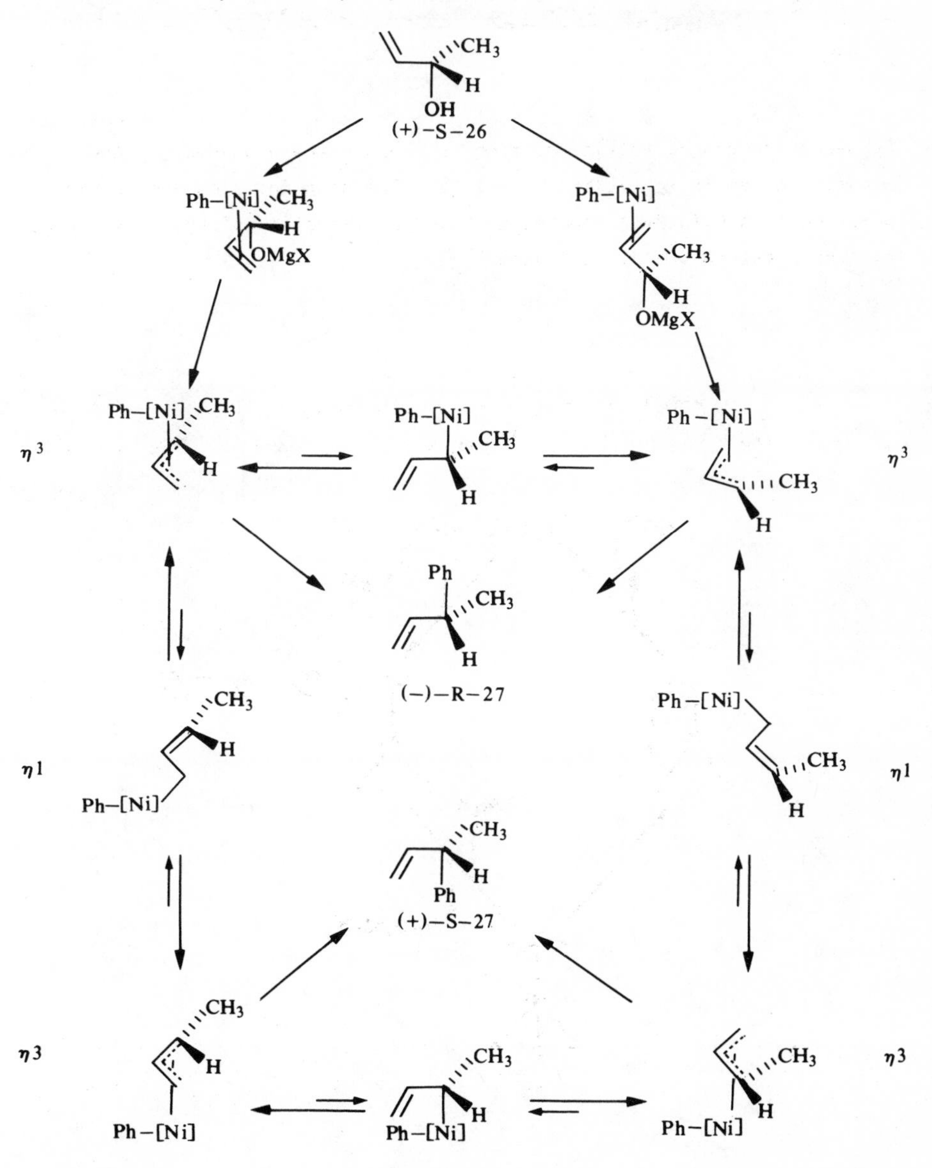

The Felkin reaction is potentially very useful for synthesis starting from unsymmetrical allyl alcohols because the L_2Ni species binds preferentially

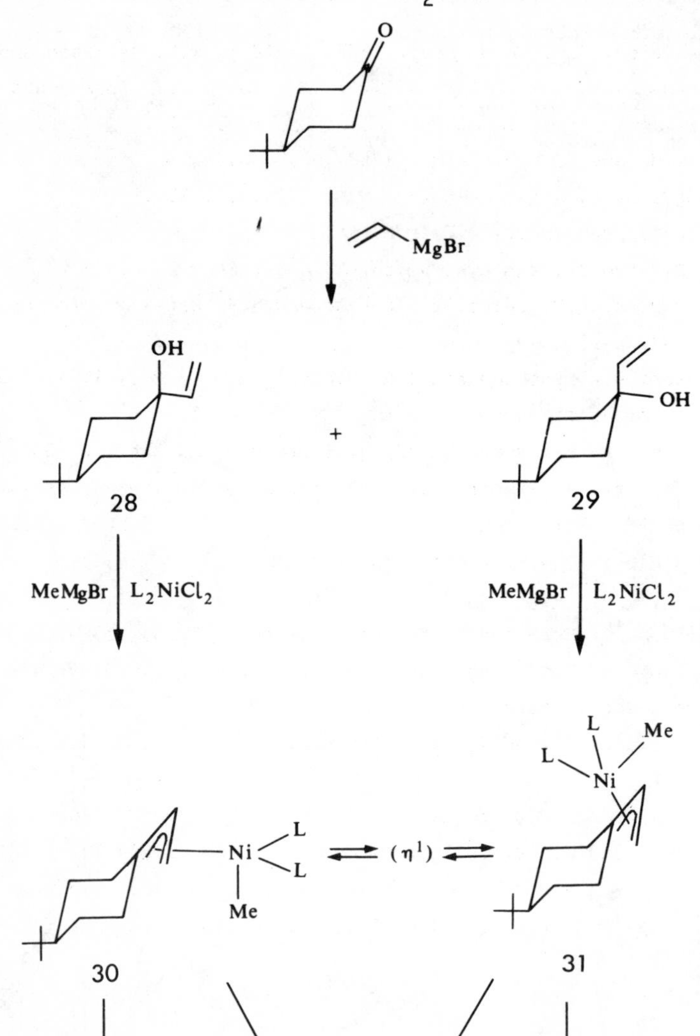

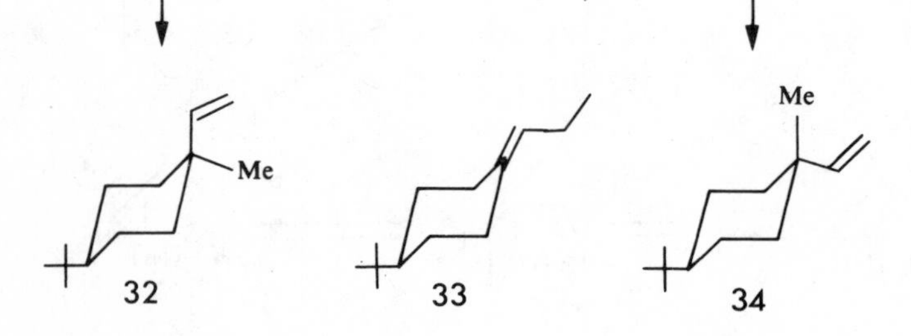

to the most accessible face of the allyl ligand in the intermediate 25. Also the new carbon-carbon bond is formed to the most substituted end of the allyl ligand to give the most stable (olefin)Ni complex (i.e. the transition state for C-C bond formation resembles the products). Reaction of either of the alcohols 28 or 29 with MeMgBr/$(Ph_3P)_2NiCl_2$ produces the olefins 32, 33 and 34 in the ratio 77:19:4. In this case the interconversion of the two η^3-allyl complexes 30 and 31 via the η^1-complex is very rapid because disubstitution facilitates formation of the η^1-complex. It is reasonable that the large L_2NiMe group will prefer the less hindered pseudoequatorial position in 30 over the pseudoaxial position in 31. Methyl migration from the Ni to the allyl ligand thus explains the stereoselective preference of this reaction for the formation of 32 with the Me group equatorial.[64]

This stereoselectivity in the formation of quaternary centres has been employed in a synthesis of some diterpenes from manool.[64]

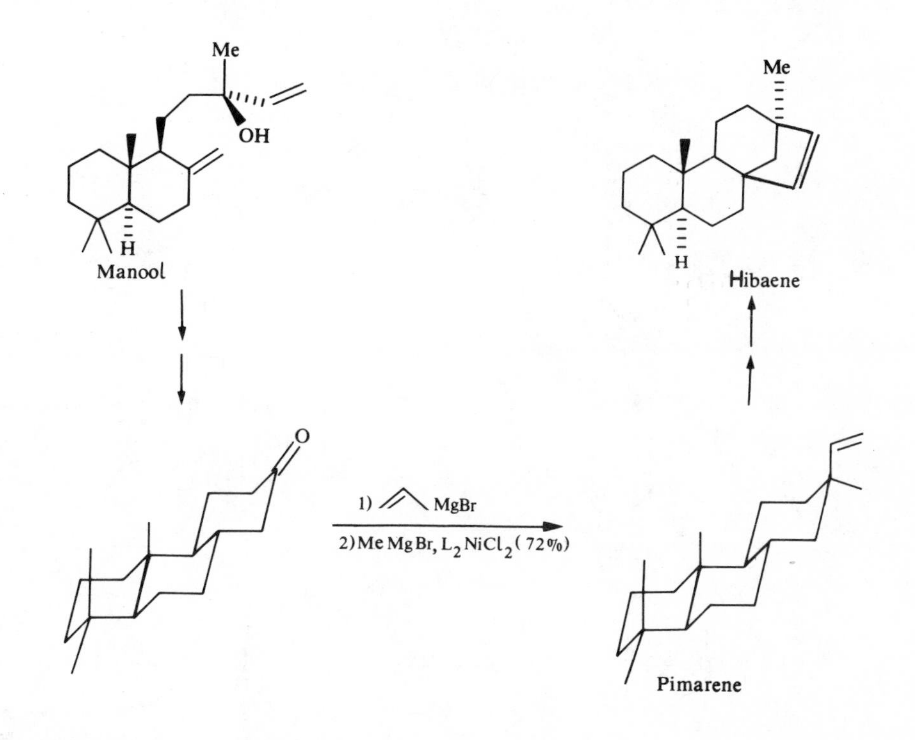

The use of optically active diphosphines as ligands for Ni in the Felkin reaction leads to the formation of optically active olefins.[65,66]

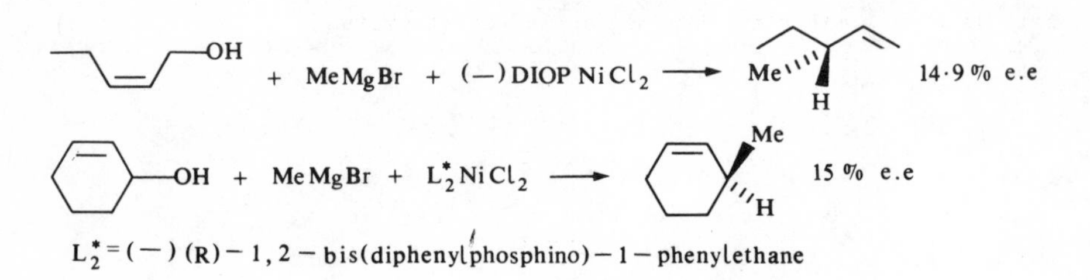

6.4 COUPLING REACTIONS INVOLVING OLEFINS AND ACETYLENES

The coupling of olefins to give dimers and polymers is a very important synthetic process. Olefins may be stereospecifically polymerised by Ziegler-Natta catalysts. The mechanism of this polymerisation is believed to involve transition metal carbene complexes as intermediates formed by an α-elimination mechanism.[67]

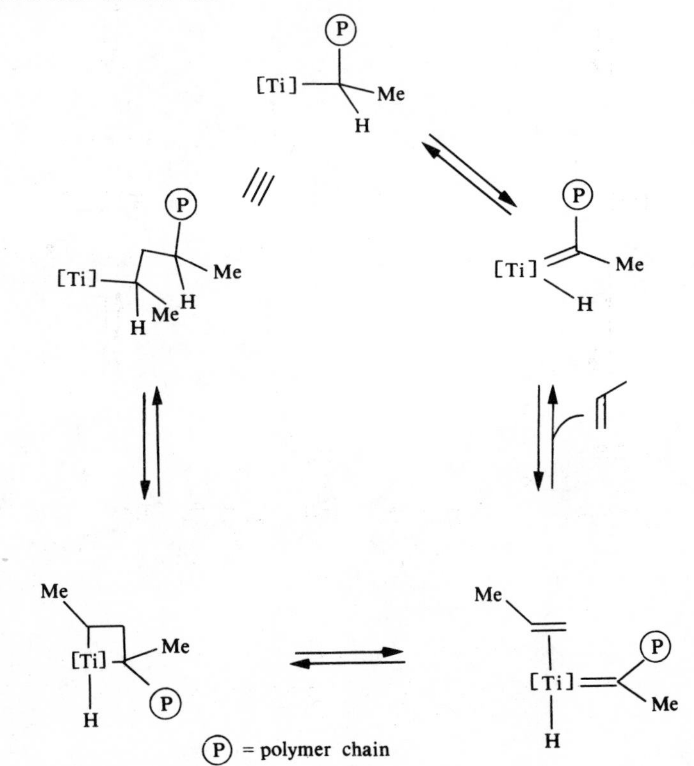

Carbene intermediates have also been invoked for the olefin metathesis reaction, which is catalysed by a variety of transition metal systems, many of them similar to the Ziegler-Natta catalysts.[68]

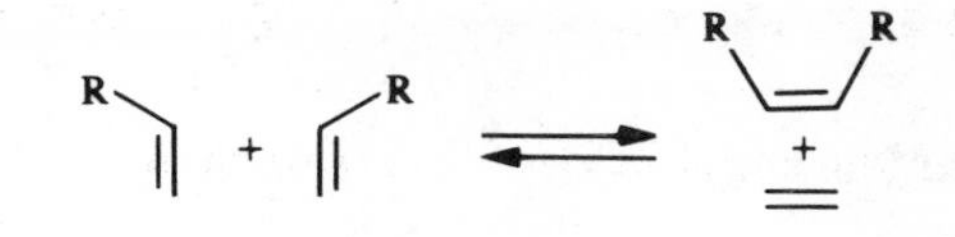

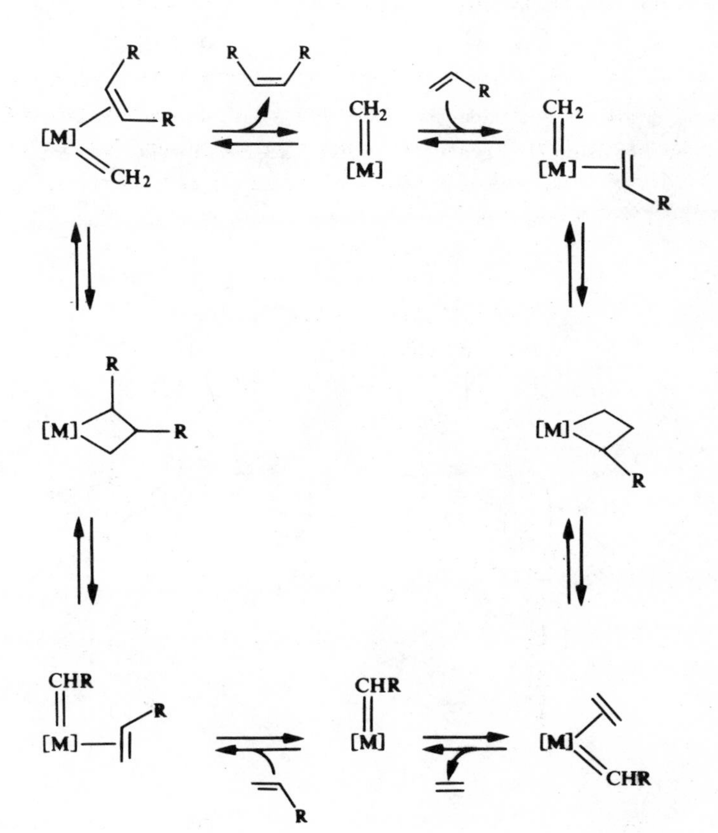

The main problem with utilising polymerisation and metathesis reactions for synthesis is that the initial products are themselves susceptible to the same reactions as the starting materials. However these reactions have found synthetic uses where some control of the products is possible. For example, the C_{16} musk compound 35 may be prepared by the metathesis of cyclo-octene.[69]

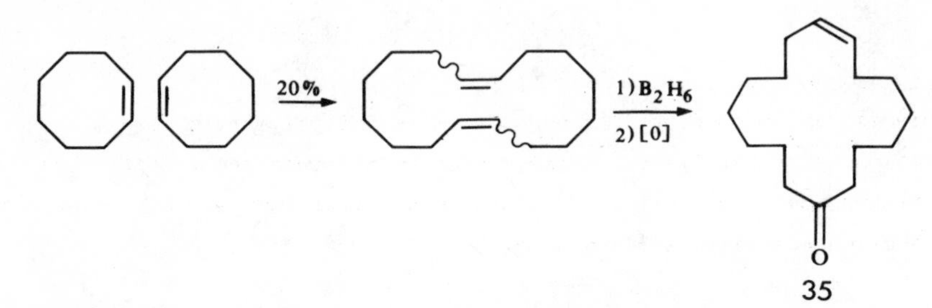

Norbornene and ethylene undergo a coupling reaction in the presence of a phosphine-containing Ni catalyst to give exo-2-vinyl norbornane.

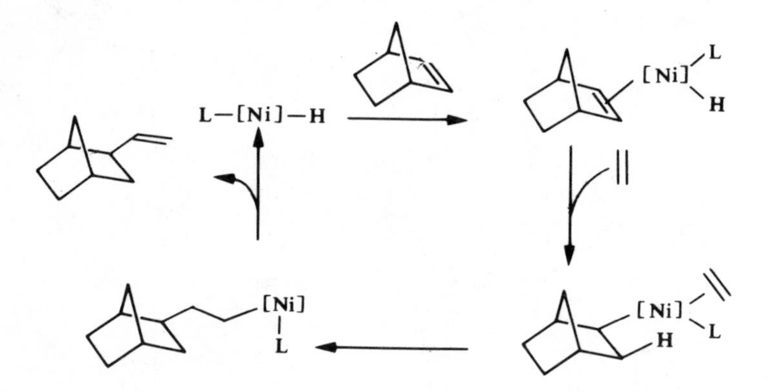

In the presence of chiral phosphines an asymmetric synthesis of 2-vinyl norbornane via this method is observed with a high optical yield of 80.6%.[70]

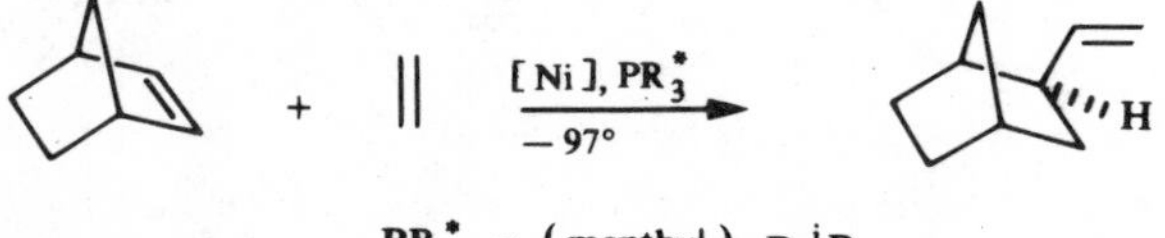

In a similar fashion norbornadiene is converted to *exo*-5-vinyl norbornene with an optical yield of 77.5%.[70]

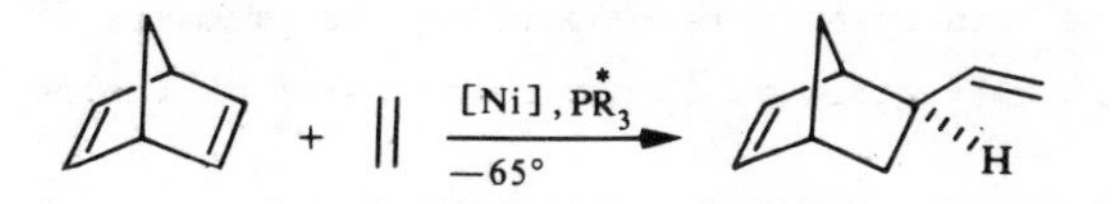

1,3-Dienes can be coupled to dimers, trimers and polymers by many transition metal catalysts.[71] The mechanism involves the addition of a metal hydride to a diene to generate an η^3-allyl complex that undergoes the coupling reaction as shown below.

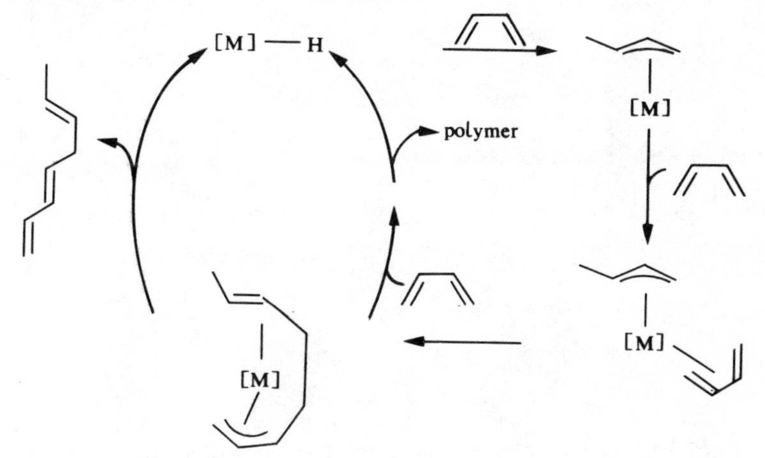

For example, butadiene can be converted to 1,3,6-octatriene with $(Ph_3P)_2NiBr_2/NaBH_4$.[72]

Isoprene can be reductively dimerised by Pd catalysts in the presence of formic acid and Et_3N to give 79% head to tail dimer, which can be used for the synthesis of monoterpenes.[73]

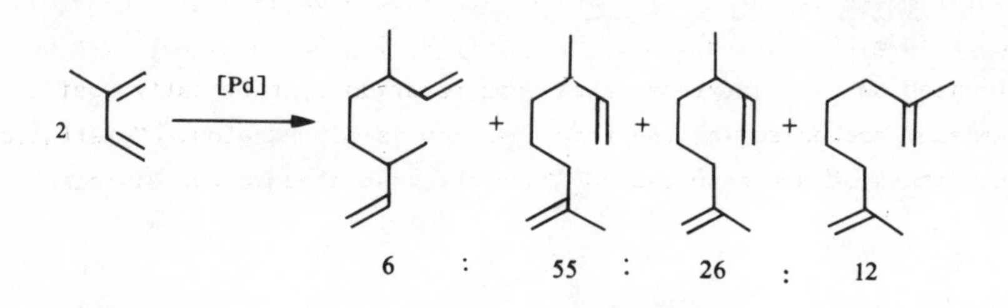

The reaction of three molecules of butadiene with Ni(O) complexes produces the bis(allyl)Ni complex 36, which can be converted to a variety of organic compounds.[71]

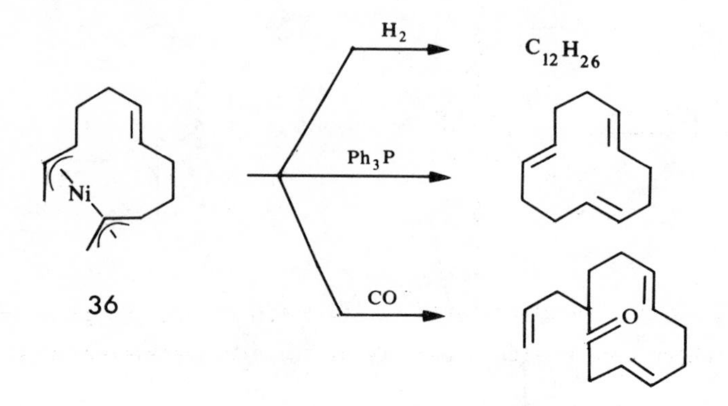

Treatment of 36 first with allene and then ^tBuNC produces the imine 37 which is readily convertible to (±)muscone.[74]

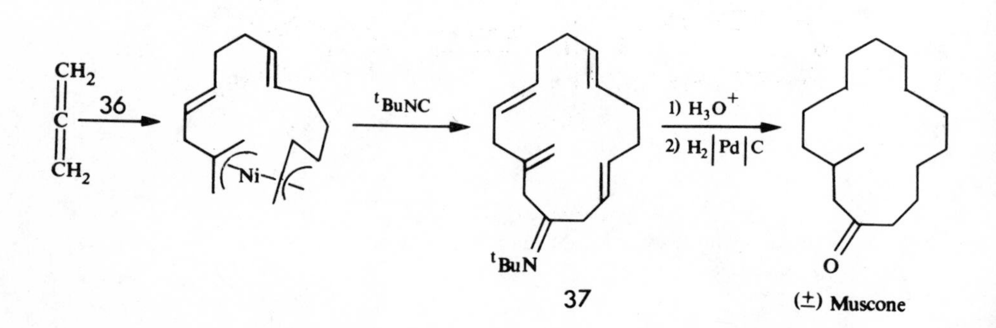

The trimerisation of acetylenes to arenes can be effected by many transition metal catalysts.[75] The fact that a variety of functional groups (e.g. Ph, $CH{=}CH_2$, C=CR, CO_2H, CO_2R, OH, OR, NR_2, R_3Si) can be tolerated on the acetylene makes this reaction synthetically useful. A general mechanism for the trimerisation is given below. Metallacyclopentadienes 38 are involved and have been isolated in certain cases.

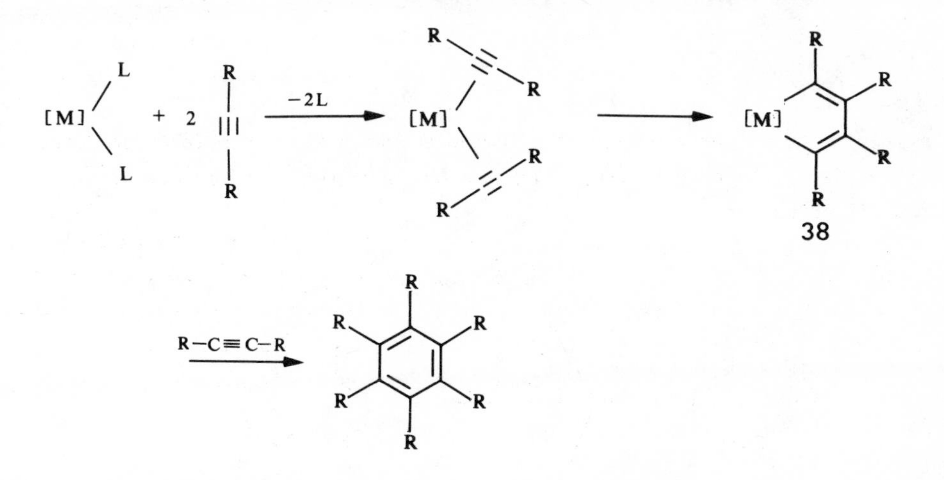

Diacetylenes react with Wilkinson's catalyst $(Ph_3P)_3RhCl$ to give compounds of type 38, which can undergo a variety of further transformations.[76]

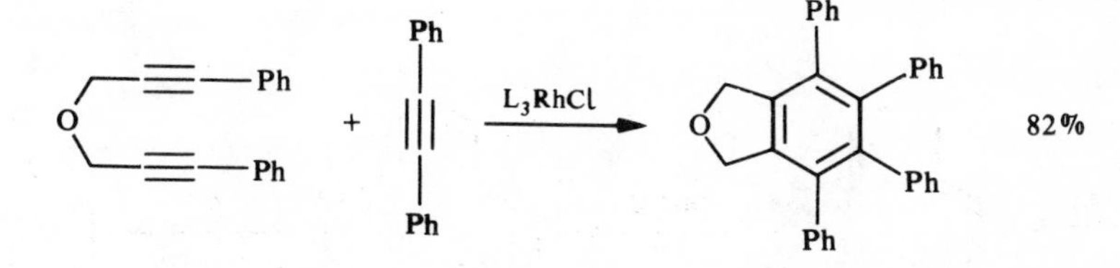

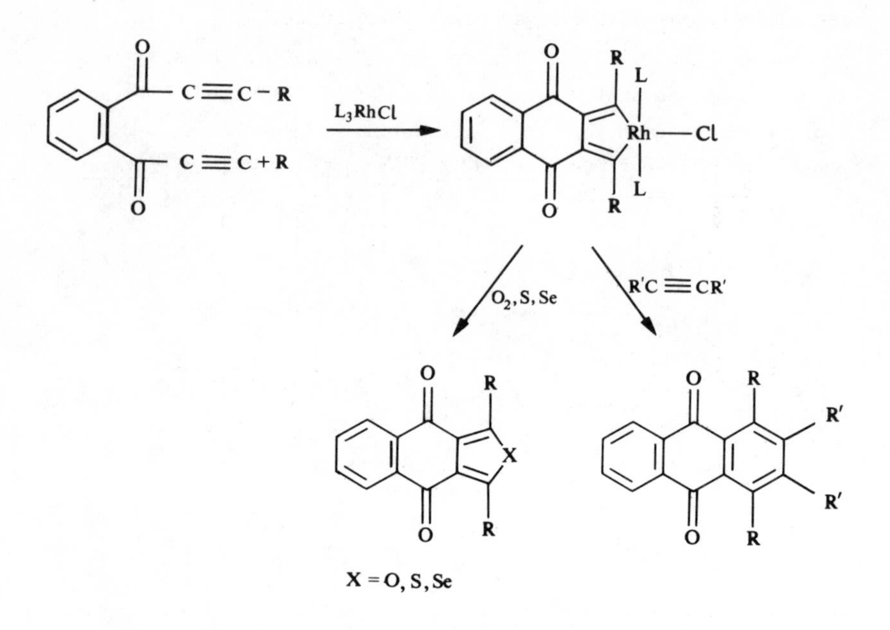
O
C≡C−R
C≡C+R
O
L₃RhCl
O
R
L
Rh—Cl
L
O
R
O₂, S, Se
R'C≡CR'
O
R
X
O
R
X = O, S, Se
O
R
R'
R'
O
R

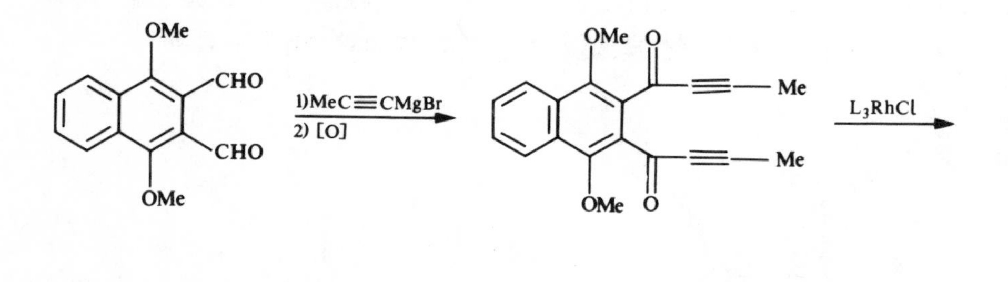
OMe
CHO
CHO
OMe
1)MeC≡CMgBr
2) [O]
OMe O
Me
Me
OMe O
L₃RhCl

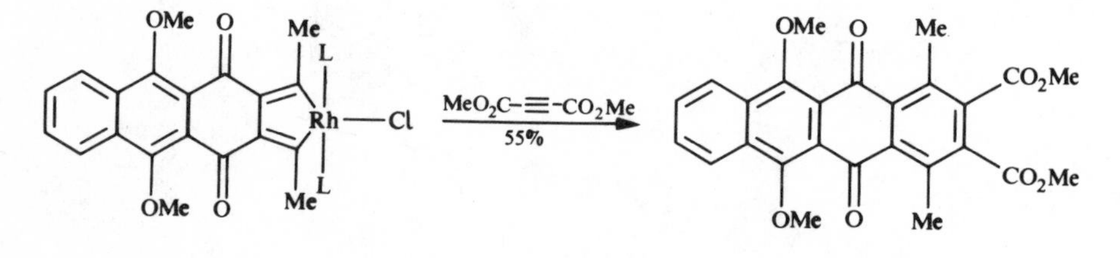
OMe O
Me
L
Rh—Cl
L
Me
OMe O
MeO₂C-≡-CO₂Me
55%
OMe O
Me
CO₂Me
CO₂Me
OMe O
Me

Polysubstituted naphthalenes may be obtained from the cyclobutene 39 and an acetylene in the presence of $(Ph_3P)_3RhCl$.[77]

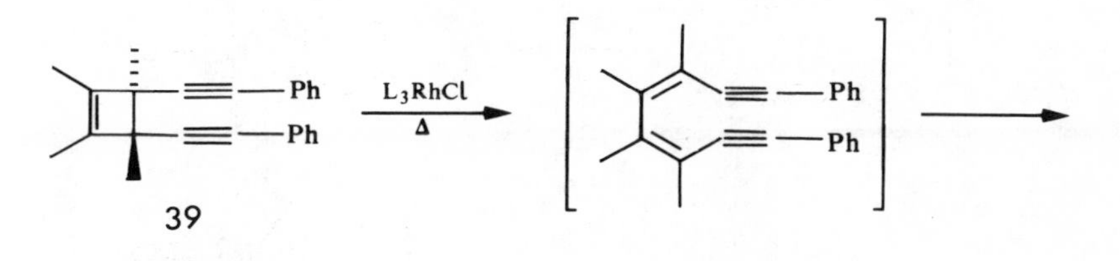

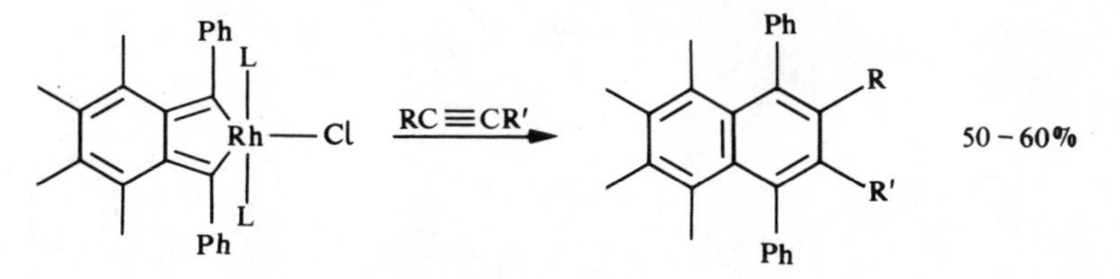

$CpCo(CO)_2$ is a very efficient catalyst for the formation of arenes by coupling diacetylenes with acetylenes.[75]

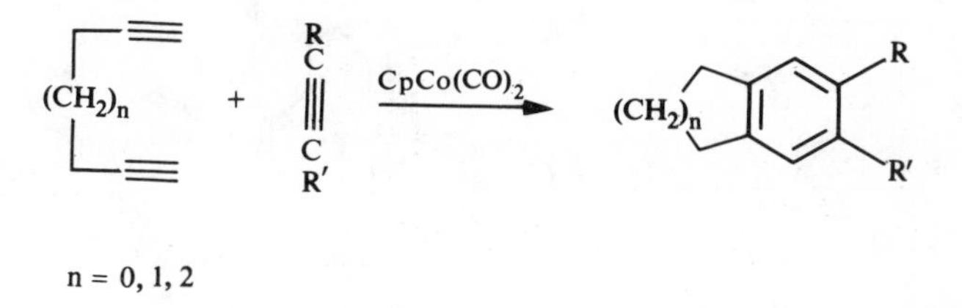

Bis(trimethylsilyl) acetylene is a very useful substrate for this reaction because the Me_3Si groups can be readily converted to other functional groups after formation of the desired arene.[78,79]

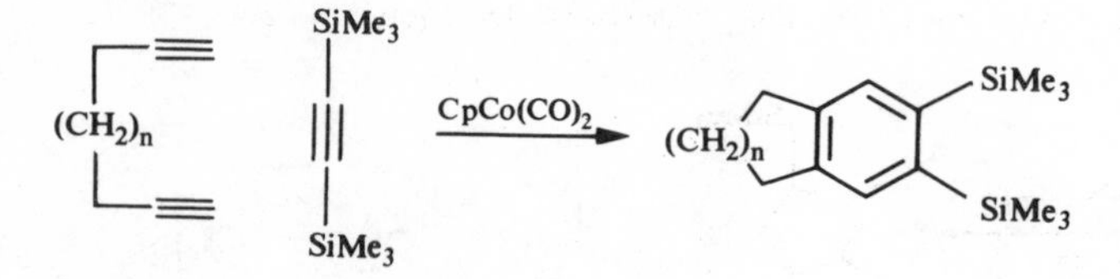

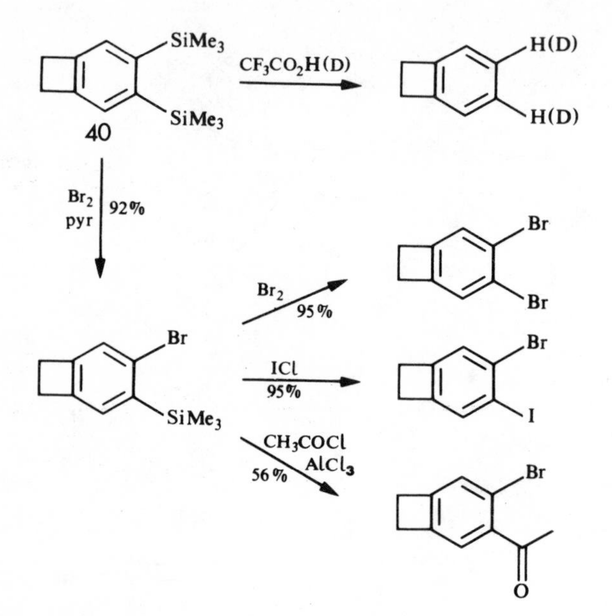

Benzocyclobutanes such as 40 are a useful source of o-xylylenes which can be trapped intramolecularly by olefin.[78,80]

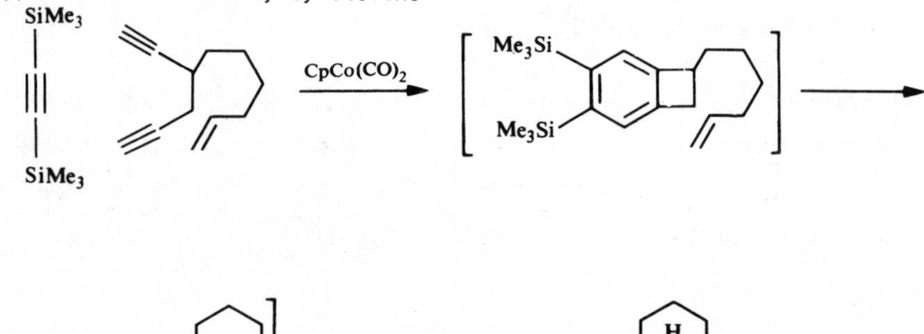

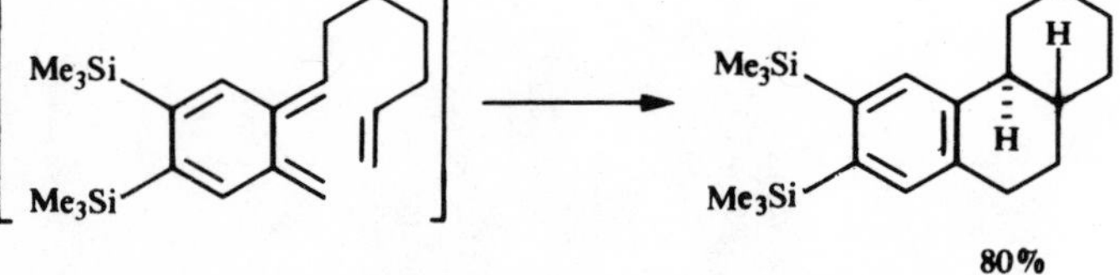

This cyclisation reaction has been employed in the synthesis of (±)estrone **41** with an overall yield of 24% from methylcyclopentenone.[78,81]

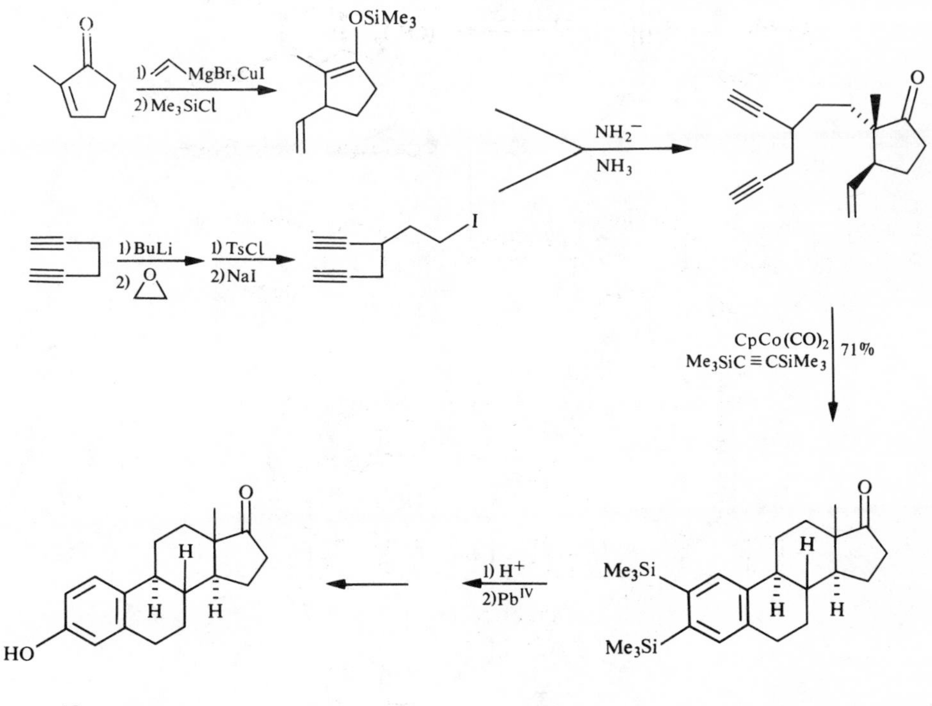

41 (±) Estrone

Ni complexes may also be used for the trimerisation of acetylenes. $(Ph_3P)_2Ni(CO)_2$ is used to prepare ^{14}C labelled DL-phenylalanine from labelled acetylene and **42**.[82]

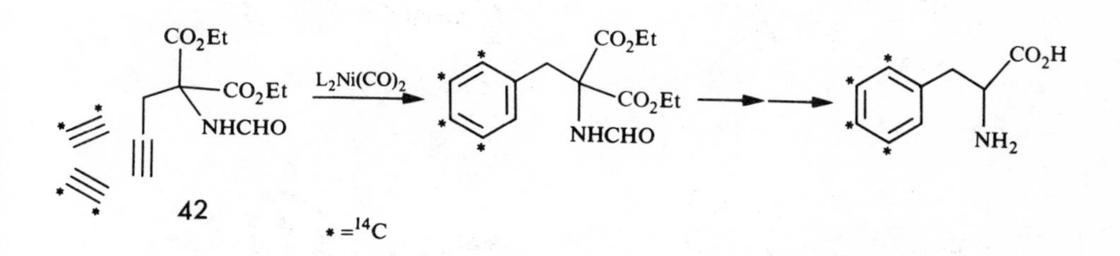

In certain cases the metallacyclopentadiene intermediate may be trapped by olefin to give cyclohexa-1,3-dienes.[83]

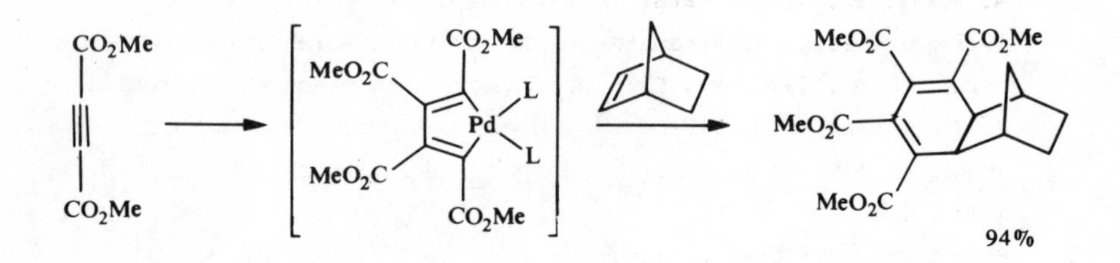

$CpCo(CO)_2$ catalyses the coupling of acetylenes with nitriles to generate pyridines.[84] Monosubstituted and unsymmetrical acetylenes give mixtures of isomeric products.

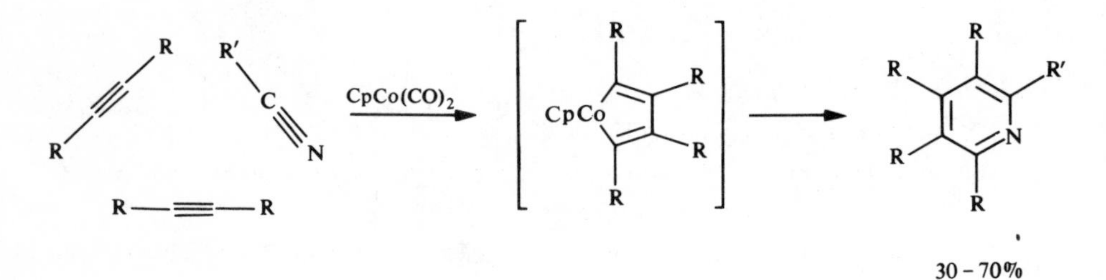

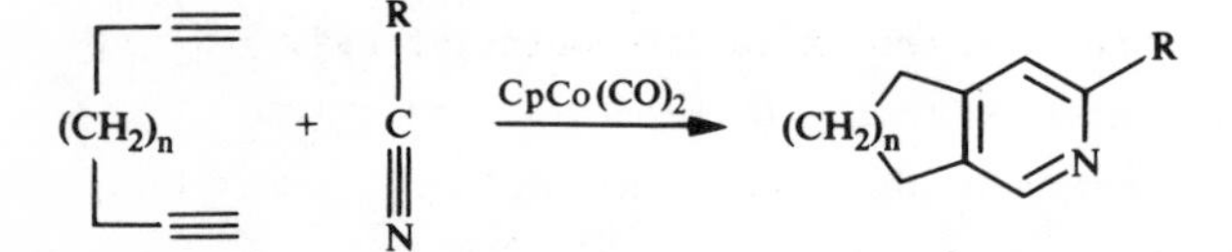

n = 1, 2, 3

2-Pyridones may be synthesised from acetylenes and isocyanates in this way.[85]

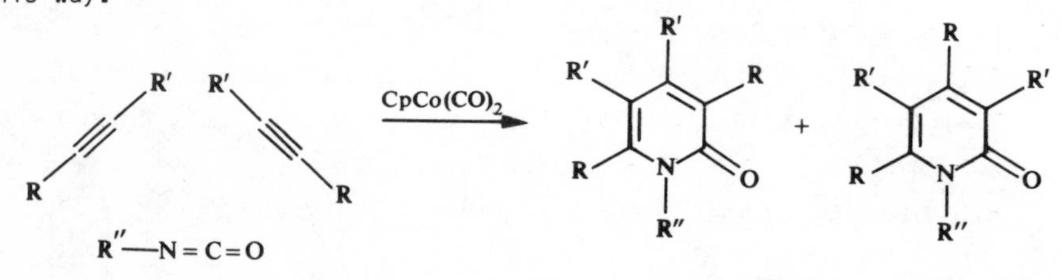

6.5 REFERENCES

1. R.A. Kretchmer and R. Glowinski, *J. Org. Chem.*, 1976, *41*, 2661.
2. A. Sekiya and N. Ishikawa, *J. Organometal. Chem.*, 1976, *118*, 349; 1977, *125*, 281; M. Yamamura, I. Moritani, S. Murahashi, *J. Organometal. Chem.*, 1975, *91*, C39; K. Tamao, K. Sumitani and M. Kumada, *J. Amer. Chem. Soc.*, 1972, *94*, 4374; M. Kumada, *Pure and Appl. Chem.*, 1980, *52*, 669; A.O. King, N. Okukado and E. Negeshi, *Chem. Comm.*, 1977, 683.
3. M.F. Semmelhack and L. Ryono, *Tet. Letters*, 1973, 2967.
4. A. Minato, K. Tamao, T. Hayashi, K. Suzuki and M. Kumada, *Tet. Letters*, 1980, 845.
5. M.F. Semmelhack, B.P. Chong, R.D. Stauffer, T.D. Rogerson, A. Chong and L.D. Jones, *J. Amer. Chem. Soc.*, 1975, *97*, 2507.
6. G. Consiglio and C. Botteghi, *Helv. Chim. Acta*, 1973, *56*, 460; Y. Kiso, K. Tamao, N. Miyake, K. Yamamoto and M. Kumada, *Tet. Letters*, 1974, 3.
7. R.J.P. Corriu and J.P. Masse, *Chem. Comm.*, 1972, 144.
8. H.P. Dang and G. Linstrumelle, *Tet. Letters*, 1978, 191; S.I. Murahashi, M. Yamamura, K. Yanagisawa, N. Mita and K Kondo, *J. Org. Chem.*, 1979, *44*, 2408.
9. E. Negishi, L.F. Valente and M. Kobayashi, *J. Amer. Chem. Soc.*, 1980, *102*, 3298.
10. T. Hayashi, M. Tajika, K. Tamao and M. Kumada, *J. Amer. Chem. Soc.*, 1976, *98*, 3718; K. Tamao, T. Hayashi, H. Matsumoto, H. Yamamoto and M. Kumada, *Tet. Letters*, 1979, 2155.
11. K. Sonogashira, Y. Tohida and N. Hagihara, *Tet. Letters*, 1975, 4467; H.A. Dieck and R.F. Heck, *J. Organometal. Chem.*, 1975, *93*, 259; K. Edo, H. Yamanaka and T. Sakamoto, *Heterocycles*, 1978, *9*, 271; A. Ohsawa, Y. Abe and H. Igeta, *Chem Letters*, 1979, 241.
12. M. Zembayashi, K. Tamao, T. Hayashi, T. Mise and M. Kumada, *Tet. Letters*, 1977, 1799.
13. B. Akermark, L. Eberson, E. Jonsson and E. Petterson, *J. Org. Chem.*, 1975, *40*, 1365; T. Itahara and T. Sakakibora, *Synthesis*,1978, 607; T. Itahara, *Synthesis*, 1979, 151.
14. R.F. Heck, *Acc. Chem. Res.*, 1979, *12*, 146.
15. R.F. Heck, *J. Amer. Chem. Soc.*, 1968, *90*, 5518; 1969, *91*, 6707.

16. R.F. Heck, *Pure Appl. Chem.*, 1978, *50*, 691.
17. R.F. Heck, *J. Amer. Chem. Soc.*, 1968, *90*, 5526.
18. R.F. Heck, *J. Amer. Chem. Soc.*, 1968, *90*, 5531.
19. R.C. Larock, J.C. Bernhardt and R.J. Driggs, *J. Organometal. Chem.*, 1978, *156*, 45; R.C. Larock and B. Riefling, *J. Org. Chem.*, 1978, *43*, 1468.
20. A. Kasahara, T. Izumi, K. Endo, T. Takeda and M. Ookita, *Bull. Chem. Soc. Jap.*, 1974, *47*, 1967; H. Horino, M. Arai and N. Inoue, *Bull. Chem. Soc. Jap.*, 1974, *47*, 1683.
21. H. Horino and N. Inoue, *Chem. Comm.*, 1976, 500.
22. R.C. Larock, *Chem. Abs.*, 1979, *90*, 71826t.
23. D.E. Bergstom and J.L. Ruth, *J. Amer. Chem. Soc.*, 1976, *98*, 1587; D.E. Bergstom and M.K. Ogawa, *J. Amer. Chem. Soc.*, 1978, *100*, 8106.
23a. R.A. Holton, *J. Amer. Chem. Soc.*, 1977, *99*, 8083.
24. M.J. Loots and J. Schwartz, *J. Amer. Chem. Soc.*, 1977, *99*, 8045; *Tet. Letters*, 1978, 4381; J. Schwartz, M.J. Loots and H. Kosugi, *J. Amer. Chem. Soc.*, 1980, *102*, 1333.
25. Y. Ito, H. Aoyama, T. Hirao, A Mochizuki and T. Saegusa, *J. Amer. Chem. Soc.*, 1979, *101*, 494.
26. R.P. Hughes and J. Powell, *J. Organometal. Chem.*, 1971, *30*, C45; M.C. Gallazzi, L. Porri and G. Vitulli, *J. Organometal. Chem.*, *97*, 131.
26a. G.P. Chiusoli, G. Salerno and F. Dallatomatsina, *Chem. Comm.*, 1977, 793;
26b. E. Wenkert, T.W. Ferreira and E.L. Michelotti, *Chem. Comm.*, 1979, 637; E. Wenkert, E.L. Michelotti and C.S. Swindell, *J. Amer. Chem. Soc.*, 1979, *101*, 2246.
27. B.A. Patel, C.B. Ziegler, N.A. Cortese, J.E. Plevyak, T.C. Zebovitz, M. Terpko and R.F. Heck, *J. Org. Chem.*, 1977, *42*, 3903; N.A. Cortese, C.B. Ziegler, Jr., B.J. Hrnjez and R.F. Heck, *J. Org. Chem.*, 1978, *43*, 2952; M. Mori, S. Kudo and Y. Ban, *J.C.S. Perkin I*, 1979, 771; M. Mori, K. Chiba and Y. Ban, *Tet. Letters*, 1977, 1037; M. Mori and Y. Ban, *Tet. Letters*, 1979, 1133.
28. M. Mori and Y. Ban, *Tet. Letters*, 1976, 1803, 1807.
29. B.A. Patel, L.C. Kao, N.A. Cortese, J.V. Minkiewicz and R.F. Heck, *J. Org. Chem.*, 1979, *44*, 918.
30. J.B. Melpolder and R.F. Heck, *J. Org. Chem.*, 1976, *41*, 265;

A.J. Chalk and S.A. Magennis, *J. Org. Chem.*, 1976, *41*, 273, 1206.

31. Y. Tamaru, Y. Yamada and Z. Yoshida, *Tet. Letters*, 1977, 3365; *Chem. Letters*, 1977, 423; *Tet. Letters*,1978, 919; *Tetrahedron*, 1979, *35*, 329.
32. W.C. Frank, Y.C. Kim and R.F. Heck, *J. Org. Chem.*, 1978, *43*, 2947.
33. Y. Tamaru, Y. Yamada and Z. Yoshida, *J. Org. Chem.*, 1978, *43*, 3396.
34. Y. Fujiwara, R. Asano, I. Moritani and S. Teranishi, *Chem. Letters*, 1975, 1061.
35. M. Julia, M. Duteil and J.Y. Lallemand, *J. Organometal. Chem.*, 1975, *102*, 239; R.A. Holton, *Tet. Letters*, 1977, 355; M.I. Bruce, *Ang. Chem. Int. Ed.*, 1977, *16*, 73.
36. B.M. Trost and J.P. Genêt, *J. Amer. Chem. Soc.*, 1976, *98*, 8516;
37. B.M. Trost, S.A. Godleski and J.P. Genêt, *J. Amer. Chem. Soc.*, 1978, *100*, 3930.
38. B.M. Trost, S.A. Godleski and J.L. Belletire, *J. Org. Chem.*, 1979, *44*, 2052.
39. R. Grigg, T.R.B. Michell and A. Ramasubbu, *Chem. Comm.*, 1979, 669; 1980, 27.
40. E.J. Corey and M.F. Semmelhack, *J. Amer. Chem. Soc.*, 1967, *89*, 2755.
41. J.A. Marshall and P.G.M. Wuts, *J. Amer. Chem. Soc.*, 1978, *100*, 1627.
42. L.S. Hegedus and R.K. Stiverson, *J. Amer. Chem. Soc.*, 1974, *96*, 3250.
43. L.S. Hegedus, G.F. Allen and E.L. Waterman, *J. Amer. Chem. Soc.*, 1976, *98*, 2674; L.S. Hegedus, G.F. Allen, J.J. Bozell and E.L. Waterman, *J. Amer. Chem. Soc.*, 1978, *100*, 5800.
44. D.E. Korte, L.S. Hegedus and R.K. Wirth, *J. Org. Chem.*, 1977, *42*, 1329.
45. K. Sato, S. Inoue, S. Ota and Y. Fujita, *J. Org. Chem.*, 1972, *37*, 462.
46. E.J. Corey and E. Hamanaka, *J. Amer. Chem. Soc.*, 1967, *89*, 2758; *J. Amer. Chem. Soc.*, 1964, *86*, 1641; E.J. Corey and E.K.W. Wat, *J. Amer. Chem. Soc.*, 1967, *89*, 2757.
47. E.J. Corey and H.A. Kirst, *J. Amer. Chem. Soc.*, 1972, *94*, 667.
48. L.S. Hegedus, S.D. Wagner, E.L. Waterman and K. Siirala-Hansen, *J. Org. Chem.*, 1975, *40*, 593.
49. M.F. Semmelhack and E.S.C. Wu, *J. Amer. Chem. Soc.*, 1976, *98*, 3384.
50. M.F. Semmelhack, A. Yamashita, J.C. Tomesch and K. Hirotsu, *J. Amer. Chem. Soc.*, 1978, *100*, 5565.

51. L.S. Hegedus, B.R. Evans, D.E. Korte, E.L. Waterman and K. Sjöberg, *J. Amer. Chem. Soc.*, 1976, *98*, 3900.

52. R. Noyori, K. Yokoyama and Y. Hayakawa, *J. Amer. Chem. Soc.*, 1973, *95*, 2722; *J. Amer. Chem. Soc.*, 1978, *100*, 1791; R. Noyori, *Acc. Chem. Res.*, 1979, *12*, 61.

53. Y. Hayakawa, F. Shimizu and R. Noyori, *Tet. Letters*, 1978, 993; R. Noyori, N. Nishizawa, F. Shimizu, Y. Hagakawa, K. Maruoka, S. Hashimoto, H. Yamamoto and M. Nozaki, *J. Amer. Chem. Soc.*, 1979, *101*, 220.

54. Y. Hayakawa, K. Yokoyama and R. Noyori, *J. Amer. Chem. Soc.*, 1978, *100*, 1799.

55. M. Takaya, S. Makino, Y. Hayakawa and R. Noyori, *J. Amer. Chem. Soc.*, 1978, *100*, 1765.

56. H. Takaya, Y. Hayakawa, S. Markino and R. Noyori, *J. Amer. Chem. Soc.*, 1978, *100*, 1778; R. Noyori, S. Makino, T. Okita and Y. Hayakawa, *J. Org. Chem.*, 1975, *40*, 806.

57. Y. Hayakawa, M. Sakai and R. Noyori, *Chem. Letters*, 1975, 509; Y. Baba, Y. Hayakawa, S. Makino and R. Noyori, *J. Amer. Chem. Soc.*, 1978, *100*, 1786.

58. R. Noyori, T. Sato and Y. Hayakawa, *J. Amer. Chem. Soc.*, 1978, *100*, 2561.

59. T. Sato, R. Ito, Y. Hayakawa and R. Noyori, *Tet. Letters*, 1978, 1829.

60. R. Noyori, Y. Baba and Y. Hayakawa, *J. Amer. Chem. Soc.*, 1974, *96*, 3336.

61. R. Noyori, Y. Hayakawa, S. Makino, N. Hayakawa and H. Takaya, *J. Amer. Chem. Soc.*, 1973, *95*, 4103.

62. H. Felkin and G. Swierczewski, Tetrahedron Report No.5, *Tetrahedron*, 1975, *31*, 2735.

63. H. Felkin and M. Joly-Goudket, personal communication.

64. B.L. Buckwalter, I.R. Burfitt, H. Felkin, M. Joly-Goudket, K. Naemura, M.F. Salomon, E. Wenkert and P.M. Woukulich, *J. Amer. Chem. Soc.*, 1978, *100*, 6445.

65. M. Cherest, S.G. Davies and H. Felkin, *Chem. Comm.*, 1981, 682.

66. G. Consiglio, F. Morardini and O. Piccolo, *Helv. Chem. Acta*, 1980, *63*, 987.

67. K.J. Ivin, J.J. Rooney, C.D. Stewart, M.L.H. Green and R. Mahtab,

Chem. Comm., 1978, 604; M.L.H. Green, *Pure and Appl. Chem.*, 1978, *50*, 27.

68. N. Calderon, *Acc. Chem. Res.*, 1972, *5*, 127; J.L. Herisson and Y. Chauvin, *Makromol. Chem.*, 1970, *141*, 161; J.J. Rooney and A. Stewart, *Catalysis J. Chem. Soc. Sp. Rep.*, 1977, *1*, 277.
69. L.G. Wideman, *J. Org. Chem.*, 1968, *33*, 4541.
70. B. Bogdanovic, *Ang. Chem. Int. Ed.*, 1973, *12*, 954.
71. R. Baker, *Chem. Rev.*, 1973, *73*, 487.
72. C.U. Pittman Jr. and L.R. Smith, *J. Amer. Chem. Soc.*, 1975, *97*, 341.
73. J.P. Neilan, R.M. Laine, N. Cortese and R.F. Heck, *J. Org. Chem.*, 1976, *41*, 3455.
74. R. Baker, R.C. Cookson and J.R. Vinson, *Chem. Comm.*, 1974, 515.
75. K.P.C. Vollhardt, *Acc. Chem. Res.*, 1977, *10*, 1.
76. E. Müller, C. Beissner, H. Jäkle, E. Langer, H. Muhm, G. Odenigbo, M. Sauerbier, A. Segmitz, D. Streichfuss and R. Thomas, *Ann.*, 1971, *754*, 64; E. Müller and A. Segmitz, *Chem. Ber.*, 1973, *106*, 35; A. Scheller, W. Winter and E. Müller, *Ann.*, 1976, 1448; J. Hambrecht and E. Müller, *Ann.*, 1977, 387.
77. H. Straub, A. Huth and E. Müller, *Synthesis*, 1973, 783.
78. R.L. Funk and K.P.C. Vollhardt, *Chem. Soc. Rev.*, 1980, *9*, 41 and references therein.
79. R.L. Hillard III and K.P.C. Vollhardt, *J. Amer. Chem. Soc.*, 1977, *99*, 4058.
80. R.L. Funk and K.P.C. Vollardt, *J. Amer. Chem. Soc.*, 1976, *98*, 6755.
81. R.L. Funk and K.P.C. Vollhardt, *J. Amer. Chem. Soc.*, 1977, *99*, 5483; R.L. Funk and K.P.C. Vollhardt, *J. Amer. Chem. Soc.*, 1979, *101*, 215.
82. L. Pichat, P.N. Liem and J.P. Guermont, *Bull. Soc. chim. France*, 1972, 4224.
83. M. Suzuki, K. Itoh, Y. Ishii, K. Simon and J.A. Ibers, *J. Amer. Chem. Soc.*, 1976, *98*, 8494.
84. Y. Wakatsuki and H. Yamazaki, *J.C.S. Dalton*, 1978, 1278; *Synthesis*, 1976, 26, *Tet. Letters*, 1973, 3383; A. Naiman and K.P.C. Vollhardt, *Ang. Chem. Int. Ed.*, 1977, *16*, 708.
85. P. Hong and H. Yamazaki, *Tet. Letters*, 1977, 1333; *Synthesis*, 1977, 50.

CHAPTER 7

ISOMERISATION REACTIONS

7.1 ISOMERISATION OF OLEFINS AND ACETYLENES

There are two general mechanisms for the isomerisation of olefins by transition metal complexes. The first mechanism shown in equation (I) involves the reversible addition of a metal hydride across a double bond to generate a transition metal σ-alkyl species. Olefin is regenerated by loss of metal hydride. Removal of a different hydrogen atom in the second step to the one initially added results in overall isomerisation. This is the mechanism commonly found to be occurring in many transition metal catalysed reductions of olefins by hydrogen. The metal hydride isomerisation catalysts may be stable metal hydride species or they may be generated in the reaction media. For example olefins may be isomerised by compounds of Rh, Pd, Pt, Ni or Fe in the presence of various co-catalysts.[1,2]

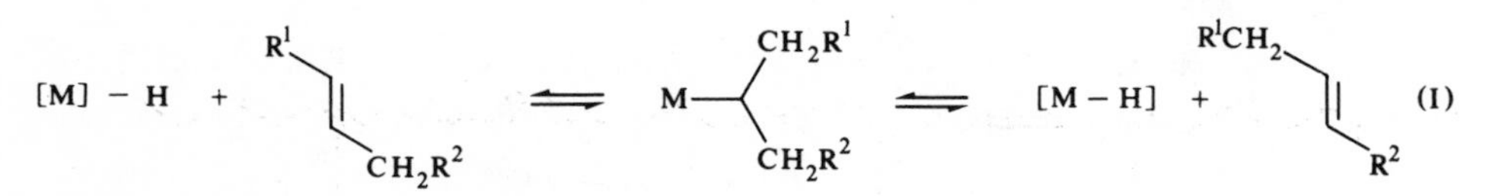

The second general mechanism is shown in equation (II) and involves

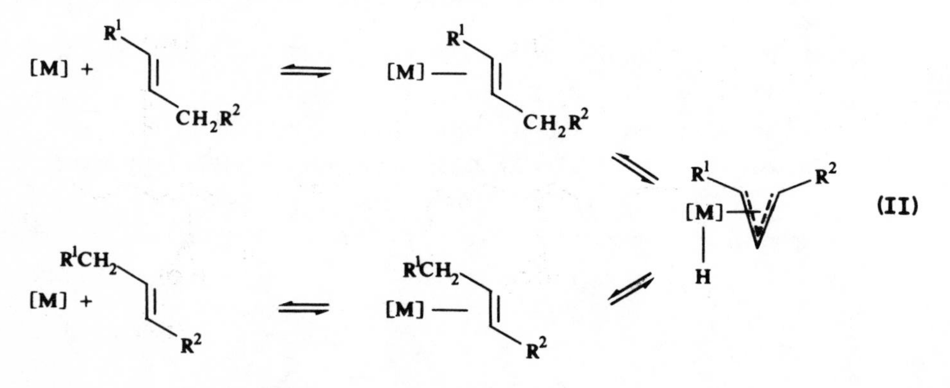

coordination of the olefin to the transition metal followed by insertion of the metal into an allylic carbon-hydrogen bond to generate a π-allyl metal hydride species. Reversal of this process by putting the hydrogen onto the other end of the π-allyl ligand leads to overall isomerisation.

It should be noted that in mechanism (I) it is necessary for the [M]-H catalytic species to be ≤16e to allow coordination of the olefin whereas in mechanism (II) the [M] species must be 14e to allow coordination of an η^3-allyl ligand and a H atom.

Olefins may be isomerised to their more thermodynamically stable isomers by a variety of catalysts. For example, allyl benzene may be isomerised to 1-phenyl propene by $HCo(CO)_4$[3] or by $(Ph_3P)_4Ru(MeCN)$.[4] In the case of the Ru complex an intermediate η^3-allyl ruthenium hydride complex can be isolated.

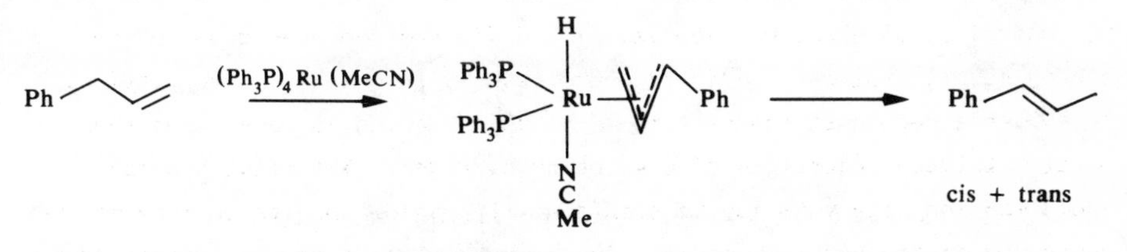

1-Hexene is catalytically isomerised into a mixture of 2-hexenes and 3-hexenes by iron carbonyls.[6]

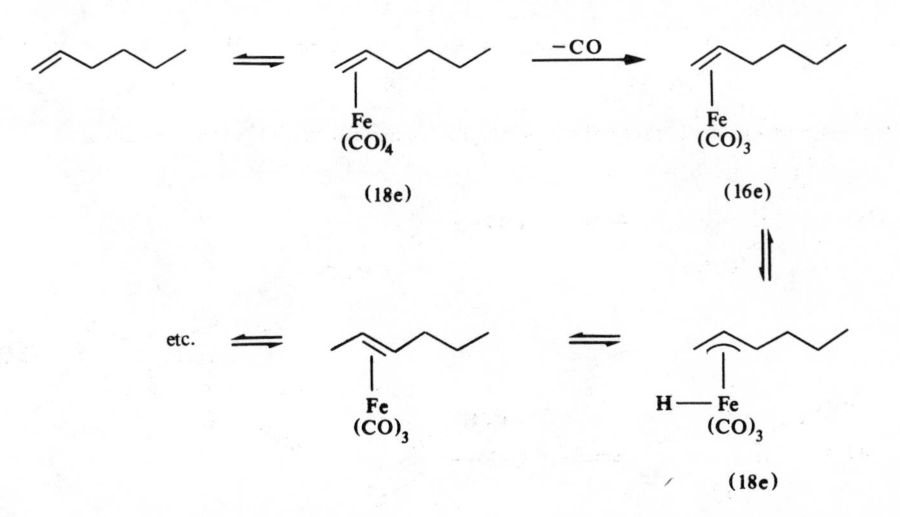

A number of terminal olefins have been isomerised to internal olefins using a bisphosphite nickel ethylene complex.[6] A variety of functional groups may be present in the molecule without causing further rearrangement.

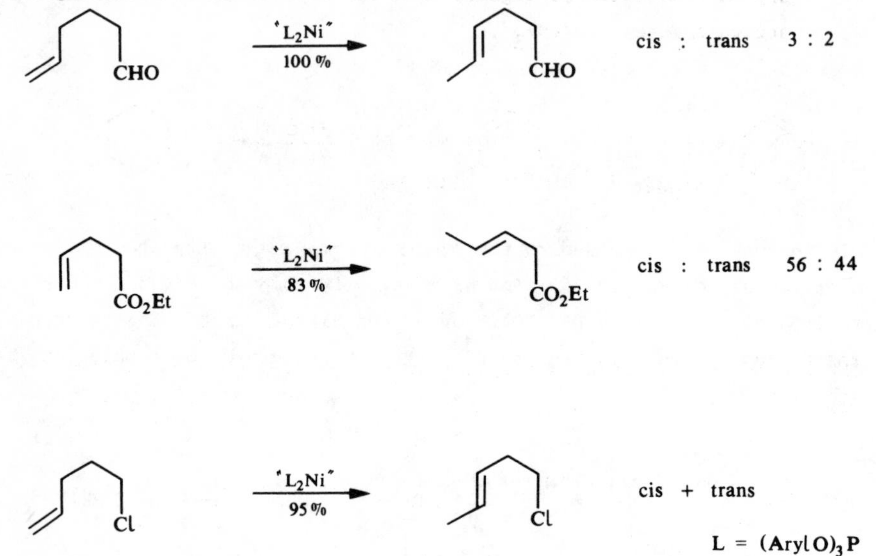

The isomerisation of the anti-Bredt olefin 1 into a mixture of four isomers is achieved by heating its $(Ph_3P)_2Pt$ complex 2 which is stable at 20°.[7]

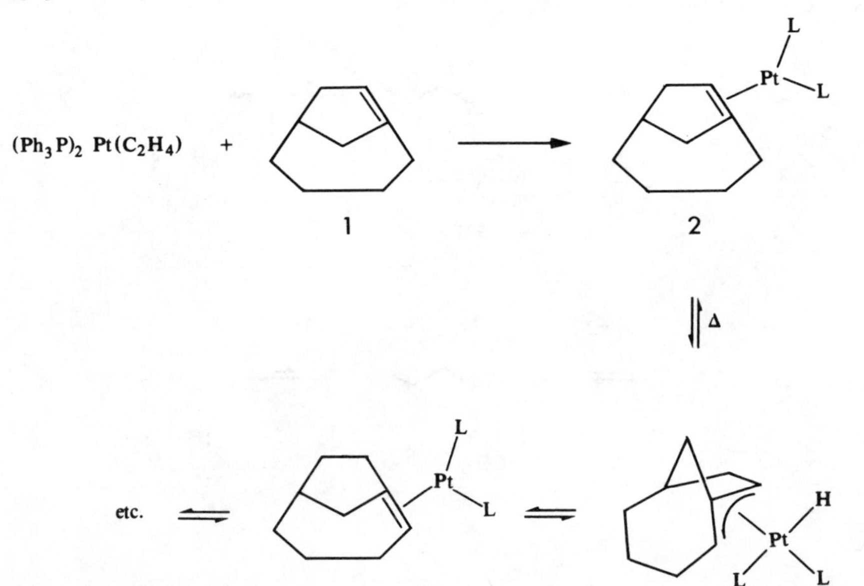

Non-conjugated unsaturated esters may be isomerised to the more stable α,β-unsaturated esters using $Fe(CO)_5$ either thermally at 150° or photochemically at 20°.[8] The thermolysis or photolysis conditions are necessary to remove two CO ligands from the iron to generate the catalytically active species "$Fe(CO)_3$".

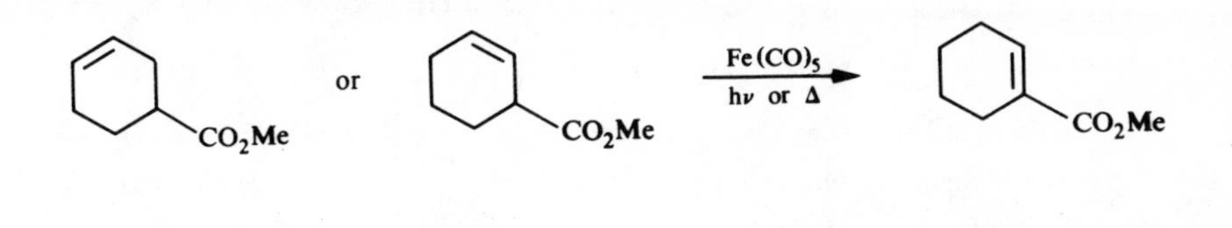

The synthetically useful isomerisation of enones such as 3 to their more stable isomers 4 have been achieved in high yields using $RhCl_3.3H_2O$ as catalyst.[9]

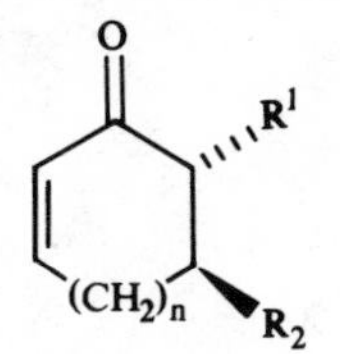

$RhCl_3.3H_2O$

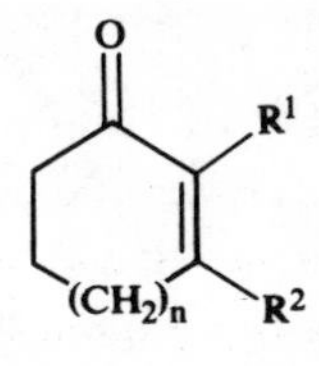

3
n = 1, 2, 3

4

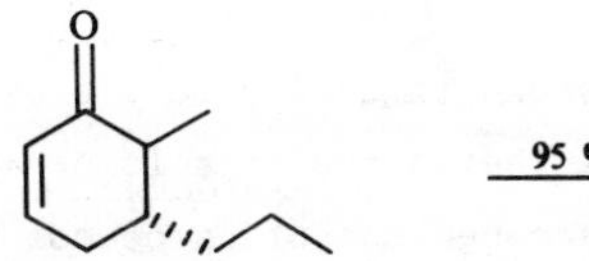

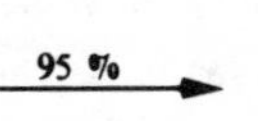

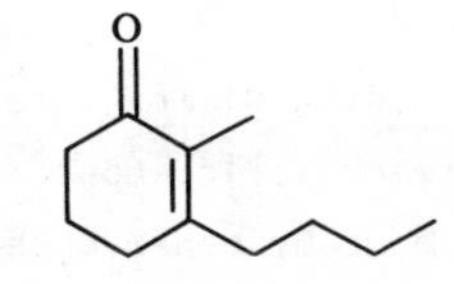

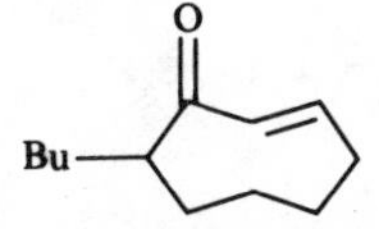

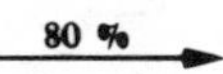

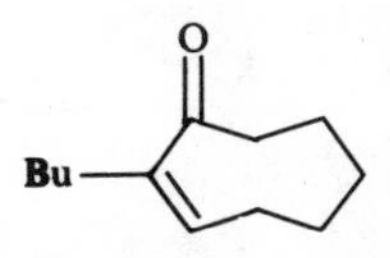

This isomerisation involves migration of the double bond around the ring and it is therefore blocked if a tertiary carbon atom is present in the chain. For example it is not possible to isomerise 5 into 6.

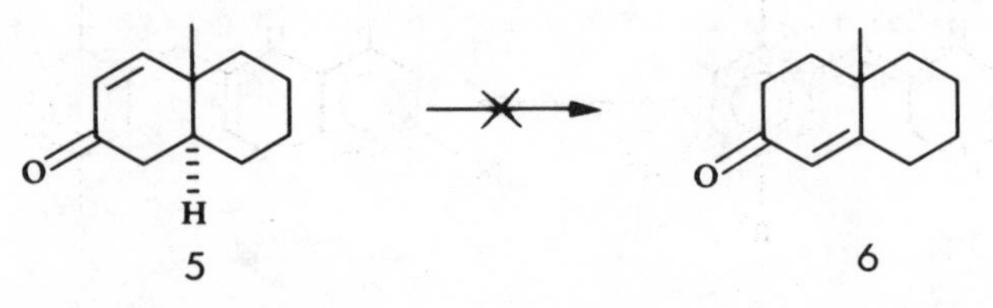

Non-conjugated enones are also isomerised under the same conditions. For instance dihydrocarvone gives a mixture of 7 and 8 in the ratio of 7:1

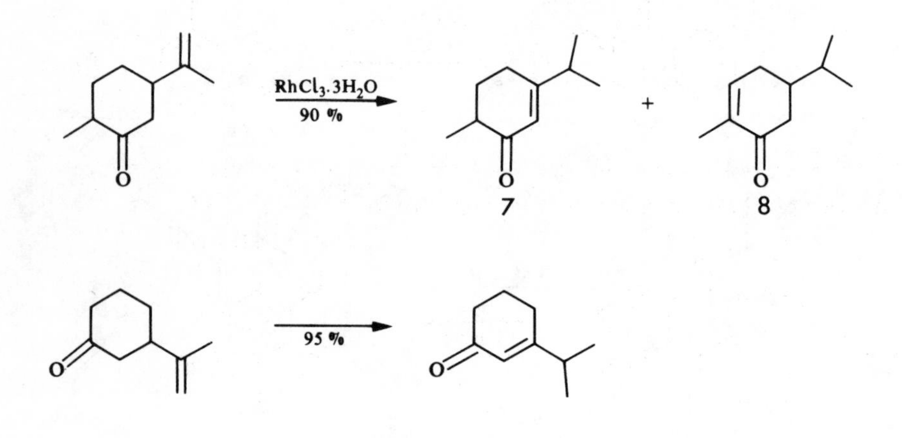

Rhodium trichloride trihydrate has been shown to give good yields in exocyclic-endocyclic enone isomerisations that are difficult or impossible to bring about by other methods. The rearrangement of compounds 9 to 10 provides a convenient route to 2-substituted cyclopentenones which are otherwise difficult to prepare.[10]

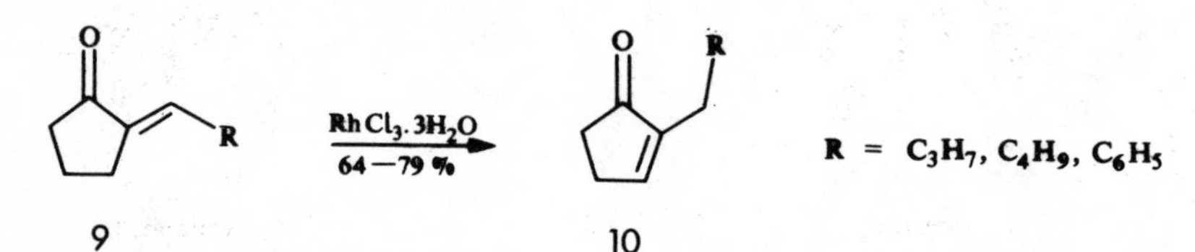

The quantitative rearrangement of the arylmethylene chroman-4-ones 11 allows access into the homoisoflavone skeleton.

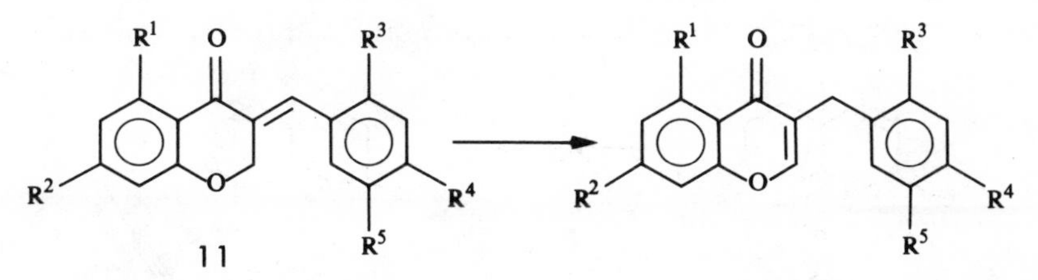

Benzylidene tetralone is converted quantitatively into 2-benzyl-1-naphthol by $RhCl_3.3H_2O$[10] and by $(Ph_3P)_2Ir(CO)Cl$ at 250°.[11]

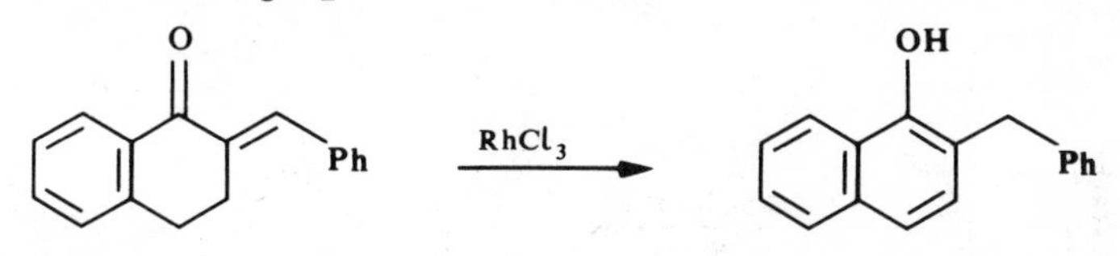

The dimethylene ketones 12 are converted by $(Ph_3P)_2M(CO)Cl$ (M = Rh, Ir) into the corresponding phenols in high yield.[11]

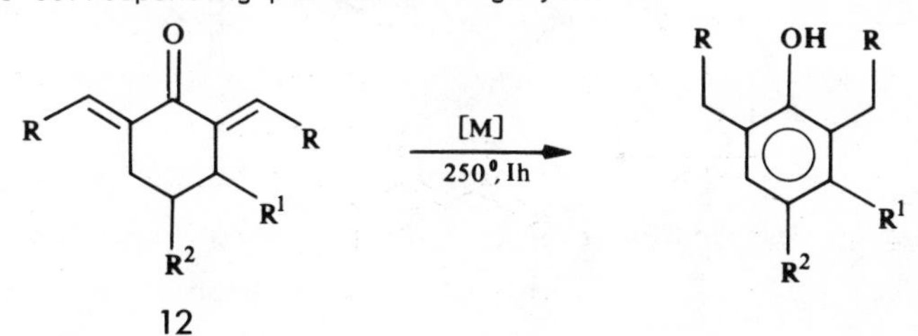

12

Damsin is isomerised to isodamsin by $(Ph_3P)_2RhH_2Cl$. This reaction probably proceeds via reversible Rh-H addition to the double bond since $(Ph_3P)_3RhCl$ is inactive.[12]

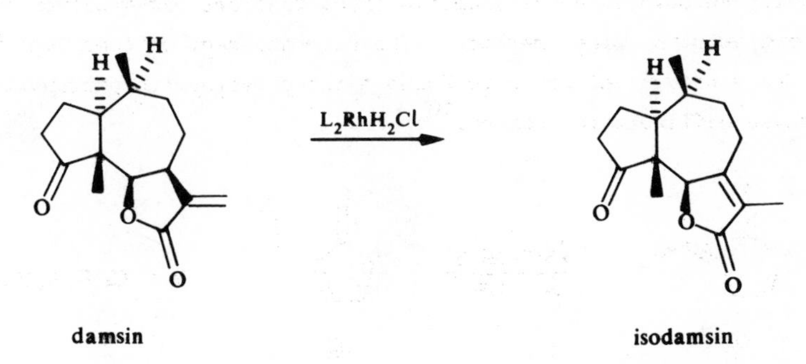

The $(MeC_5H_4)Mn(CO)_2$ complexes of electron poor acetylenes easily isomerise to the corresponding allene complexes in the presence of acids or bases.[12a]

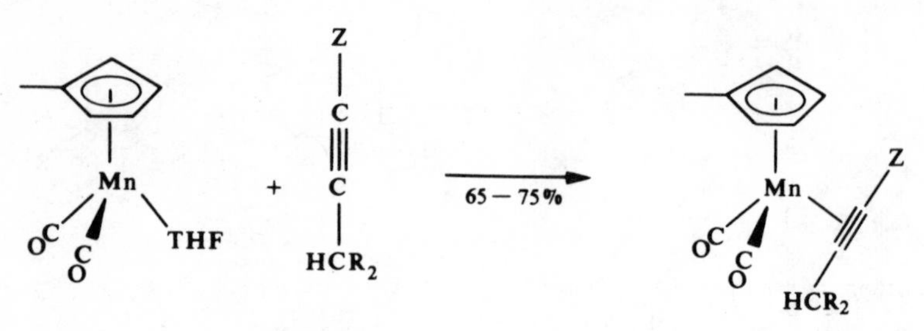

$Z = CO_2Me, CHO$

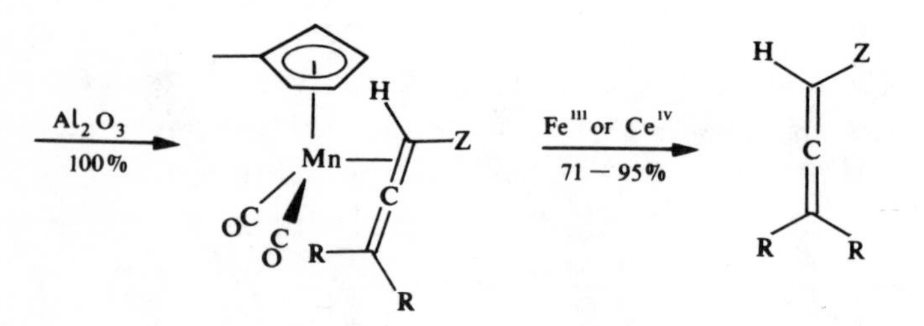

This isomerisation reaction has been used to synthesise an insect pheromone where the $(MeC_5H_4)Mn(CO)_2$ group not only promotes an acetylene

$$RCH_2C{\equiv}CCHO \xrightarrow[75\%]{[Mn]\ THF} RCH_2C{\equiv}C{-}CHO\ ([Mn]) \xrightarrow[100\%]{Al_2O_3} RCH{=}C{=}CHCHO\ ([Mn])$$

$$RCH{=}C{=}CHCHO\ ([Mn]) \xrightarrow[95\%]{^{-}CH(CO_2Me)PO(OMe)_2} RCH{=}C{=}CH.CH\overset{E}{=}CHCO_2Me\ ([M])$$

$$RCH{=}C{=}CH.CH\overset{E}{=}CHCO_2Me\ ([M]) \longrightarrow RCH{=}C{=}CH.CH\overset{E}{=}CHCO_2Me$$

$R = C_8H_{17}$ $[Mn] = (MeC_5H_4)Mn(CO)_2$

to allene rearrangement but also serves as a protecting group for the allene once formed.

7.2 ISOMERISATION OF DIENES

Nonconjugated dienes may be isomerised by a variety of catalysts to their thermodynamically more stable conjugated isomers. For example, 1,5-cyclooctadiene (1,5-COD) is isomerised by $Fe(CO)_5$ to 1,3-cyclooctadiene by the π-allyl iron hydride mechanism described above.[13]

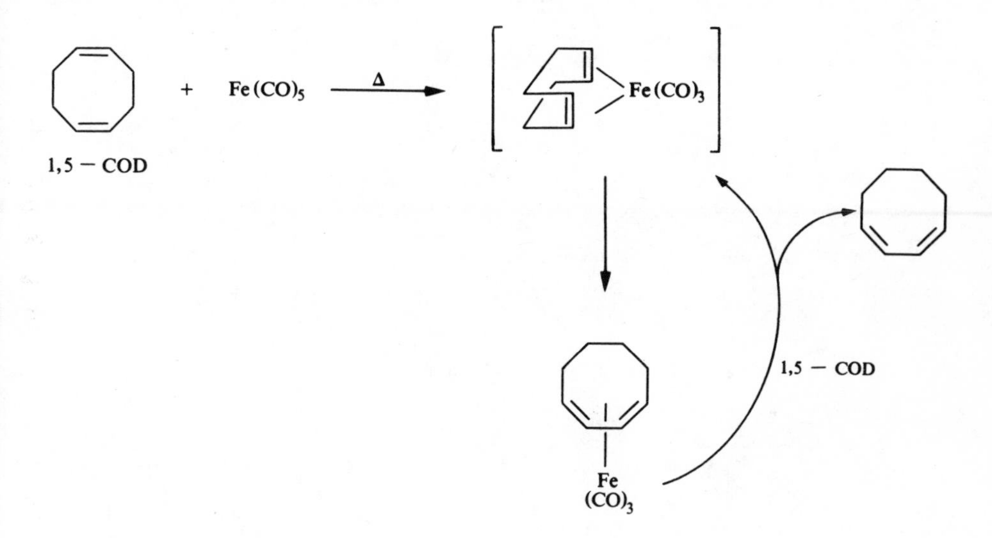

The same transformation can be achieved with $(Ph_3P)_2Ir_2(1,5\text{-COD})H_2Cl_2$ as catalyst. The mechanism in this case is reversible addition of Ir-H to one of the double bonds and the reaction is believed to involve the equilibrium between 1,5-COD and 1,4-COD the latter then going irreversibly to 1,3-COD.[14]

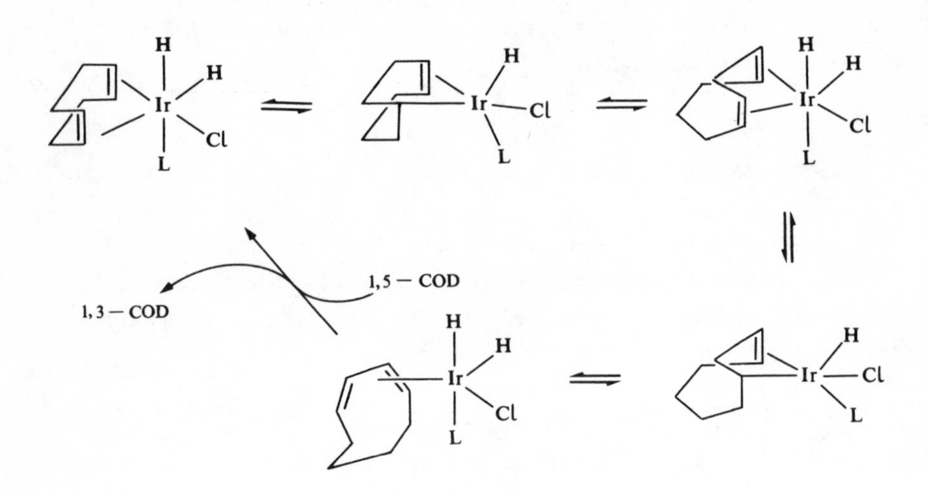

Unlike $Fe(CO)_3$ which prefers to coordinate to 1,3-dienes, complexes of Pd[15] and Rh[16] prefer to complex 1,5-dienes and this effect has been used to bring about the thermodynamically unfavourable rearrangement of 1,3-COD to 1,5-COD. In this case however the reactions are necessarily stoicheiometric and involve the formation of 1,5-COD complexes.

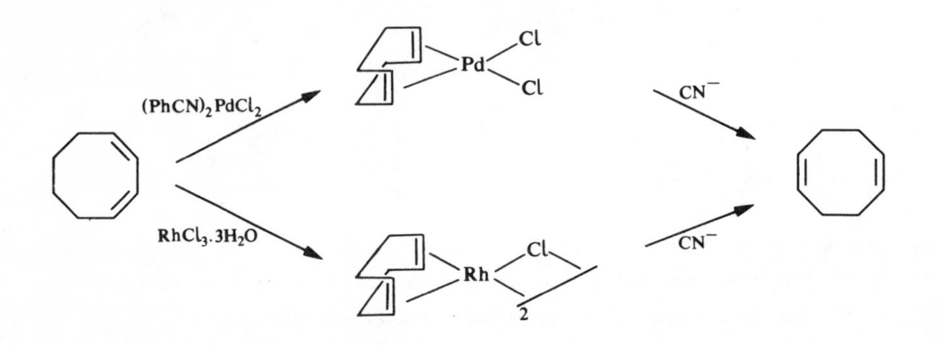

Conjugated and non-conjugated dienes react with $Fe(CO)_5$ to produce (diene)$Fe(CO)_3$ complexes (section 2.5.1). This reaction is accompanied in many cases by rearrangement of the diene complex.[17]

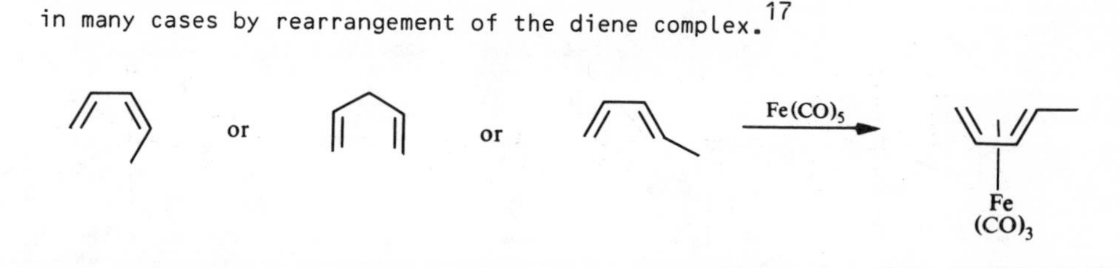

Cyclohexa-1,4-dienes readily available from the Birch reaction react with $Fe(CO)_5$ to produce 1,3-cyclohexadiene iron tricarbonyl complexes.[18]

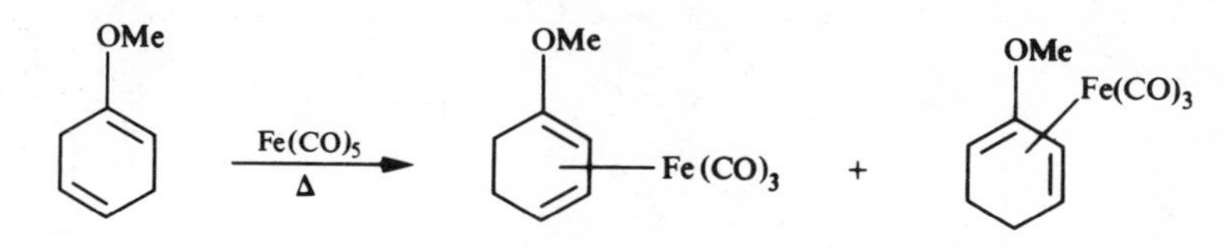

4-Vinylcyclohexene rearranges to 1-ethyl and 2-ethyl cyclohexa-1,3-diene iron tricarbonyl complexes. The ratio depends on the conditions.[19]

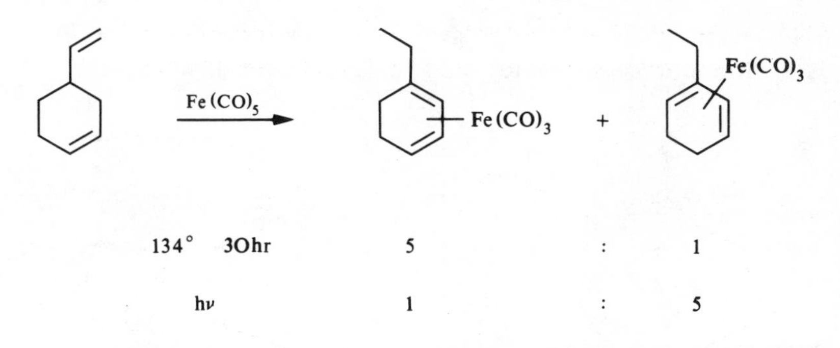

$(Ph_3P)Fe(CO)_4$ is a more selective reagent leading only to the production of the 2-ethyl cyclohexa-1,3-diene iron tricarbonyl complex.

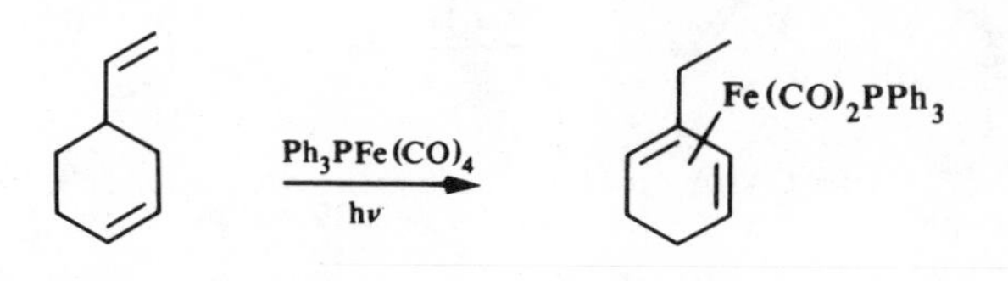

Cyclohexa-1,4-dienes may also be rearranged catalytically to 1,3-dienes with $(Ph_3P)_3RhCl$.[20]

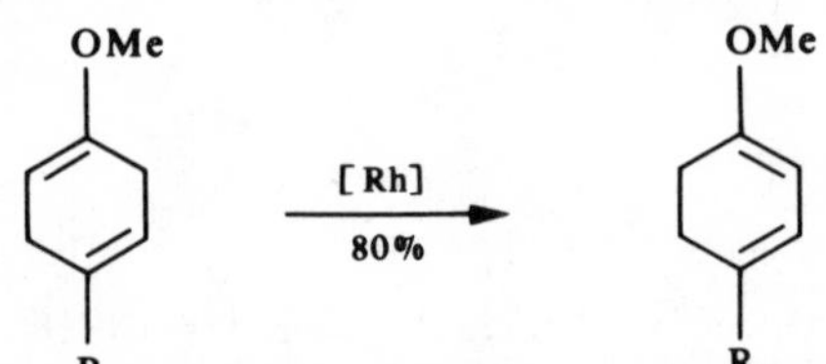

$(Ph_3P)_3RhCl$ isomerises 1-methoxycyclohexa-1,4-dienes to transoid 1-methoxy-dienes where possible. This rearrangement has proved useful for the synthesis of steroid derivatives.

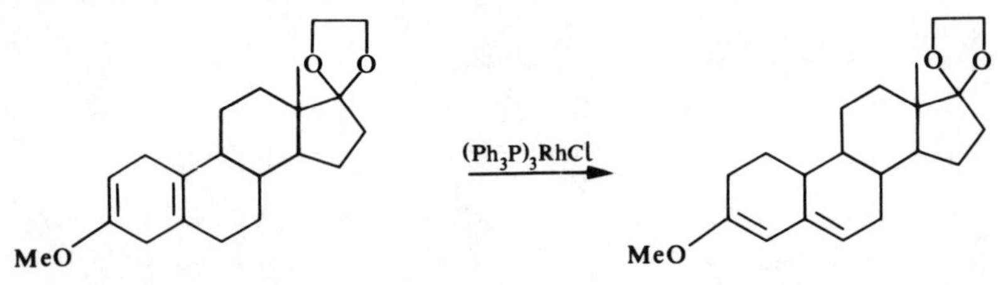

The diene 13 is isomerised by the Rh catalyst to the transoid diene 14 whereas base catalysed isomerisation stops at the cisoid diene 15 .

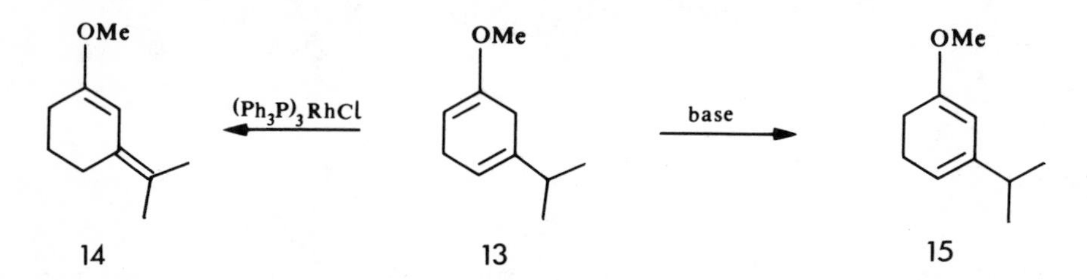

The diene 16 is rearranged quantitatively by $Cp_2TiCl_2/LiAlH_4$ to the diene 17.[21]

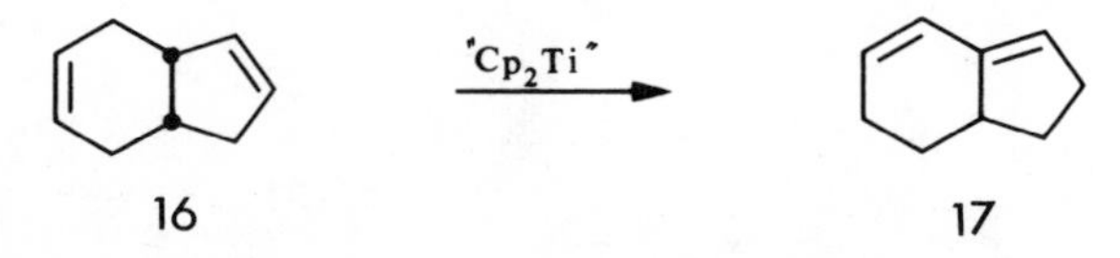

The rearrangement of the dienones 18 and 19 by $RhCl_3.3H_2O$ into the corresponding phenols provides a convenient synthesis of substituted aromatic systems.[22]

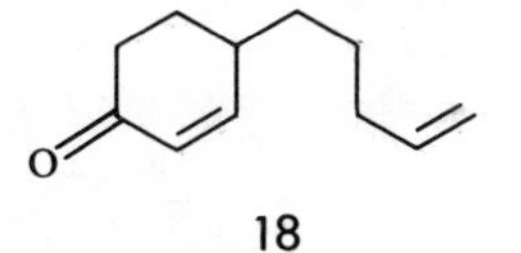

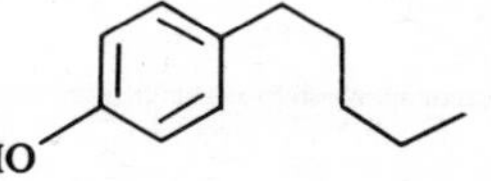

18

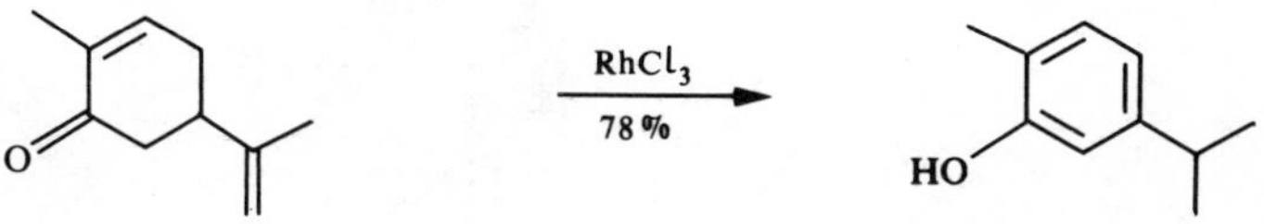

O

RhCl$_3$

78%

HO

19 carvone

The same rearrangement of the corresponding imines provides a convenient synthesis of diphenylamines.

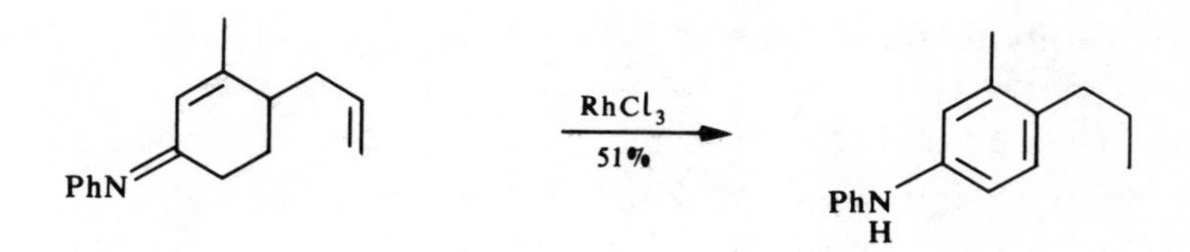

Iron carbonyl isomerisation of the nonconjugated diene lactol 20, a readily available synthetic precursor for prostaglandins A, allows entry into the prostaglandin C series.[23]

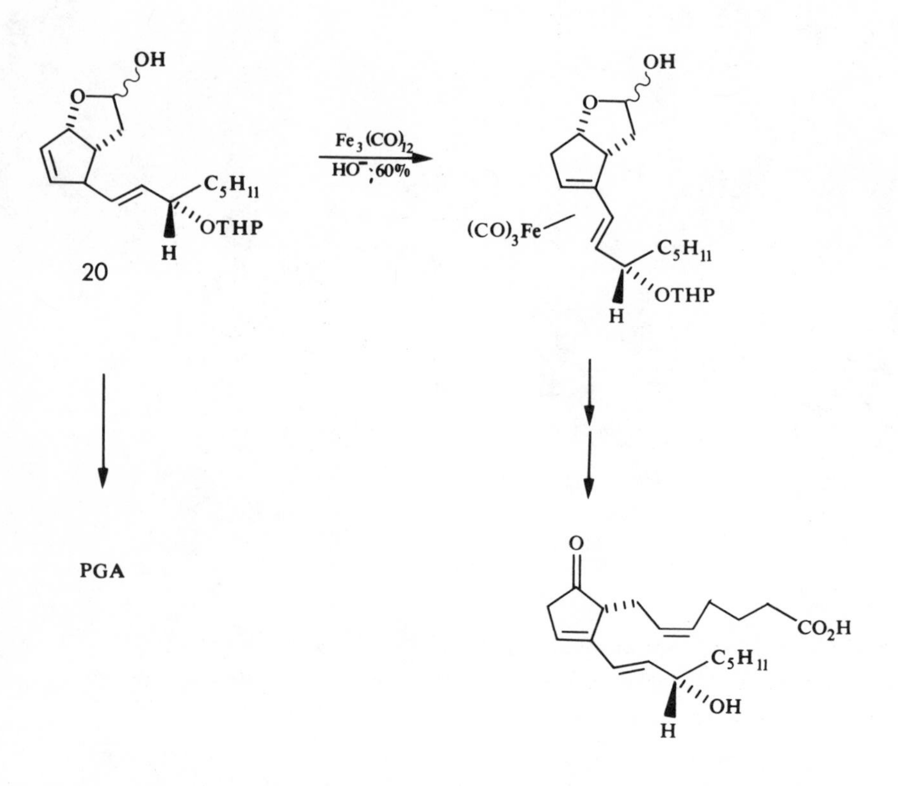

Conjugated steroid transoid dienes may be converted in high yield to cisoid diene iron tricarbonyl complexes on treatment with $Fe(CO)_5$. For example the dienes 22 (R = H), 23 (R = H, Me, OMe) and 24 (R = H) are all converted to the complex 21.[24]

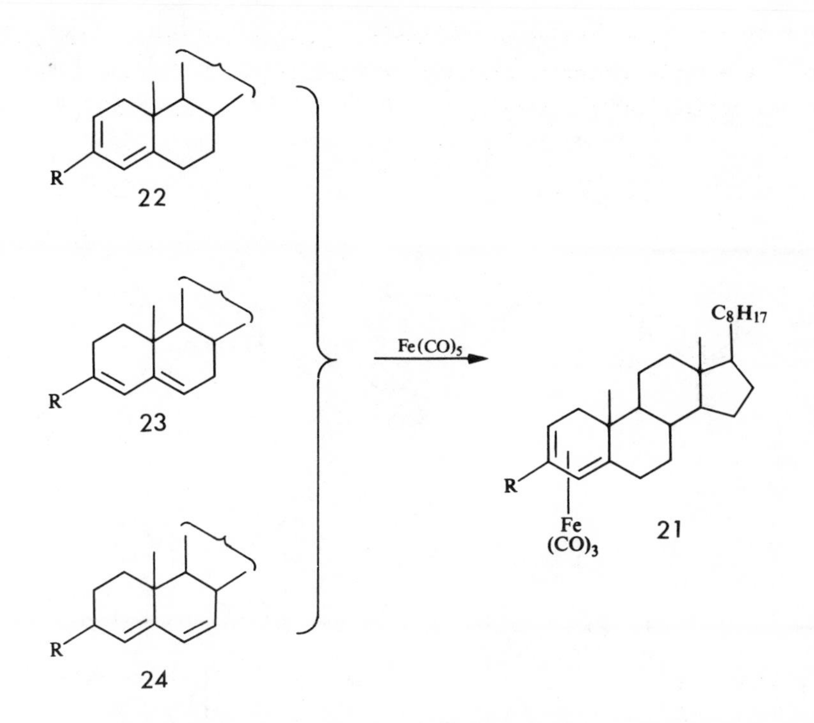

Thus formation of complexes 21 followed by decomplexation (with $FeCl_3$) of the diene allows the conversion of heteroannular steroidal dienes into their thermodynamically less stable homoannular isomers.

Acid promoted rearrangement of ergosteryl acetate 25 yields a mixture of ergosteryl B_1 26, B_2 27 and B_3 28 acetates. However treatment of ergosteryl acetate 25 with $Cr(CO)_6$ in refluxing n-octane promoted a smooth isomerisation into ergosteryl B_2 acetate 27.[25]

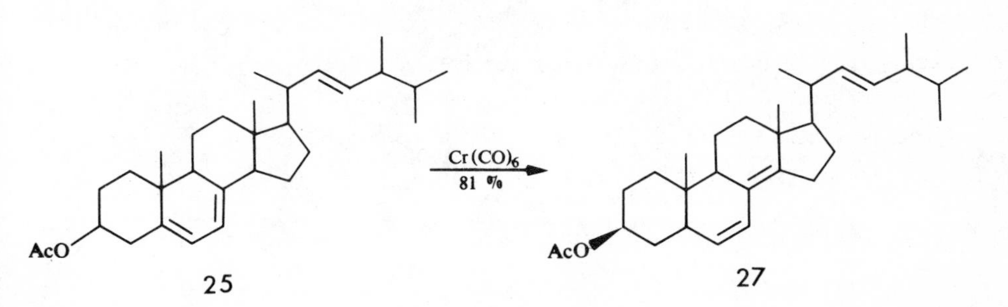

A probable mechanism for this rearrangement involves coordination of the cisoid diene to $Cr(CO)_4$ followed by hydrogen migration via the metal. The product, a transoid diene, cannot function as a four electron ligand for $Cr(CO)_4$ and hence the reaction is irreversible.

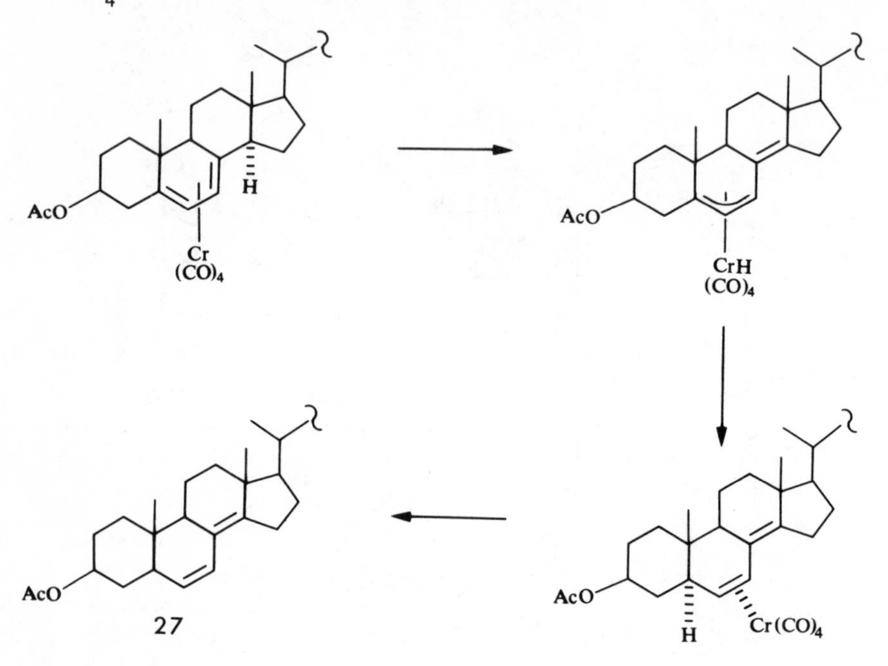

In agreement with this mechanism isomerisation of ergosteryl B_3 acetate 28 with $Cr(CO)_6$ gave only ergosteryl B_1 acetate 26.

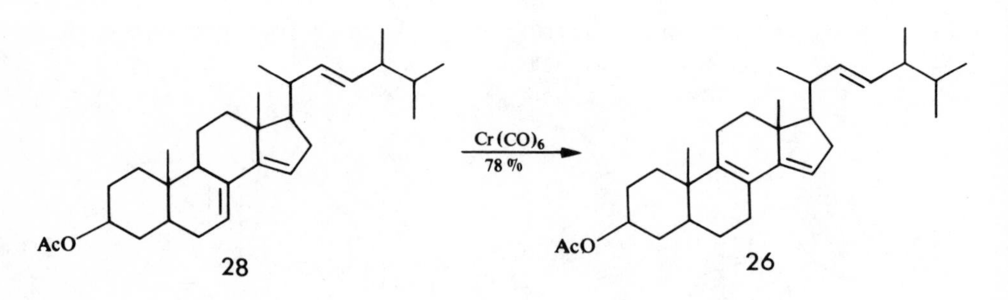

Thus, in suitable cases $Cr(CO)_6$ offers a highly regioselective method for the isomerisation of a cisoid diene into a transoid isomer. This isomerisation complements the conversion of transoid dienes into cisoid dienes by $Fe(CO)_5$ described above.

$(PhCN)_2PdCl_2$ can be used to catalyse the Cope rearrangement of many unstrained conformationally flexible acyclic 1,5-dienes at 20° whereas without the catalyst, high temperatures and long reaction times are needed.[26]

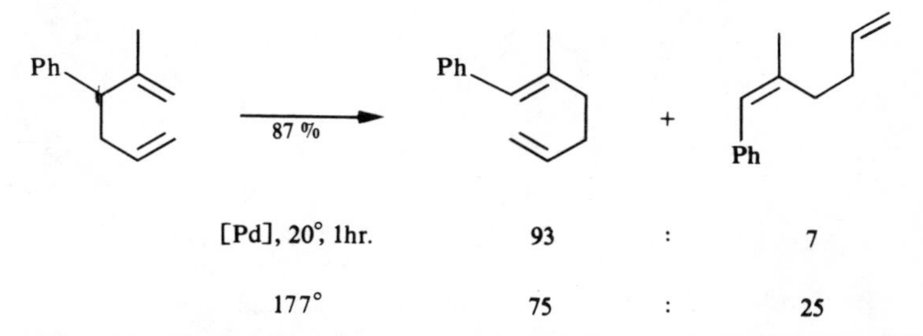

The restrictions on this reaction are that one or other but not both positions 2 and 5 must have an alkyl substituent. This is presumably to stabilise a carbonium-like intermediate. If both are substituted, initial complexation is hindered.

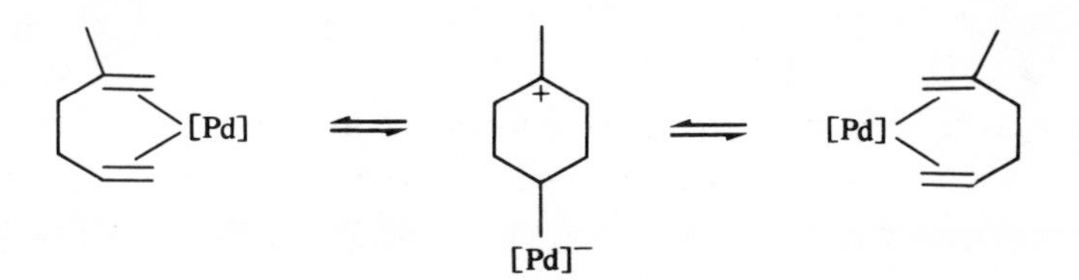

This reaction is non-catalytic if the diene $PdCl_2$ complexes are too stable or are insoluble in the solution. In these cases stoicheiometric quantities of $(PhCN)_2PdCl_2$ are needed to promote the Cope rearrangements. The rearranged dienes may be liberated from the Pd with cyanide ion or dimethyl sulphoxide.[27,28]

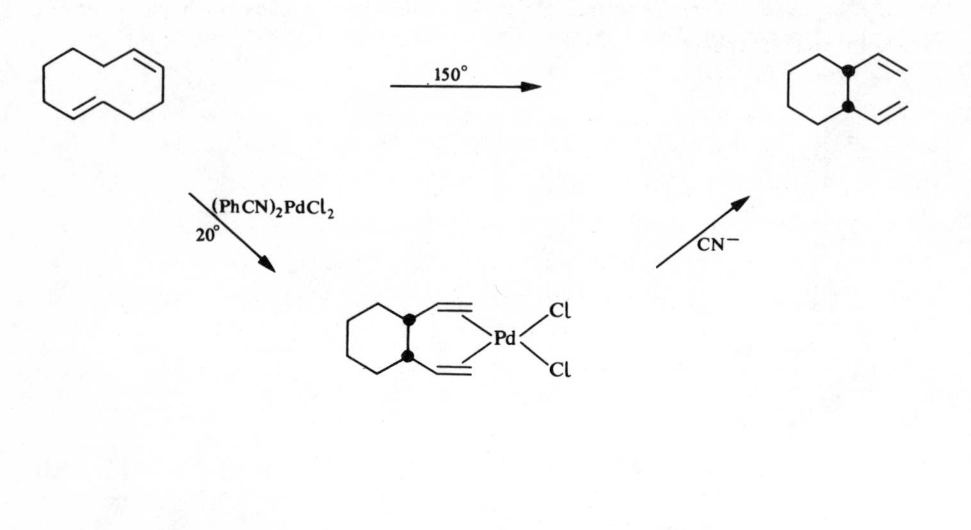

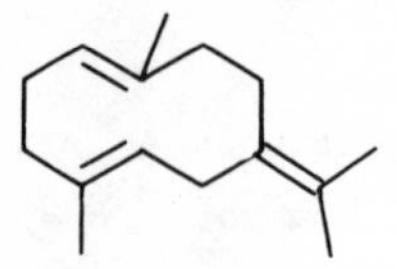
germacratriene

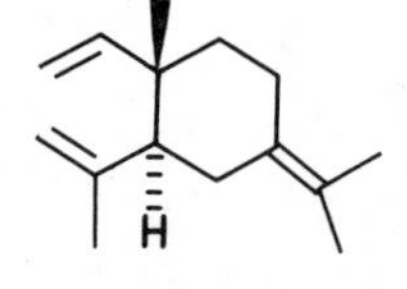

γ — elemene

1)(PhCN)$_2$PdCl$_2$
2)DMSO

7.3 ISOMERISATION OF ALLYLIC ALCOHOLS, ETHERS, AMINES, ETC

Allyl alcohol is isomerised to propionaldehyde by a variety of homogeneous catalysts such as $Fe(CO)_5$[29] or $HCo(CO)_4$.[30]

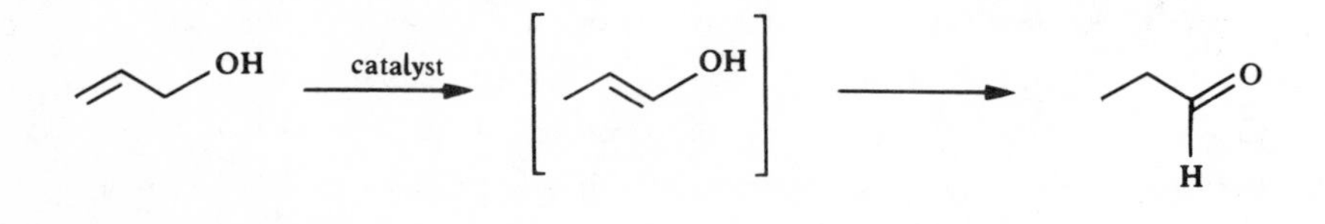

$Fe(CO)_5$ has been used to isomerise primary and secondary allylic alcohols and can be used in the presence of other functional groups such as esters, alcohols and ketones.[31]

The mechanism of this isomerisation is believed to be via a π-allyl iron hydride species 29. Evidence in favour of this mechanism comes from deuterium labelling experiments (e.g. 30 gives 31)[33] and the fact that the *endo* alcohol 32 isomerises with $Fe(CO)_5$ whereas the *exo* alcohol 33 does not. The latter experiment indicates that the iron carbonyl must bond to the same face as the hydrogen to be isomerised; *endo* bonding to alcohol 33 is sterically unfavourable.

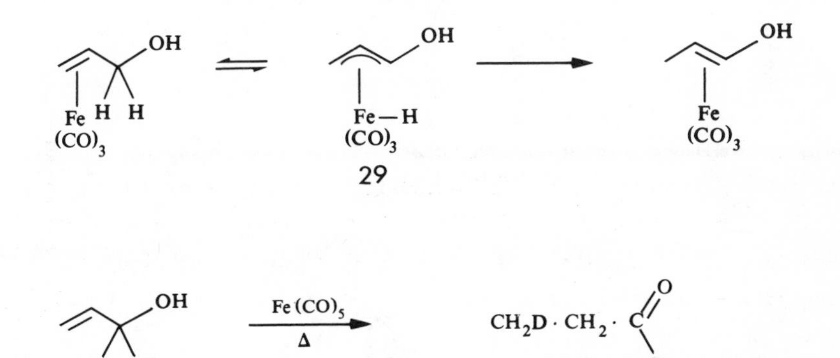

29

OH

D D

30

$Fe(CO)_5$, Δ

$CH_2D \cdot CH_2 \cdot C(=O)D$

31

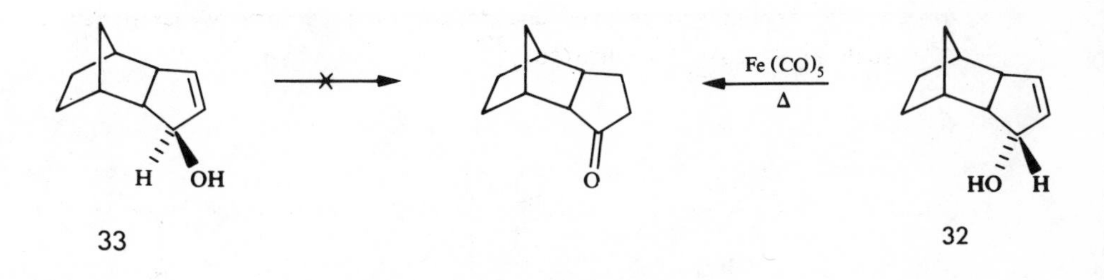

33 32

Methallyl alcohol is isomerised quantitatively to isobutyraldehyde with $(Ph_3P)_3RhH(CO)$ (70°, 3 hrs).[32] $RhCl_3$ and $IrCl_3$ may also be used but lead to the production of by-products.

Secondary allylic alcohols may be isomerised to ketones by bisphosphite nickel ethylene.[6] However higher yields are obtained with $(Ph_3P)_3RuHCl$.[35] The Rh catalyst may not be used for primary alcohols because the initially formed aldehyde is rapidly decarbonylated and the rhodium carbonyl species thus formed is inactive in the isomerisation reaction.

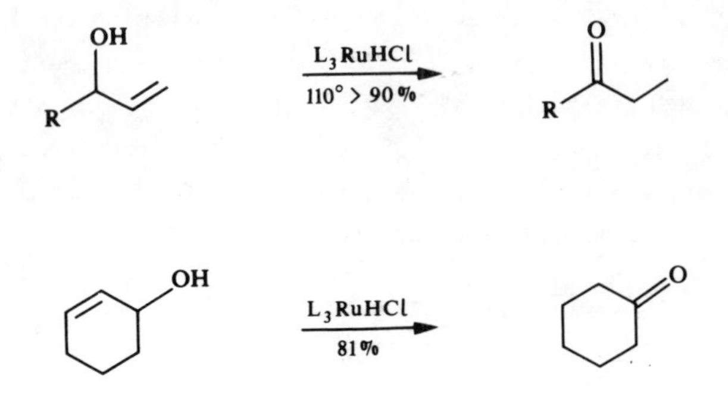

One of the mildest isomerisation catalysts for allylic alcohols is the cation$[(Ph_2MeP)_2Ir(1,5\text{-}COD)]^+PF_6^-$ activated by H_2.[36] Many simple allylic alcohols are isomerised in high yield at 20°. However the catalyst is rather susceptible to the substitution pattern on the allyl alcohol presumably because of steric constraints either in the initial olefin complex or in the π-allyl intermediate.

Alcohols isomerised (>98%) by $[(R_3P)_2Ir(1,5\text{-}COD)]^+$ in the presence of H_2.

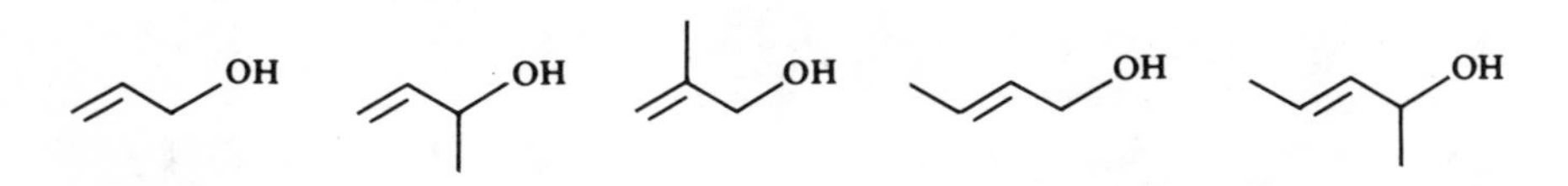

Alcohols not readily susceptible to isomerisation by $[(PR_3)_2Ir(1,5\text{-}COD)]^+$ in the presence of H_2.

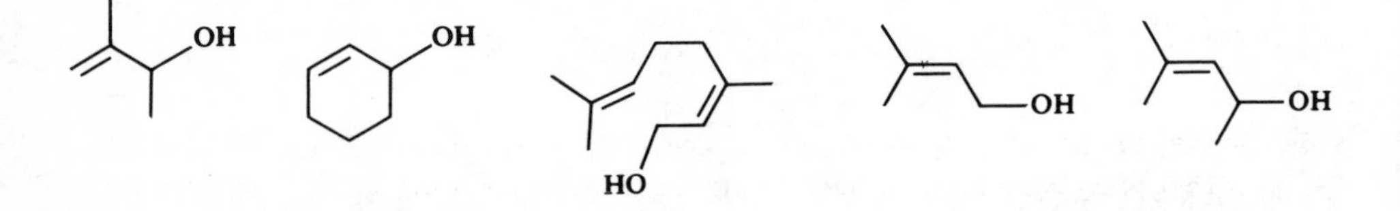

The catalysts described above for the isomerisation of allylic alcohols may also be used to isomerise allylic ethers to vinyl ethers. For example $Fe(CO)_5/h\nu$ [37] and $(Ph_3P)_3RhCl$ [38] give very high yields of isomerised products.

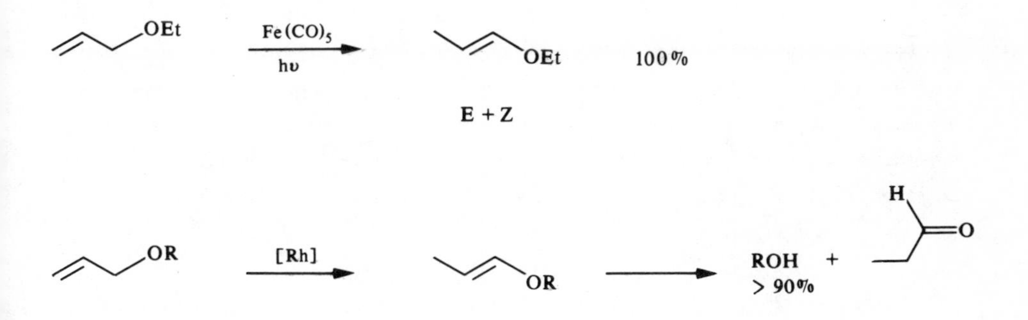

R = cholesteryl, menthyl, decyl

Aryl allyl ethers are rearranged quantitatively to vinyl ethers by $(PhCN)_2PdCl_2$ or by $(Ph_3P)_3RuCl_2$.[39] The reaction is insensitive to substituents on the aryl ring but is impeded by substituents on the allyl group. Trimethylsilyl enol ethers may be synthesised by rearrangement of trimethylsilyl allyl ethers catalysed by $(Ph_3P)_4RuH_2$.[39a]

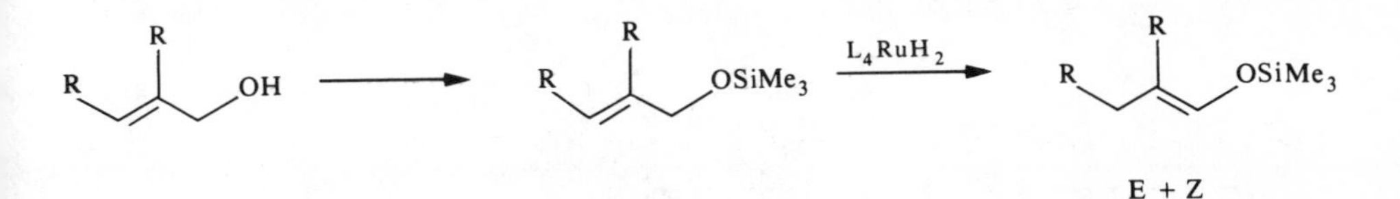

E + Z

Many of the catalytic systems for the conversion of allyl ethers to vinyl ethers lead to mixtures of *cis* and *trans* propenyl ethers. However allyl ethers may be converted stereoselectively into *cis*-propenyl ethers using $(Ph_3P)_2PtH(ClO_4)$ [40] or into *trans*-propenyl ethers using $[(Ph_2MeP)_2Ir(1,5\text{-}COD)]^+PF_6^-$ activated by H_2.[41]

The rearrangement of relatively inert allyl ethers into easily hydrolysable vinyl ethers allows allyl groups to be used as protecting groups in organic synthesis. Allyl ether protecting groups have proved particularly useful in carbohydrate chemistry. They are good temporary blocking groups in the presence of acetate groups and can be easily removed by $(Ph_3P)_3RhCl$.[42-44]

The effect of substituents on the allyl group on the rate of isomerisation has been used to selectively remove allyl protecting groups in sugars containing several O-allyl groups. For example O-allyl groups are isomerised very quickly by $ClRh(PPh_3)_3RhCl$ (10 min - 1hr) whereas but-2-enyl groups isomerise only slowly (∿24 hr) and 3-methyl but-2-enyl groups are essentially inert. These rate differences allow O-allyl groups to be removed in preference to both but-2-enyl[43] and 3-methyl but-2-enyl groups[44] and allow the selective protection of alcohol groups.

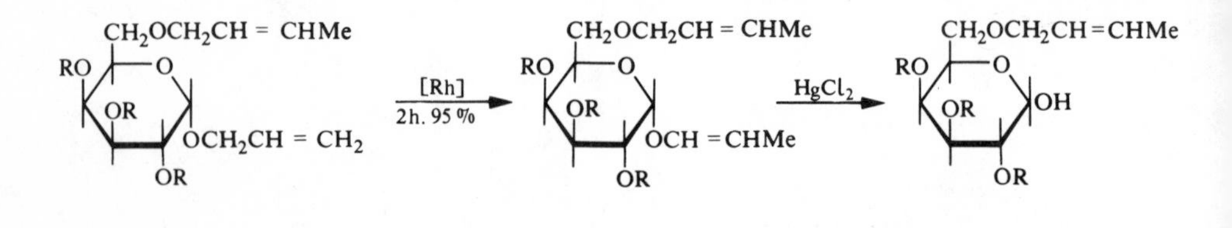

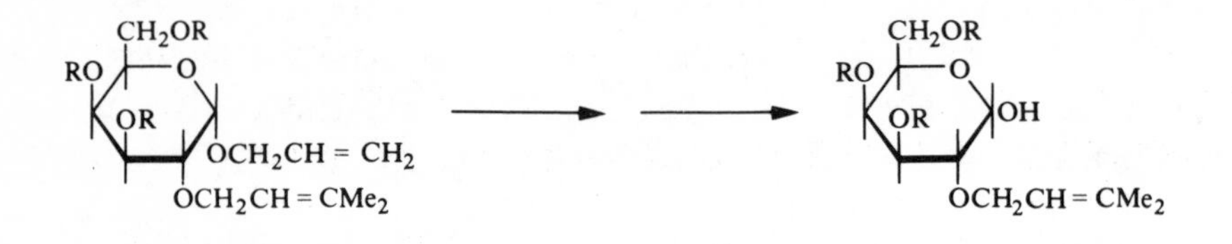

$R = CH_2Ph$

In a similar manner 2-methylene-1,3-propane diol may be used as a protecting group for aldehydes and ketones.[45]

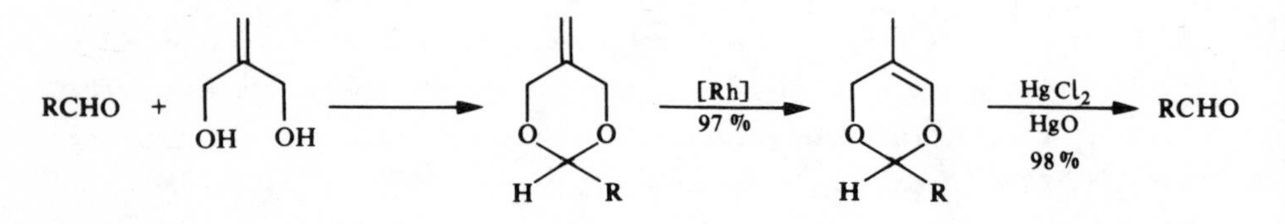

Unsymmetrical diallyl ethers are rearranged to γ,δ-unsaturated aldehydes and ketones by $(Ph_3P)_3RuCl_2$. The ruthenium catalyst selectively isomerises the least sterically hindered allyl group to produce an allyl vinyl ether which undergoes a Claisen rearrangement.[46]

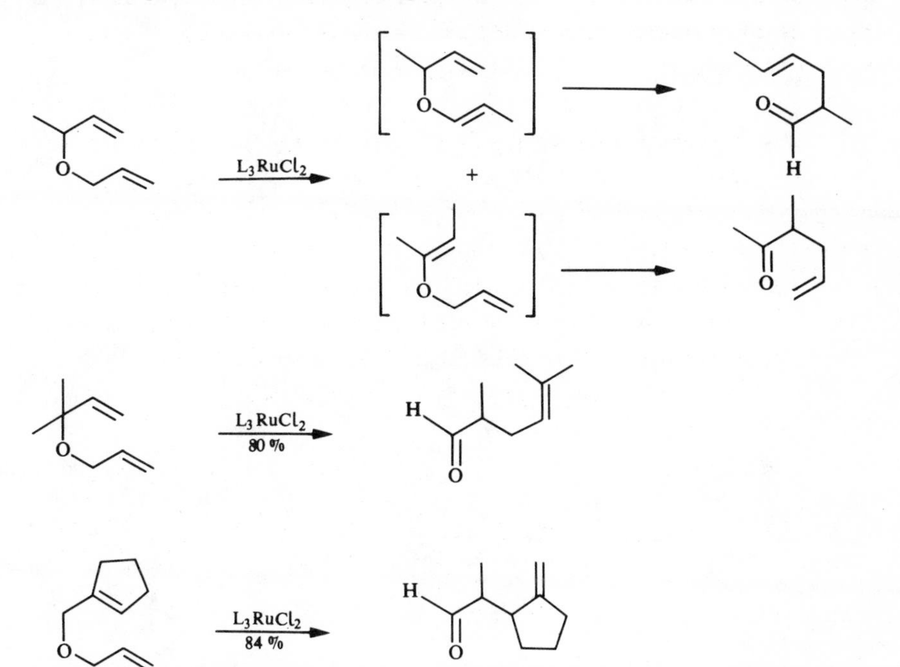

The Claisen rearrangement of allyl vinyl ethers to give 4,5-unsaturated aldehydes and ketones occurs with inversion of the allyl group. Rearrangement of allyl vinyl ethers in the presence of $(Ph_3P)_4Pd$, however, similarly gives 4,5-unsaturated aldehydes and ketones but without the allyl inversion.[46a] The mechanism presumably involves the formation of $(\eta^3$-allyl)Pd complexes.

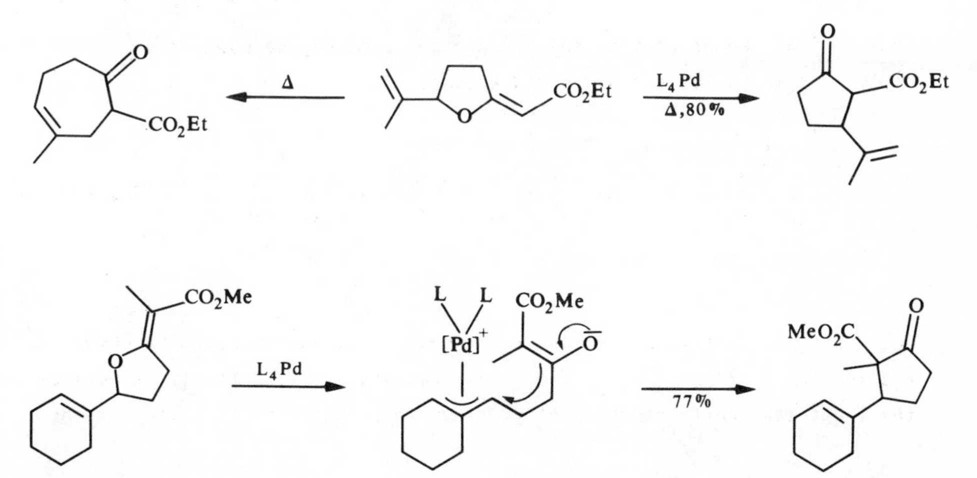

$(PhCN)_2PdCl_2$ catalyses the allylic rearrangement of allyl acetates via formation of intermediate η^3-allyl complexes.[46b] The ratio of starting allyl acetate to transposed allyl acetate is thermodynamically controlled.

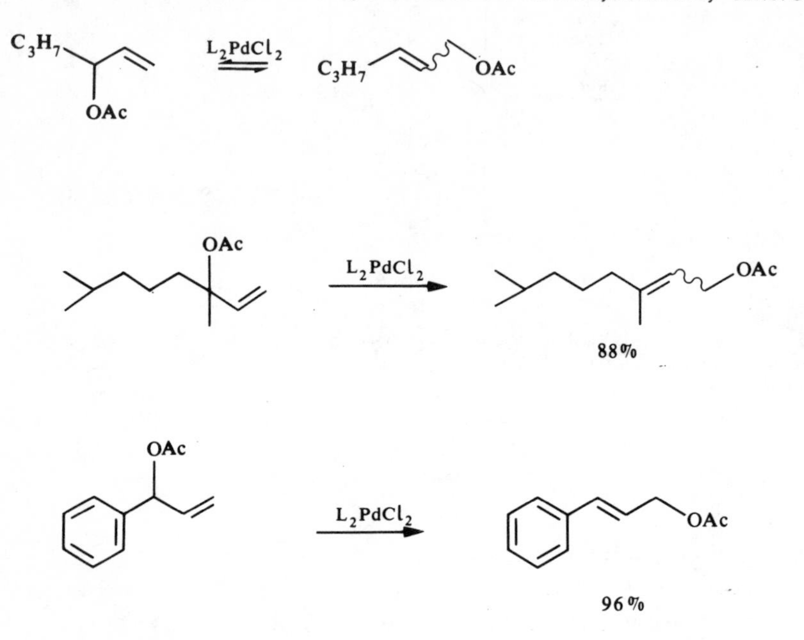

The allyl ester 34 rearranges in the presence of $[(^iPrO)_3P]_4Ni$ to give a mixture of acids 36 and 37. The mechanism is believed to involve the π-allyl complex 35, (L = $(^iPrO)_3P$).[47]

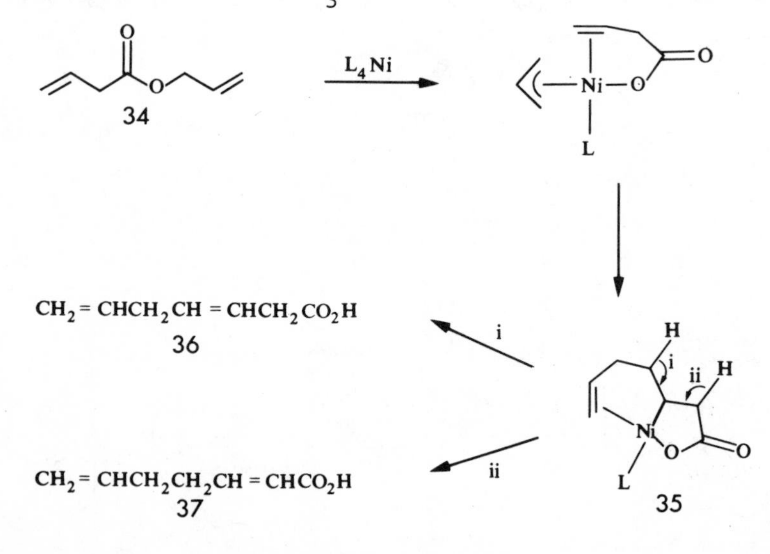

Allyl amines may be isomerised to enamines with $(Ph_3P)_3Co(N_2)H$.[47a] This catalyst is not effective for the isomerisation of allyl ethers or allyl alcohols.

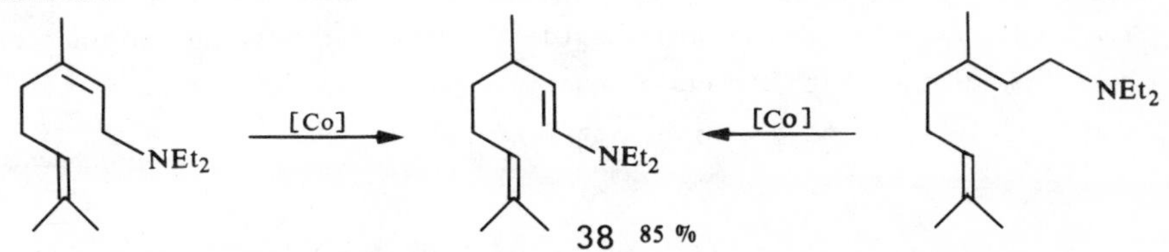

The use of Co catalysts containing chiral phosphines, for the above reaction, results in the formation of chiral enamines.

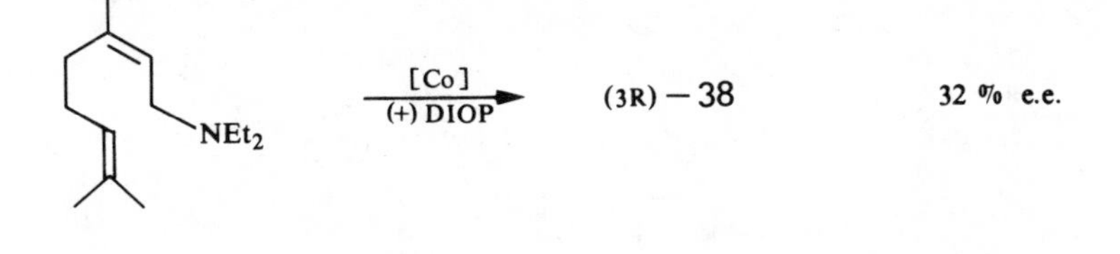

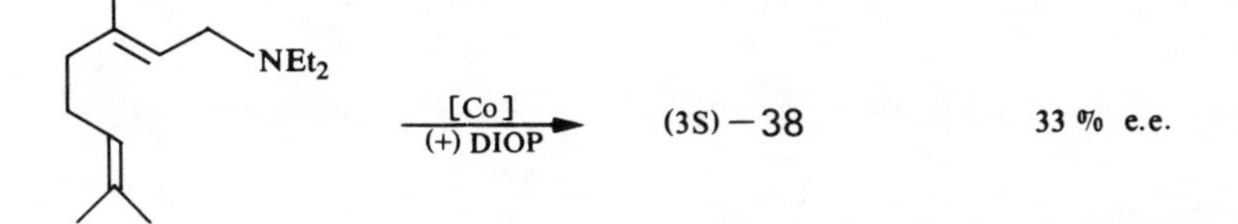

N-Allyl amides are rearranged to prop-2-enyl amides by photolysis in the presence of $Fe(CO)_5$ in yields of 50-80%.[48]

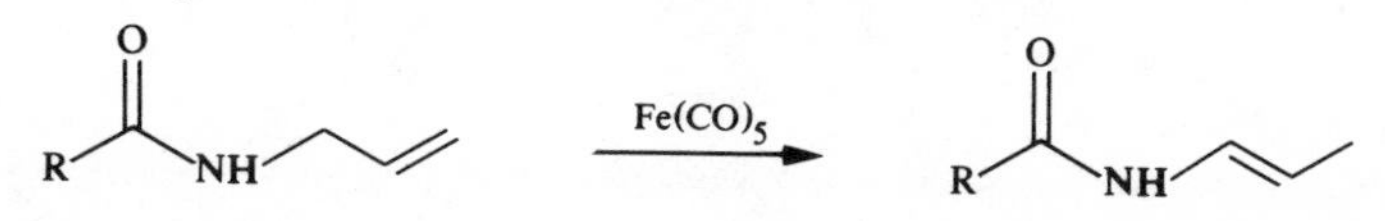

R = Me, NH_2, PhNH, PhO

Protection of amides in this way has been used in the synthesis of biotin.[48a] In this case $(Ph_3P)_3RhCl$ is the best isomerisation catalyst.

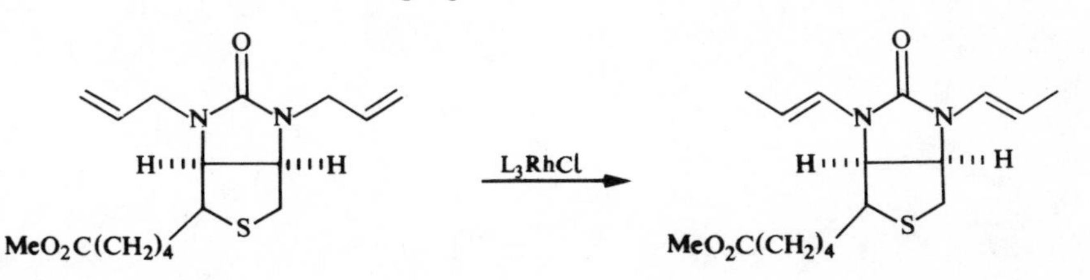

N,N-Diallyl groups have been used to protect primary amines against acids, bases and nucleophiles, for example during Li/liquid NH_3 reductions.[49] The N-allyl groups can be removed by using Wilkinson's catalyst $(Ph_3P)_3RhCl$ in acetonitrile-water. The propionaldehyde produced must be continuously removed by distillation to avoid deactivation of the catalyst due to decarbonylation of the aldehyde with formation of $(Ph_3P)_2RhCOCl$.

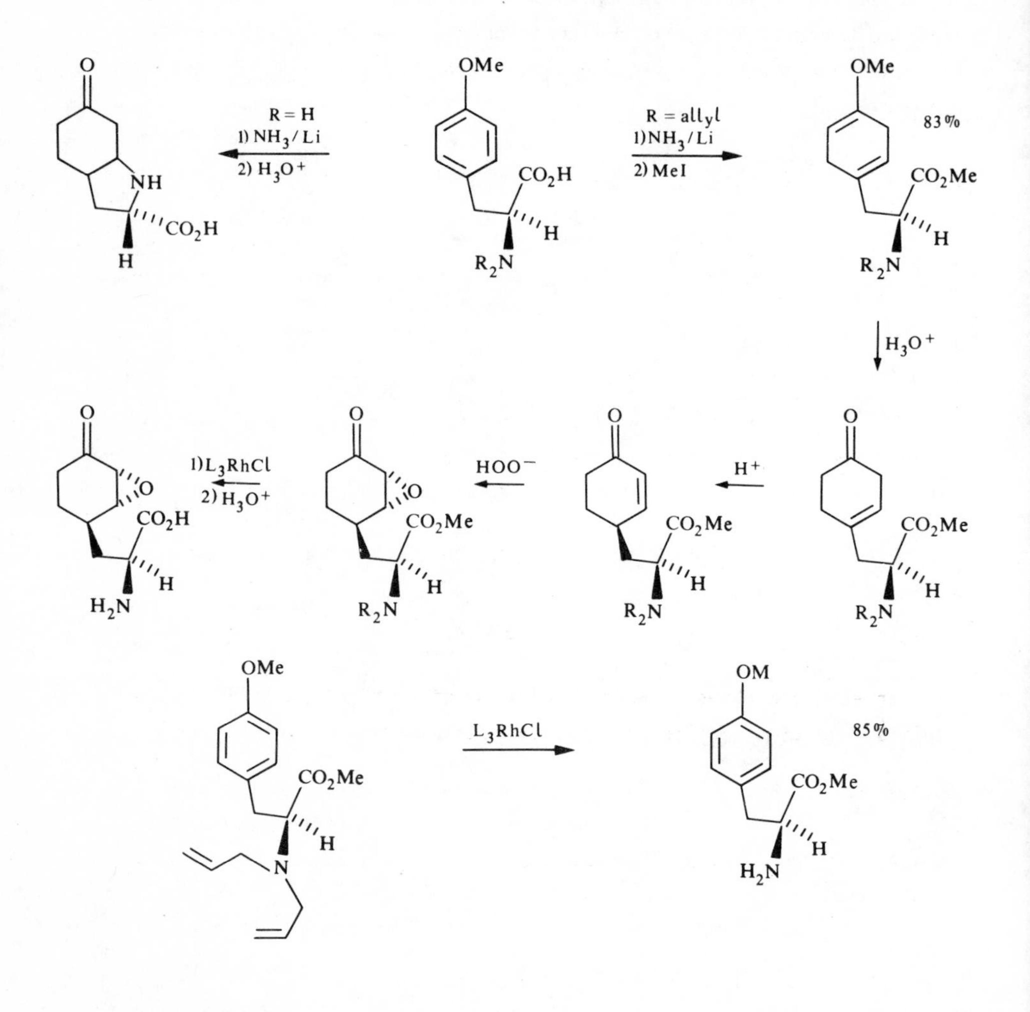

7.4 REARRANGEMENTS OF SMALL RING HYDROCARBONS

Small ring hydrocarbons (cyclopropanes, cyclobutanes and cyclobutenes) undergo a variety of rearrangement reactions in the presence of transition metal compounds.[50] The interaction of cyclopropanes and cyclobutanes with transition metals can be regarded as analogous to the interaction between a metal and an olefin. The highest occupied molecular orbitals on the small ring hydrocarbons are believed to be mainly p-orbital in character and can donate electrons into an empty metal orbital. Small ring hydrocarbons will, however, be poorer electron donors than olefins and therefore electron poor transition metal systems will interact best with them.

Bicyclo[2.1.0]pentane rearranges to cyclopentene with $[Rh(CO)_2Cl]_2$

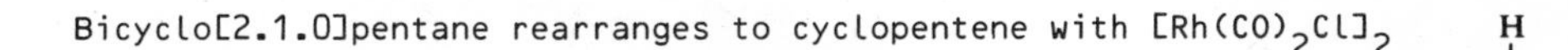

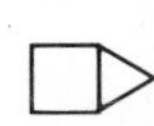

Evidence for the above mechanism comes partly from the isolation of 40 in the isomerisation of 39 by $(Ph_3P)_3RhCl$.[51]

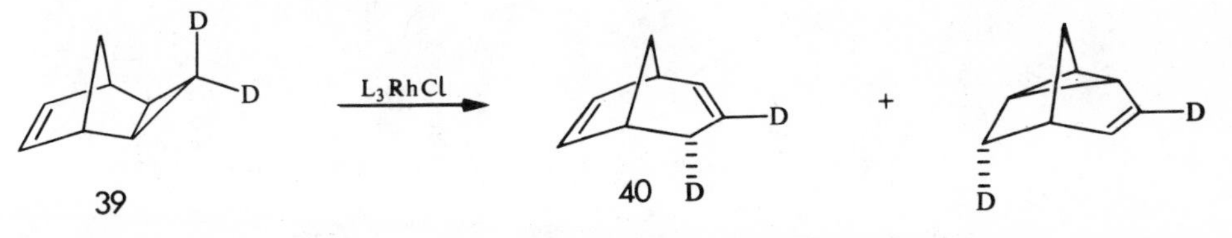

Further evidence comes from the isolation of an η^3-allyl iridium hydride species from the reaction of phenylcyclopropane and $(Ph_3P)_2IrN_2Cl$.[52]

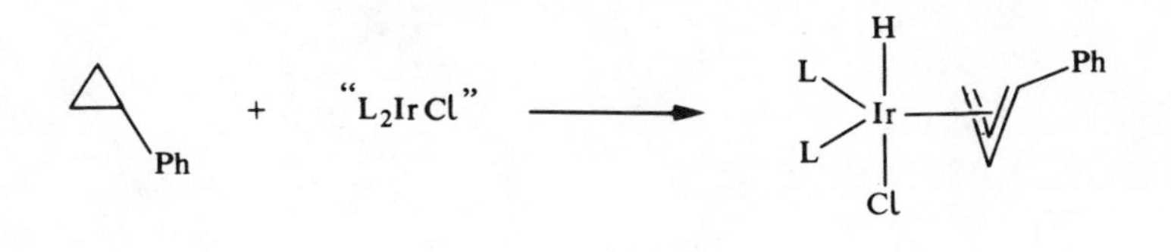

The *cis*-cyclopropyl derivative 41 is slowly rearranged to 1,5-cyclononadiene by $PdCl_2(PhCN)_2$. If, however, the reaction is quenched by CN^- after 18 min the *trans*-cyclopropyl isomer 42 can be isolated. This provides a convenient method of synthesis of 42.[53]

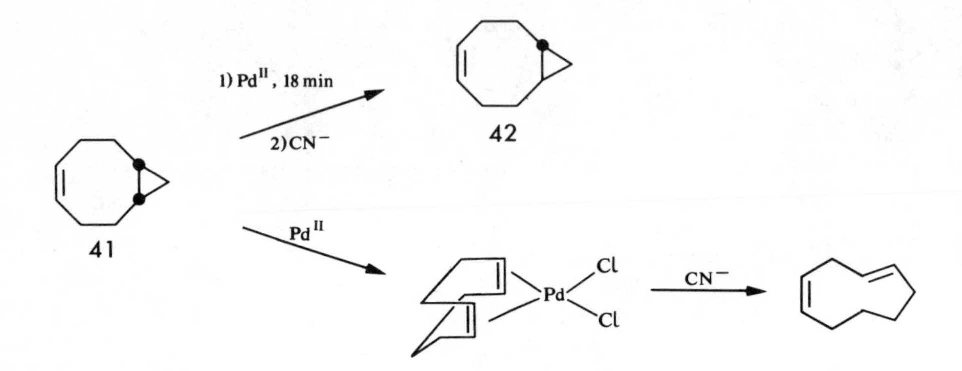

Catalytic isomerisation of 43 gives only the *cis*-dihydroindene 44 whereas thermolysis alone produces a mixture of *cis* and *trans* isomers in the ratio of 9:1.[54]

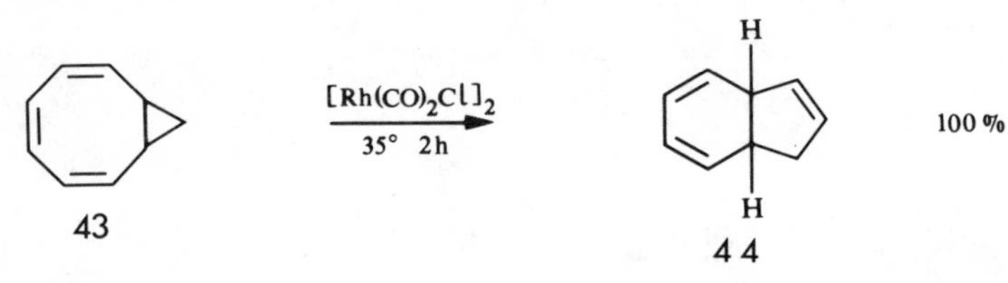

It is interesting that thermolysis of the $Mo(CO)_3$ complex of 43 provokes a different type of rearrangement to 45.[55]

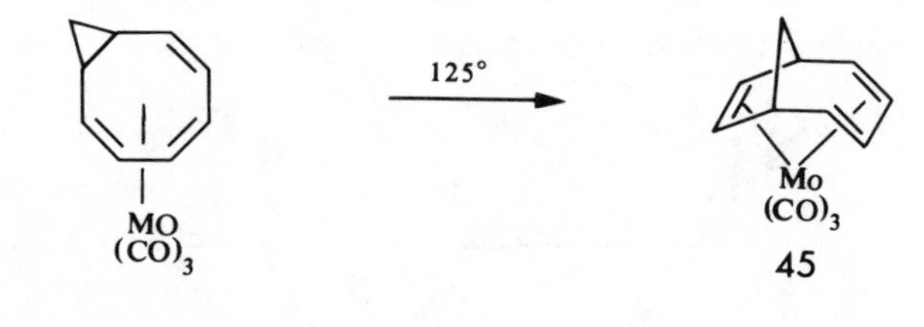

The bicyclononatriene in 45 can also be prepared by catalytic isomerisation of the tricyclononadiene 46 with $[Rh(CO)_2Cl]_2$.[55]

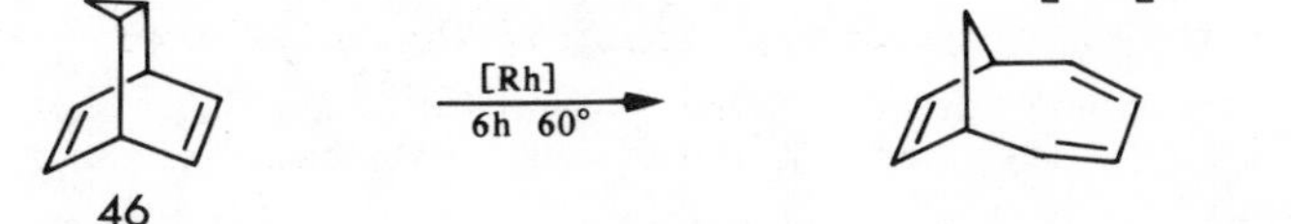

The *exo* isomer 47 rearranges in the presence of $[IrCl(CO)(PPh_3]_2$ to give 49 presumably via 48. The *endo* isomer 50 which cannot form an intermediate analogous to 49 was inert.[57] The tetracyclononane 51 was, however, susceptible to rearrangement.

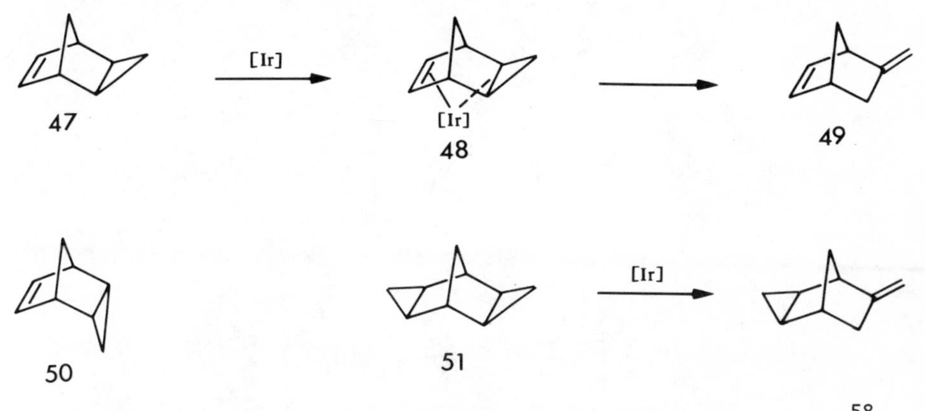

Cyclobutenes are also readily isomerised by transition metal species.[58]

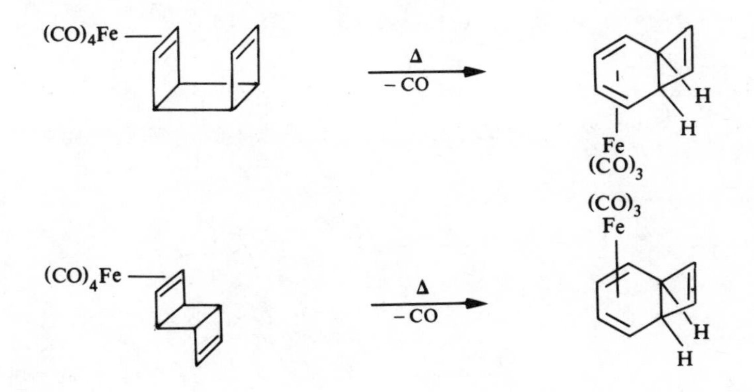

7.5 ISOMERISATION OF SMALL RING HETEROCYCLES

7.5.1 Oxygen heterocycles

Epoxides undergo a wide variety of rearrangements catalysed by transition metals. Simple epoxides are generally isomerised to ketones by $Co_2(CO)_8$[59] or by $(Ph_3P)_3RhCl$.[60]

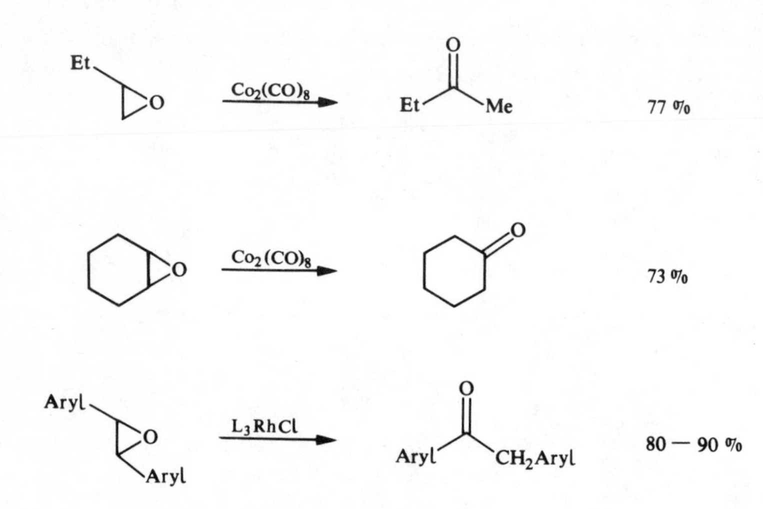

The epoxides derived from α,β-unsaturated ketones are readily isomerised to 1,3-diketones by $(Ph_3P)_4Pd$ or $(diphos)_2Pd$ at 80-140°.[60a]

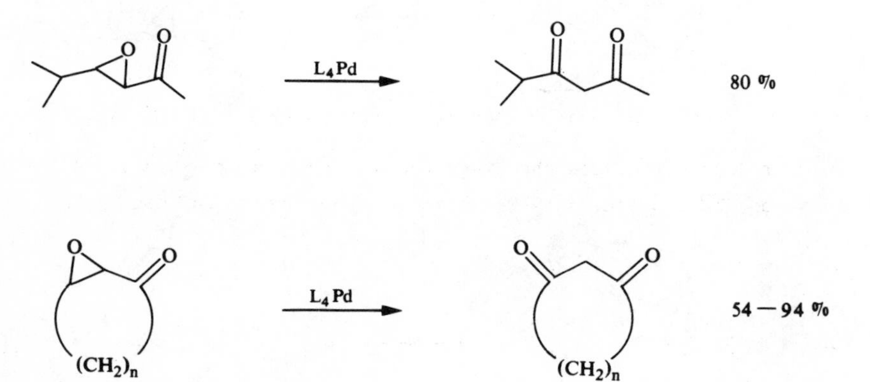

n = 2 — 5,9

When monoepoxides of 1,3-dienes are isomerised in the presence of Pd complexes, the type of product is dependent on the substitution pattern. If there is an alkyl substituent in the 2 or 4 positions then allylic alcohols tend to be produced.[61]

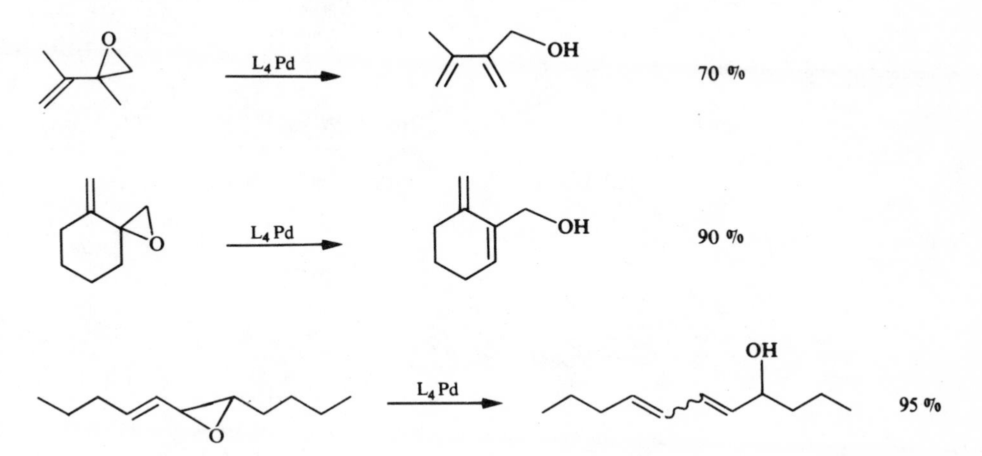

Monoepoxides of medium ring 1,3-dienes however give non-conjugated enones with $(Ph_3P)_4Pd$.[61]

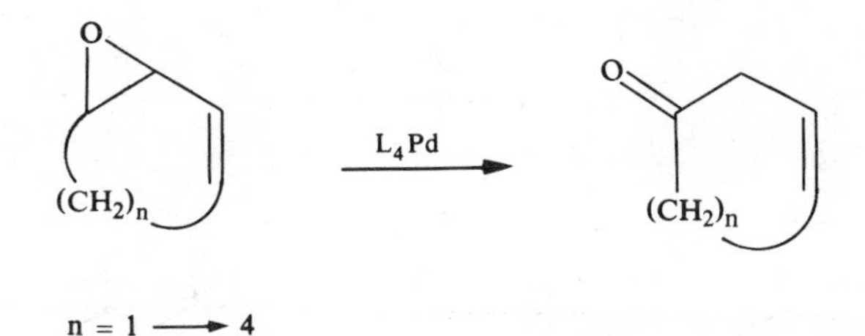

This procedure provides the basis for a synthesis of 4-hydroxy-2-cyclopentenone 52 a useful intermediate for the synthesis of prostaglandins.

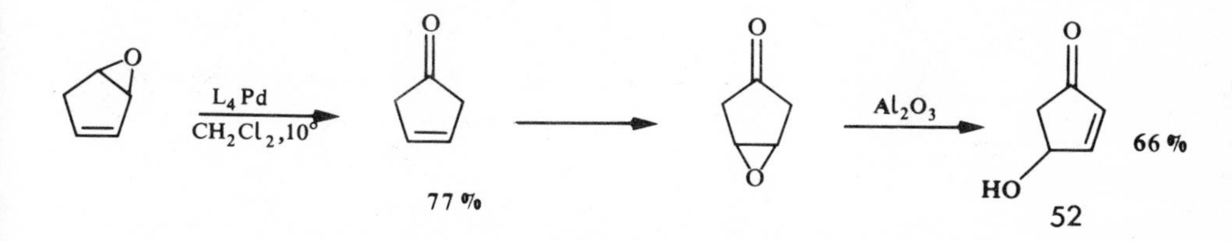

Butadiene monoepoxide rearranges in the presence of $[Rh(CO)_2Cl]_2$ to give *trans*-crotonaldehyde.[62]

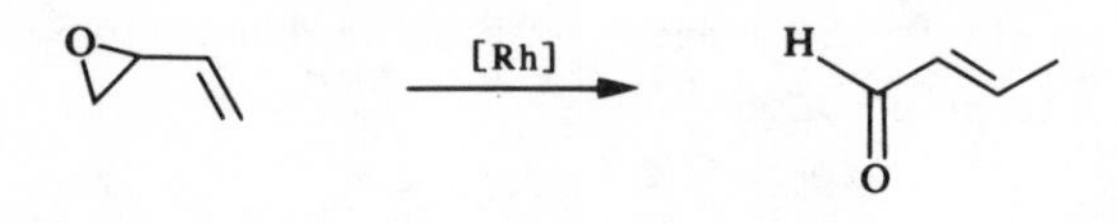

The monoepoxide of cyclooctatetraene 53 is isomerised by $Fe_3(CO)_{12}$ in benzene[63] or by $[Rh(CO)_2Cl]_2$ at -50°.[54] The epoxide alone is thermally stable up to $\sim 400^o$.

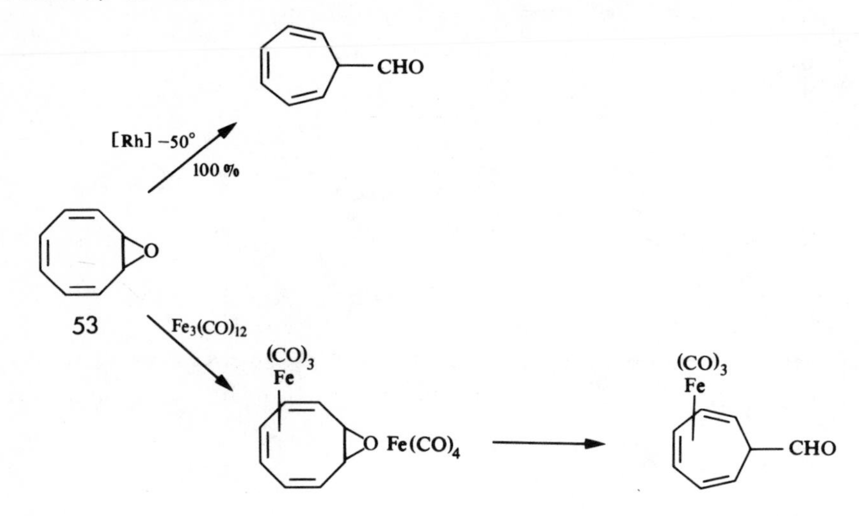

The monoepoxide of cyclohexa-1,3-diene is isomerised to the *endo*-5-hydroxycyclohexadiene iron tricarbonyl complex 54 on photolysis with $Fe(CO)_5$.[65]

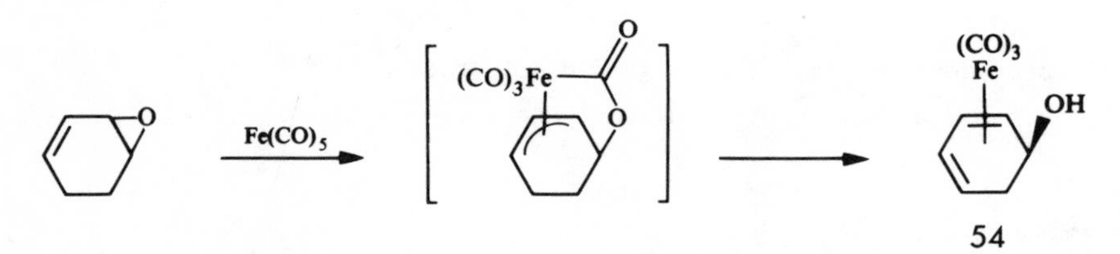

Photolysis of $Fe(CO)_5$ in the presence of 53 results in the formation of the $Fe_2(CO)_7$ complex 55 from which the previously unknown 9-oxabicyclo [4.2.1]nona-2,4,7-triene 56 can be released with Me_3NO.[64]

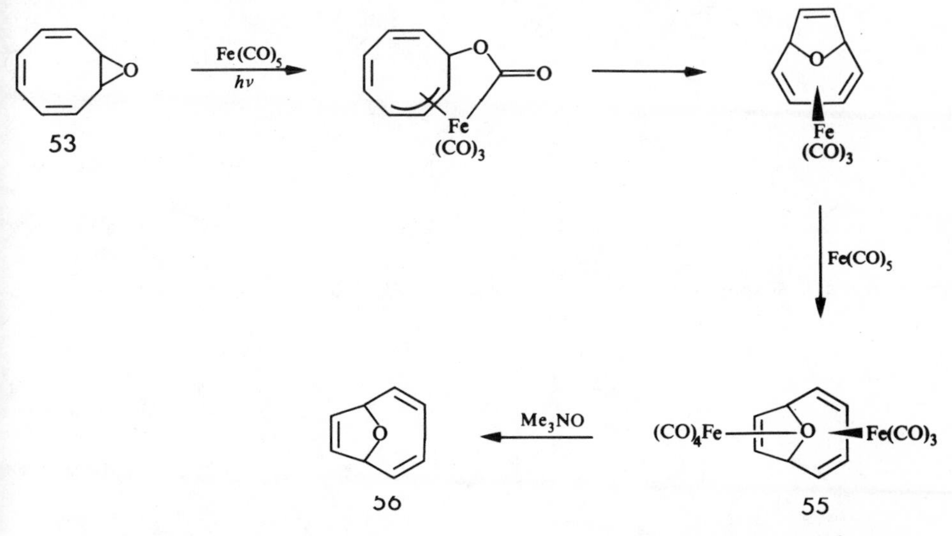

The conversion of an epoxide to an allylic alcohol normally requires strong base and forcing conditions. Treatment of the epoxides 57, 58 and 59 with $Fe_2(CO)_9$ at 50°, however, smoothly converts them to the allylic alcohols shown.[66] The cyclopropane rings in 57 and 58 are

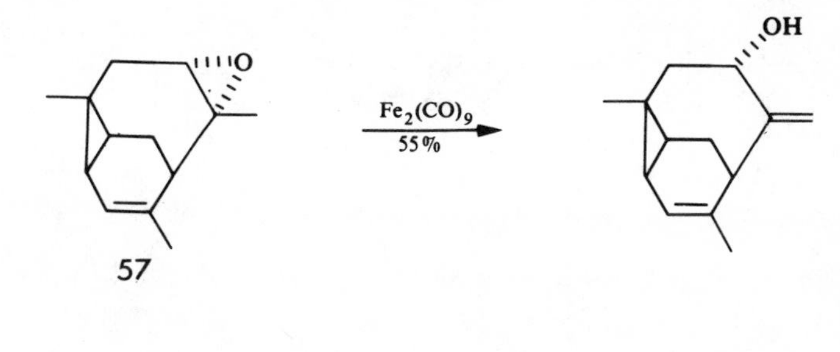

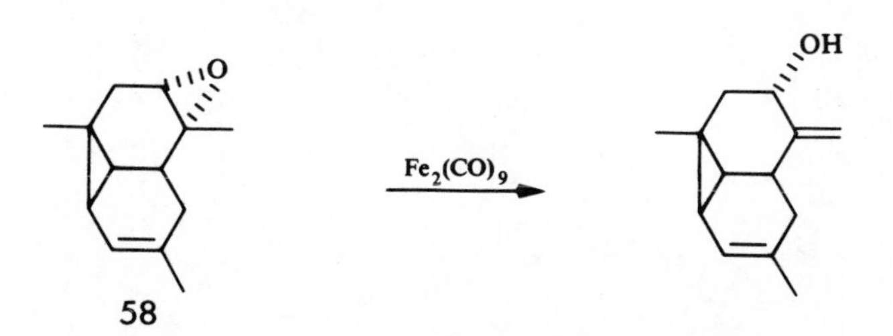

unaffected under conditions that cause isomerisation of the epoxides.

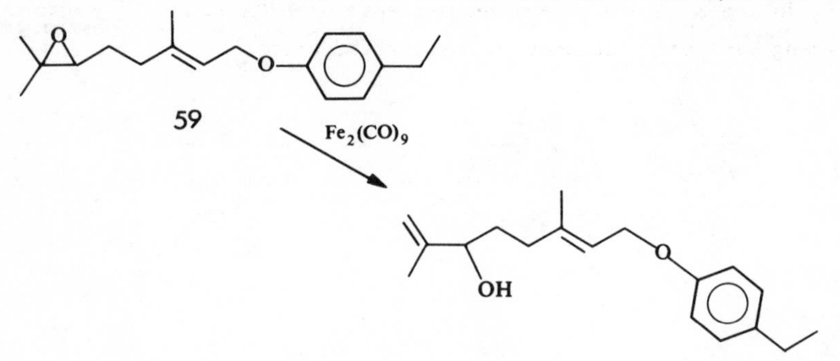

Oxetanes are susceptible to rearrangement by $[Rh(CO)_2Cl]_2$ and generally give olefins and aldehydes in high yields.[64]

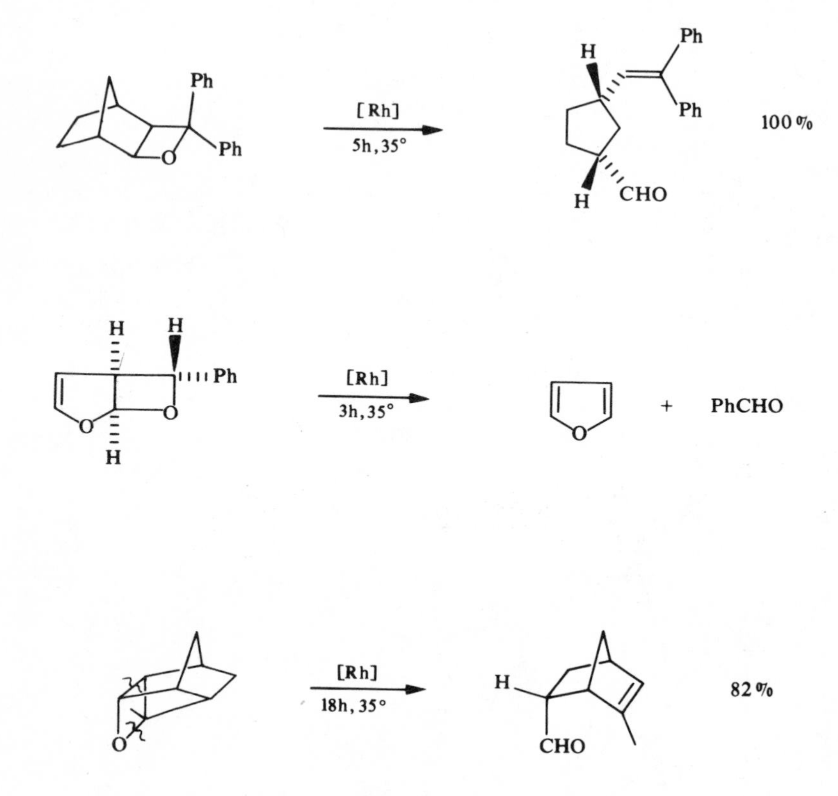

$PdCl_2$ in the presence of four equivalents of $P(OMe)_3$ at 20° catalyses the rearrangement of β-lactones to unsaturated acids.[67]

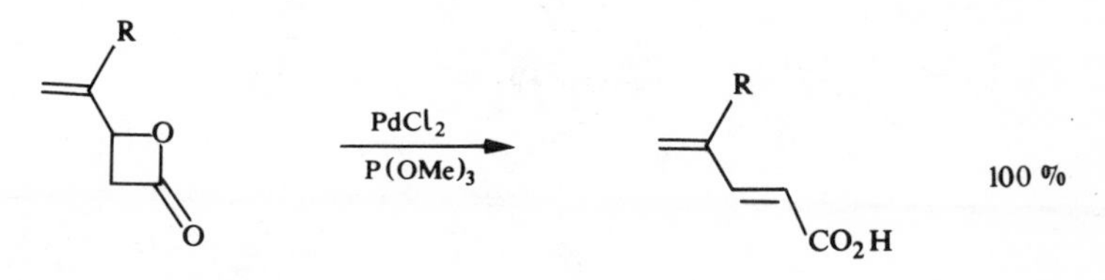

7.5.2 *Nitrogen heterocycles*

$(PhCN)_2PdCl_2$ catalytically rearranges the aziridine 60 to 61 in high yield.[68]

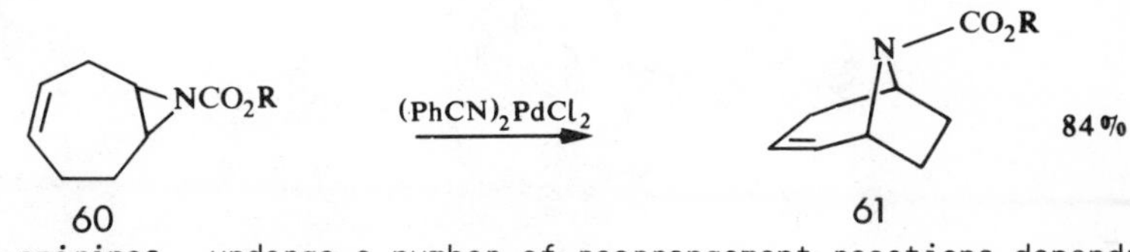

2-Aryl **azirines** undergo a number of rearrangement reactions depending on the metal carbonyl complex present.[69]

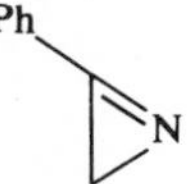

$Co_2(CO)_8$

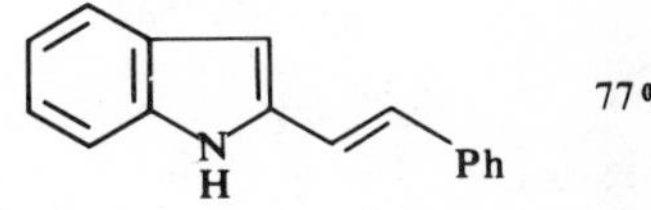

77 %

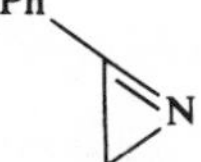

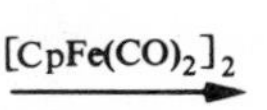

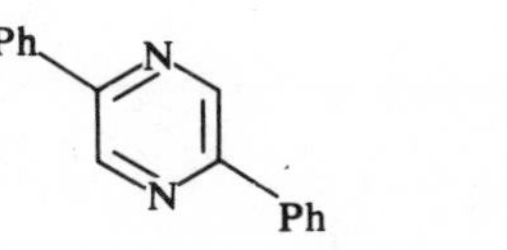

39 %

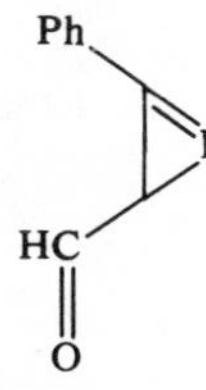

$Mo(CO)_6$

Ph

O–N

81 %

Finally the diazatriene 62 rearranges to the pyrrole 63 on treatment with $Fe_2(CO)_9$.[70]

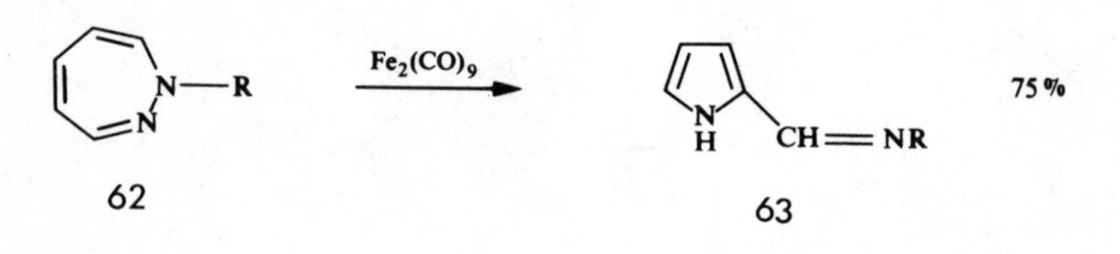

7.6 REFERENCES

1. R. Cramer and R.V. Lindsey Jr., *J. Amer. Chem. Soc.*, 1966, *88*, 3534.
2. R. Cramer, *Acc. Chem. Res.*, 1968, *1*, 186.
3. L. Roos and M. Orchin, *J. Amer. Chem. Soc.*, 1965, *87*, 5502.
4. E.O. Sherman Jr. and M. Olson, *J. Organometal. Chem.*, 1979, *172*, C13.
5. T.A. Manuel, *J. Org. Chem.*, 1962, *27*, 3941.
6. C.F. Lochow and R.G. Miller, *J. Org. Chem.*, 1976, *41*, 3020.
7. E. Stamm, K.B. Becker, P. Engel and R. Keese, *Helv. Chim. Acta*, 1979, *62*, 2181.
8. R. Damico, *J. Org. Chem.*, 1968, *33*, 1550.
9. P.A. Grieco, M. Nichizawa, N. Marinovic and W.J. Ehmann, *J. Amer. Chem. Soc.*, 1976, *98*, 7102.
10. J. Andrieux, D.H.R. Barton and H. Patin, *J.C.S. Perkin I*, 1977, 359.
11. Y. Pickholtz, Y. Sasson, J. Blum, *Tet. Letters*, 1974, 1263.
12. J.F. Biellmann and M.J. Jung, *J. Amer. Chem. Soc.*, 1968, *90*, 1673.

12a. M. Franck-Neumann and F. Brion, *Ang. Chem. Int. Ed.*, 1979, *18*, 688.

13. J.E. Arnet and R. Pettit, *J. Amer. Chem. Soc.*, 1961, *83*, 2954.
14. M. Gargano, P. Giannoccaro, and H. Rossi, *J. Organometal.Chem.*, 1975, *84*, 389.
15. M. Frye, E. Kuljian and J. Viebrock, *Inorg. Chem.*, 1965, *4*, 1499.
16. R.E. Rinehart and J.S. Lasky, *J. Amer. Chem. Soc.*, 1964, *86*, 2516.
17. G.F. Emerson, J.E. Mahler, R. Kochmar and R. Pettit, *J. Org. Chem.*, 1964, *29*, 3620.
18. A.J. Birch and K.B. Chamberlain, *Org. Synth.*, 1977, *57*, p.16.
19. P. McArdle and T. Higgins, *Inorg. Chim. Acta*, 1978, *30*, L303.
20. A.J. Birch and G.S.R. Subba Rao, *Tet. Letters*, 1968, 3797.
21. F. Turecek, H. Antropiusova, K. Mach, V. Manus and P. Sedemera, *Tet. Letters*, 1980, 637.
22. P.A. Grieco and N.Marinovic, *Tet. Letters*, 1978, 2545.
23. E.J. Corey and G. Moinet, *J. Amer. Chem. Soc.*, 1973, *95*, 7185.
24. H. Alper and J.T. Edward, *J. Organometal. Chem.*, 1968, *14*, 411.
25. D.H.R. Barton, S.G. Davies and W.B. Motherwell, *Synthesis*, 1979, 265.
26. L.E. Overman, and F.M. Knoll, *J. Amer. Chem. Soc.*, 1980, *102*, 865.
27. J.C. Trebellas, J.R. Olechowski and H.B. Jonassen, *J. Organometal. Chem.*, 1966, *6*, 412 (see also: P. Heimbach and M. Molin, *J. Organometal. Chem.*, 1973, *49*, 477).

28. E.D. Brown, T.W. Sam, J.K. Sutherland and A. Torre, *J.C.S. Perkin I*, 1975, 2326.

29. G.F. Emerson and R. Pettit, *J. Amer. Chem. Soc.*, 1962, *84*, 4591.

30. R.W. Goetz and M. Orchin, *J. Amer. Chem. Soc.*, 1963, *85*, 1549.

31. R. Damico and T.J. Logan, *J. Org. Chem.*, 1967, *32*, 2356.

32. W. Strohmeier and L. Weigelt, *J. Organometal. Chem.*, 1975, *86*, C17.

33. W.T. Hendrix, F.G. Cowherd and J.L. von Rosenberg, *Chem. Comm.*, 1968, 97.

34. F.G. Cowherd and J.L. von Rosenberg, *J. Amer. Chem. Soc.*, 1969, *91*, 2157.

35. Y. Sasson and G.L. Rempel, *Tet. Letters*, 1974, 4133.

36. D. Baudry, M. Ephritikhine and H. Felkin, *Nouveau J. Chimie*, 1978, *2*, 355.

37. P.W. Jolly, F.G.A. Stone, and K. Mackenzie, *J. Chem. Soc.*, 1965, 6416.

38. E.J. Corey and J.W. Suggs, *J. Org. Chem.*, 1973, *38*, 3224.

39. P. Golborn and F. Scheinmann, *J.C.S. Perkin I*, 1973, 2870.

39a. H. Suzuki, Y. Koyama, Y. Moro-Oka and T. Ikawa, *Tet. Letters*, 1979, 1415.

40. H.C. Clark and H. Kurosawa, *Inorg. Chem.*, 1973, *12*, 357, 1566.

41. D. Baudry, M. Ephritikhine and H. Felkin, *Chem. Comm.*, 1978, 694.

42. C.D. Warren and R.W. Jeanloz, *Carbohydrate Research*, 1977, *53*, 67; C. Augé and A. Veyrières, *Carbohydrate Research*, 1977, *54*, 45; J-C. Jacquinet and P. Sinay, *J.C.S. Perkin I*, 1979, 314, 319.

43. P.A. Gent and R. Gigg, *Chem. Comm.*, 1974, 277.

44. R. Gigg, *J.C.S. Perkin I*, 1980, 738.

45. E.J. Corey and J.W. Suggs, *Tet. Letters*, 1975, 3775.

46. J. Tsuji, Y. Kobayashi and I. Shimizu, *Tet. Letters*, 1979, 39; J.M. Reuter and R.G. Salomon, *J. Org. Chem.*, 1977, *42*, 3360.

46a. B.M. Trost, T.A. Runge and L.N. Jungheim, *J. Amer. Chem. Soc.*, 1980, *102*, 2840.

46b. L.E. Overman and F.M. Knoll, *Tet. Letters*, 1979, 321; B.M. Trost, T.R. Verhoeven and J.M. Fortunak, *Tet. Letters*, 1979, 2301.

47. G.P. Chiusoli, G. Salerno and F. Dallatomasina, *Chem. Comm.*, 1977, 793.

47a. H. Kumobayashi, S. Akutagawa and S. Otsuka, *J. Amer. Chem. Soc.*, 1978, *100*, 3949.

48. A.J. Hubert, P. Moniotte, G. Goebbells, R. Warin and P. Teyssié, *J.C.S. Perkin II*, 1973, 1954.

48a. B. Moreau, S. Lavielle and A. Marquet, *Tet. Letters*, 1977, 2591.

49. B.C.Laguzza and B.Ganem, personal communication,*Tet. Letters*,1981,1483.

50. K.C. Bishop, *Chem.Rev.*, 1976, *76*, 461 and references therein.

51. T.J. Katz and S.A. Cerefice, *J. Amer. Chem. Soc.*, 1971, *93*, 1049.

52. T.H. Tulip and J.A. Ibers, *J. Amer. Chem. Soc.*, 1979, *101*, 4201.

53. G. Albelo and M.F. Rettig, *J. Organometal. Chem.*, 1972, *42*, 183.

54. R. Grigg, R. Hayes and A. Sweeney, *Chem. Comm.*, 1971, 1248.

55. W. Grimme, *Chem. Ber.*, 1967, *100*, 113.

56. T.J. Katz and S.A. Cerefice, *Tet. Letters*, 1969, 2561.

57. H.C. Volger, H. Hogeveen and M.M.P. Gaasbeek, *J. Amer. Chem. Soc.*, 1969, *91*, 218, 2137.

58. W.Slegeir, R. Case, J.S. McKennis and R. Pettit, *J. Amer. Chem. Soc.*, 1974, *96*, 287.

59. J.L. Eisenmann, *J. Org. Chem.*, 1962, *27*, 2706.

60. D. Milstein, O. Buchman and J. Blum, *J. Org. Chem.*, 1977, *42*, 2299.

60a. M. Suzuki, A. Watanabe and R. Noyori, *J. Amer. Chem. Soc.*, 1980, *102*, 2095.

61. M. Suzuki, Y. Oda and R. Noyori, *J. Amer. Chem. Soc.*, 1979, *101*, 1623.

62. G. Adames, C. Bibby and R. Grigg, *Chem. Comm.*, 1972, 491.

63. H. Maltz and G. Deganello, *J. Organometal. Chem.*, 1971, *27*, 383.

64. R. Aumann and H. Averbeck, *J. Organometal. Chem.*, 1975, *85*, C4.

65. R. Aumann, K. Fröhlich and H. Ring, *Angew. Chem. Int. Ed.*, 1974, *13*, 275.

66. K. Hayakawa and H. Scmid, *Helv. Chim. Acta*, 1977, *60*, 1942.

67. A. Noels and P. Lefebure, *Tet. Letters*, 1973, 3035.

68. G.R. Wiger and M.F. Rettig, *J. Amer. Chem. Soc.*, 1976, *98*, 4168.

69. F. Bellamy, J-L. Schuppiser and J. Streith, *Heterocycles*, 1978, *11*, 461.

70. H. Alper and T. Sakakibara, *Can. J. Chem.*, 1979, *57*, 1541.

CHAPTER 8

OXIDATION AND REDUCTION

8.1 OXIDATION

8.1.1 The Wacker Process and Related Reactions

The Wacker process is the one pot conversion of ethylene into acetaldehyde. This oxidation is performed by Pd^{II} which is reduced to $Pd.^{0}$ The reaction is rendered catalytic in Pd by $Cu^{II}Cl_2$ which reoxidises the Pd^0 to Pd^{II}. The Cu^I thus produced is in turn reoxidised to Cu^{II} by molecular oxygen;

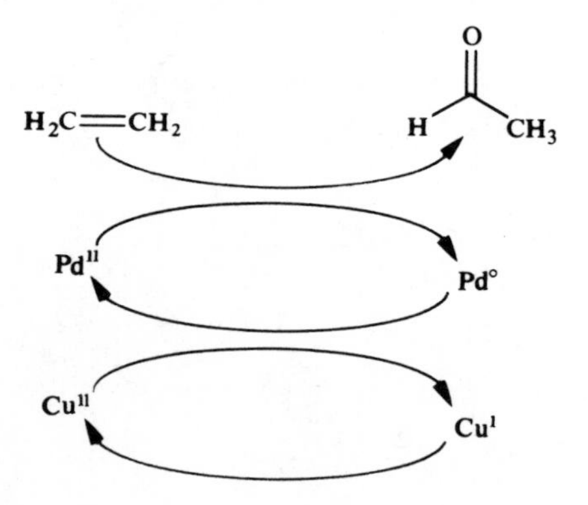

The mechanism of the Wacker process is shown below. It is believed to involve formation of a Pd(ethylene) complex 1 which is susceptible to nucleophilic attack by OH^- (see Chap. 4) to give the alkyl complex 2. β-Elimination and readdition gives the complex 3 which decomposes to acetaldehyde and Pd^0.

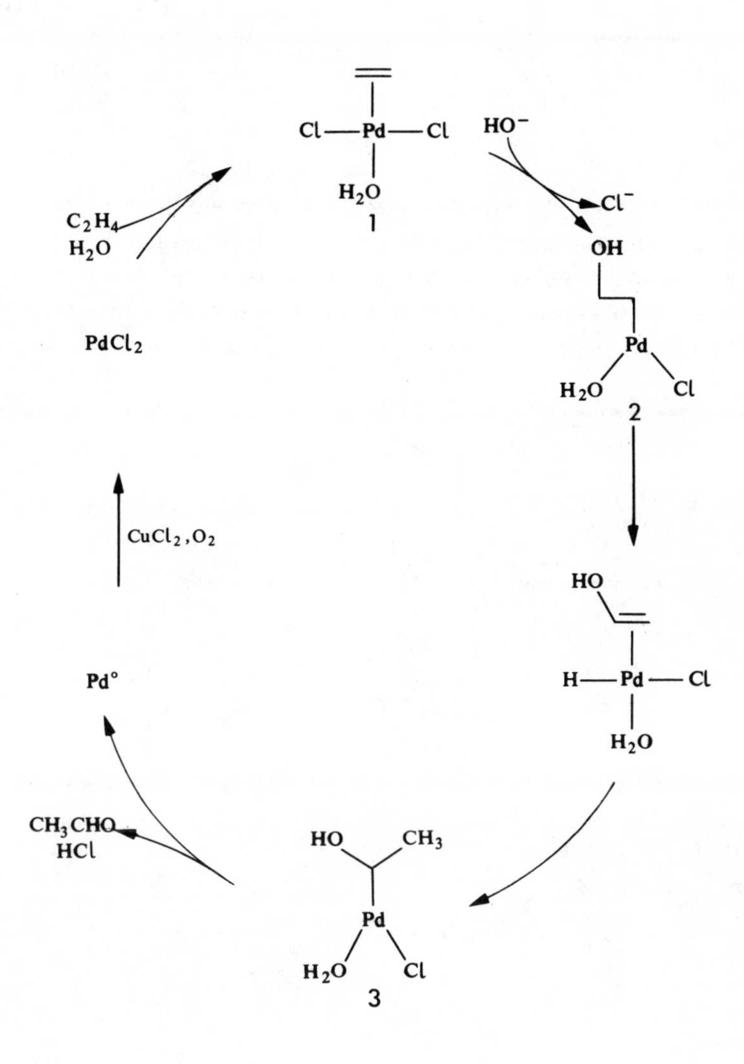

In the presence of excess chloride or acetate ions vinyl chloride and vinyl acetate are produced respectively.

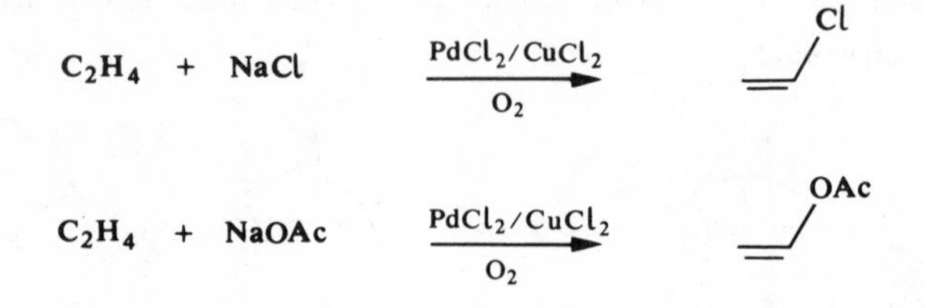

When this reaction is performed on monosubstituted olefins methyl ketones are obtained in high yields. For example 1-dodecene is converted to 2-dodecanone in 87% yield.[1] If the use of Cu^{II} or oxygen is disadvantageous then p-benzoquinone may be used to reoxidise the Pd^0 to Pd^{II}. 1,2-disubstituted olefins may also be oxidised to ketones.[2]

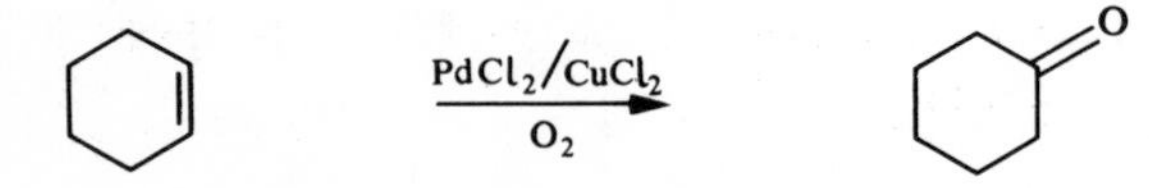

If the reaction is carried out in anhydrous alcohol as solvent then acetals are formed.[2,3] In this case Cu^{II} salts interfer in the reaction and stoicheiometric amounts of $PdCl_2$ give rise to the best yields.

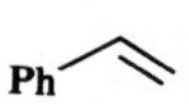

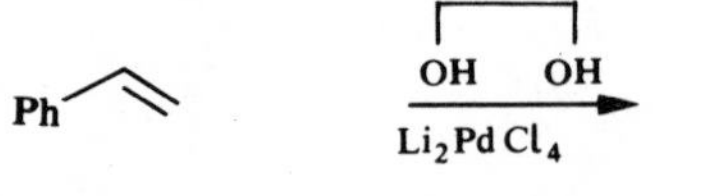

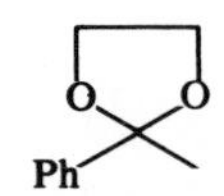

95%

Terminal double bonds may be selectively oxidised in the presence of internal double bonds.[4]

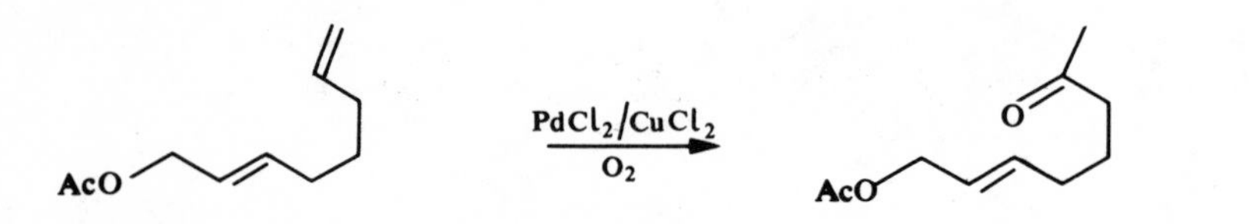

1,1-Disubstituted olefins generally fail to undergo this oxidation, unless alkyl migration is particularly favoured, due either to steric effects making coordination to Pd unfavourable or to the formation of stable π-allyl Pd complexes.[2,5]

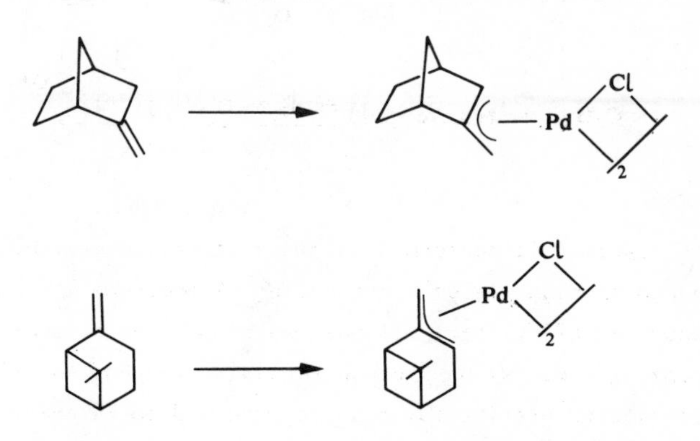

Methylene cyclobutanes, however, undergo oxidation accompanied by ring expansion to cyclopentanones.[5]

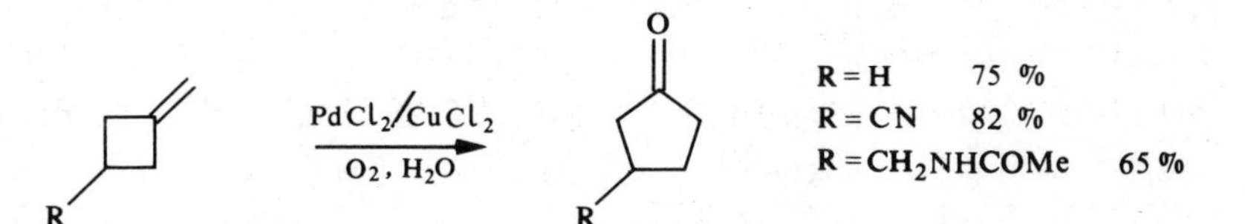

A reaction related to the Wacker process is the aromatic substitution reaction shown below. A variety of nucleophiles (OAc^-, N_3^-, Cl^-, NO_2^-, Br^-, CN^-, SCN^-) and oxidants (CrO_3, $Pb(OAc)_4$, $NaClO_3$, $KMnO_4$, $NaNO_3$, $NaNO_2$)

$$\text{Aryl—H} + X^- \xrightarrow[\text{Oxidant}]{Pd^{II}} \text{Aryl—X}$$

have been successfully used.[6]

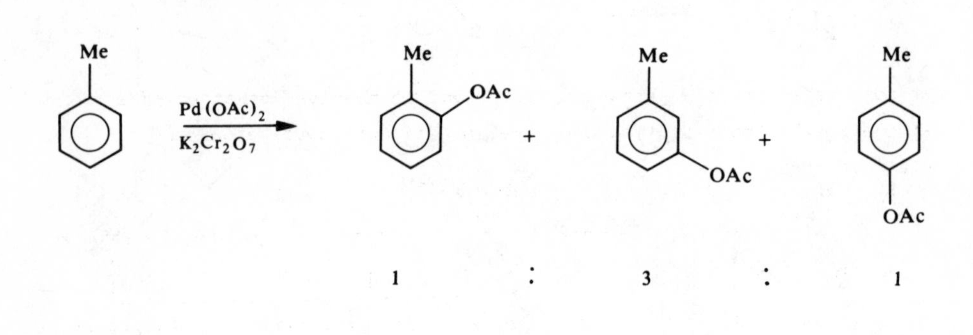

8.1.2. Dehydrogenation Reactions

Primary and secondary alcohols may be dehydrogenated to aldehydes and ketones at room temperature using $(Ph_3P)_3RuCl_2$ as catalyst with N-methylmorpholine-N-oxide as reoxidant for the Ru.[7] For simple alcohols and allylic alcohols high yields of product are obtained. However, the oxidation fails for homoallylic alcohols (e.g. cholesterol) presumably because of the formation of inert Ru alkoxy–olefin complexes.

Oxidation by $(Ph_3P)_3RuCl_2$ and N-methylmorpholine N-oxide

Cyclododecanol	100%
d-Carveol	94%
5α-cholester-3β-ol	87%
testosterone	94%
17-Hydroxy-5α-Δ^2-androstene	83%
3α-Hydroxy-5α-Δ^1-cholestene	87%
Cholesterol	0%
1-Nonene-4-ol	7%

Palladium chloride may also be used to oxidise secondary alcohols to ketones. In this case O_2 is used as reoxidant.[8] Olefins and unhindered

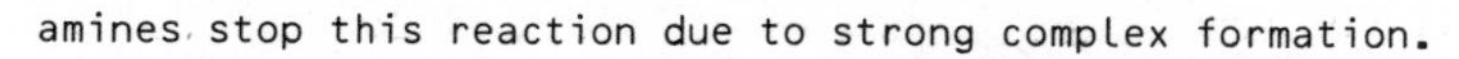

amines stop this reaction due to strong complex formation.

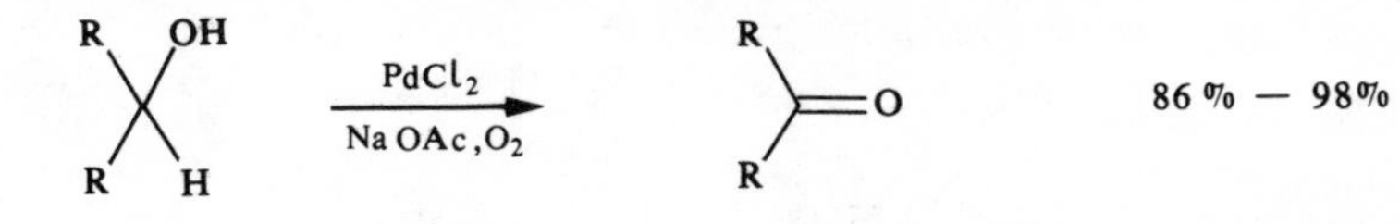

For both catalysts the mechanism is thought to involve formation of a metal-alkoxyl complex followed by β-elimination to give ketone and metal-hydride.

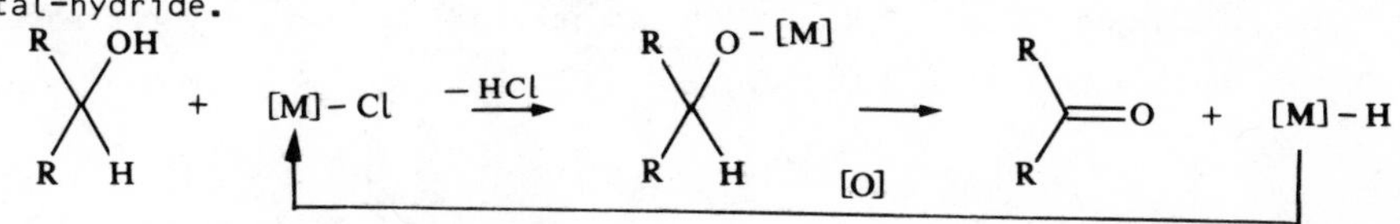

Amines may be oxidised to aldehydes or ketones using the cation $[(C_5H_5)_2$-

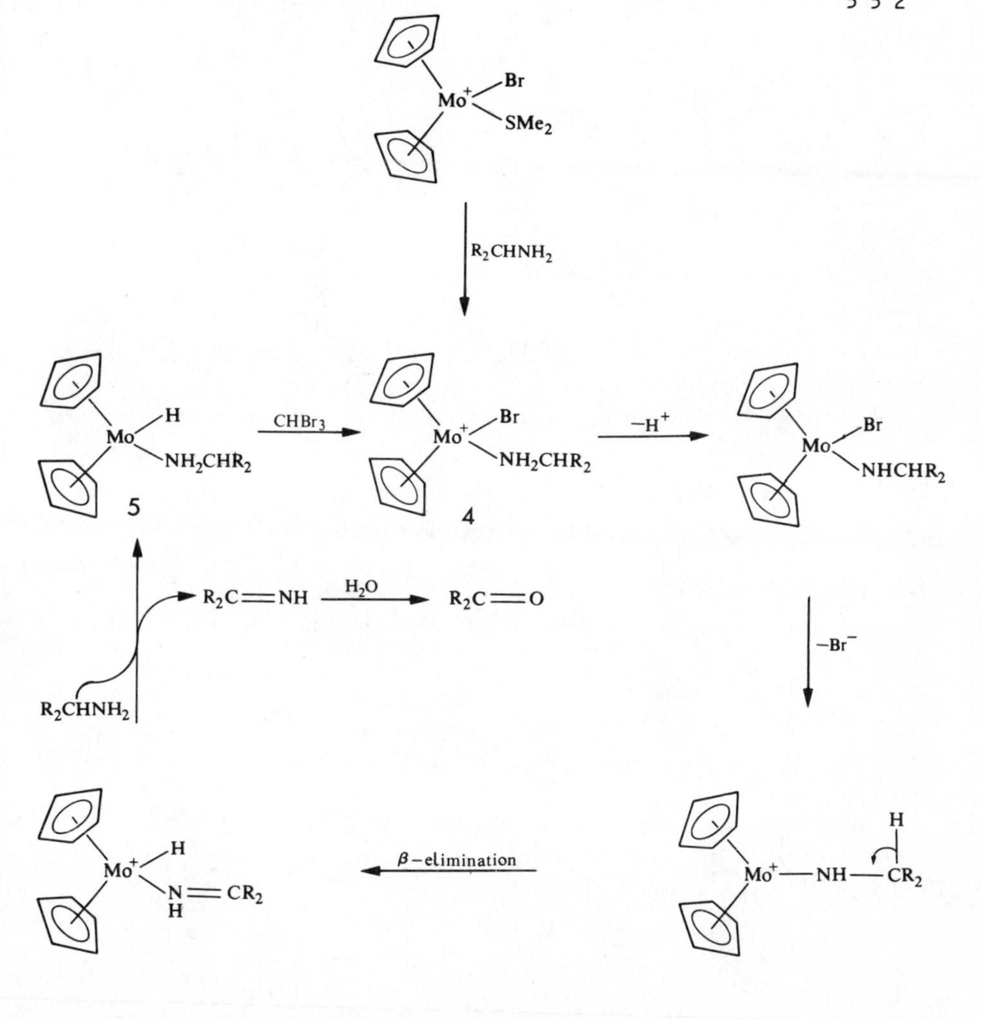

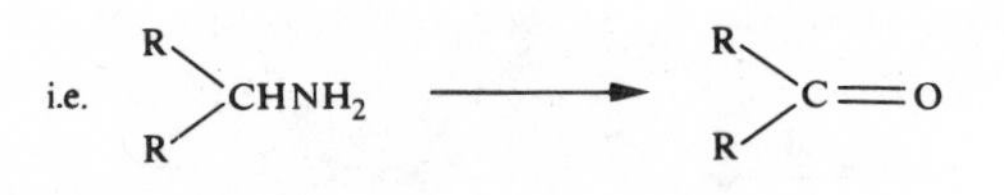

$Mo(SMe_2)Br]^+PF_6^-$.[9] The reaction has been shown to proceed through the intermediates 4 and 5 which have been isolated.

$PdCl_2$, $(PhCN)_2PdCl_2$ and $(C_6H_{10})_2PdCl_2$ stoicheiometrically dehydrogenate cyclohexanones to cyclohexenones.[10]

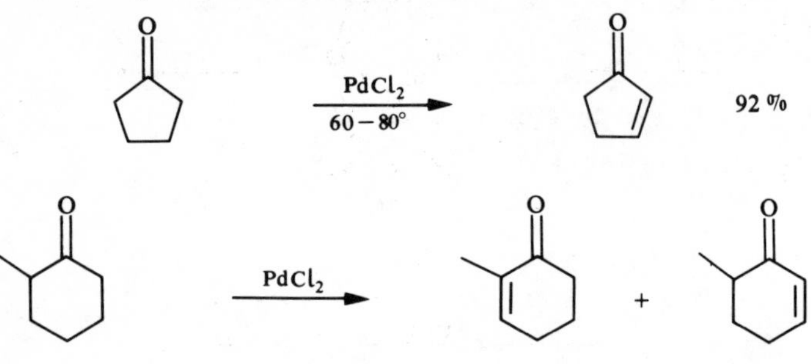

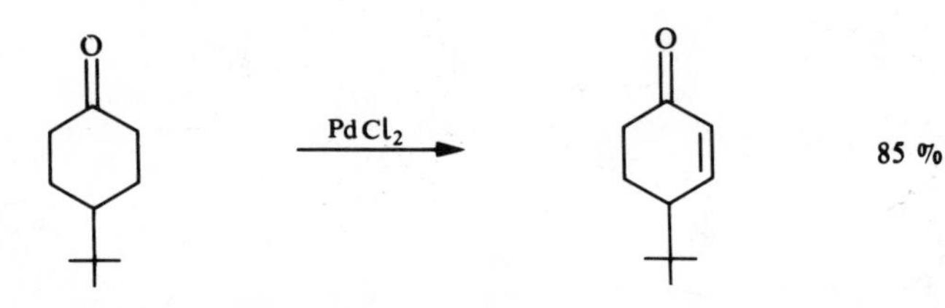

$(Ph_3P)_3RhCl$ catalyses hydrogen transfer between cycloheptene and aromatic cyclic amines, such as indoline and tetrahydroquinoline.[11]

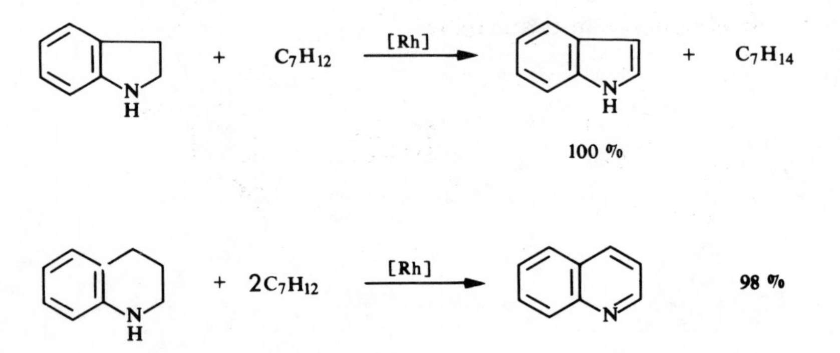

8.1.3 Epoxidation reactions with ROOH

Olefins may be selectively epoxidised by alkyl hydroperoxides (usually t-butyl, t-amyl or cumyl) in the presence of transition metal catalysts. This reaction has many characteristics similar to the reactions of organic peracids.

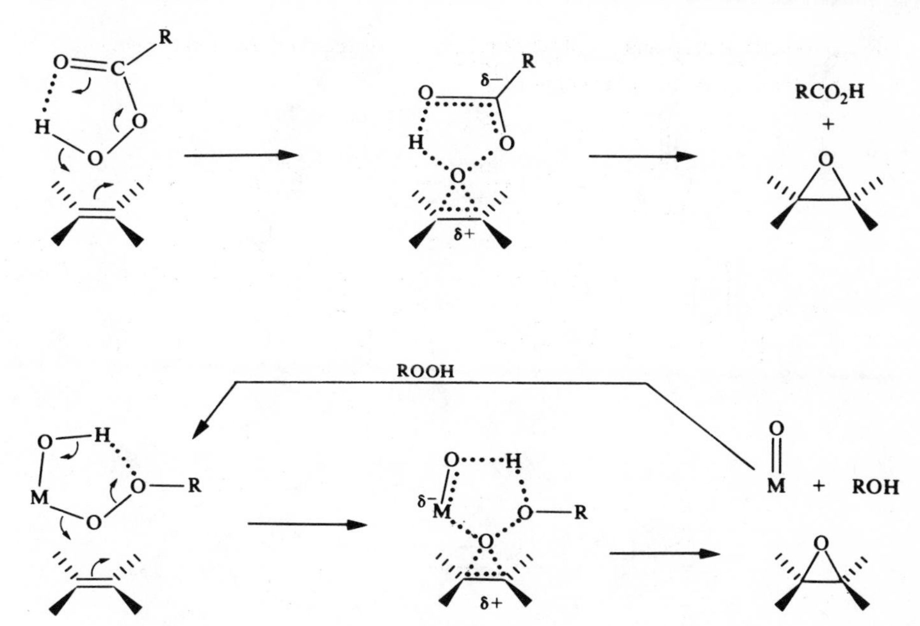

As in the case of organic peracids, the rate of epoxidation increases with increasing olefin substitution.[12]

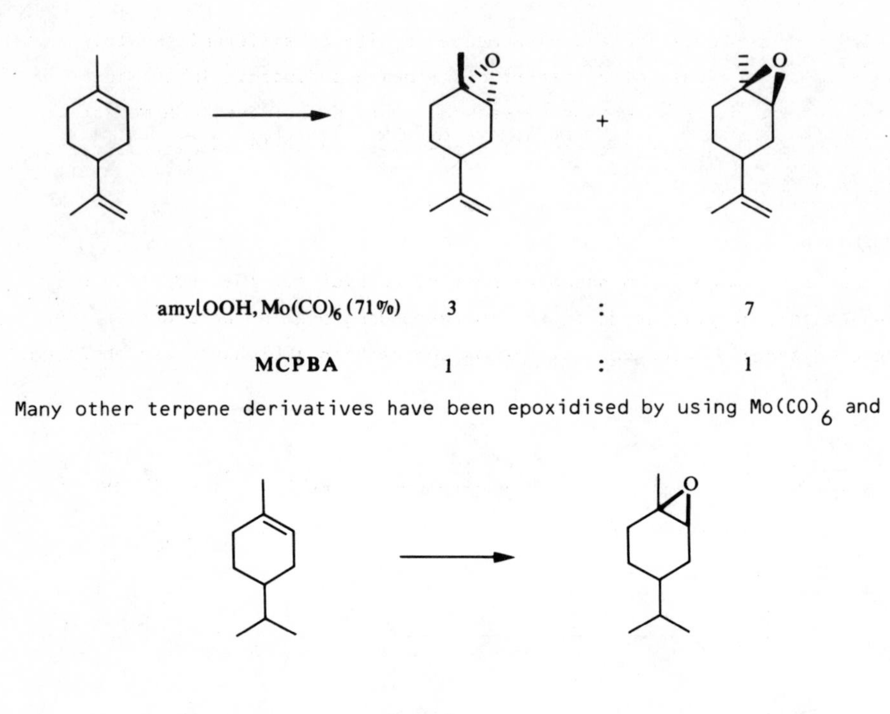

Many other terpene derivatives have been epoxidised by using $Mo(CO)_6$ and

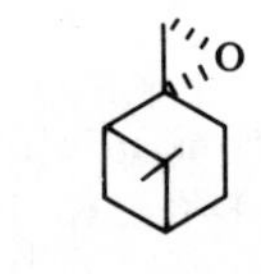

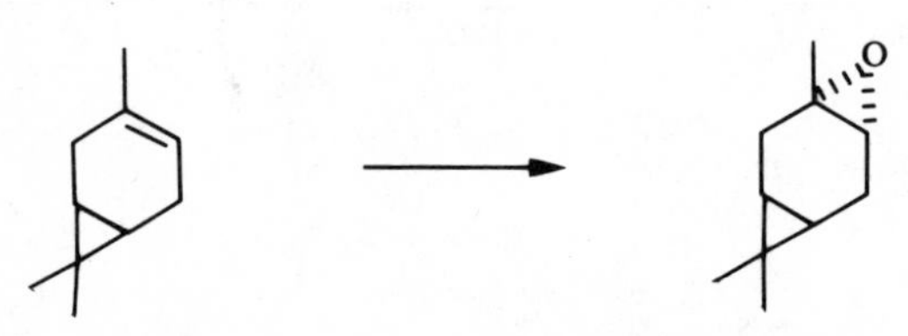

alkyl hydroperoxide usually with a high degree of stereoselectivity.[13]

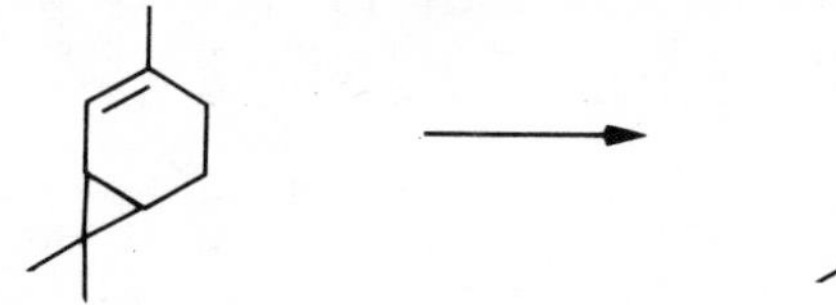

For epoxidations with $VO(acac)_2$ and t-BuOOH, the presence of an allylic or homoallylic hydroxyl group increases the rate and causes preferential *syn* epoxidation. This effect is due to coordination of the OH function to the transition metal catalyst. Cyclohexen -3-ol is epoxidised 200 times faster than cyclohexene.[14] The monosubstituted double bond with an allylic OH is epoxidised faster than the trisubstituted double bond in lanalool 4.[14]

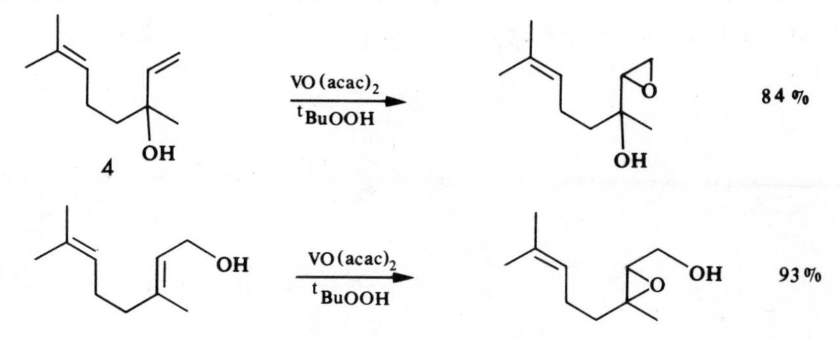

The epoxidation of small and medium ring cyclic allylic alcohols with $VO(acac)_2$-^{t}BuOH produces stereoselectively *syn*-epoxy alcohols with very high yields.[15,16] This is in contrast to peracids where, for the allylic alcohols n = 5 and 6, the *anti*-epoxy alcohol is produced.[17]

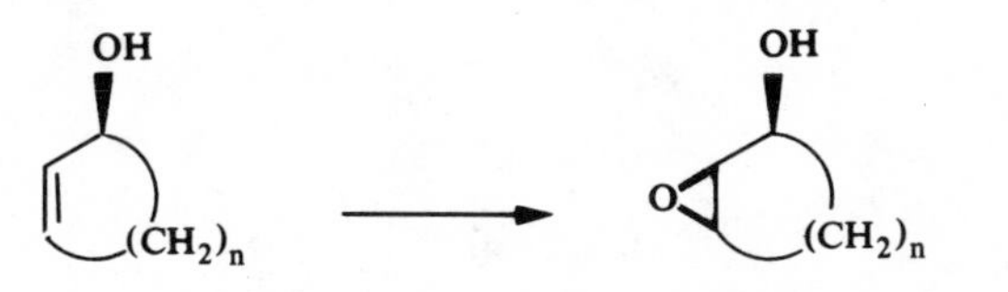

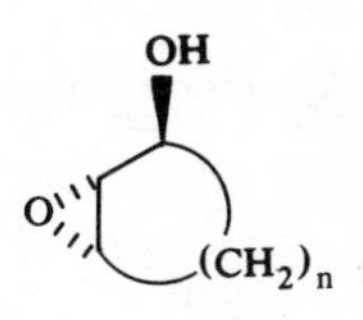

syn anti

	^tBuOOH, VO(acac)$_2$			MCPBA		
	yield	% syn	% anti	Yield	% syn	% anti
n = 3	86	99.7	0.3	83	95	5
4	75	99.6	0.4	95	61	39
5	83	97	3	81	0.2	99.8
6	78	91	9	89	0.2	99.8

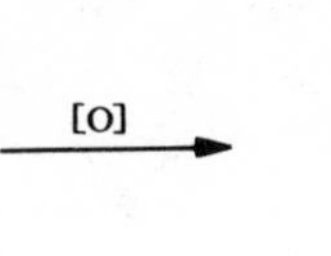

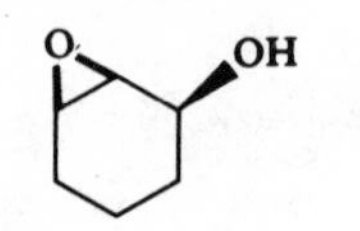

98 %

[O] = **MCPBA** or VO(acac)$_2$ / ^{t}BuOOH

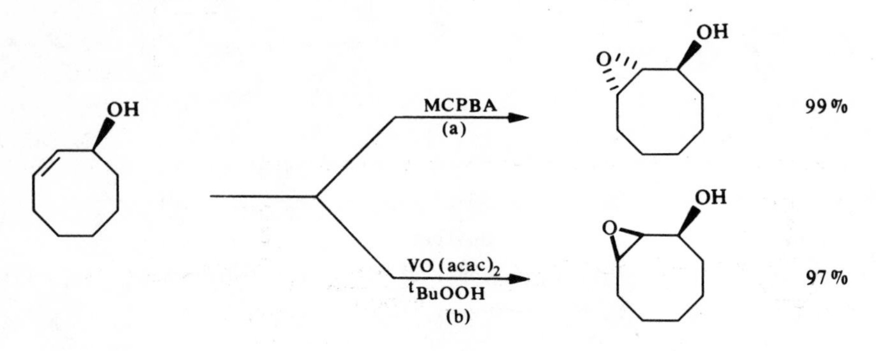

(a)

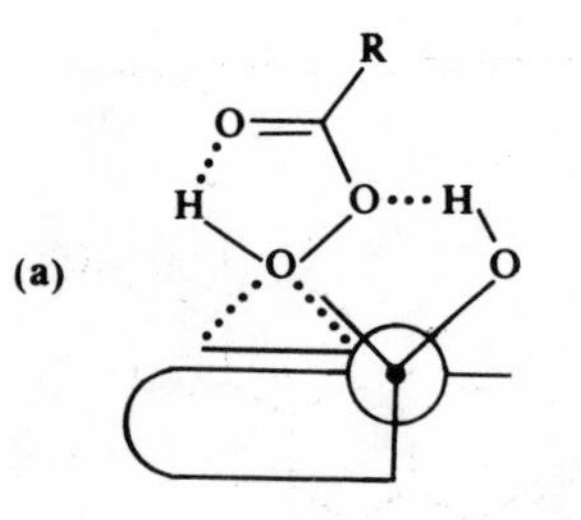

(b)

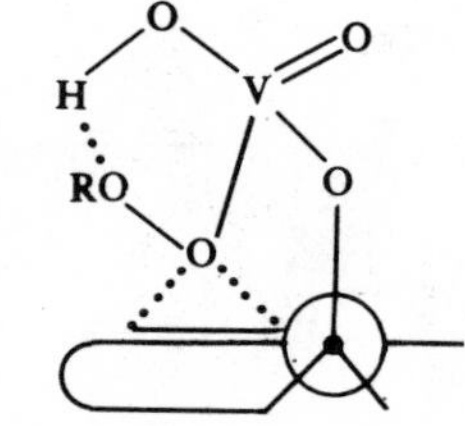

A competing reaction, oxidation of the allylic alcohol to the enone, occurs where the relative orientation of the alcohol and olefin are such that the transition state for intramolecular epoxidation is disfavoured. Thus the axial alcohol 5 yields 92% epoxy alcohol and 8% enone whereas the equatorial alcohol 6 gives 91% enone.[18]

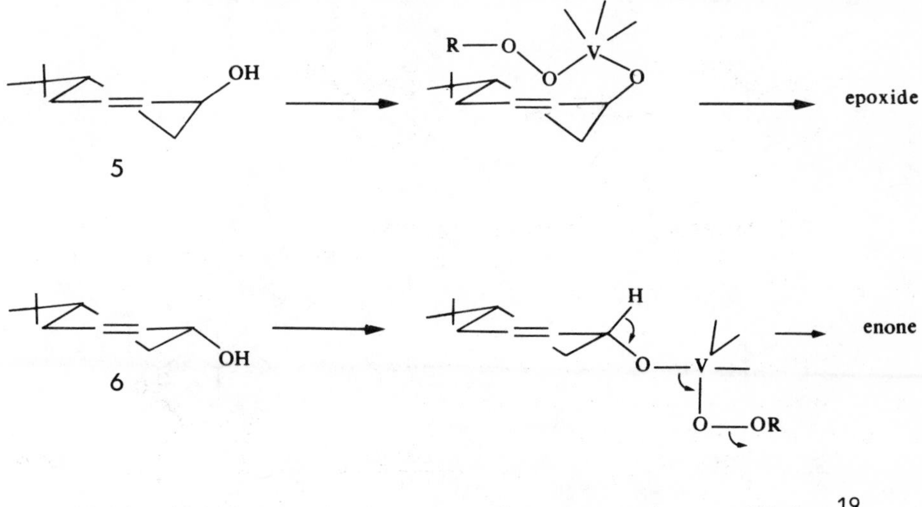

Homoallylic alcohols also lead to rate enhancement and *syn* addition.[19]

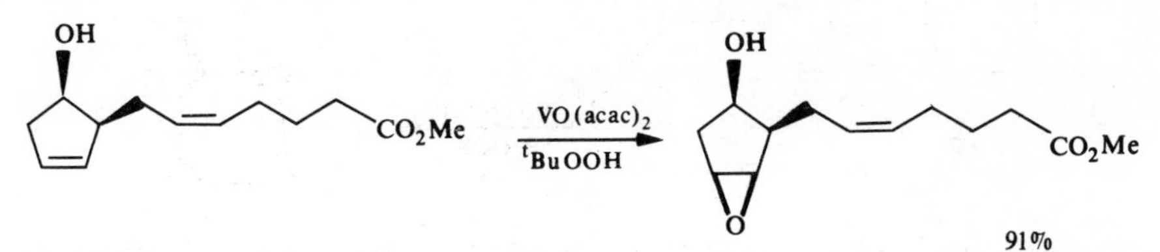

Remote epoxidation can be controlled in certain steroids by the alcohol directing effect.[20]

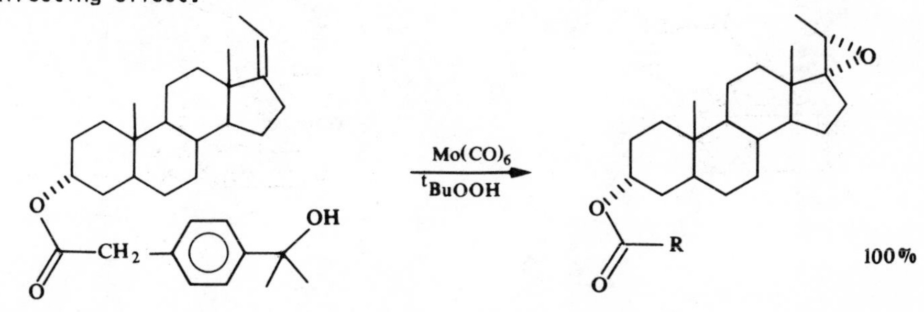

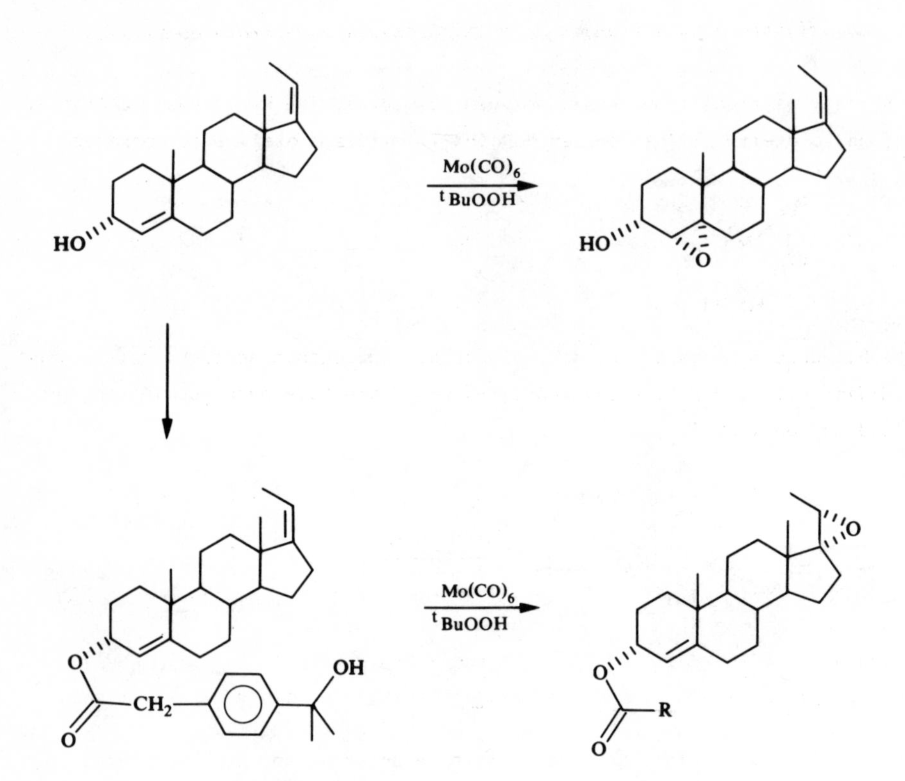

Asymmetric epoxidation has been achieved by using chiral ligands coordinated to Mo[21] and V.[22] Enantiomeric excesses of up to 33 and 50% respectively were accomplished.

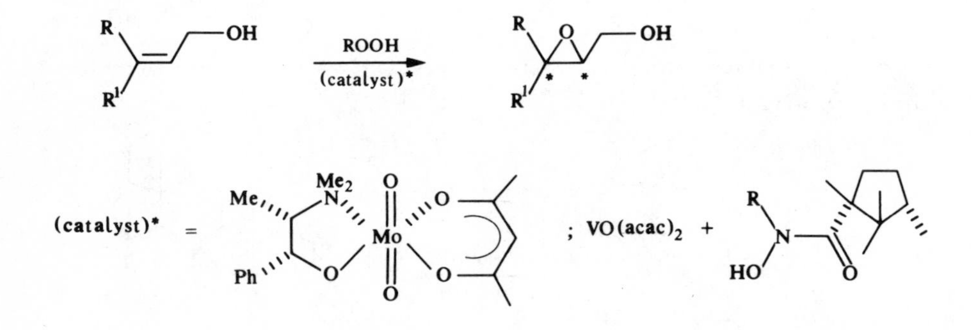

8.1.4 Oxidations with transition metal dioxygen and peroxy complexes

A variety of transition metal dioxygen complexes have been shown to transfer oxygen to olefins. The complex MoO_5(HMPT) oxidises olefins to epoxides.[23]

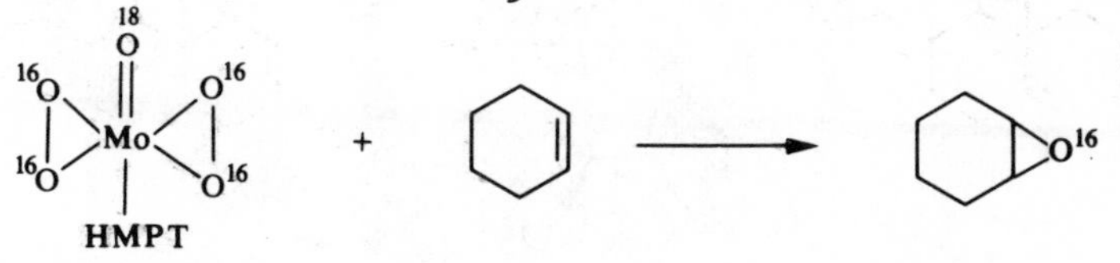

The 5,6-double bonds of cholesterol acetate, diosgenin acetate, solasodine-O,N-diacetate and 3β-acetoxy androst-5-en-17-one have been epoxidised in yields of 80-100%.[24]

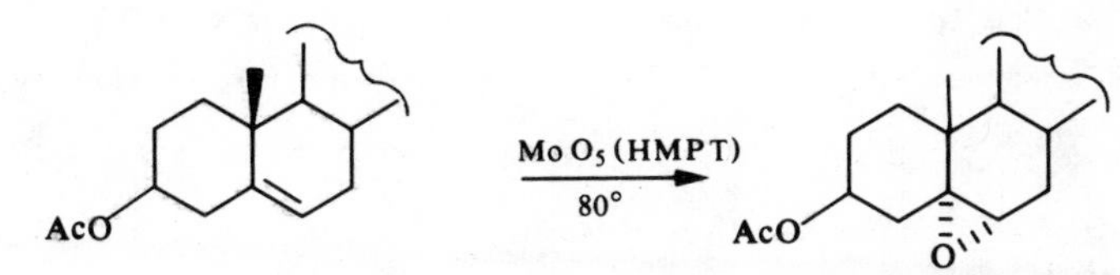

3,4-dihydro-2H-pyran is oxidised to the aldehyde ester 7 by MoO_5(HMPT) in high yield. This is the same product that is obtained by ozonolysis.[25]

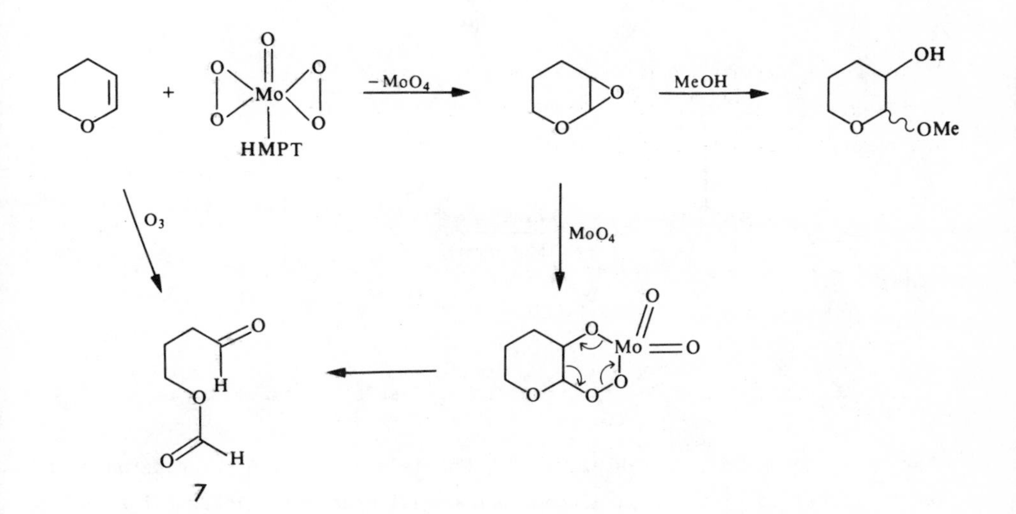

Asymmetric epoxidation has been achieved using an optically active MoO_5L species. Enantiomeric excesses of up to 30% were observed.[26]

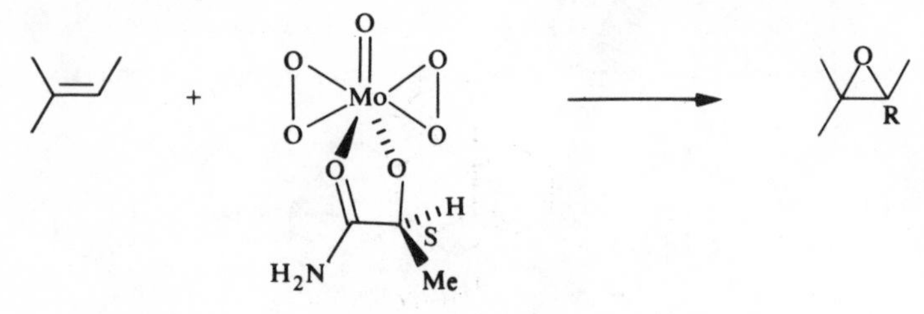

The complex $(Ph_3P)_2PtO_2$ in the presence of PhCOCl epoxidises cyclohexene and norbornene.[27]

The complex MoO_5(pyridine) reacts with enolate anions to give α-hydroxy ketones.[28] For example, it is possible to prepare the labile acyloin 8 by this method.

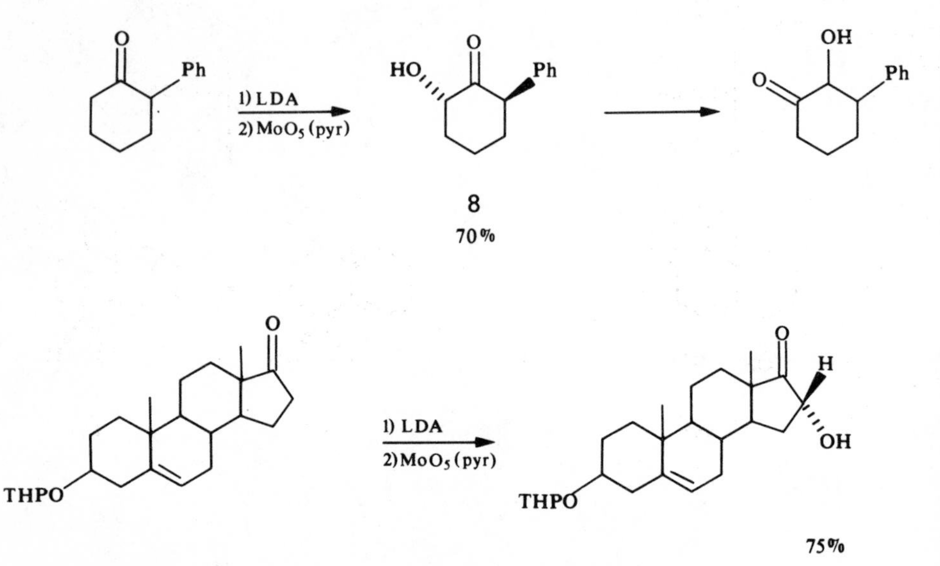

8.1.5 Oxidative decomplexation reactions. (see section 2.1.8)

The Pt-carbon σ-bond in $(Ph_3P)_2PtCl(CH_2Ph)$ is cleaved by m-chloroperbenzoic acid (mcpba) to give benzyl alcohol and benzyl m-chlorobenzoate. The Pt-carbon bond is cleaved stereospecifically with retention of configuration.[29]

Steroidal η^3-allyl palladium complexes undergo oxidation to allylic alcohols with MCPBA[30] and oxidation to enones with Collins reagent.[31]

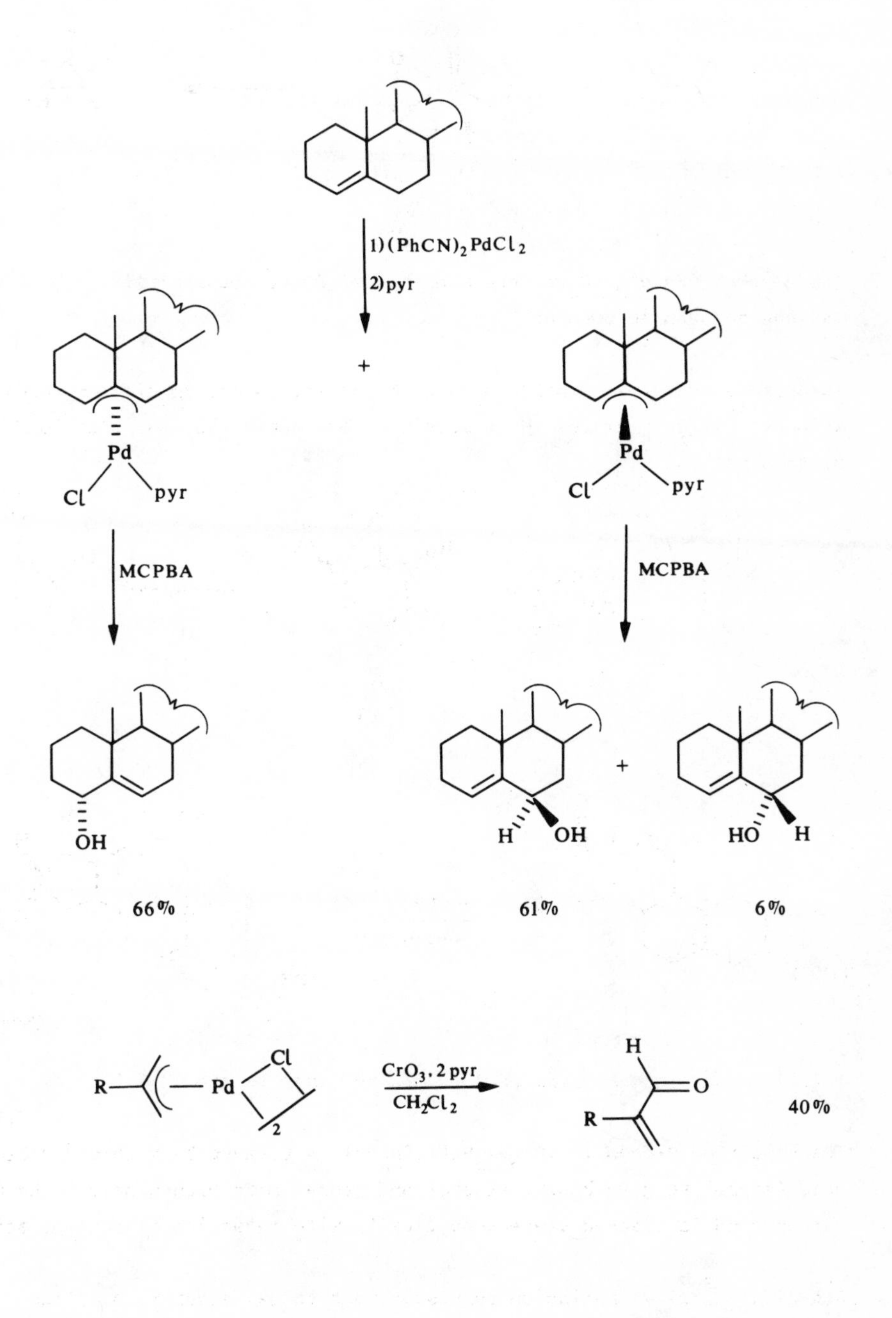

8.2 REDUCTIONS

8.2.1 Addition of H_2 and related reactions

Hydrogenation of olefins with heterogeneous and homogeneous catalysts is one of the most studied organometallic reactions.[32] One of the most widely used homogeneous catalysts is Wilkinson's catalyst, $(Ph_3P)_3RhCl$. The catalytic species is $(Ph_3P)_2RhCl$ (14e) which reacts with H_2 and olefin to give a (18e) Rh dihydro olefin complex which decomposes via a σ-alkyl RhH to hydrocarbon and $(Ph_3P)_2RhCl$. Hydrogen is added *cis* across the double bond. The catalytic cycle has been described in section 1.5.1.

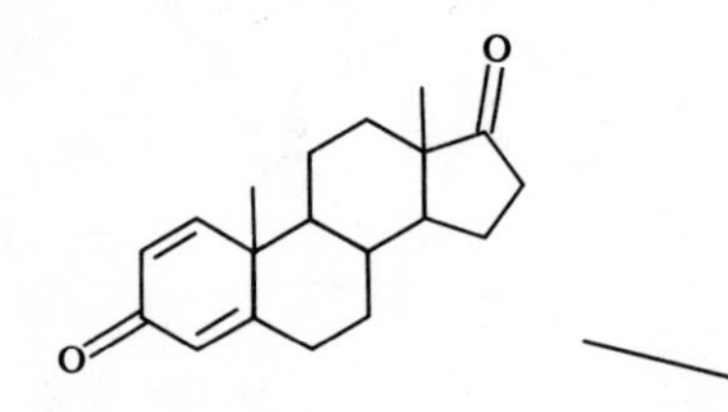

$(Ph_3P)_3RhCl/H_2$

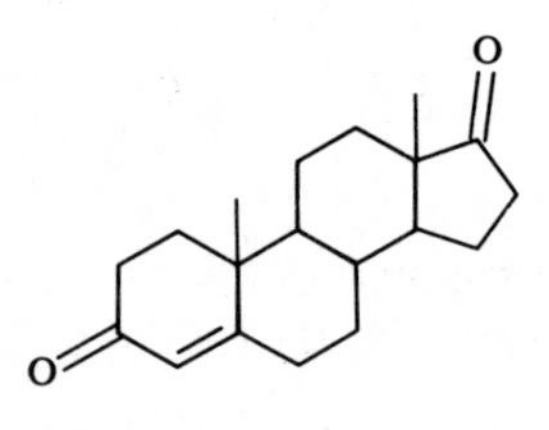

75 – 80 %

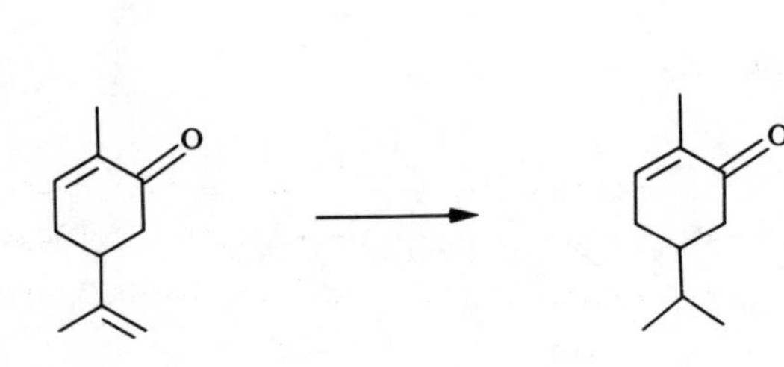

Hydrogenation of olefins and acetylenes can be carried out in the presence of functional groups such as RCHO, R_2CO, OH, CN, NO_2, Cl, ROR, CO_2R, CO_2H. Normally, less sterically hindered double bonds are hydrogenated faster because they form complexes with the catalyst more readily.[33]

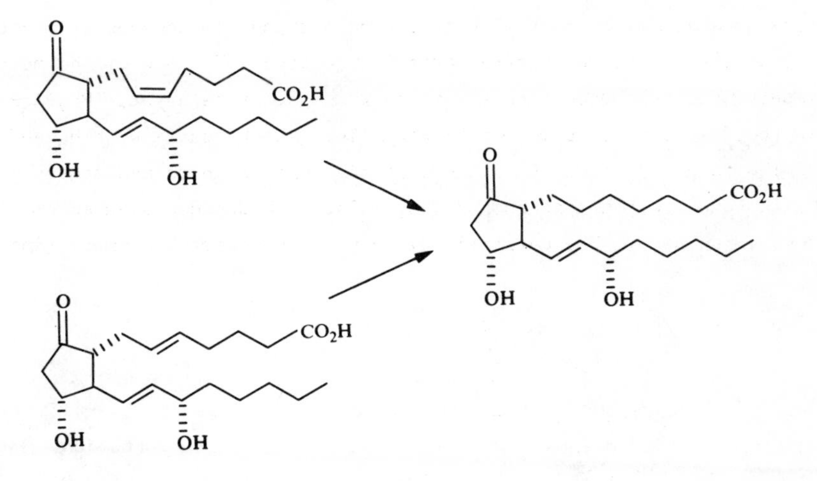

Many other homogeneous catalysts have been investigated. Some examples are $(Ph_3P)_2IrCl(CO)$, H_2IrCl_6, $HCo(CO)_4$. Variation of the phosphines and the solvent allow rates and selectivities to be controlled.[32] Many hydrogenation catalysts also catalyse the hydrosilation reaction shown below.

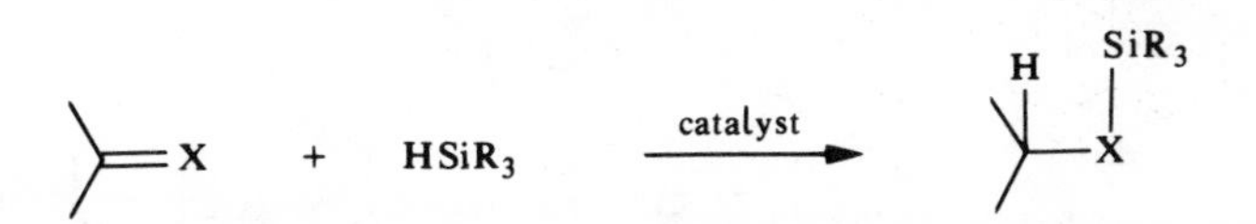

$X = O, NR, CR_2$

Cp_2MoH_2 behaves as a selective catalyst for the partial hydrogenation of 1,3- or 1,4-dienes. Cyclopentadiene, 1,3- and 1,4-cyclohexadiene, 1,3-cycloheptadiene, 1,3-cyclooctadiene and norbornadiene are reduced to their corresponding monoenes at 140-180°.[134]

$(CH_2)_n$ + Cp_2MoH_2 + H_2 ⟶ $(CH_2)_n$

n = 1, 2, 3

1,3-Dienes are also partially reduced catalytically by $Cr(CO)_6$ and H_2. The reaction is photo-induced to remove the CO ligands from the chromium and generate the catalytic species. Only 1,4-hydrogenation occurs and only with dienes that can achieve *s-cis* conformation.[35]

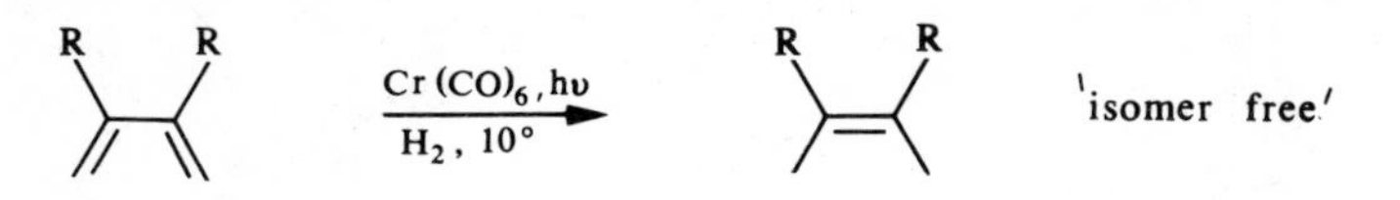

Methyl benzoate chromium tricarbonyl also catalyses the 1,4-addition of hydrogen to 1,3-dienes that can attain the *s-cis* conformation. *Trans,trans*-1,3-dienes give *cis* olefins as products.[36]

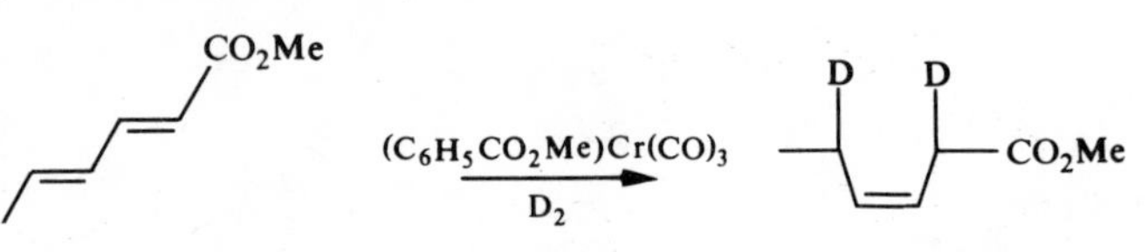

$(Ph_3P)_2(CO)_2RuCl_2$ is inert towards monoolefins and can be used to hydrogenate polyenes to monoenes.[37]

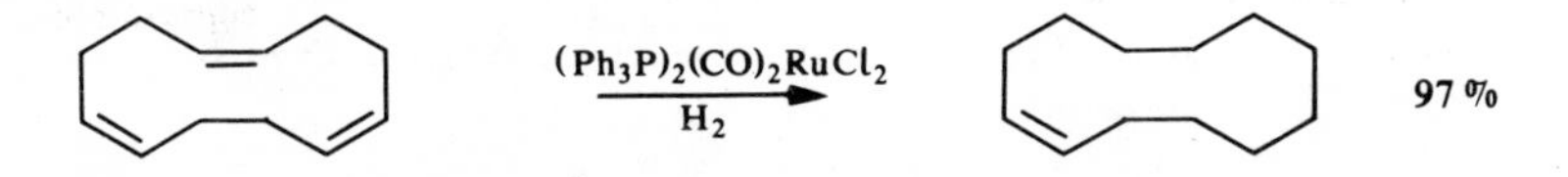

Hydrogenation of 7-dehydrostigmasterylacetate 9 with $(Ph_3P)_3RhCl$ gave α-spinasterylacetate 10.[38]

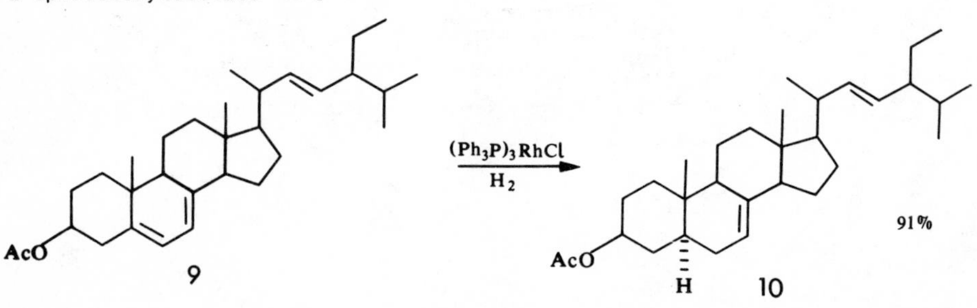

The reaction of $Fe(CO)_5$ with aqueous base allows the generation *in situ* of hydrido iron carbonyl complexes that selectively reduce α,β-unsaturated carbonyl compounds (aldehydes, ketones, esters, lactones etc.) to the corresponding saturated carbonyl compounds.[39]

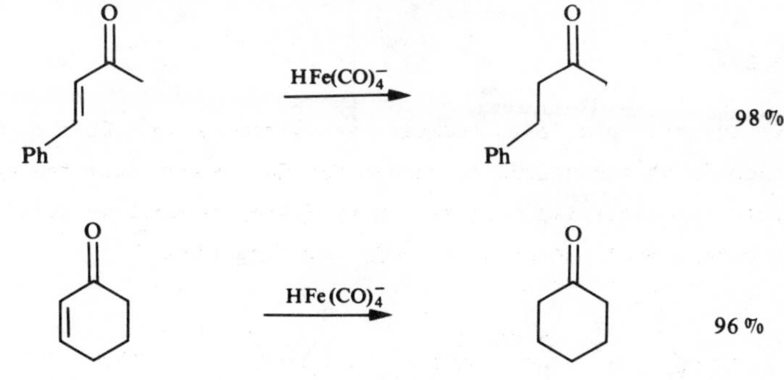

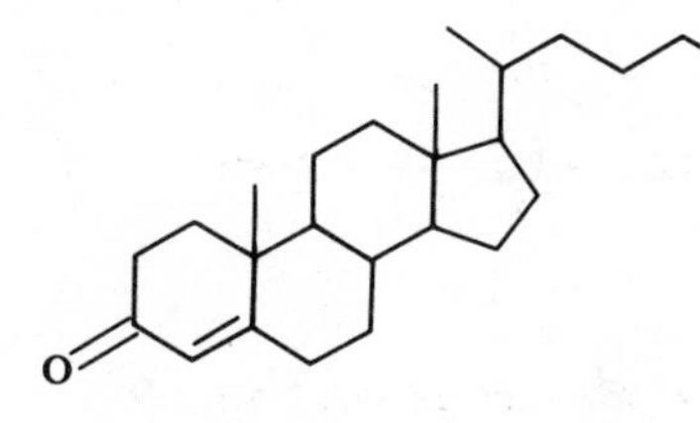

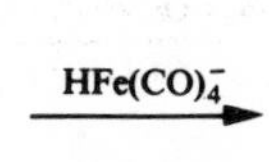

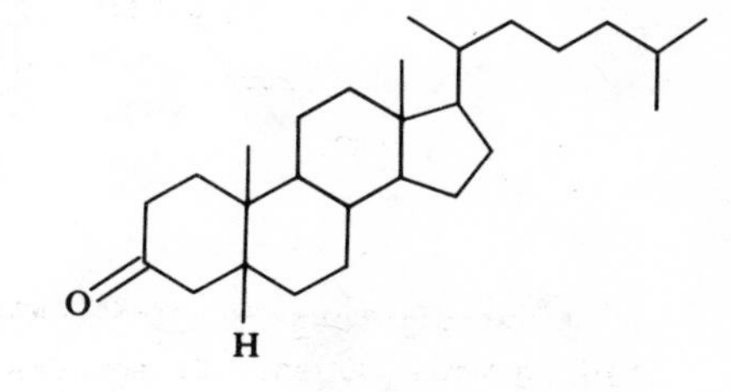

4-cholesten-3-one

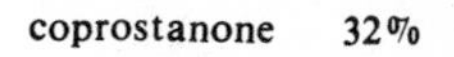

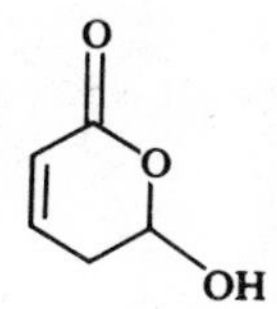

$HFe(CO)_4^-$

O O OH 90 %

α,β-unsaturated ketones are reduced to triethyl silyl enol ethers by Et_3SiH in the presence of $(Ph_3P)_3RhCl$. Hydrolysis of the silyl enol ethers gives the corresponding ketones.[40]

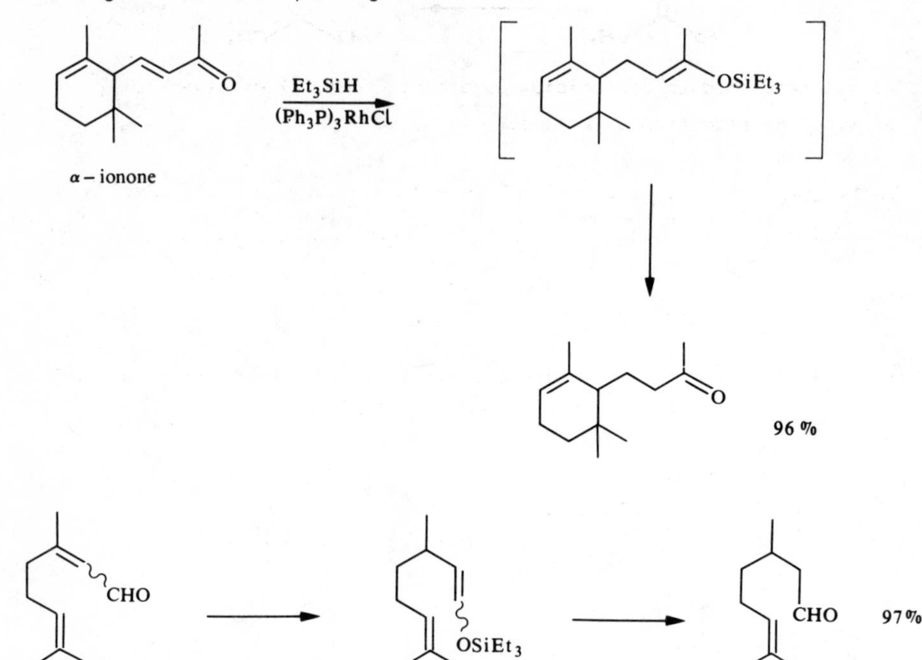

Reduction of the carbonyl group to alcohol may also occur. The ratio of ketone to allylic alcohol produced depends upon the silane employed.

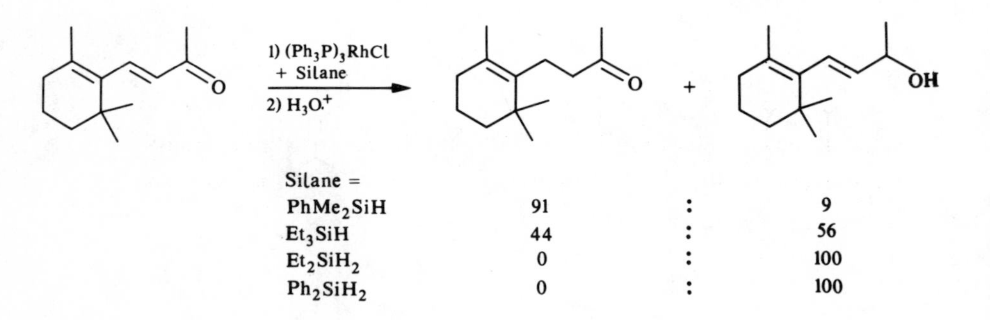

Silane =			
$PhMe_2SiH$	91	:	9
Et_3SiH	44	:	56
Et_2SiH_2	0	:	100
Ph_2SiH_2	0	:	100

Ketones may be hydrogenated to alcohols by $(Ph_3P)_3RhCl$ in the presence of Et_3N.[41]

The cationic rhodium and iridium complexes $[(PhMe_2P)_2M(solvent)_2H_2]^+$ catalyse the reduction of ketones.[42]

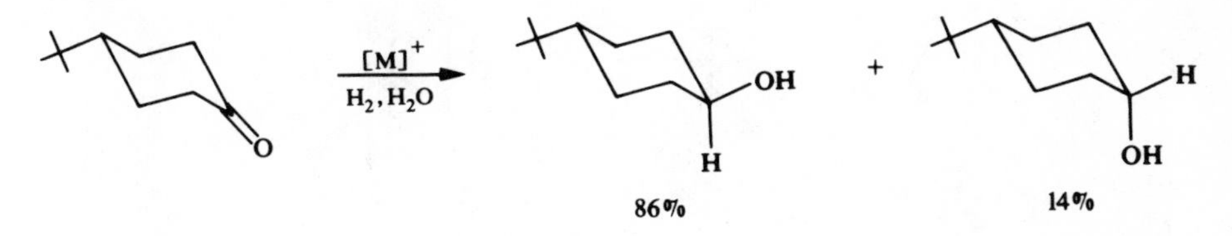

$(Ph_3P)_3RuCl_2$ catalyses the hydrogenation of anhydrides to γ-lactones.[43]

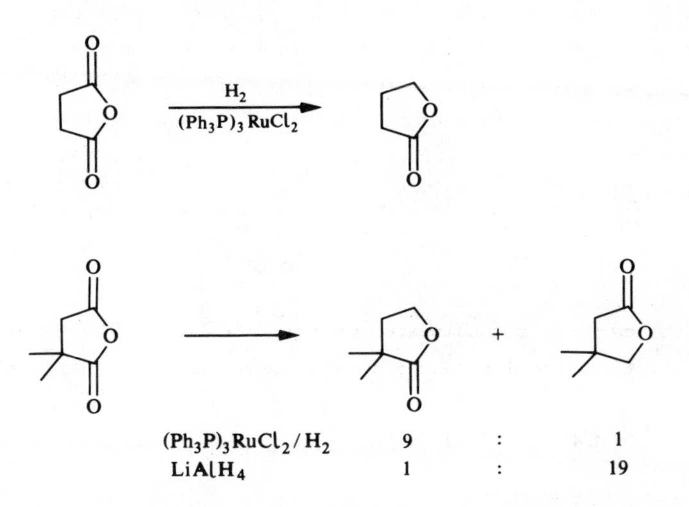

Arenes can be hydrogenated to cyclohexanes faster than alkenes are hydrogenated with η^3-allyl $Co[P(OMe)_3]_3$ as catalyst. The addition of all six hydrogens is to the same face of the arene.[44]

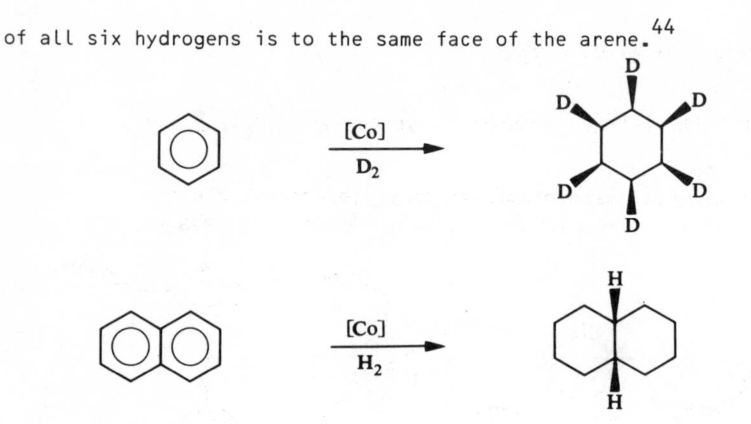

8.2.2 Asymmetric hydrogenation and hydrosilylation reactions

An asymmetric synthesis is a reaction that converts a prochiral centre into a chiral centre in such a way that the enantiomers are produced in unequal amounts.[45] Asymmetric synthesis occurs because a chiral reagent and a prochiral substrate form diastereomeric transition states which differ in energy. The magnitude of this energy difference determines the enantiomeric excess. Two catalytic processes that have been studied extensively are hydrogenation and hydrosilyation.

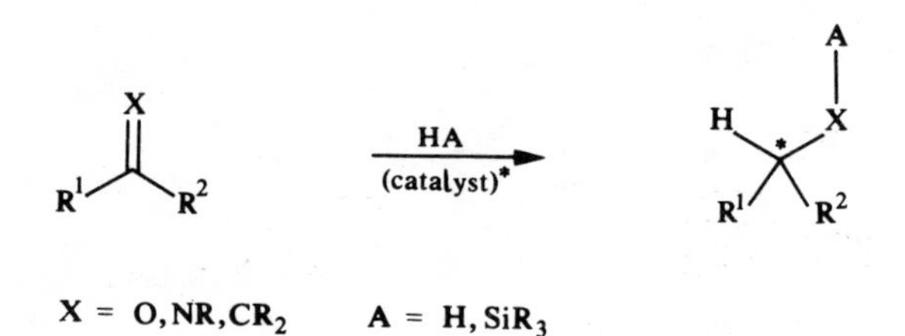

The catalysts contain optically active ligands coordinated to a transition metal. The majority of reactions studied use Rh as the metal although reactions with Co, Ni, Pd, Cu,etc., are also known. The chiral ligands are generally phosphines with some examples of amines, amino alcohols, etc., being known. The optically active phosphines can be chiral on phosphorus or on the alkyl groups.

$R_1R_2R_3P^*$ R = Ph, $PhCH_2$, *o*-anisyl, cyclohexyl, Me, Et, nBu, etc.

R_2R^*P R = menthyl, neomenthyl, myrtanyl ferrocenyl etc.

Some of the most successful ligands are bidentate bisphosphines.

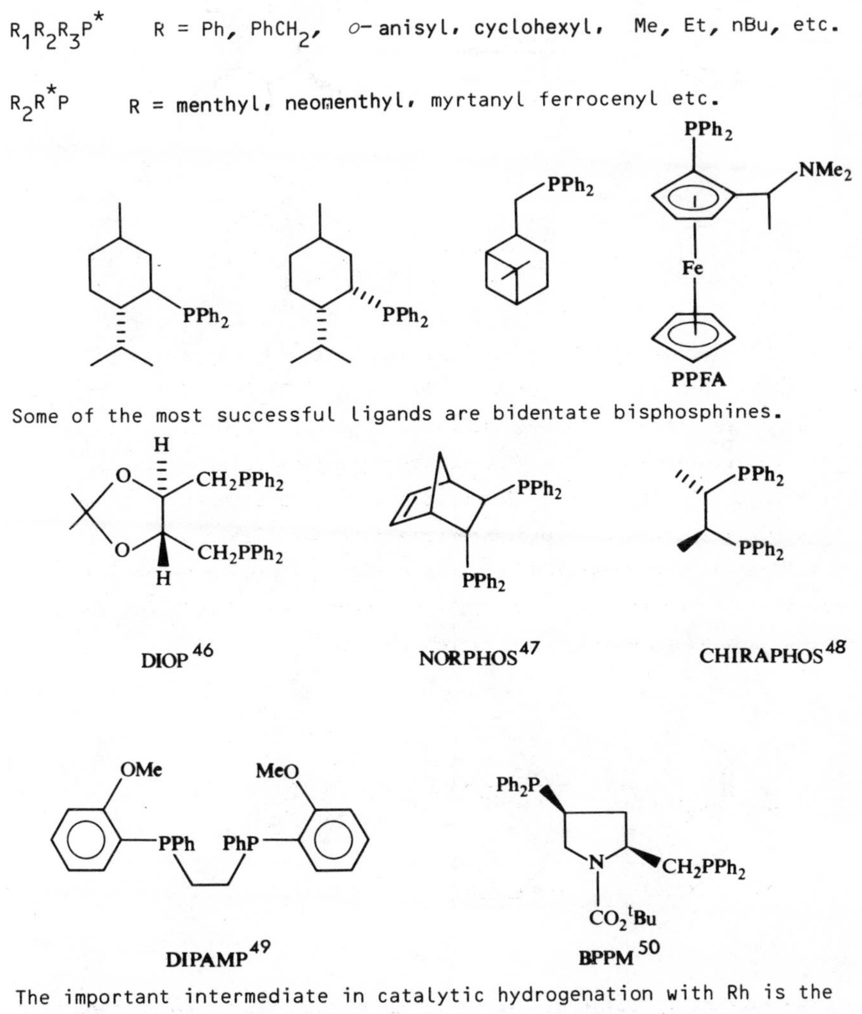

The important intermediate in catalytic hydrogenation with Rh is the complex where two hydrogens and the olefin are bound to the metal. When only achiral ligands (L) are present, bonding to both faces of the double bond is equally favourable (11 and 12 are enantiomers and therefore have equal energy) and racemic products are produced.

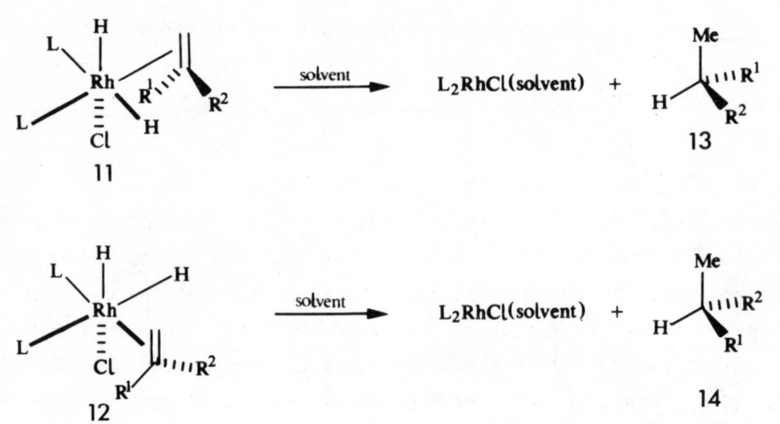

When chiral ligands (L) are used the intermediates 11 and 12 are diastereomeric and are no longer of equal energy. Their stabilities,and

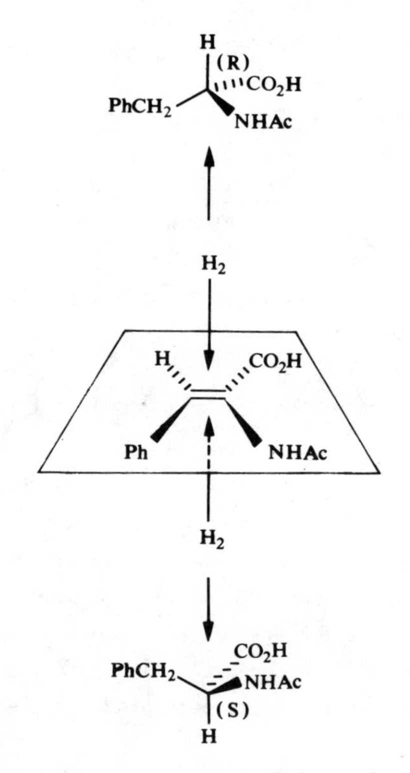

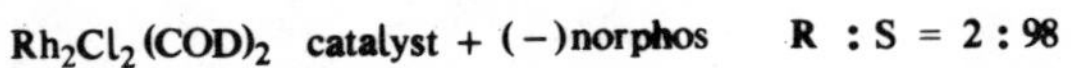
$Rh_2Cl_2(COD)_2$ catalyst + (−)norphos R : S = 2 : 98

their rates of formation and disappearance, will be different and this may lead to an excess of either 13 or 14 being produced. In the ideal situation the difference is so great that only one of the intermediates 11 or 12 and thus only one product enantiomer, 13 or 14 respectively, is obtained (100% e.e.).

Asymmetric hydrogenation has been most successful for the reduction of[46,50] α-N-acylaminoacrylic acids to the corresponding α-amino acid derivatives, where in certain cases optical yields approaching 100% have been obtained.

$$RCH=C(NHCOR')CO_2H \xrightarrow[L_2^*]{Rh_2Cl_2(COD)_2} RCH_2-\overset{*}{C}H(NHCOR')-CO_2H$$

R	ligand(L_2^*)	e.e%
H	CHIRAPHOS	91
	DIPAMP	95
Ph	CHIRAPHOS	99
	NAPHOS	96
	DIOP	81
	BPPM	91
^iPr	CHIRAPHOS	100
p-HOC_6H_4	"	92
	DIOP	80
m,p$(MeO)_2C_6H_3$	"	83

β-Keto acids and esters and α-keto esters and α-diketones also undergo catalytic hydrogenation reactions.[45,51]

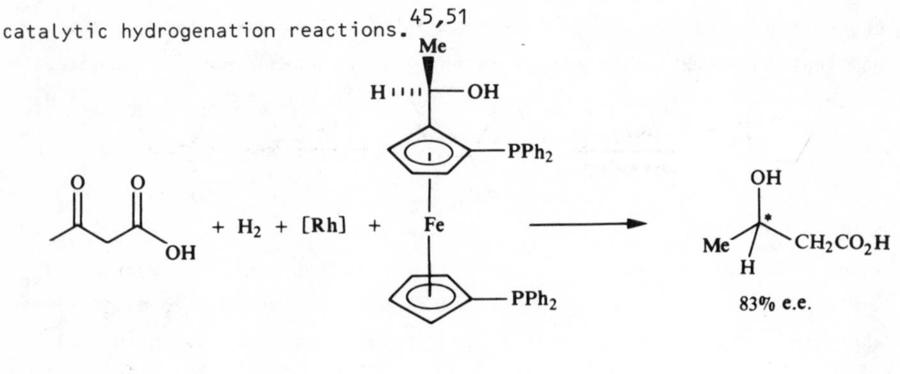

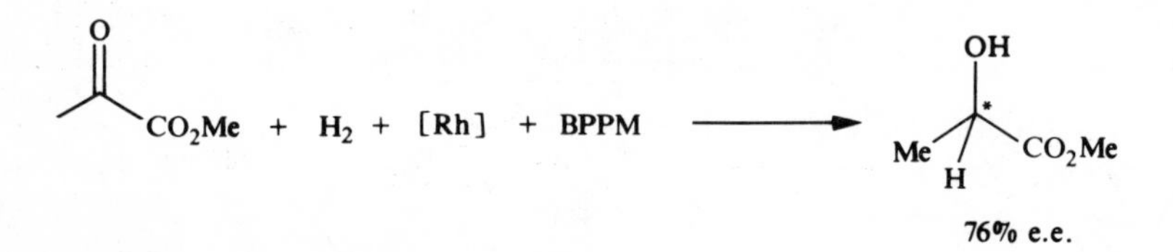

Asymmetric hydrosilylation of ketones is catalysed by rhodium complexes of chiral phosphines.[52]

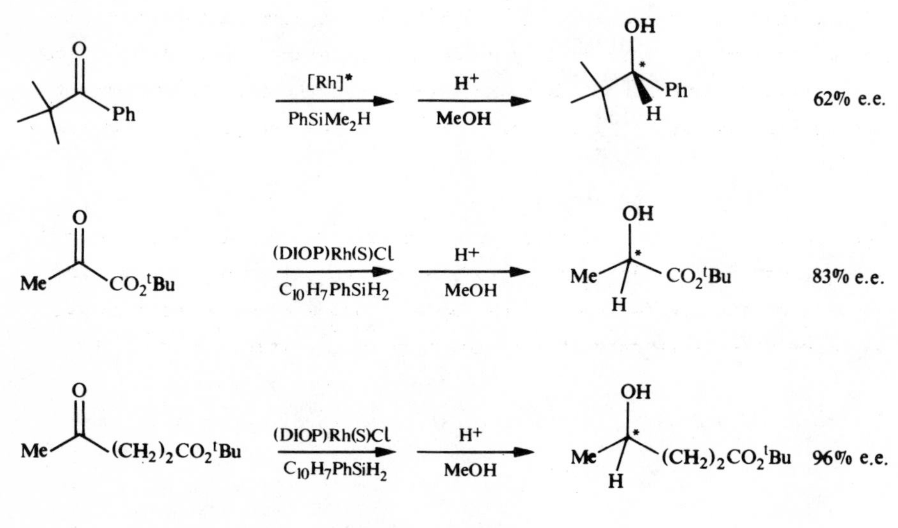

Asymmetric hydrosilylation of olefins followed by reoxidation allows the preparation of optically active alcohols in up to 50% e.e.[53]

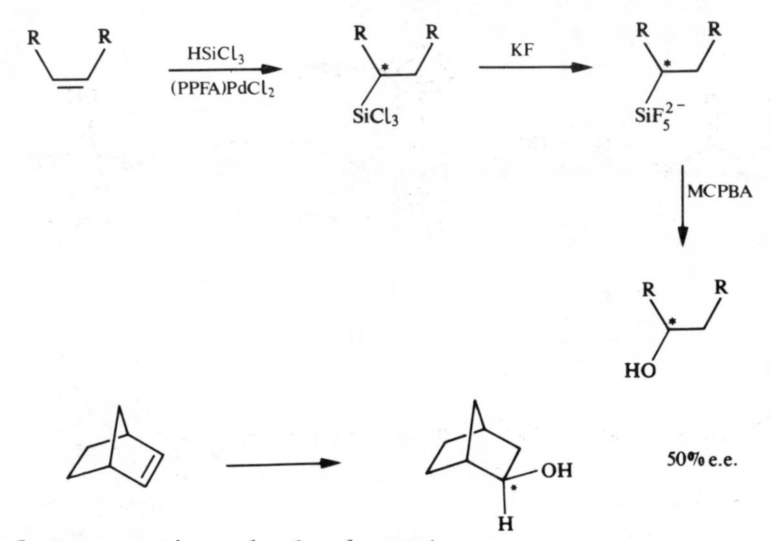

8.2.3 Deoxygenation and related reactions

The deoxygenation of epoxides to the corresponding olefins, the reverse of the epoxidation reaction, is synthetically a very useful reaction. Since epoxidations with peracid proceed with retention of olefin stereochemistry, deoxygenation reactions that invert the stereochemistry of the epoxide/olefin, lead to overall inversion of the geometry for the olefin.

Nucleophilic opening of epoxides by $CpFe(CO)_2^-Na^+$ gives the alkoxide **15** which on treatment with acid gives the $CpFe(CO)_2(olefin)^+$ cation **16**. Release of the olefin from the cation with NaI/acetone gives free olefin with the same stereochemistry as the epoxide.[54] Thermolysis of the alkoxides **15** gives olefins with inversion of stereochemistry i.e. *cis* elimination of [Fe]=O.[55]

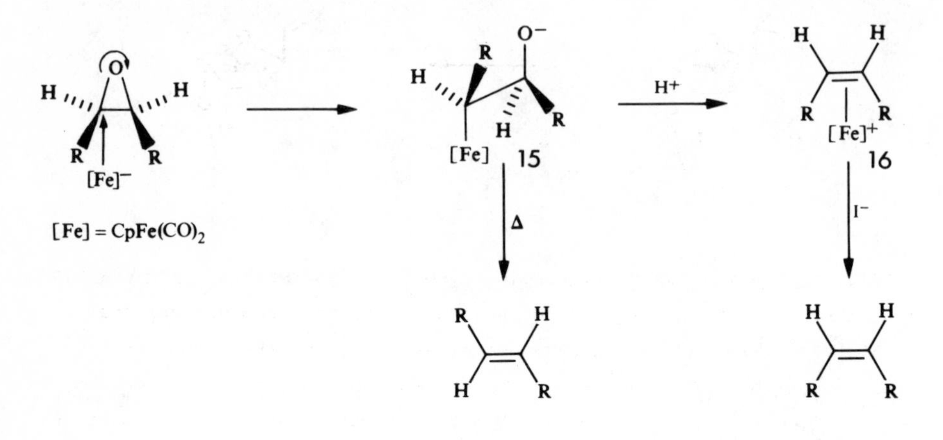

epoxide	olefin after H^+/I^-		olefin after thermolysis	
cis-but-2-ene oxide	*cis*-but-2-ene	(64%)	*trans*-but-2-ene	(86%)
trans-but-2-ene oxide	*trans*-but-2-ene	(50%)	*cis*-but-2-ene	(69%)
cis-stilbene oxide	*cis*-stilbene	(82%)	*trans*-stilbene	(96%)
trans-stilbene oxide	*trans*-stilbene	(83%)	*cis*-stilbene	(92%)
trans-ethylcrotonate oxide	*trans*-ethylcrotonate	(96%)	*cis*:*trans* 61:39	(81%)

The stereoselectivity for both methods is generally very high although the thermolysis method is less stereoselective for αβ-unsaturated esters. Monosubstituted epoxides are opened by $[CpFe(CO)_2]^-$ faster than disubstituted epoxides and this allows diepoxides such as **17** to be selectivity deoxygenated.

17

1) $CpFe(CO)_2Na$
2) H^+ 3) I^-

50%

The sugar epoxide **18** may also be deoxygenated in high yield.[56]

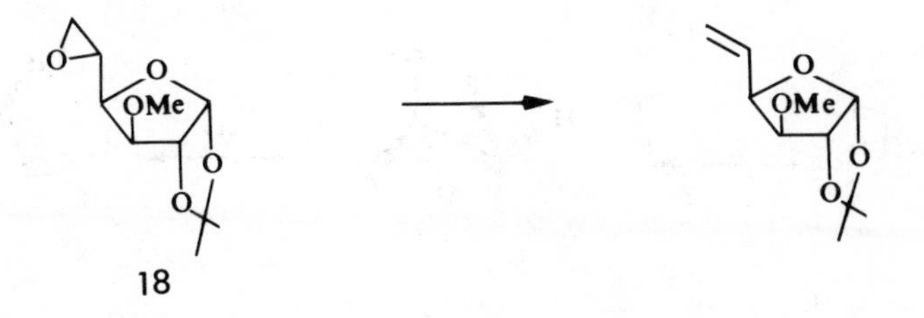

$Co_2(CO)_8$ is an effective deoxygenating agent for epoxyesters. The reaction has been shown to be highly stereospecific for certain epoxyesters that have electron withdrawing groups on both carbon atoms; the resulting olefin having the opposite stereochemistry to the epoxide. This implies initial S_N2 ring opening of the epoxide by a cobalt species. The reaction is much less stereospecific for epoxides where Lewis acid catalysed ring opening to a carbonium ion by a Co species is not disfavoured.[57]

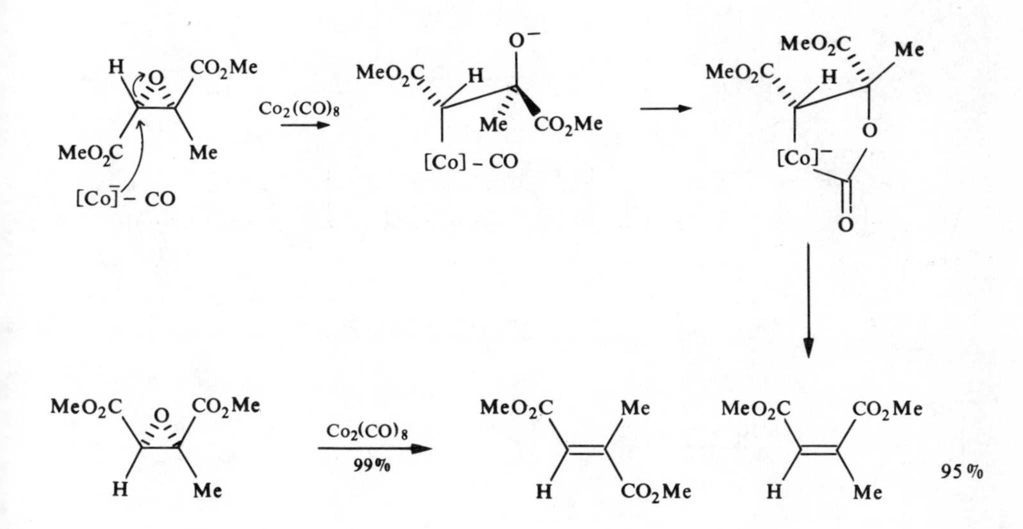

Epoxides may be deoxygenated non-stereospecifically with $Fe(CO)_5$[58] or $Mo(CO)_6$.[59] For example *trans* - stilbene oxide gives *trans*-(56%) and *cis* - stilbene (22%) on reaction with $Fe(CO)_5$ in tetramethyl urea. Some other examples are shown below.

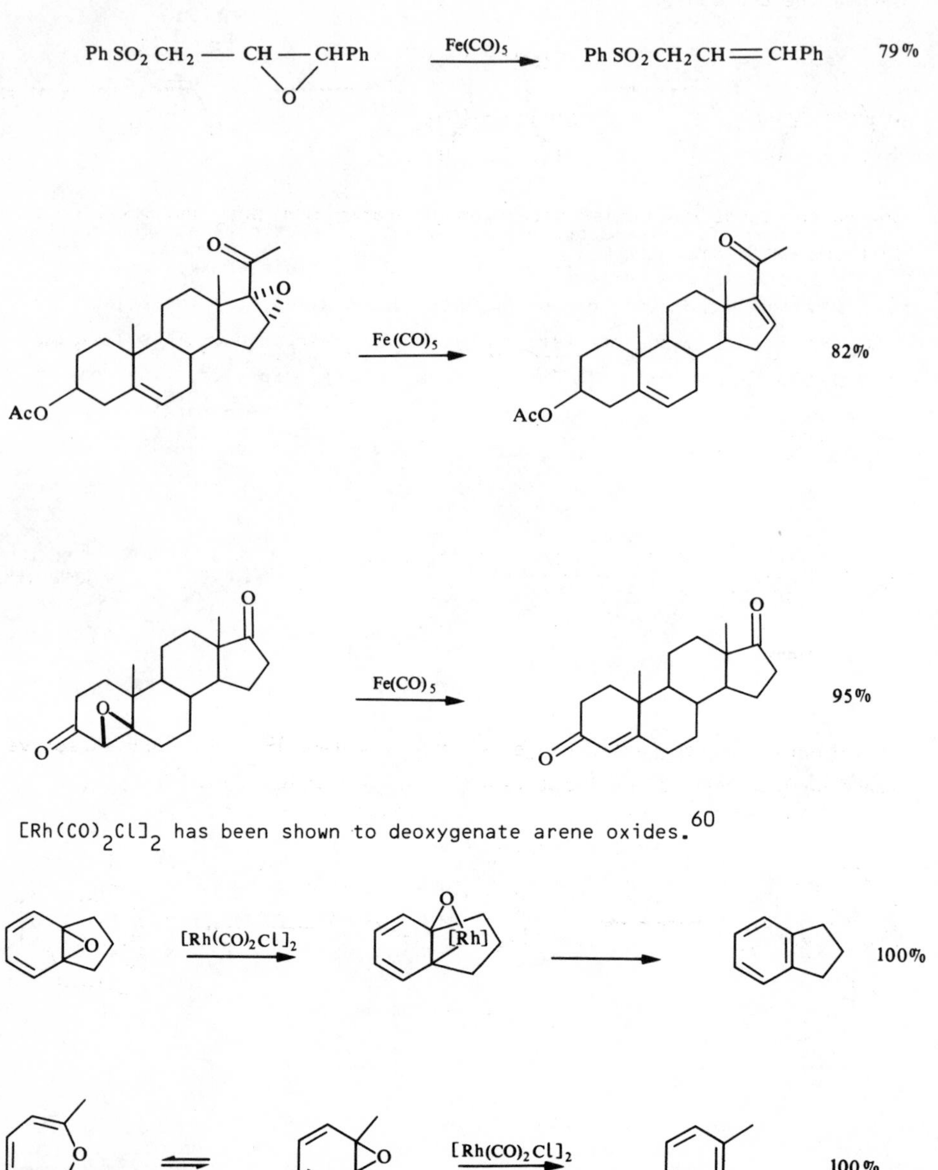

$[Rh(CO)_2Cl]_2$ has been shown to deoxygenate arene oxides.[60]

Low valent electron poor organometallic species deoxygenate epoxides in high yield. The reactions generally go stereospecifically with retention of stereochemistry. The initial step is believed to be insertion into one of the C-O bonds.

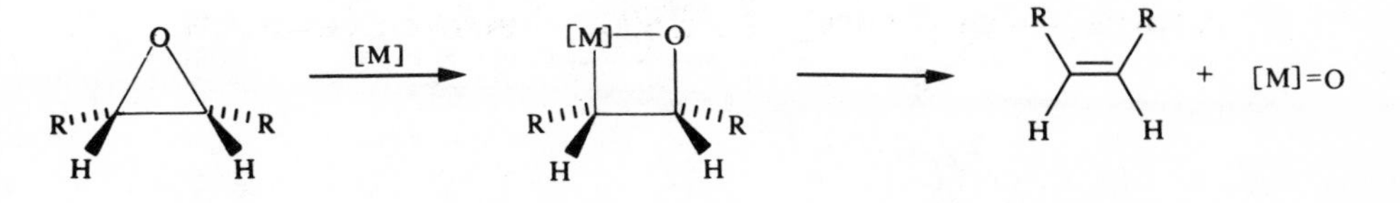

Low valent tungsten species have been generated from WCl_6 and alkyl lithiums or lithium metal.[61]

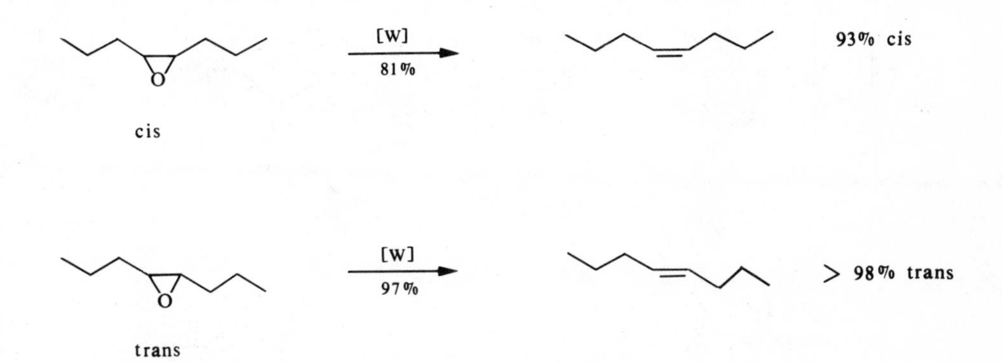

Oxidation and subsequent deoxygenation of humulene 19 allows the selective functionalisation of the least reactive ($\Delta^{4,5}$) double bond.[62]

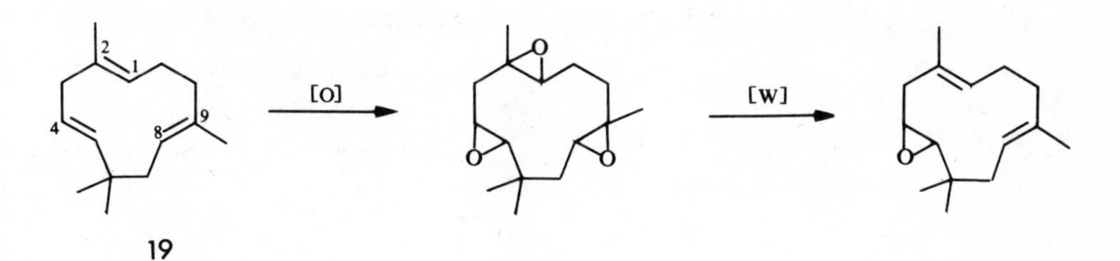

Bisbenzene titanium may be used as a source of titanium atoms to deoxygenate propylene oxide.[63]

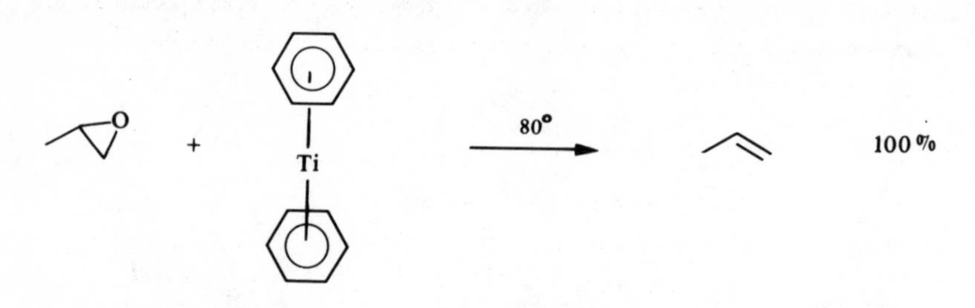

The metallocenes $[Cp_2M]$, M = Ti, Zr, Mo, W generated by a variety of methods also deoxygenate epoxides. When M = Mo or W, Cp_2MCl_2 can be used catalytically in the presence of Na/Hg, while for M = Ti and Zr a stoicheiometric amount is required. This is attributed to the solubility of $Cp_2M{=}O$ M = Mo and W which allows regeneration of "Cp_2M".[64]

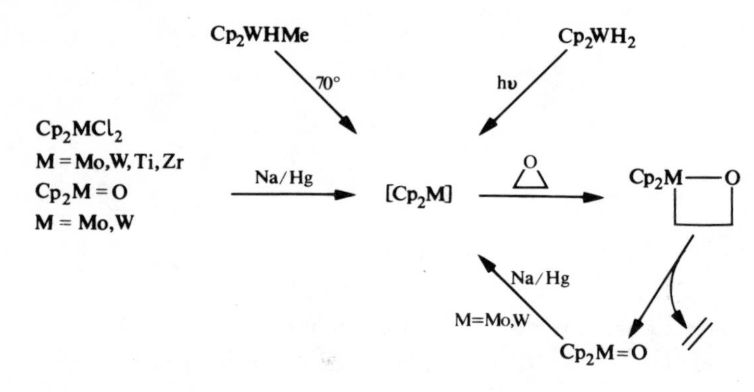

Nucleophilic oxidants ($X^{+}—O^{-}$) can be deoxygenated by certain transition metal carbonyls. The intermediate 20 thus formed can decompose with loss of X and CO_2. This reaction can be used to remove coordinated CO from transition metal complexes or to reduce the nucleophilic oxidant. Sulphoxides and amine oxides are deoxygenated by $Fe(CO)_5$ to sulphides[65]

and amines[66] respectively.

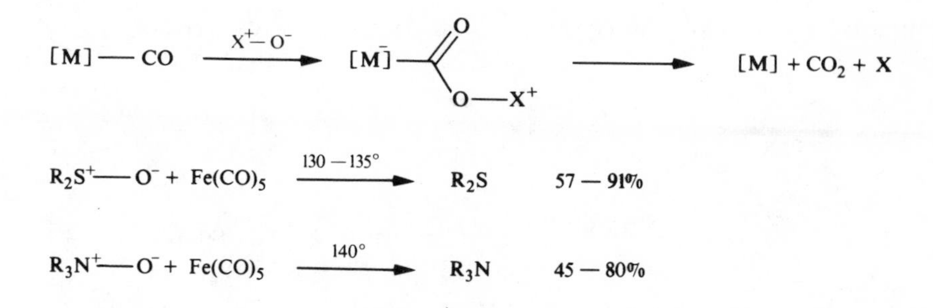

$Fe(CO)_5$ also reduces amide oximes to amidines and oximes to imines.[67] This latter reaction can be used to regenerate carbonyl compounds from their oximes.[66,68]

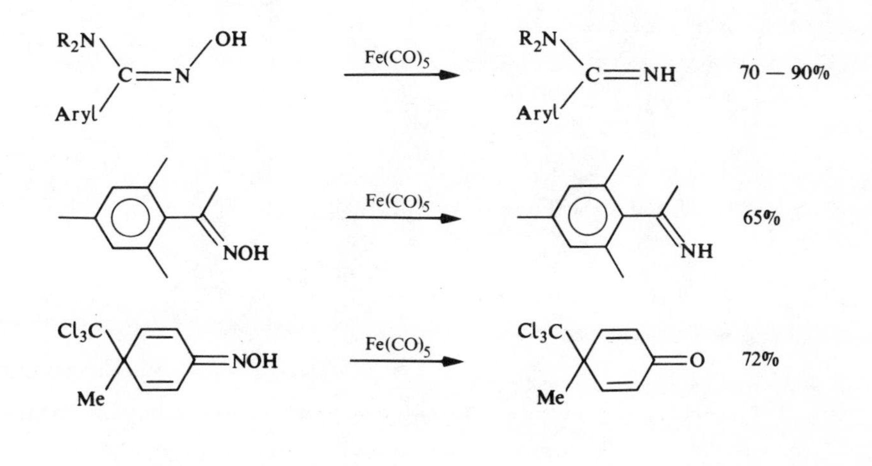

Alcohols that can give rise to stable carbanions are deoxygenated via their alcoholates to the corresponding hydrocarbons by $Fe(CO)_5$ and HCl.[69]

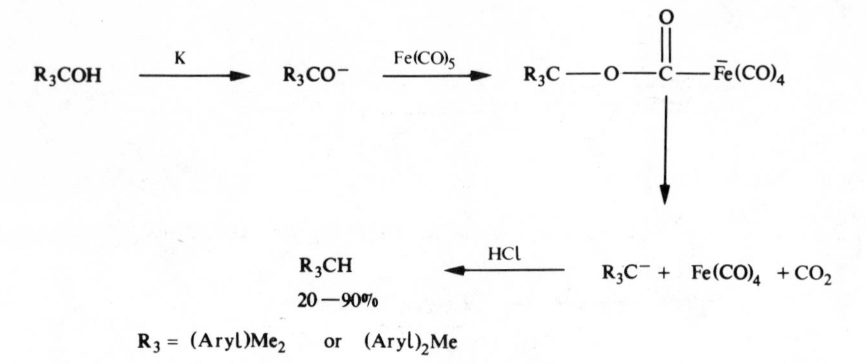

8.2.4 *Miscellaneous reductions*

It is possible to desulphurise thiocarbonyl compounds with a variety of transition metal reagents. Thioketones may be reduced to alkanes with $Fe(CO)_5$/KOH (i.e. $HFe(CO)_4^-$) or reductively dimerised with $Co_2(CO)_8$, $[CpFe(CO)_2]_2$ or $Mn_2(CO)_{10}$.

$$R_2C{=}CR_2 \xleftarrow{Co_2(CO)_8} R_2C{=}S \xrightarrow[KOH]{Fe(CO)_5} R_2CH_2$$

45 — 83%   60 — 81%

Thioamides are reduced to amines by $Fe(CO)_5$/KOH.[70]

$Fe(CO)_5$ or $(1,5\text{-}COD)_2Ni$ provide useful alternative desulphurisation agents to $P(OMe)_3$ in the Corey-Winter olefin synthesis especially for the preparation of thermally labile olefins.[71]

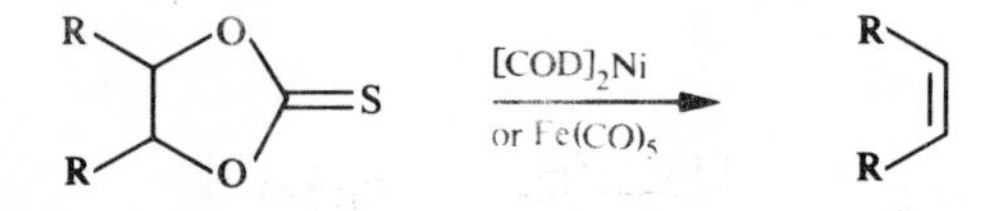

Thioanhydrides may be converted to olefins by desulphurisation with $(Ph_3P)_2Ni(CO)_2$, $Fe_2(CO)_9$ or $(Ph_3P)_3RhCl$.[72]

Episulphides are reduced to olefins by $Fe_2(CO)_9$ or $(CO)_5MnH$.[75]

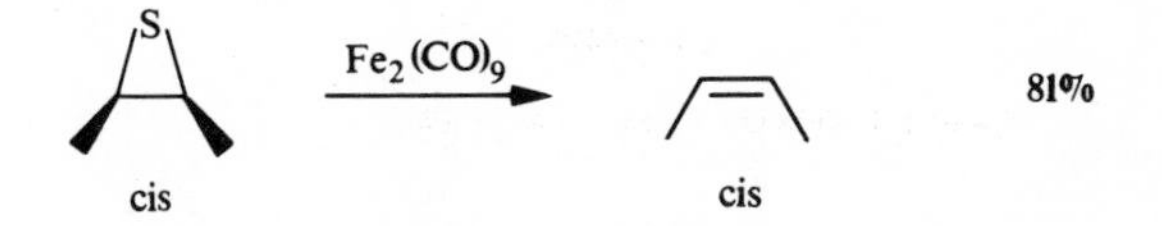

$Mo(CO)_6$/AcOH reduces thiols to alkanes.[74]

$$\text{RSH} \xrightarrow[\text{AcOH}]{\text{Mo(CO)}_6} \text{RH} \qquad 54-92\%$$

The nucleophilic $Fe(CO)_4^{2-}$ can be used to convert anhydrides to aldehydes and acids.[75] Acid chlorides are similarly reduced to aldehydes.[76]

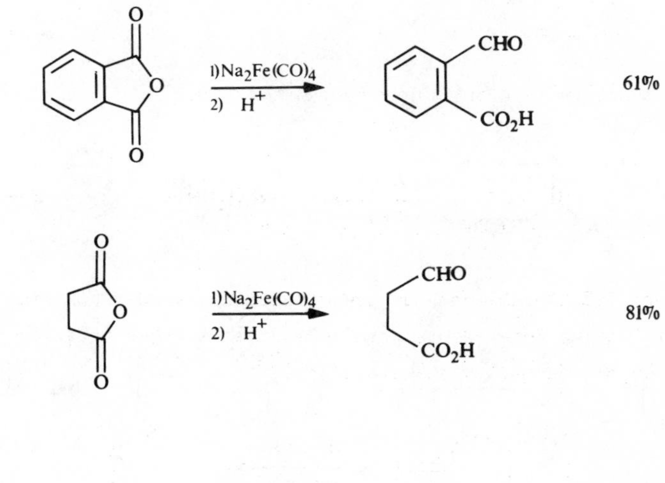

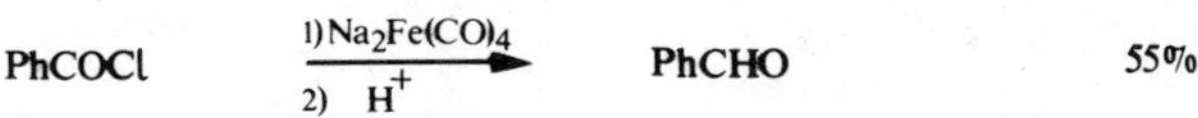

Amines may be alkylated by aldehydes in the presence of $(CO)_4FeH^-$. Presumably the reaction proceeds by reduction of the initially formed imines. Primary amines may be selectively mono or di-alkylated.[77]

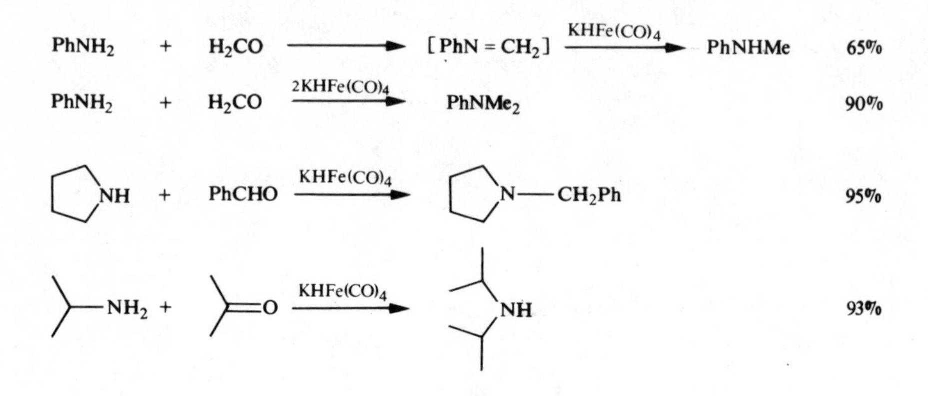

α,β-Unsaturated ketones are reduced selectively to saturated ketones in very high yields by $Fe(CO)_5/NaOH$ ($(CO)_4FeH^-$) in MeOH or by $NaHCr_2(CO)_{10}$.[78]

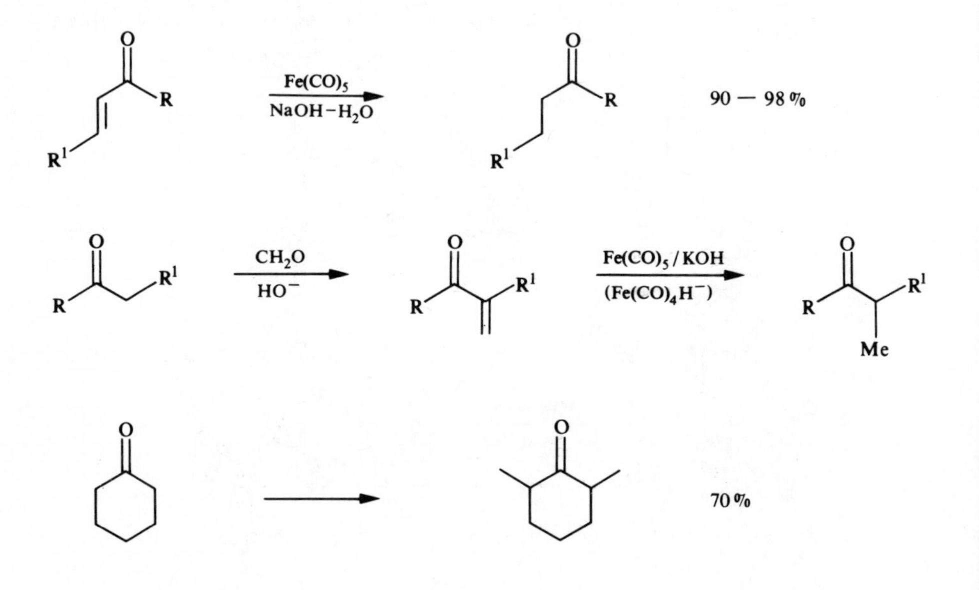

This reduction provides a method for the methylation of aldehydes and ketones.[79]

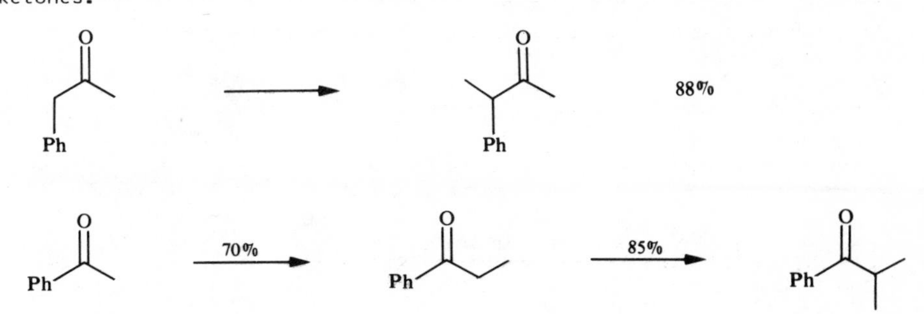

KHFe(CO)$_4$ dehalogenates alkyl halides but not aryl halides at 20°. The initial step is probably nucleophilic substitution of the halide by [Fe] since the reaction has been shown to proceed with inversion of configuration.[80]

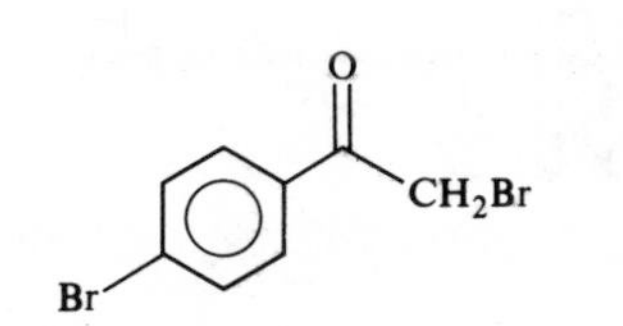

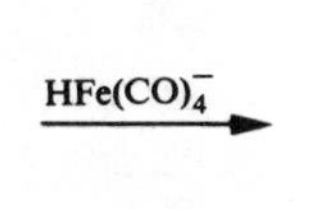

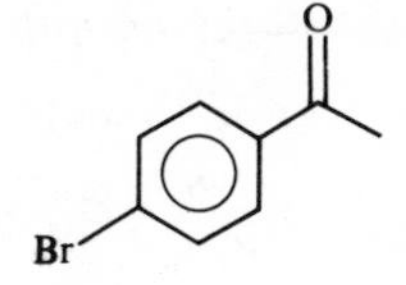

66 %

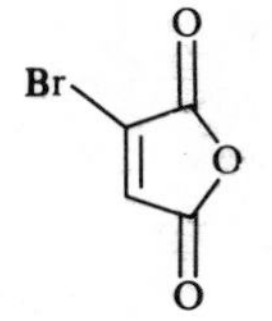

HFe(CO)$_4^-$

84 %

DFe(CO)$_4^-$

60%

Dehalogenation reactions via free radical intermediates may be effected using many organotransition metal reagents. The anion $CpV(CO)_3H^-$ reduces a variety of alkyl halides to the corresponding alkanes via free radical intermediates.[81]

$$RX + CpV(CO)_3H^- \longrightarrow CpV(CO)_3X^- + RH$$

$$X = Cl, Br, I$$

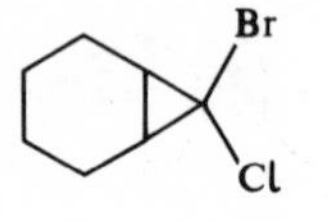

$$\xrightarrow{CpV(CO)_3H^-}$$

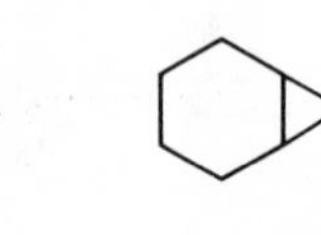

$$(-)\text{-}PhCH(Br)Me \xrightarrow{CpV(CO)_3D^-} (\pm)\text{-}PhCH(D)Me$$

The reactions of $CpV(CO)_3H^-$ show many similarities to the reactions of Bu_3SnH. A significant difference, however, occurs in the hydrogen transfer step which is very much faster for $CpV(CO)_3H^-$ presumably because the radical generating species contains a hydrogen atom for $CpV(CO)_3H^-$ but not for Bu_3SnH. This results in the suppression of the radical rearrangement and decomposition reactions for $CpV(CO)_3H^-$ that are commonly found for the reactions of Bu_3SnH.

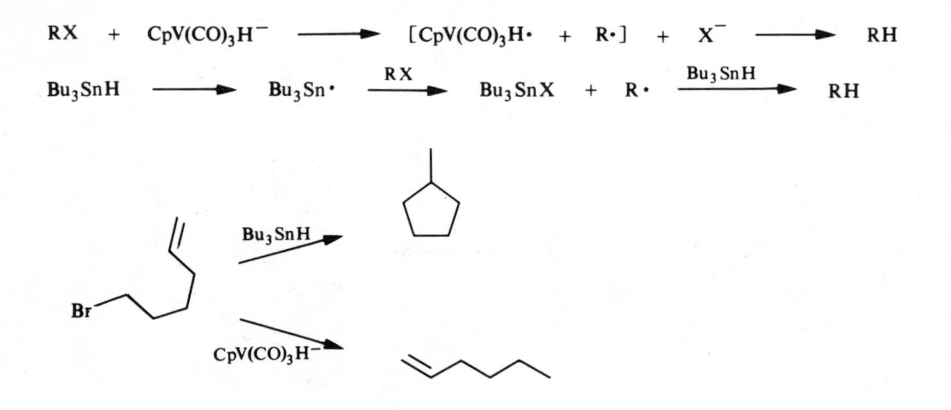

Cp_2MoH_2 appears to be a useful reagent for the monodehalogenation of gem-dihalo compounds. The least sterically hindered halide is reduced.[82] α-Halo ketones are also reduced to methyl ketones.

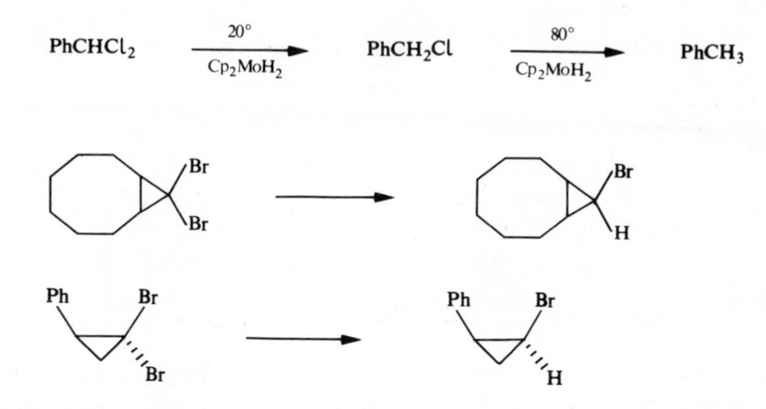

Vinyl halides and enol acetates may be reduced to olefins by $Fe(CO)_5$. α-Acetoxy and α-halo ketones are reduced to the respective ketones.[83]

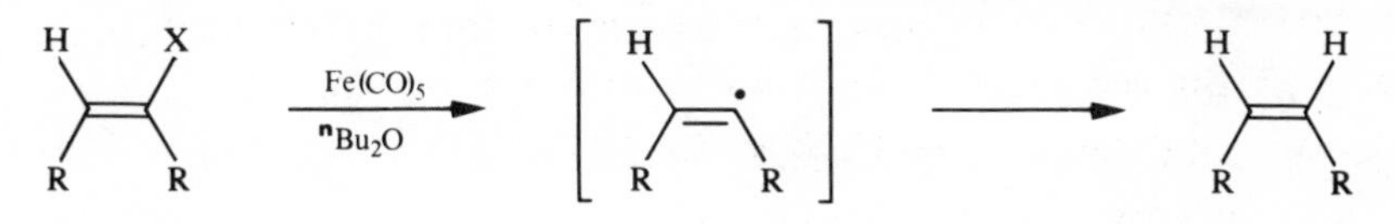

X = halide, OAc

Treatment of organometallic species with acid or hydride may also lead to selective reductions.[84]

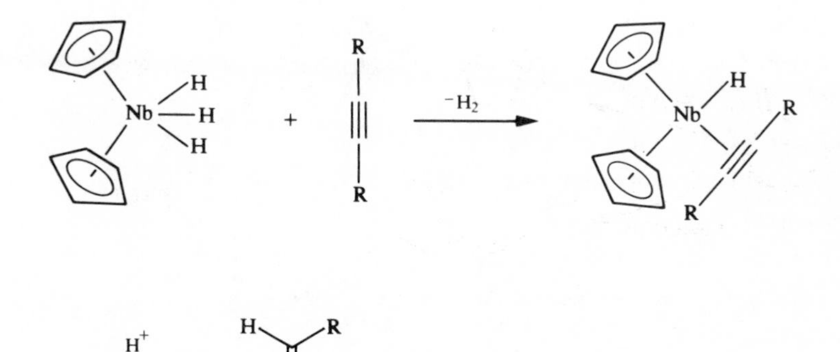

8.3. References

1. W.H. Clement and C.M. Selwitz, *J. Org. Chem.*, 1964, *29*, 241.
2. W.G. Lloyd and B.J. Luberoff, *J. Org. Chem.*, 1969, *34*, 3949.
3. D.F. Hunt and G.T. Rodeheaver, *Tet. Letters*, 1972, 3595.
4. J. Tsuji and T. Mandai, *Tet. Letters*, 1978, 1817.
5. P. Boontanonda and R. Grigg, *Chem. Comm.*, 1977, 583.
6. P.M. Henry, *J. Org. Chem.*, 1971, *36*, 1886.
7. K.B. Sharpless, K. Akashi and K. Oshima, *Tet. Letters*, 1976, 2503.
8. T.F. Blackburn and J. Schwartz, *Chem. Comm.*, 1977, 157.
9. F.W.S. Benfield and M.L.H. Green, *Chem. Comm.*, 1971, 1274.
10. V.B. Bierling, K. Kirschke and H. Oberender, *J. Prakt. Chem.*, 1972, *314* , 170.
11. T. Nishiguchi, K. Tachi and K. Fukuzumi, *J. Org. Chem.*, 1975, *40*, 237, 240.
12. M.N. Sheng and J.G. Zajacek, *J. Org. Chem.*, 1970, *35*, 1839; V.P. Yurev, I.A. Gailyunas, L.V. Spirikhin and G.A. Tolsilkov, *J. Gen. Chem. USSR*, 1975, *45*, 2269.
13. G.A. Tolsilkov, U.V. Dzhemilev, V.P. Yurev and S.F. Rafikov, *Proc. Acad. Sci. USSR. Chem.*, 1973, *208*, 45; V.P. Yurev, I.A. Gailynas, Z.G. Isaeva and G.A. Tolsilkov, *Bull. Acad. Sci. USSR Div. Chem. Sci.*, 1974, *23*, 885.
14. K.B. Sharpless and R.C. Michaelson, *J. Amer. Chem. Soc.*, 1973, *95*, 6136.
15. T. Itoh, K. Kaneda and S. Teranishi, *Chem. Comm.*, 1976, 421.
16. R.B. Dehnel and G.H. Whitham, *J.C.S. Perkin I*, 1979, 953.
17. P. Chamberlain, M.L. Roberts and G.H. Whitham, *J. Chem. Soc. (B)*, 1970, 1374.
18. R.B. Dehnel and G.H. Whitham, *J.C.S. Perkin I*, 1979, 953.
19. M. Kobayshi, S. Kurozumi, T. Toru and S. Ishimoto, *Chem. Letters*, 1976, 1341.
20. R. Breslow and L.M. Maresca, *Tet. Letters*, 1977, 623.
21. S. Yamada, T. Mashiko and S. Terashima, *J. Amer. Chem. Soc.*, 1977 *99*, 1988.
22. R.C. Michaelson, R.E. Palermo and K.B. Sharpless, *J. Amer. Chem. Soc.*, 1977, *99*, 1990.
23. H. Mimoun, I. Seree de Roch and L. Sajus, *Tetrahedron*, 1970, *26*, 37.

24. G.A. Tolsilkov, U.M. Dzhemilev and V.P. Yur'ev, *Zh. Org. Khim.*, 1972, *8*, 2204 (C.A., *78*, 43834c).
25. A.A. Frimer, *Chem. Comm.*, 1977, 205.
26. H.B. Kagan, H. Mimoun, C. Mark and V. Schurig, *Ang. Chem. Int. Ed.*, 1979, *18*, 485.
27. M.J.Y. Chen and J.K. Kochi, *Chem. Comm.*, 1977, 204.
28. E. Vedejs, *J. Amer. Chem. Soc.*, 1974, *96*, 5944.
29. I.J. Harvie and F.J. McQuillin, *Chem. Comm.*, 1977, 241.
30. D.N. Jones and S.D. Knox, *Chem. Comm.*, 1975, 166.
31. E. Vedejs, M.F. Saloman and P.D. Weeks, *J. Organometal. Chem.*, 1972, *40*, 221.
32. A.J. Birch and D.H. Williamson, *Org. Reactions*, 1976, *24*, 1, and ref. therein.
33. A.J. Birch and K.A.M. Walker, *J. Chem. Soc. C*, 1966, 1894; R.E. Ireland and P. Bey, *Org. Syn.*, 1973, *53*, 63; F.H. Lincoln, W.P. Schneider and J.E. Pike, *J. Org. Chem.*, 1973, *38*, 951.
34. A. Nakamura and S. Otsuka, *Tet. Letters*, 1973, 4529.
35. J. Nasielski, P. Kirsch and L. Wilputte-Steinert, *J. Organometal. Chem.*, 1971, *27*, C13; M. Wrighton and M. Schroeder, *J. Amer. Chem. Soc.*, 1973, *95*, 5764.
36. M. Cais, E.N. Frankel and A. Rejoan, *Tet. Letters*, 1968, 1919; E.N. Frankel, E. Selke and C.A. Glass, *J. Amer. Chem. Soc.*, 1968, *90*, 2446; E.N. Frankel and R.O. Butterfield, *J. Org. Chem.*, 1969, *34*, 3930.
37. D.R. Fahey, *J. Org. Chem.*, 1973, *38*, 3343.
38. H.W. Kircher and F.U. Rosenstein, *J. Org. Chem.*, 1973, *38*, 2259.
39. R. Noyori, I. Umeda and T. Ishigami, *J. Org. Chem.*, 1972, *37*, 1542.
40. I. Ojima, T. Kogure and Y. Nagai, *Tet. Letters*, 1972, 5035.
41. B. Heil, S. Törös, J. Bakos and L. Markó, *J. Organometal. Chem.*, 1979, *175*, 229.
42. R.R. Schrock and J.A. Osborn, *Chem. Comm.*, 1970, 567.
43. J.E. Lyons, *Chem. Comm.*, 1975, 412; P. Morand and M. Kayser, *Chem. Comm.*, 1976, 314.
44. E.L. Muetterties and F.J. Hirsekorn, *J. Amer. Chem. Soc.*, 1974, *96*, 7920; F.J. Hirsekorn, M.C. Rakowski and E.L. Muetterties, *J. Amer. Chem. Soc.*, 1975, *97*, 237; E.L. Muetterties, M.C. Rakowski, F.J. Hirsekorn, W.D. Larson, V.J. Basus and F.A.L. Anet, *J. Amer. Chem. Soc.*, 1975, *97*, 1266.

45. J.D. Morrison and H.S. Mosher, *Asymmetric Organic Reactions*, Prentice Hall, 1971; D. Valentine and J.W. Scott, *Synthesis*, 1978, 329.
46. T.P. Dang and H.B. Kagan, *Chem. Comm.*, 1971, 481; C. Detellier, G. Gelbard and H.B. Kagan, *J. Amer. Chem. Soc.*, 1978, *100*, 7556.
47. H. Brunner and W. Pieronczyk, *Ang. Chem. Int. Ed.*, 1979, *18*, 620.
48. M.D. Fryzuk and B. Bosnich, *J. Amer. Chem. Soc.*, 1977, *99*, 6262.
49. B.D. Vineyard, W.S. Knowles, M.J. Sabacky, G.L. Bachman and D.J. Weinkauff, *J. Amer. Chem. Soc.*, 1977, *99*, 5946.
50. K. Achiwa, *J. Amer. Chem. Soc.*, 1976, *98*, 8265.
51. T. Hayashi, T. Mise and M. Kumada, *Tet. Letters*, 1976, 4351; Y. Ohgo, Y. Natori, S. Takeuchi and J. Yashimora, *Tet. Letters*, 1974, 1327.
52. T. Hayashi, K. Yamamoto, K. Kasuga, H. Omizu and M. Kumada, *J. Organometal. Chem.*, 1976, *113*, 127; I. Ojima, T. Kogure and M. Kumagai, *J. Org. Chem.*, 1977, *42*, 1671.
53. T. Hayashi, K. Tamao, Y. Katsuro, I. Nakae and M. Kumada, *Tet. Letters*, 1980, 1871.
54. W.P. Giering, M. Rosenblum and J. Tancrede, *J. Amer. Chem. Soc.*, 1972, *94*, 7170.
55. M. Rosenblum, M.R. Saidi and M. Madhavarao, *Tet. Letters*, 1975, 4009.
56. S.G. Davies, unpublished results.
57. P. Dowd and K. Kang, *Chem. Comm.*, 1974, 384.
58. H. Alper and D. Des Roches, *Tet. Letters*, 1977, 4155.
59. H. Alper, D. Des Roches, T. Durst and R. Legault, *J. Org. Chem.*, 1976, *41*, 3611.
60. R.W. Ashworth and G.A. Berchtold, *Tet. Letters*, 1977, 343.
61. K.B. Sharpless, M.A. Umbreit, M.T. Nieh and T.C. Flood, *J. Amer. Chem. Soc.*, 1972, *94*, 6538.
62. A. Sattar, J. Forrester, M. Moir, J.S. Roberts and W. Parker, *Tet. Letters*, 1976, 1403.
63. H. Ledon, I. Tkatchenko and D. Young, *Tet. Letters*, 1979, 173.
64. M. Berry, S.G. Davies and M.L.H. Green, *Chem. Comm.*, 1978, 99.
65. H. Alper and E.C.H. Keung, *Tet. Letters*, 1970, 53.
66. H. Alper and J.T. Edward, *Can. J. Chem.*, 1970, *48*, 1543.
67. A. Dondoni and G. Barbaro, *Chem. Comm.*, 1975, 761.
68. H. Alper and J.T. Edward, *J. Organometal. Chem.*, 1969, *16*, 342.

69. H. Alper and M. Salisova, *Tet. Letters*, 1980, 801.

70. H. Alper and Hang-Nam Paik, *J. Org. Chem.*, 1977, *42*, 3522; H. Alper *J. Org. Chem.*, 1975, *40*, 2694; H. Alper, *J. Organometal Chem.*, 1974, *73*, 359.

71. J. Dawb, V. Trantz and U. Echardt, *Tet. Letters*, 1973, 447; M.F. Semmelhack and R.D. Stauffer, *Tet. Letters*, 1973, 2667.

72. B.M. Trost and F. Chen, *Tet. Letters*, 1971, 2603.

73. B.M. Trost and S.D. Ziman, *J. Org. Chem.*, 1973, *38*, 932; W. Beck, W. Danzer and R. Hofer, *Ang. Chem. Int. Ed.*, 1973, *12*, 77.

74. H. Alper and C. Blais, *Chem. Comm.*, 1980, 169.

75. Y. Watanabe, M. Yamashita, T. Mitsudo, M. Tanaka, and Y. Takegami, *Tet. Letters*, 1973, 3535.

76. Y. Watanabe, T. Mitsudo, M. Tnaaka, K. Yamamoto, T. Okajima and Y. Takegami, *Bull. Chem. Soc. Japan*, 1971, *44*, 2569.

77. G.P. Boldrini, M. Panunzio and A. Umani-Ronchi, *Synthesis*, 1974, 733; Y. Watanabe, M. Yamashita, T. Mitsudo, M. Tanaka and Y. Takegami, *Tet. Letters*, 1974, 1879.

78. R. Noyori, I. Umeda and T. Ishigami, *J. Org. Chem.*, 1972, *37*, 1542; G.P. Boldrini and A. Umani-Ronchi, *Synthesis*, 1976, 596.

79. G. Cainelli, M. Panunzio and A. Umani-Ronchi, *Tet. Letters*, 1973, 2491; *J.C.S. Perkin I*, 1975, 1273.

80. H. Alper, *Tet. Letters*, 1975, 2257.

81. R.J. Kinney, W.D. Jones and R.G. Bergman, *J. Amer. Chem. Soc.*, 1978, *100*, 7902.

82. A. Nakamura, *J. Organometal. Chem.*, 1979, *164*, 183.

83. S.J. Nelson,G. Detre and M. Tanabe, *Tet. Letters*, 1973, 447; T.Y. Luh, C.H. Lai, K.L. Lei and S.W. Tam, *J. Org. Chem.*, 1979, *44*, 641.

84. J.A. Labinger and J. Schwartz, *J. Amer. Chem. Soc.*, 1975, *97*, 1596; M. Franck-Neumann, D. Martina and F. Brion, *Ang. Chem. Int. Ed.*, 1978, *17*, 690; R.O. Hutchins, K. Learn and R.P. Fulton, *Tet. Letters*, 1980, 27; T.R. Bosin, M.G. Raymond and A.R. Buckpitt, *Tet. Letters*, 1973, 4699.

CHAPTER 9

CARBONYLATION AND RELATED REACTIONS

Many industrial procedures have been developed for the synthesis of aldehydes, alcohols and acids from olefins and carbon monoxide. Many of these processes are catalysed by transition metal complexes particularly those of iron, cobalt, rhodium, nickel and palladium.

Hydroformylation:

$$RCH{=}CH_2 \xrightarrow{CO/H_2} RCH_2.CH_2.CHO$$

Hydrocarboxylation:

$$RCH{=}CH_2 \xrightarrow{CO/H_2O} RCH_2.CH_2.CO_2H$$

$$HC{\equiv}CH \longrightarrow CH_2{=}CH.CO_2H$$

Oxidative carboxylation

$$CH_2{=}CH_2 \longrightarrow CH_2{=}CH.CO_2H$$

All of the reactions of carbon monoxide involving transition metal catalysts proceed through the same two types of intermediates: a metal carbonyl complex followed by a metal acyl complex. It is the different ways of forming the acyl complex and its subsequent reactions that control the products of a given reaction.

The reactions of coordinated carbon monoxide that can lead to metal acyl are either (A) nucleophilic attack directly on the CO, or (B) alkyl migration from the metal to a CO ligand.

A Nucleophilic attack by N^- on coordinated CO

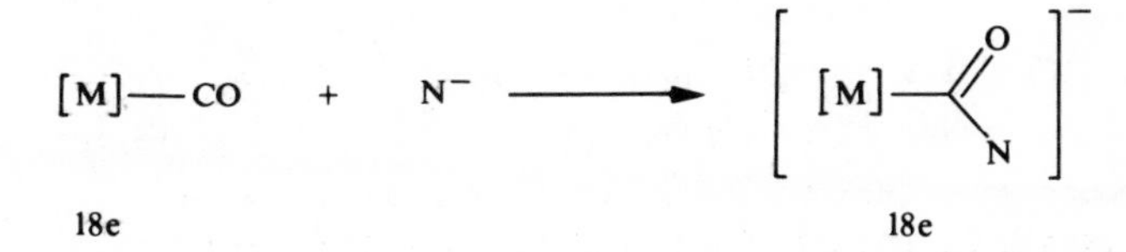

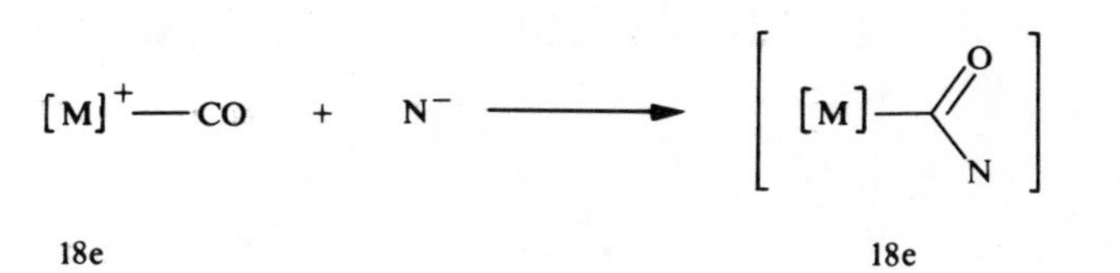

B Alkyl migration

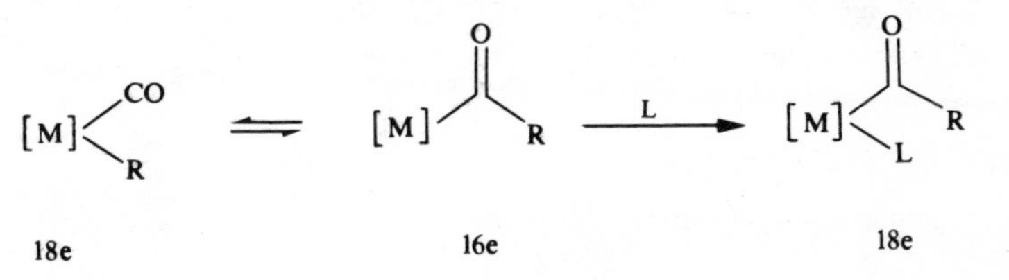

(L = 2e ligand such as CO, PR_3, solvent, etc.)

The M-alkyl group may be generated by many methods, for example, nucleophilic displacement of Cl^- from M(CO)Cl by R^-, addition of M-H across a double bond, or nucleophilic attack onto a coordinated olefin.

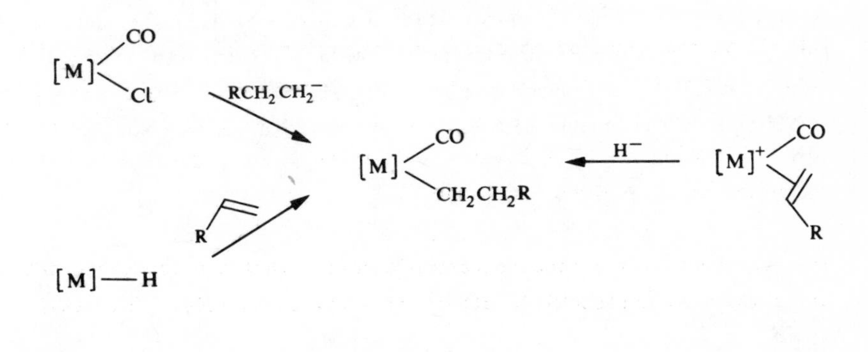

In all the transition metal systems so far studied, the migration of the alkyl group from the metal to the CO proceeds with retention of configuration at carbon.[1,2]

The acyl complexes liberate the products of the reaction either by a pericyclic elimination reaction (C), e.g. in the hydroformylation reaction, or by nucleophilic attack of an external nucleophile, e.g. H_2O, on the metal-acyl (D).

C Pericyclic elimination

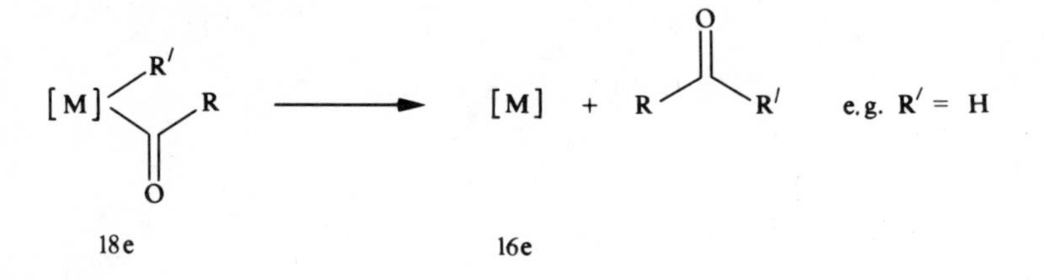

D Nucleophilic attack on M-acyl

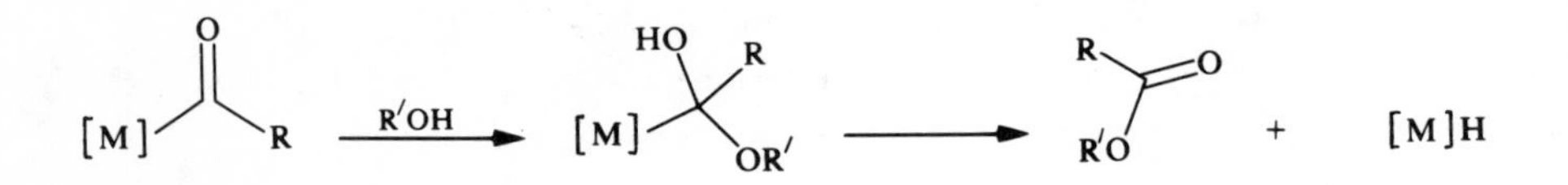

Depending on the reaction conditions the M-acyl groups may be liberated as RCHO, RCO_2H, RCOX, $RCONR'_2$, RCOR' or RCO_2COR'.

All of the reactions of coordinated carbon monoxide can be summarised in terms of the above four reactions. They are illustrated in the two catalytic cycles below for the hydroformylation of ethylene catalysed by $HCo(CO)_4$ and the carboxylation of ethylene catalysed by $PdCl_2$.

(see section 1.5.1).

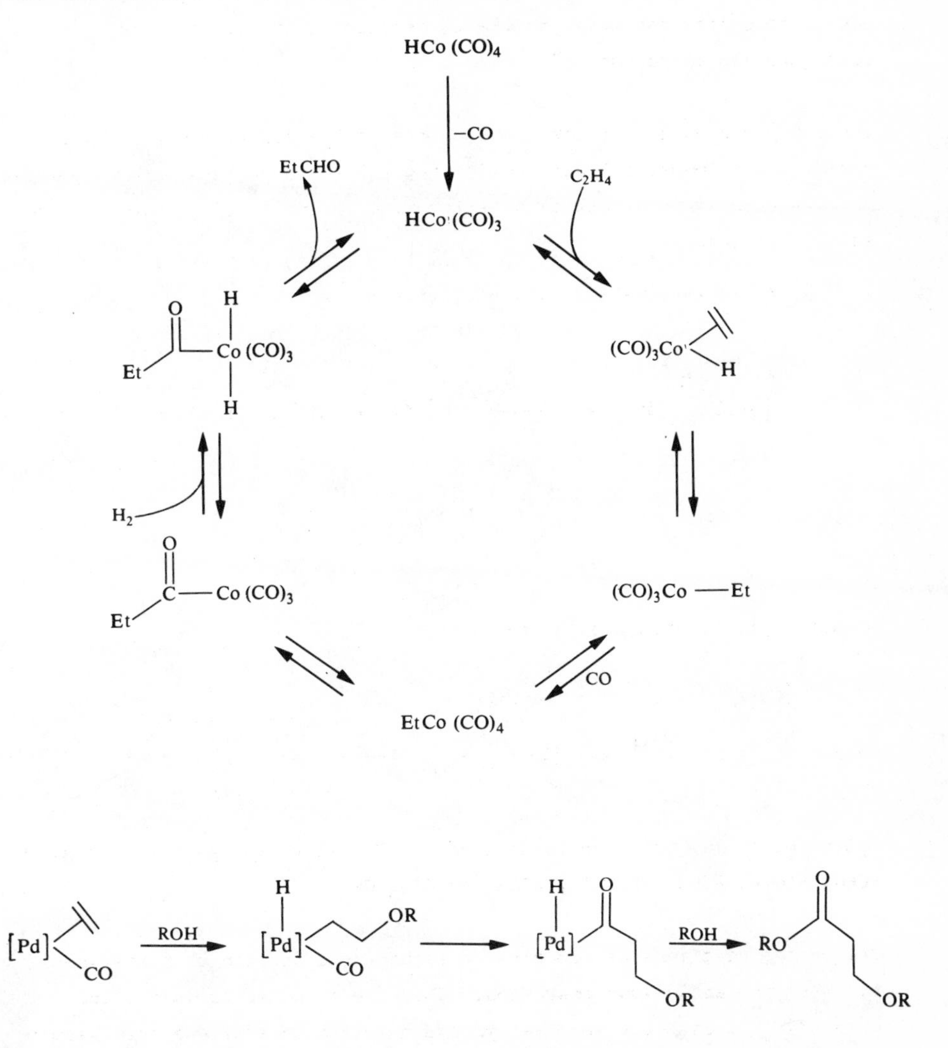

The insertion of CO into organic molecules is at the present time one of the most important uses of transition metals for organic synthesis. A large variety of functional groups (e.g. alcohols, aldehydes, ketones, esters, lactones, amides, lactams, carboxylic acids, etc) may be formed via this method. Some of the more synthetically useful reactions are given below according to the transition metal used in the synthesis.

9.1 CARBONYLATION REACTIONS WITH Zr COMPOUNDS

CO undergoes ready stoicheiometric insertion (20°; 1.5 atm.) into the Zr-C bond of many alkyl, alkenyl and γ,δ-unsaturated Cp_2ZrCl complexes to generate the corresponding stable Zr-acyl complexes. It has been demonstrated that the CO inserts into the Zr-C bond with retention of configuration at carbon.[2] Since the starting Zr complexes can be easily produced from the addition of Cp_2ZrHCl to olefins, acetylenes and conjugated 1,3-dienes, respectively, this procedure has developed into a useful CO insertion reaction. The Zr-acyls are readily converted to aldehydes, acids, esters or acid halides.

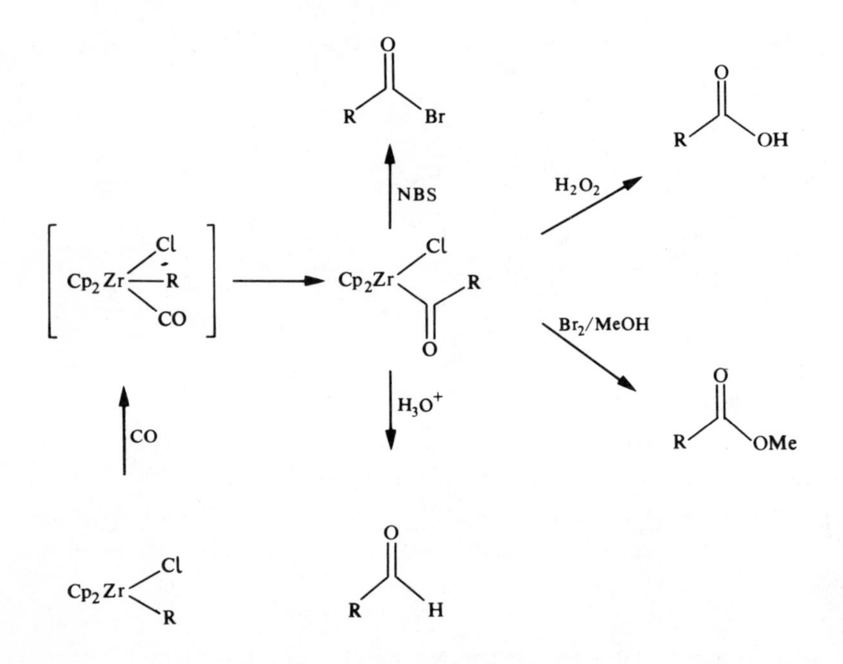

When Cp_2ZrHCl reacts with internal olefins, the Zr migrates very rapidly down the chain to produce a primary alkyl substituent on the Zr. Thus all of the hexene isomers produce the 1-hexyl Zr species.[3] Hydrozirconation followed by carbon monoxide insertion is a useful procedure for carbonylating only terminal carbon atoms. It complements many of the other transition metal procedures which give mixtures of internal-carbonylation products.[4,5]

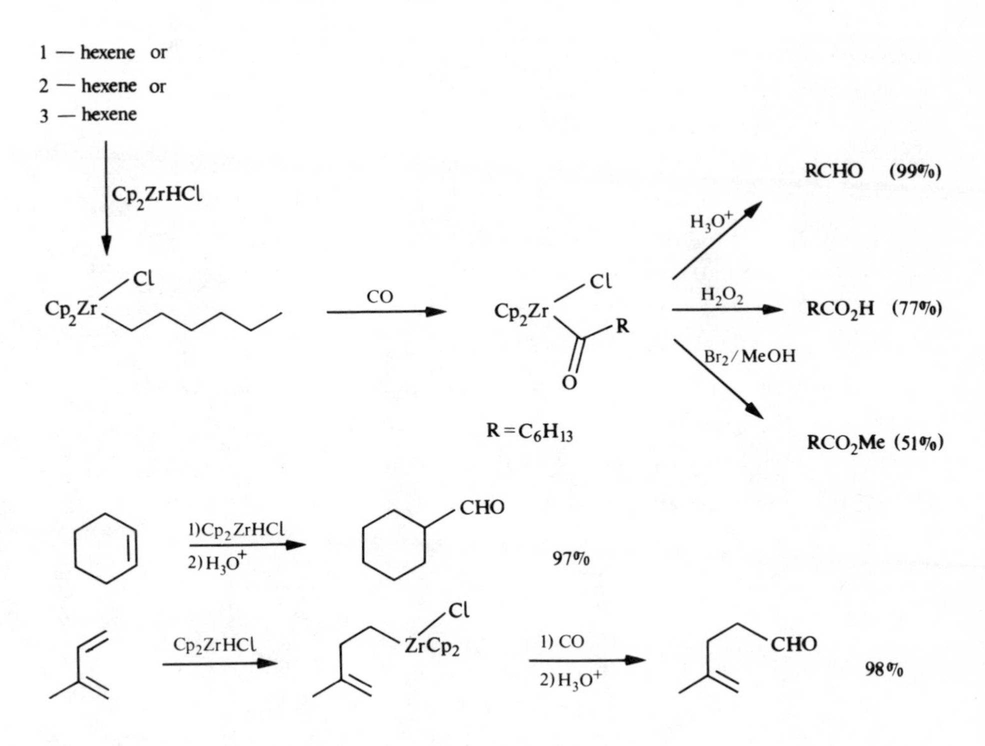

9.2 CARBONYLATION REACTIONS WITH Fe COMPOUNDS

Three types of iron complexes have found general use for the insertion of CO into organic molecules: the metal carbonyls $Fe(CO)_5$ and $Fe_2(CO)_9$, the tetracarbonyl ferrates $Fe(CO)_4^{2-}$ and $HFe(CO)_4^-$, and the $CpFe(CO)_2R$ complexes.

$Fe_2(CO)_9$ reacts with benzyl chloride to give dibenzyl ketone in 56% yield.[6]

$$PhCH_2Cl \xrightarrow[C_6H_6]{Fe_2(CO)_9} (PhCH_2)_2CO$$

Aryl azides may be converted to ureas in high yield with $Fe(CO)_5$ in acetic acid at 100°.[7]

$$\text{ArylN}_3 \xrightarrow[\text{CH}_3\text{CO}_2\text{H}]{\text{Fe(CO)}_5} (\text{ArylNH})_2\text{CO} \qquad 70-90\%$$

Cycloheptatrienone is isolated from the reaction of acetylene with $Fe_2(CO)_9$.[8]

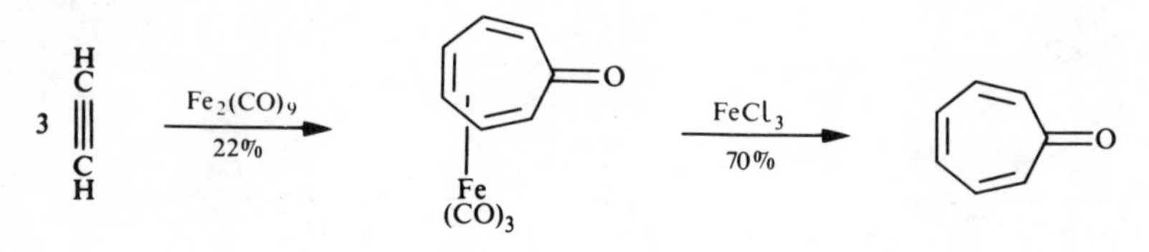

Protonation of cycloheptatriene iron tricarbonyl followed by hydride reduction generates the complex 1 in 60% yield. Treatment of 1 with CO generates the ketone 2 in 96% yield. Similarly cyclooctatetraene iron tricarbonyl gives the ketone 3 in 60% yield.[9]

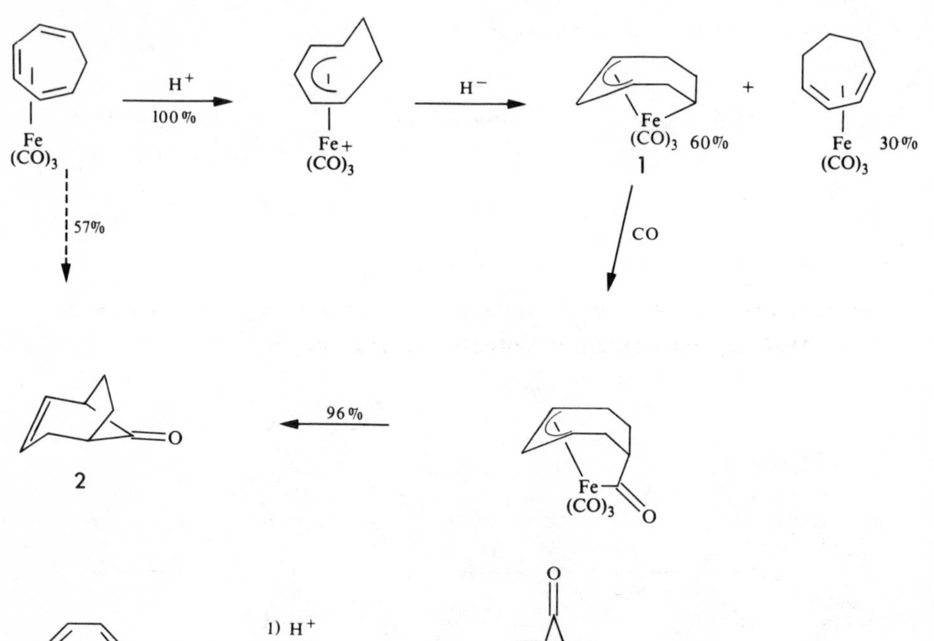

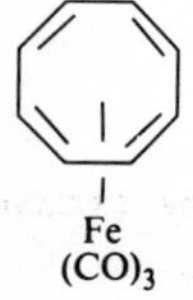

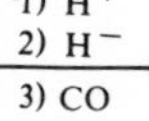

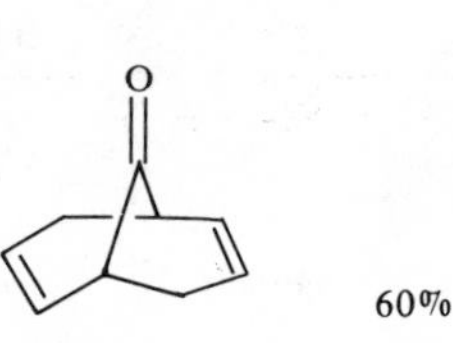

60%

The heterocycles of type 4 are first reduced and then carbonylated by $Fe_2(CO)_9$ to generate the β-lactams.[10]

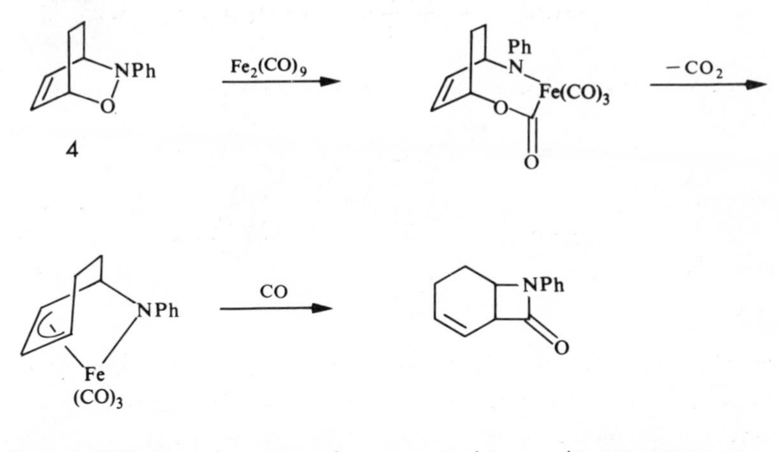

Some diene $Fe(CO)_3$ complexes can be converted to cyclopentenones on treatment with $AlCl_3$.[11] For example 2-indanone may be isolated in 48% yield from the following reaction.

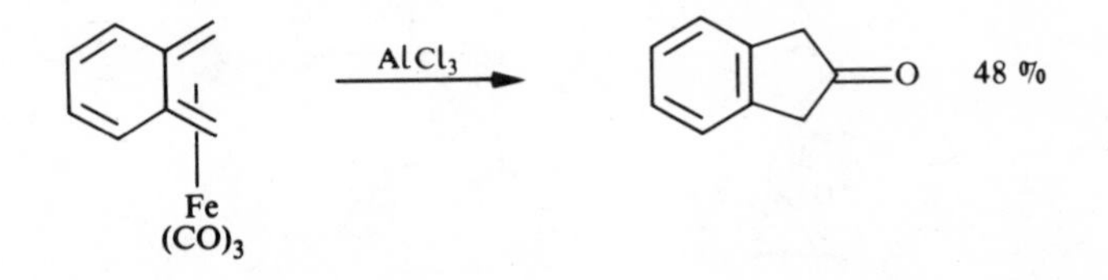

The mechanism is believed to be electrophilic attack of an activated (diene)$Fe(CO)_3$ complex on an unactivated complex.

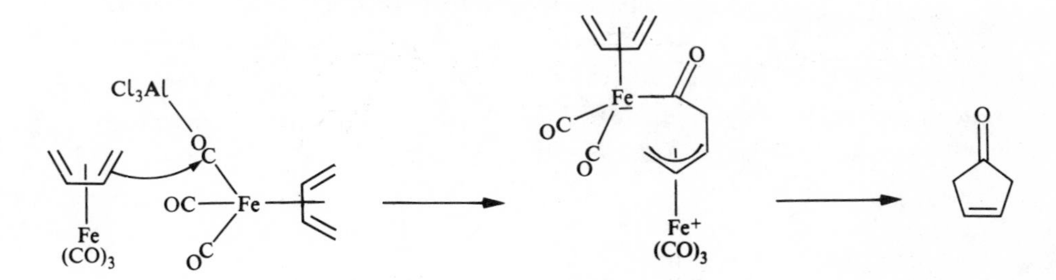

Cyclohexa-1,3-diene iron tricarbonyl reacts with $AlCl_3$ and CO to produce the ketone complex 5 which may react further with CO to give the diketone 6.[12]

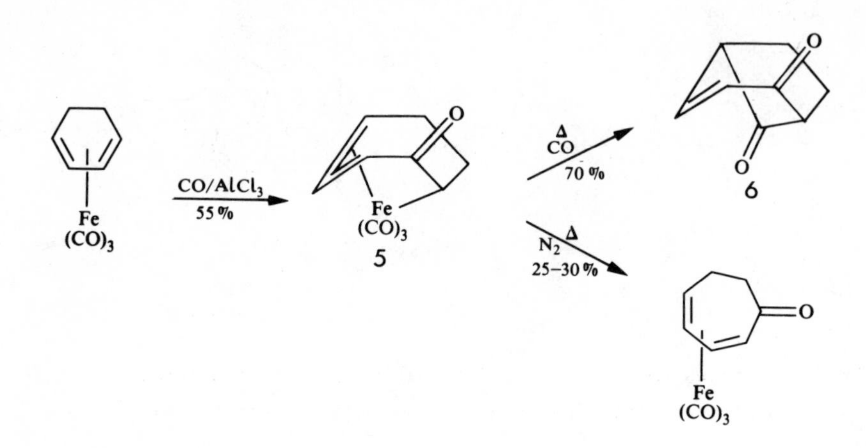

Cyclooctatetraene iron tricarbonyl reacts with $AlCl_3$ in the presence of CO to produce the complex 7 which on heating with CO gives barbaralone.[13]

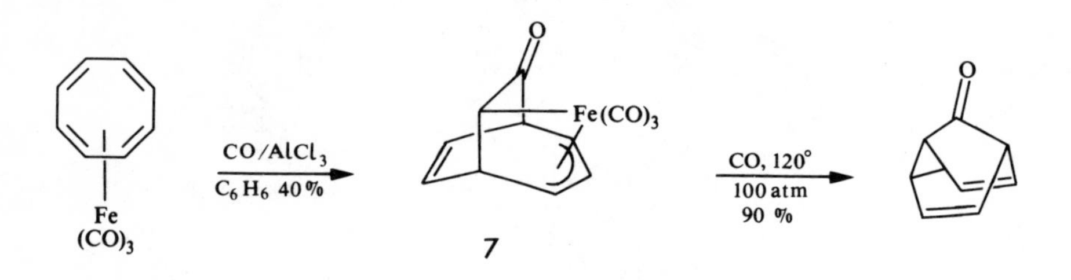

$Fe(CO)_5$ reacts with small ring compounds to give ring expansion products. The relatively inexpensive (−)β-pinene 8 can be isomerised to (−)α-pinene 9 with $Fe(CO)_5$ in 44% yield with an optical yield of 97%.[14] Other isomerisation methods give lower optical yields. The other products of this reaction are the cyclobutane to cyclopentanone ring expansion products 10 and 11.[15]

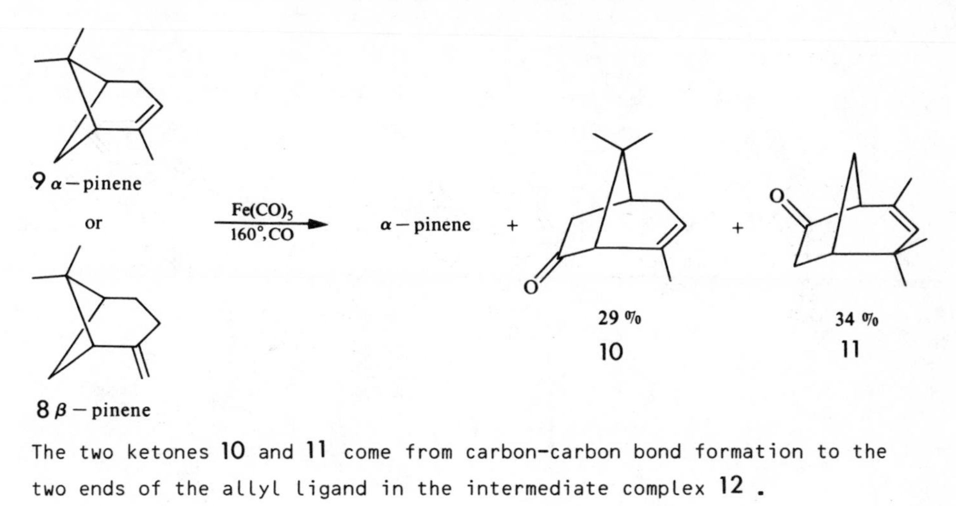

The two ketones 10 and 11 come from carbon-carbon bond formation to the two ends of the allyl ligand in the intermediate complex 12 .

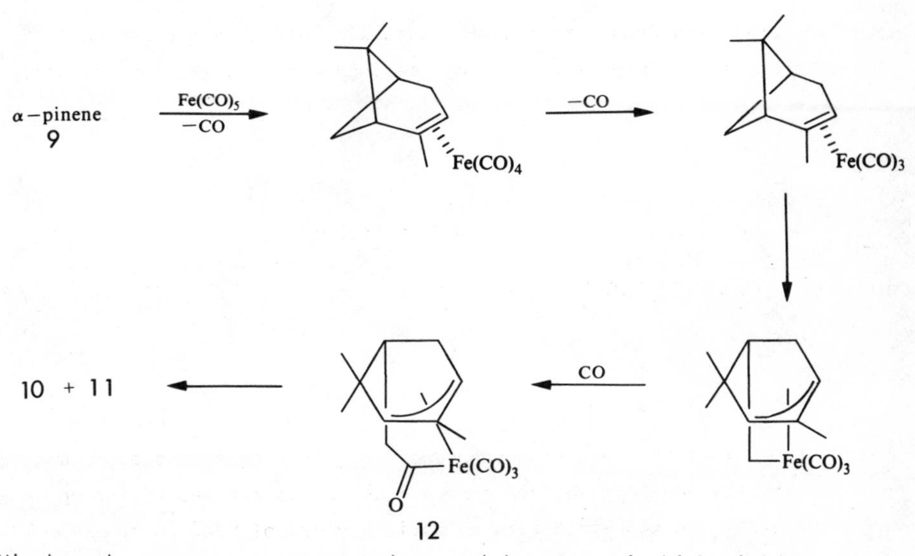

Vinyl cyclopropanes are converted to cyclohexenones in high yields on treatment with $Fe(CO)_5$ or $Fe_2(CO)_9$.[16]

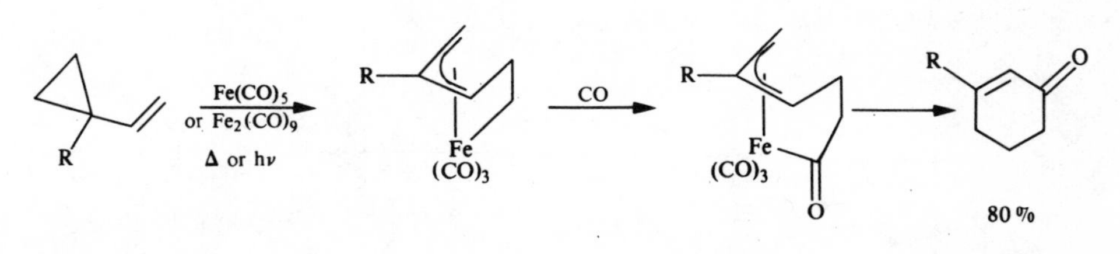

Vinyl epoxides are converted to lactones by treatment with $Fe(CO)_5$ followed by oxidation with Ce(IV).[17] Reaction of the intermediate complexes with $PhCH_2NH_2/ZnCl_2$ followed by treatment with Ce(IV) yields lactams.[17a]

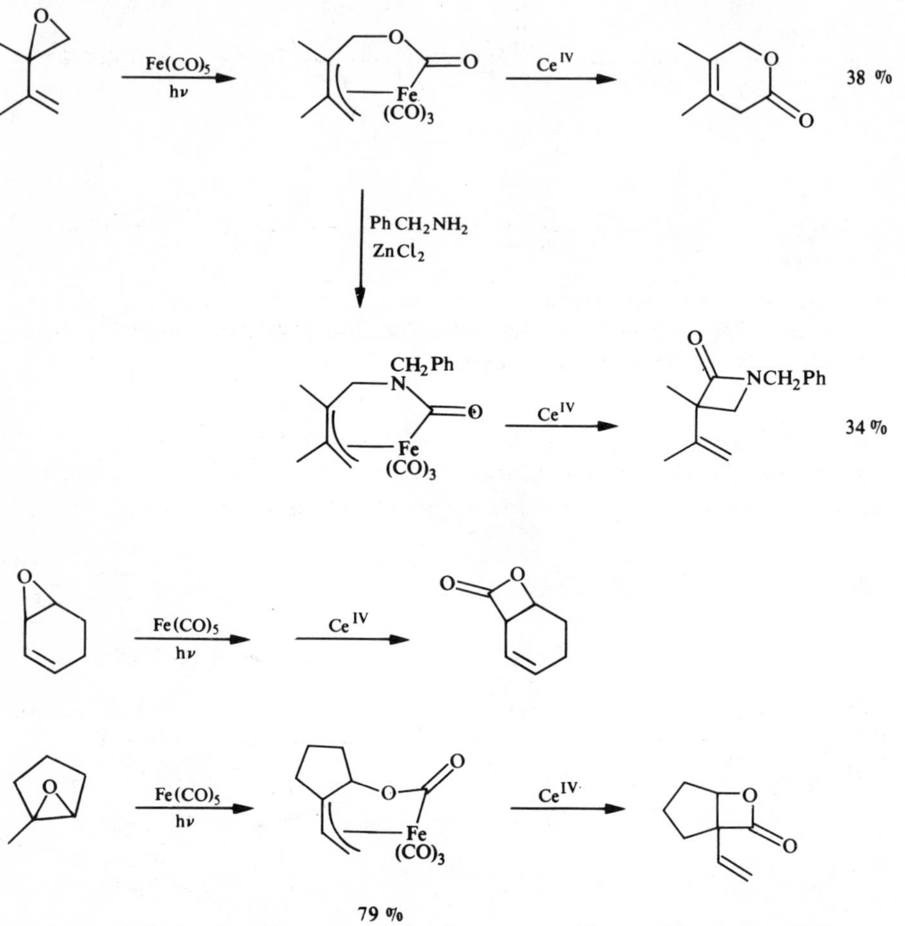

Iron and cobalt carbonyls convert isoprene monoxide to the conjugated lactone 13 whereas $[(1,5\text{-COD})RhCl]_2$ produces the unrearranged lactone 14.[18] This reaction may be regarded as the equivalent of a Diels-Alder reaction between isoprene and carbon dioxide.

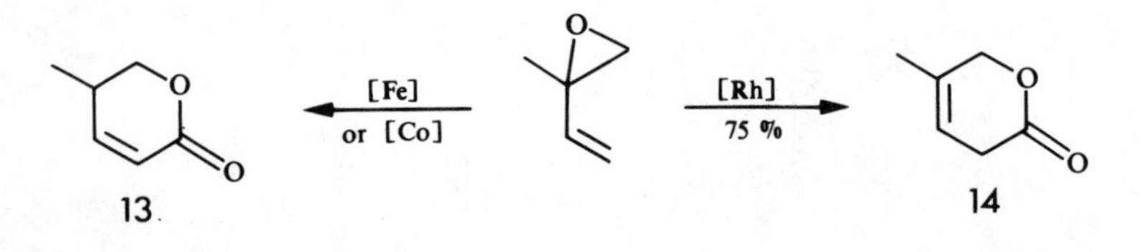

Acylferrates may be made either by the addition of an alkyl anion to $Fe(CO)_5$ or by the reaction of $Fe(CO)_4^{2-}$ with an acid chloride or alkyl halide followed by the addition of a ligand L (L = CO or PR_3).

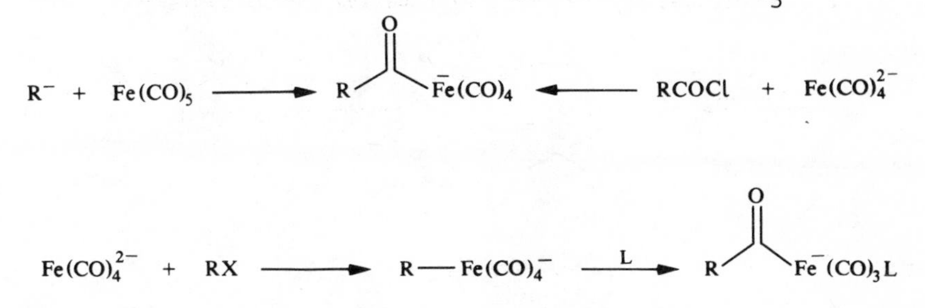

These acyl ferrates are readily converted into aldehydes, ketones, acids, amides, acid chlorides and esters.

For example, treatment of pyridine with PhLi followed by $Fe(CO)_5$ generates the complex 15 which could be converted into the 3-substituted pyridines shown.[19] 3-substitution is normally difficult to achieve for pyridine.

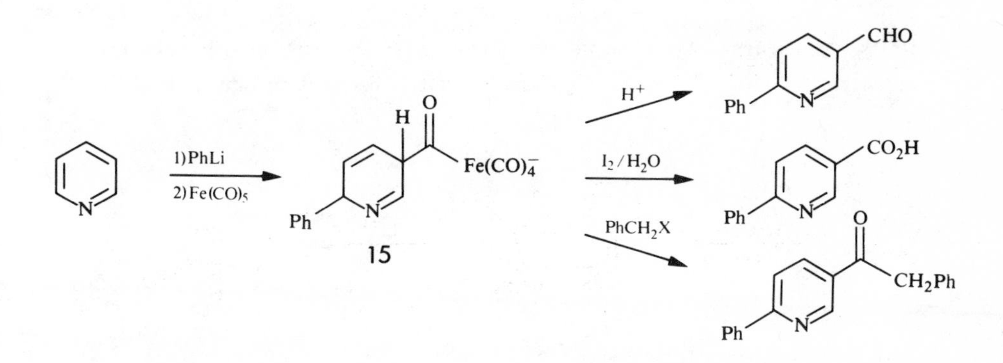

Treatment of $Fe(CO)_5$ with Grignard reagents followed by a nitro compound allows the formation of amides (70-85%).[20]

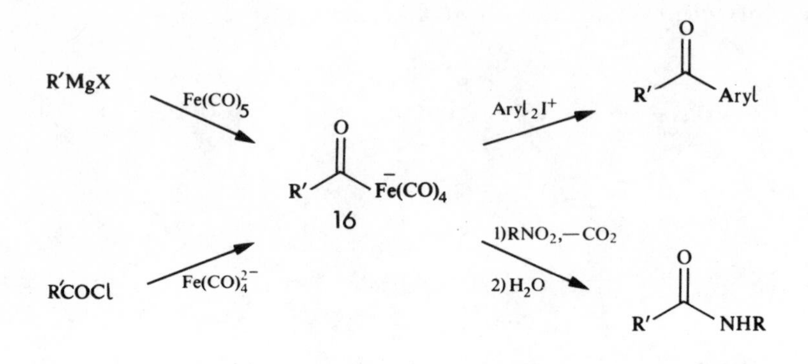

Treatment of the acyl anions **16** with a variety of diaryl iodonium salts gives aryl alkyl ketones in high yields (85-93%).[21]

$Fe(CO)_4^{2-}$ reacts with a variety of alkyl halides (RX) to generate the monoanions $RFe(CO)_4^-$ which are converted to the acyl anions with CO or PR_3.

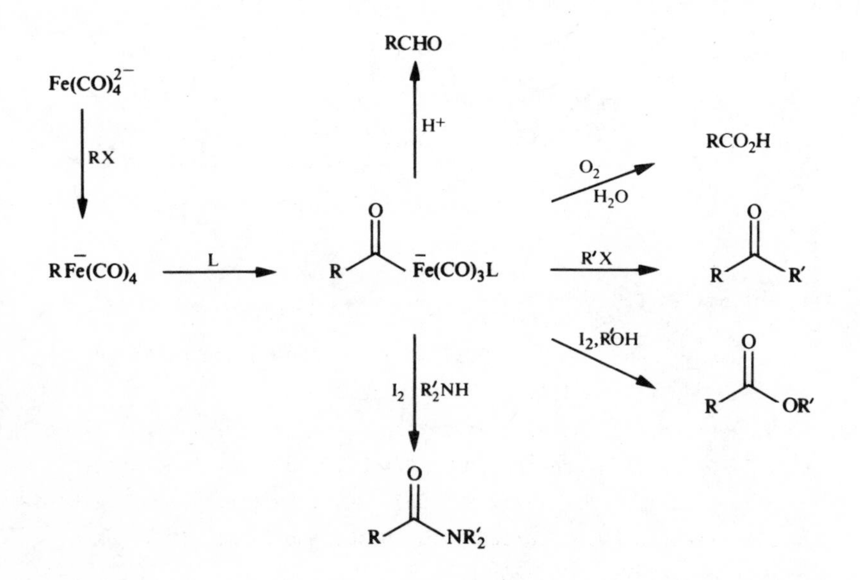

The general reactions of these acyl anions are shown below together with some specific examples.[22,23,24] Oxidation (O_2 or I_2) converts the intermediates $RFe(CO)_4^-$ directly to acids or esters (RCO_2R').

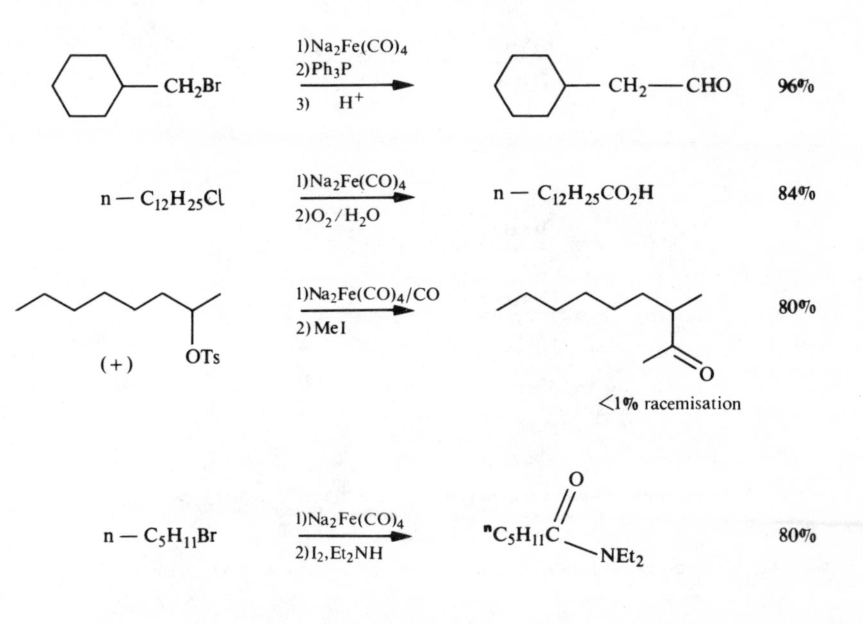

Generally for these reactions high yields (80-100%) are only obtained for primary alkyl halides. Less satisfactory yields are obtained for secondary alkyl halides. The reaction conditions described above are compatible with other functional groups in the molecule such as chloride, cyanide, ketone and ester functions.

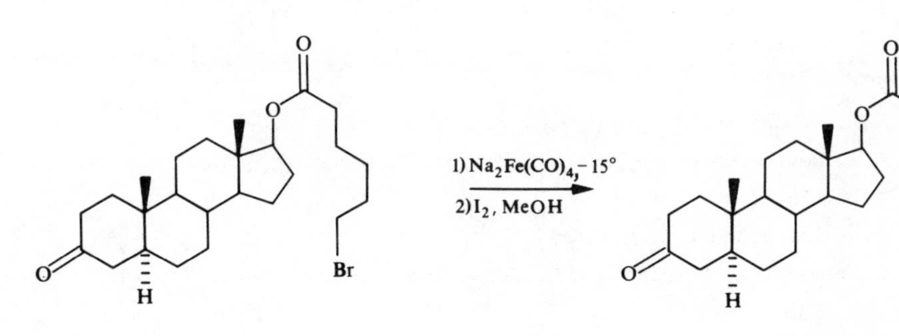

$KHFe(CO)_4$ has been used to convert ethyl acrylate to diethyl methyl malonate[25] and epoxides to β-hydroxy esters.[26]

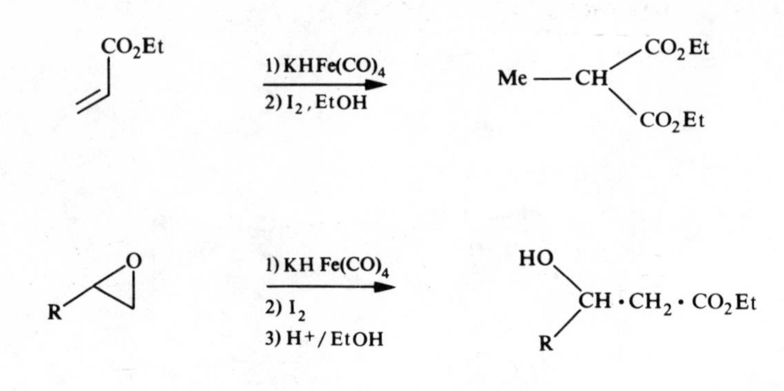

Reaction of 5-bromo-1-pentene with $Na_2Fe(CO)_4$ followed by treatment with acid produces cyclohexanone presumably via the mechanism shown below. Similarly 4-bromo-1-butene and the tosylate 17 also cyclise.[22,27]

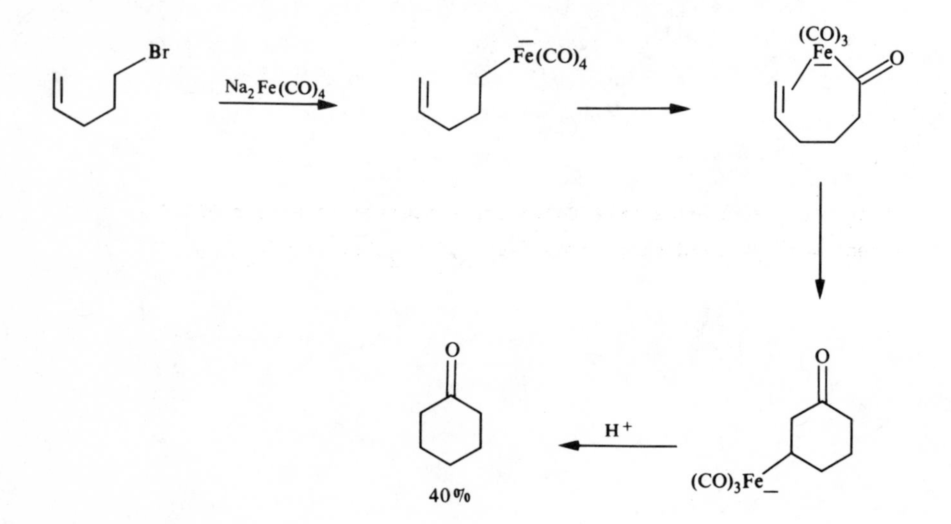

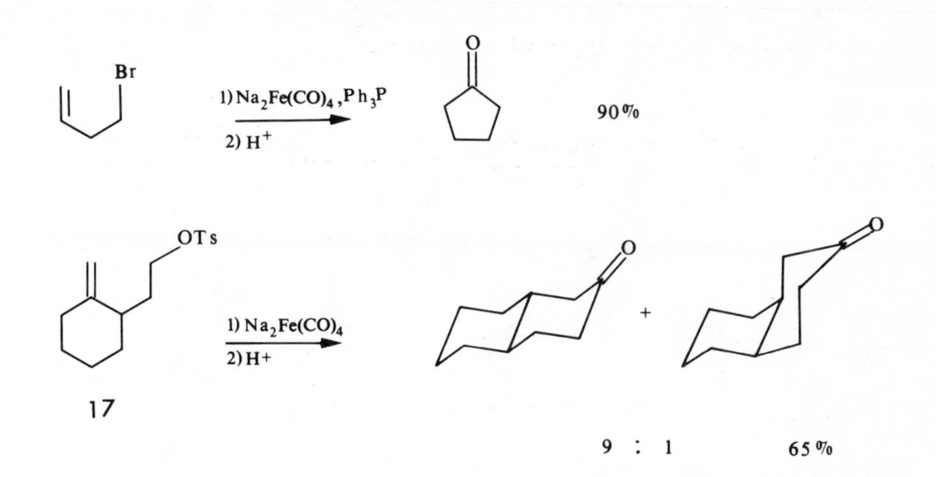

The cyclisation of 4,5-unsaturated primary tosylates to cyclohexanones forms a key step in a total synthesis of aphidicolin 18.[28]

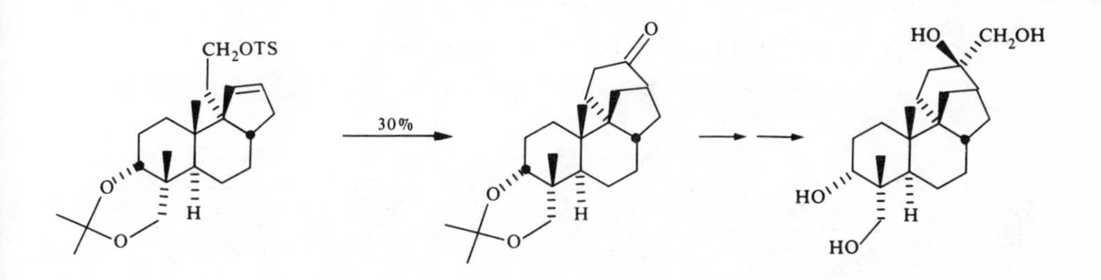

The iron alkyl complexes $CpFe(CO)_2R$ are converted to esters RCO_2R' on treatment with an oxidising agent ($FeCl_3$, $CoCl_2$, Ce(IV), Cl_2, AgO, PbO or dichlorocyanoquinone) in the presence of an alcohol R'OH.[29]. The reactions are stereospecific.[1,29]

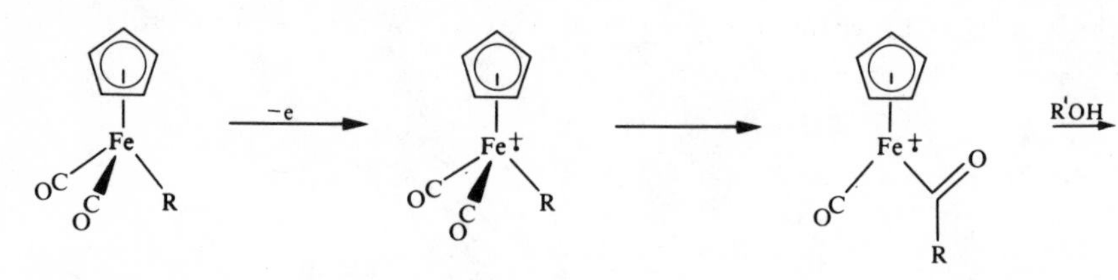

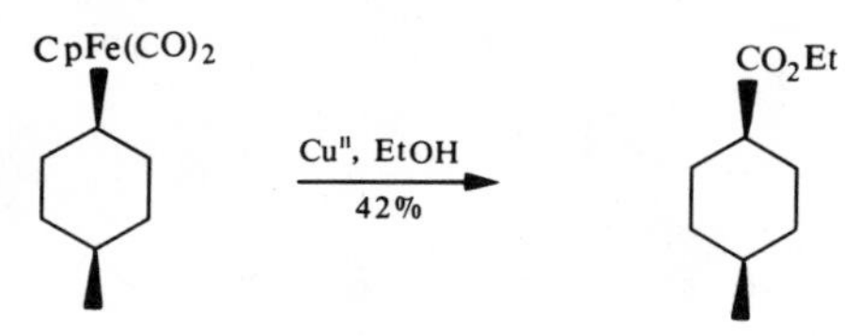

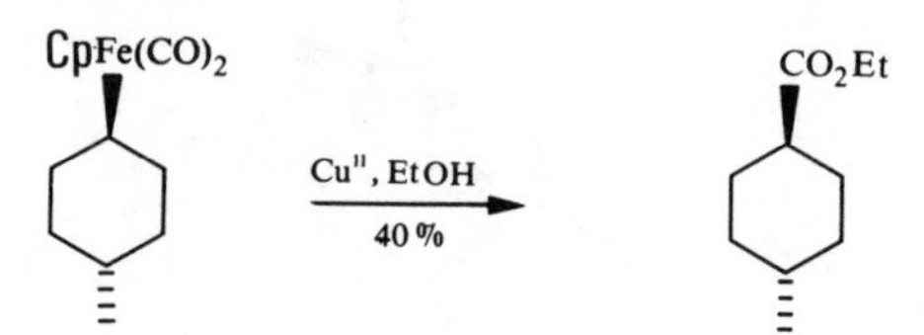

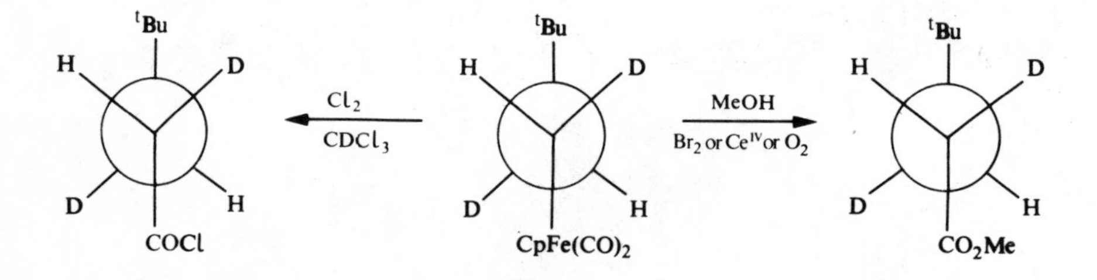

The $CpFe(CO)_2R$ complexes are readily available either from nucleophilic addition to $CpFe(CO)_2(olefin)^+$ cations or from the reaction of $CpFe(CO)_2^-$ anion with alkyl halides etc.[29] A combination of nucleophilic addition to the olefin cation by amine followed by oxidative carbonylation has provided a useful β-lactam synthesis.[30,31]

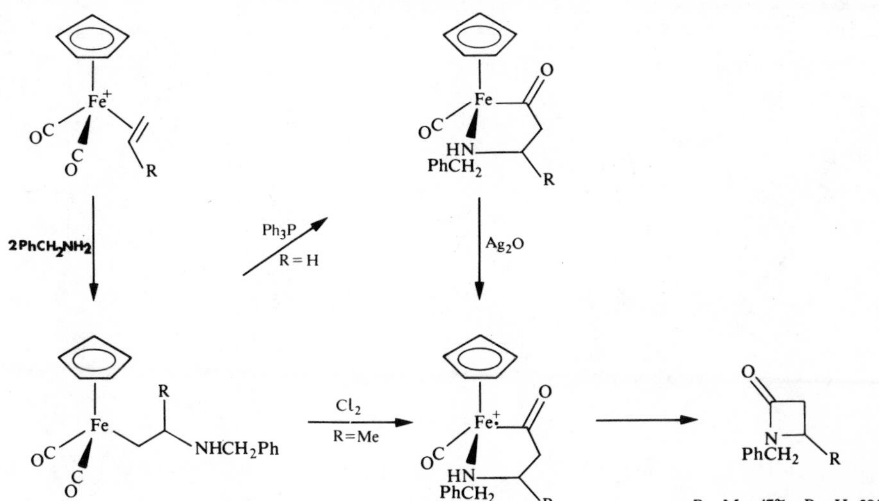

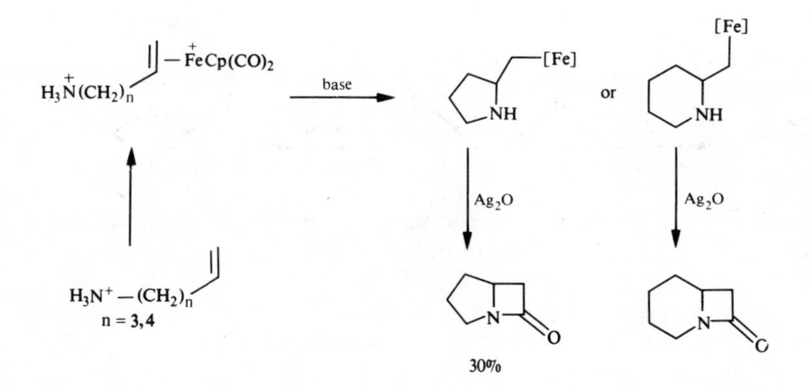

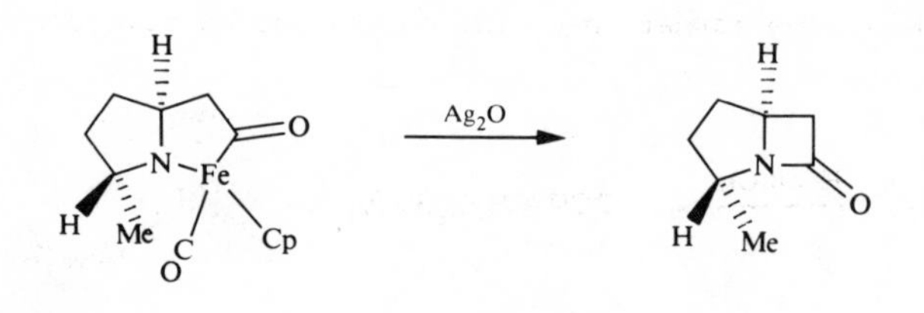

The oxidative carbonylation reaction of $CpFe(CO)_2R$ is compatible with carbonyl functions in the molecule.[32]

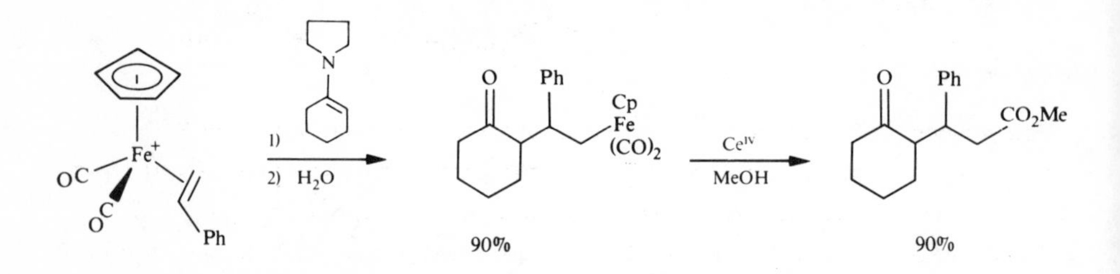

9.3. CARBONYLATION REACTIONS WITH Co AND Rh COMPOUNDS.[33]

The catalysts $HCo(CO)_4$ and $(Ph_3P)_3RhH(CO)$ have a large number of industrial applications for the hydroformylation of olefins and related reactions.[34] The active species in cobalt carbonyl carbonylations is the 16e acyl complex $RCOCo(CO)_3$. The $RCOCo(CO)_3$ may be formed from alkyl cobalt tetracarbonyl by migration of R from the metal to the ligand or by loss of CO from the acyl species $RCOCo(CO)_4$

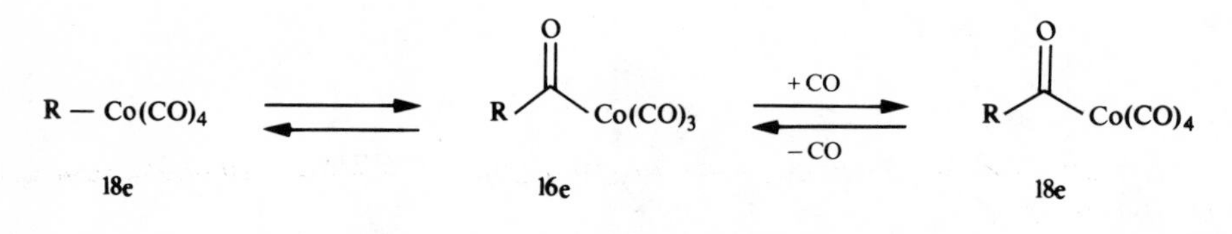

$RCo(CO)_4$ may be prepared from the reaction of $HCo(CO)_4$ with olefins, from the reaction of the anion $Co(CO)_4^-$ with RX, or from the opening of small ring heterocycles (e.g. epoxides) with $HCo(CO)_4$. Treatment of $RCo(CO)_4$

with CO leads to the formation of the corresponding acyl derivatives $RCOCo(CO)_4$.

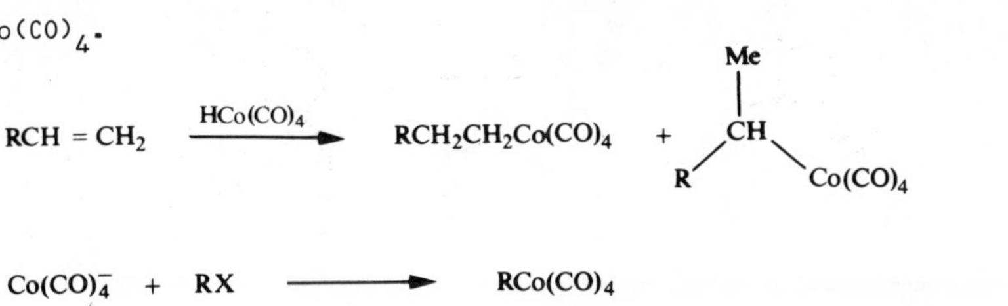

Alkyl and acyl cobalt tetracarbonyl species are often very unstable and must be generated in the presence of the other reagent necessary to form the desired final product. Although the yields of cobalt catalysed carbonylations are often very high, selectivity is a problem and mixtures of products are generally obtained. This is particularly true for the reaction of $HCo(CO)_4$ with olefins since olefins can be isomerised by addition-elimination of $HCo(CO)_4$ before carbonylation occurs.

In many reactions the cobalt compound employed is $Co_2(CO)_8$. However, there is a considerable amount of evidence available to indicate that the active catalyst in these reactions is $HCo(CO)_4$. The metal hydride is formed upon reduction of $Co_2(CO)_8$ by H_2, either deliberately added or present as an impurity in CO, or by hydrogen abstraction from the solvent or the reactants. Indeed, all of the CO insertion reactions catalysed by cobalt carbonyl complexes can be best understood in terms of the reactions of the 16e species $HCo(CO)_3$ which is formed by loss of CO from $HCo(CO)_4$.

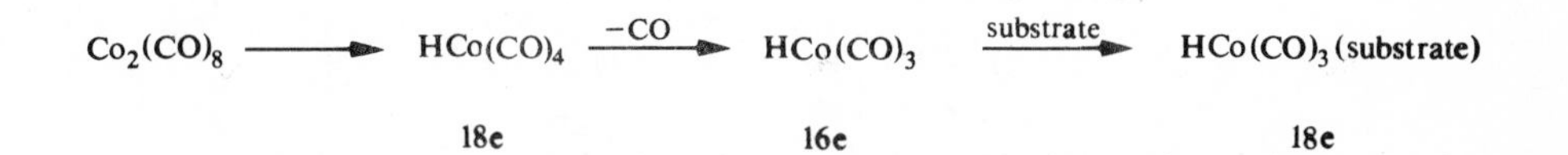

As mentioned above an intermediate in these reactions is a Co-acyl complex. The final products may be aldehydes, esters, amides, etc.

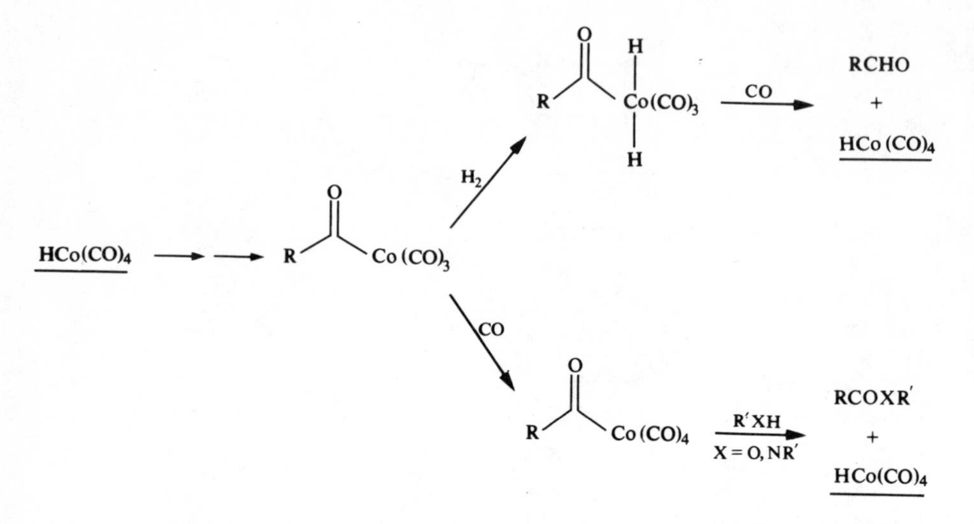

Olefins react with $HCo(CO)_4$ or $Co_2(CO)_8$ in the presence of H_2 and CO to give aldehydes as products (hydroformylation reaction).[35] Labelling experiments have shown that the addition of $HCo(CO)_4$ across a double bond is a *cis*-addition. Unfortunately this reaction is not very useful for synthesis due to the lack of regioselectivity in the addition of $HCo(CO)_4$. For example 1-pentene gives about equal amounts of the two isomeric compounds **19** and **20**.[36]

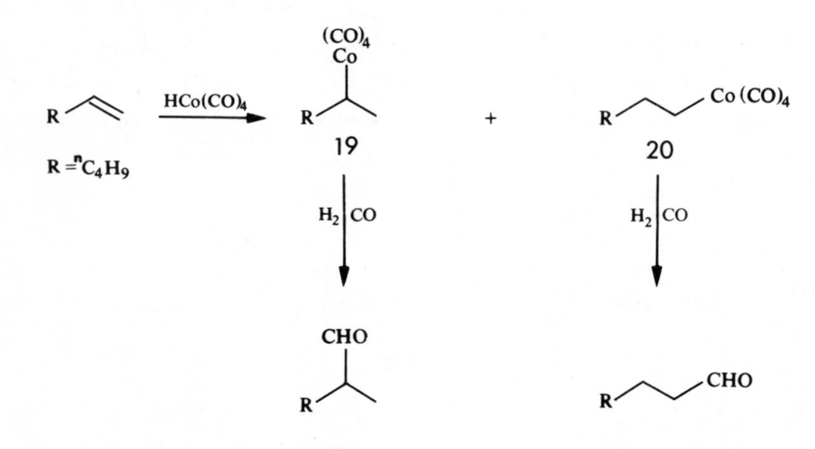

The addition of $HCo(CO)_4$ is more regioselective with vinyl ethers and vinyl acetates (the Co adding to the α-position) and for this reason the reaction has found use in carbohydrate chemistry.[37] Although in these

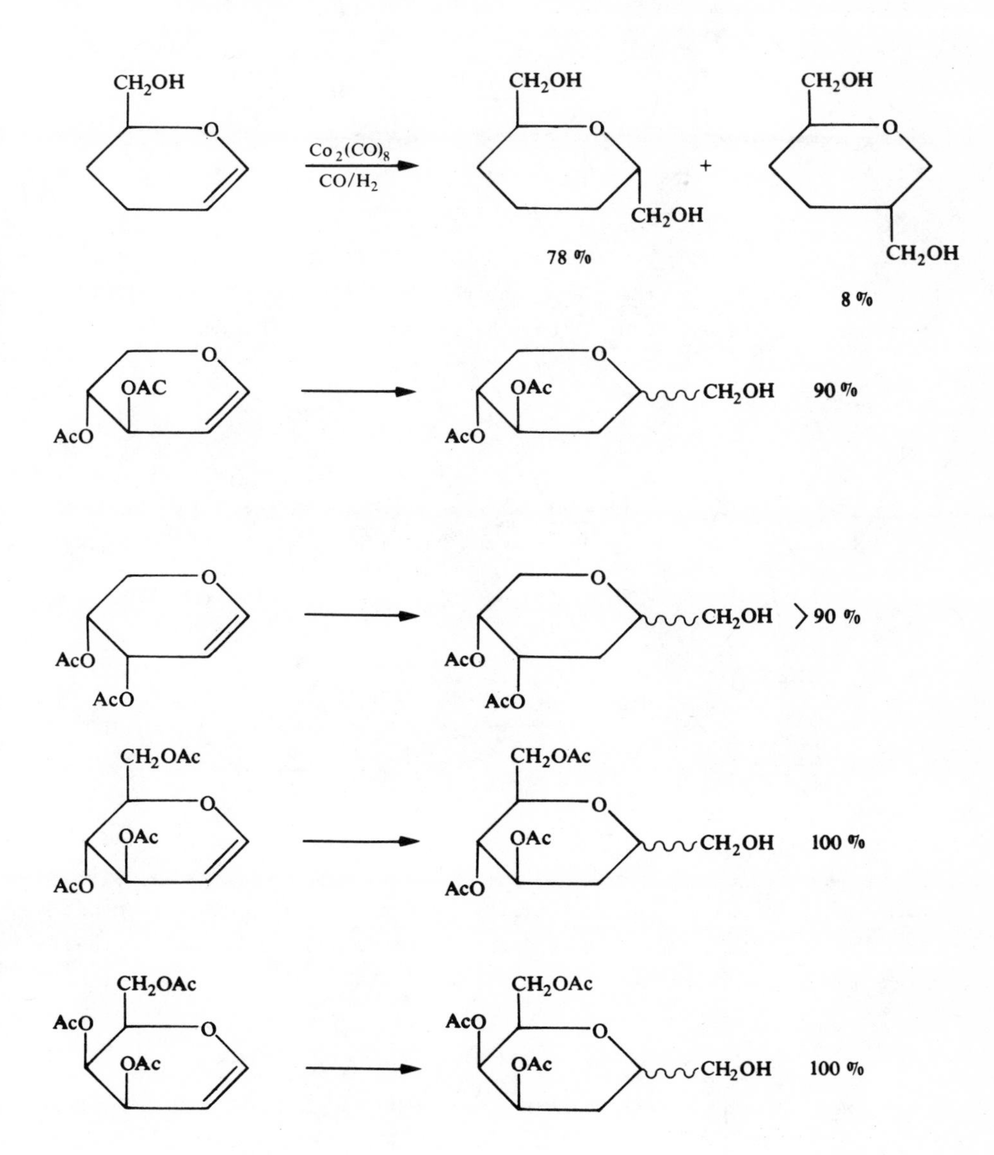

cases the regioselectivity is high there is generally little stereo-selectivity and 50:50 mixtures of α and β products are normally obtained.

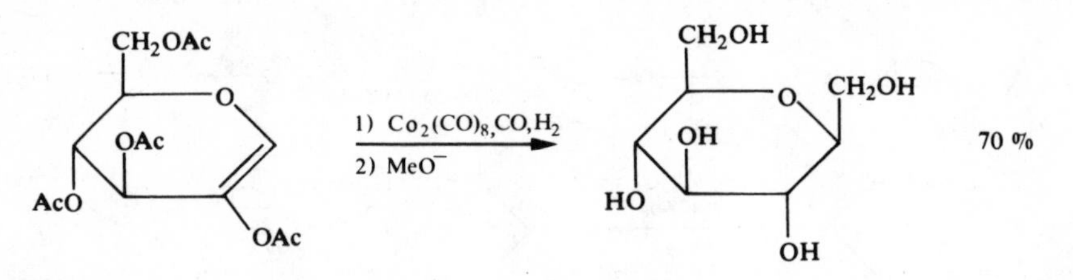

Often if an excess of H_2 is employed subsequent reduction of the aldehyde to alcohol occurs as in the example above. Utilisation of 1 equivalent of CO and 1 equivalent of H_2 allows the preparation of aldehydes.[37]

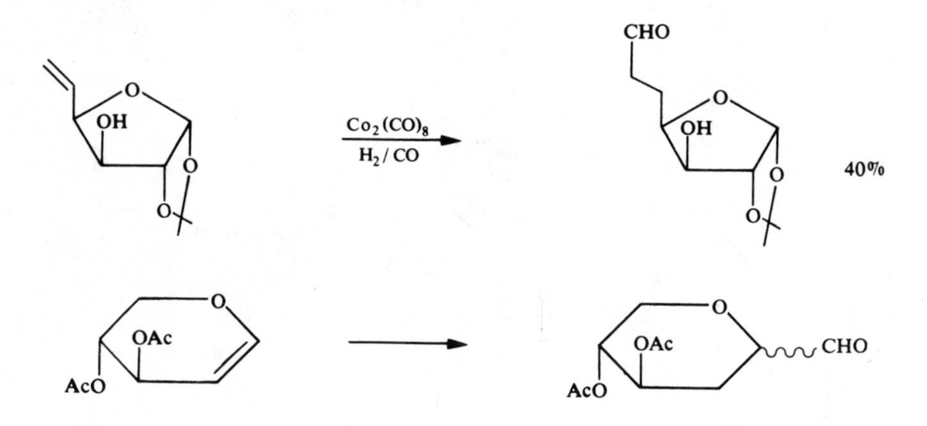

Cyclohexene, for which regioselectivity and isomerisation problems do not exist, reacts with $Co_2(CO)_8$ in the presence of CO and H_2O to give cyclo-hexane carboxylic acid.

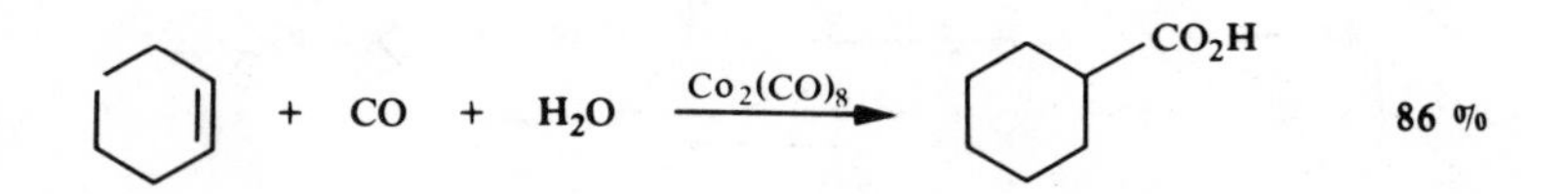

Acetylenes and α,β-unsaturated aldehydes and ketones tend to undergo rapid reduction rather than carbonylation. 1,3-dienes undergo 1,4-addition to produce stable allyl compounds.

$$CH_2{=}CH{-}CH{=}CH_2 + HCo(CO)_4 \longrightarrow CH_3CH{=}CHCH_2{-}Co(CO)_4 \xrightarrow{-CO} (\pi\text{-allyl}){-}Co(CO)_3$$

Alkylation of the nucleophilic anion $Co(CO)_4^-$ generated from $HCo(CO)_4$ and base (MeO^-, hindered amine) has proved the most useful method for the preparation of specific alkyl and acyl-$Co(CO)_4$ species. The isomerisation reaction by loss of $HCo(CO)_4$ and readdition to the olefin thus formed is relatively slow and therefore mixtures of products can be avoided by this method. The alkylating agent may be an alkyl chloride, bromide, iodide, sulphonate etc. Performing th reaction under an atmosphere of CO leads directly to the acyl derivatives and treatment of the acyl derivatives with alcohols or amines generates esters and amides respectively.[33]

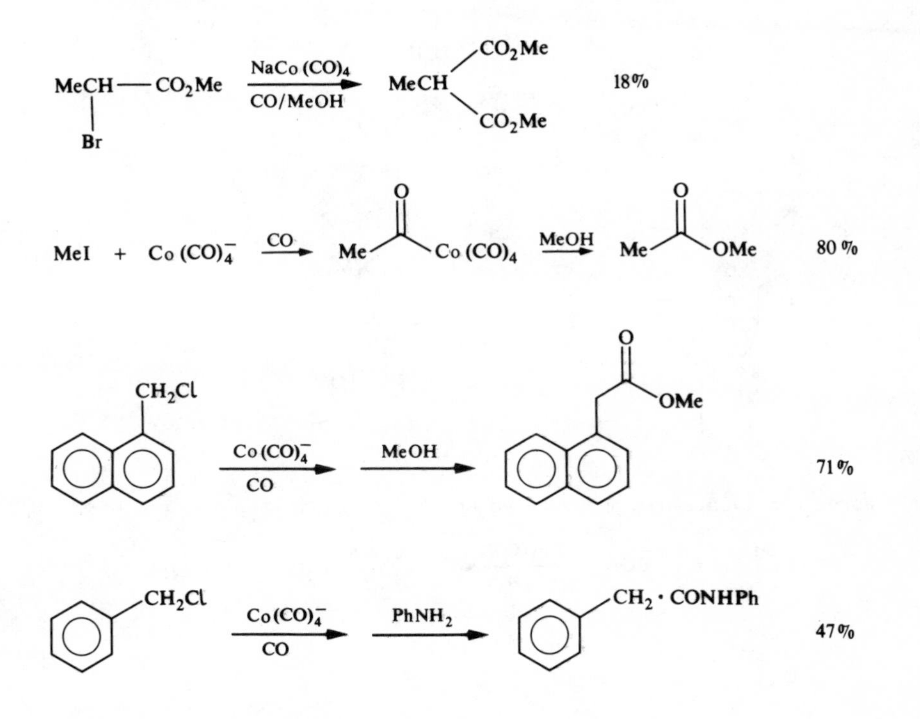

Epoxides react with $HCo(CO)_4$, $Co_2(CO)_8$ and $Co(CO)_4^-$ to generate 2-alkoxy-alkyl-$Co(CO)_4$ species. These complexes can produce a wide variety of products depending on the reaction conditions. For example, in the absence of CO, epoxides are rearranged to ketones by $Co(CO)_4^-$.[38]

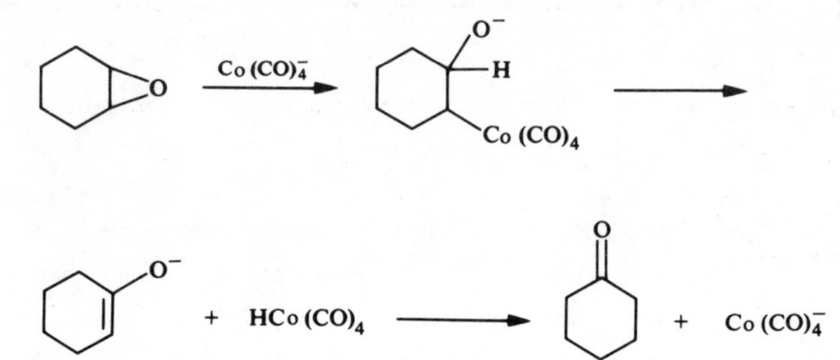

Propylene oxide in the presence of $Co_2(CO)_8$ or $Co(CO)_4^-$, CO and MeOH gives a mixture of 3-hydroxy esters.[33,37]

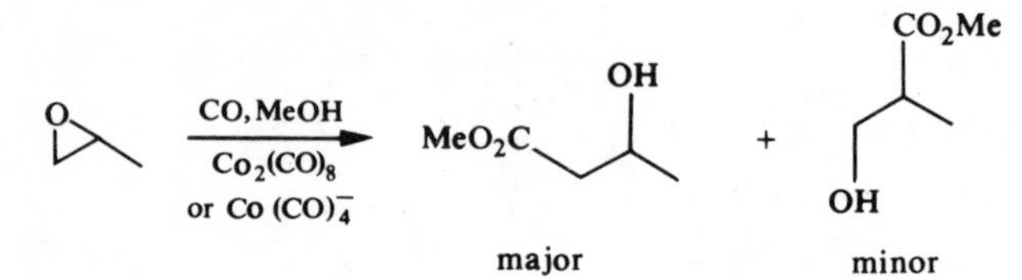

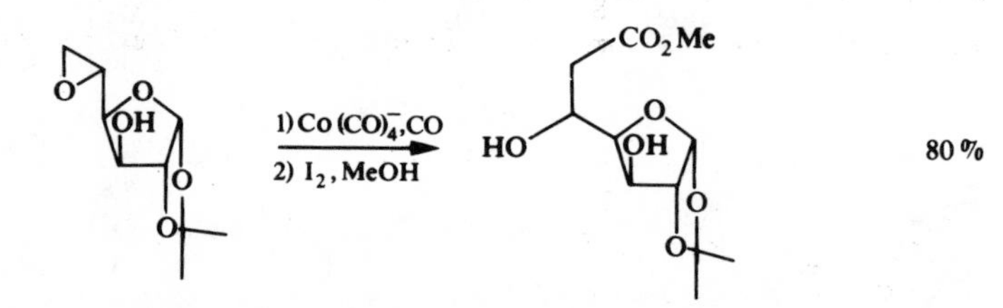

In aprotic solvents α,β unsaturated acids are obtained.

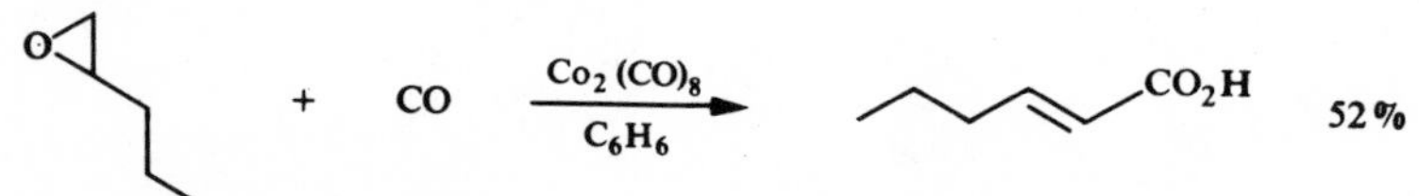

If hydrogen is also present then aldehydes or alcohols are formed.[33,37,39]

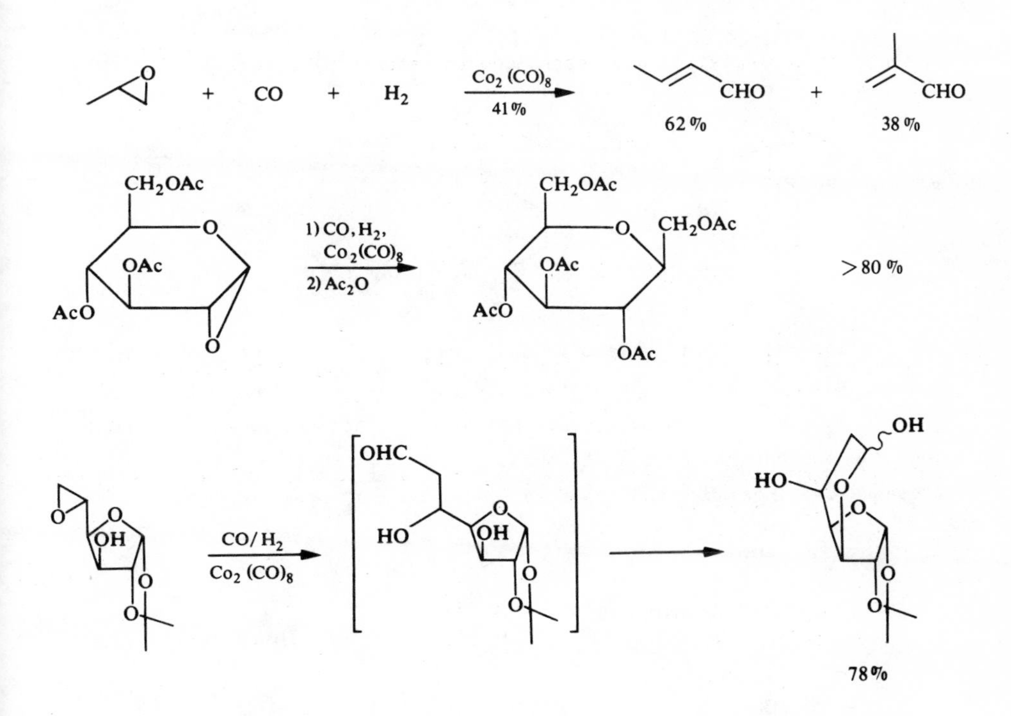

Oxetanes also react with $Co_2(CO)_8$ and CO. The primary products, 4-hydroxy acyl cobalt tetracarbonyls, decompose to give γ-lactones[33] or reduction

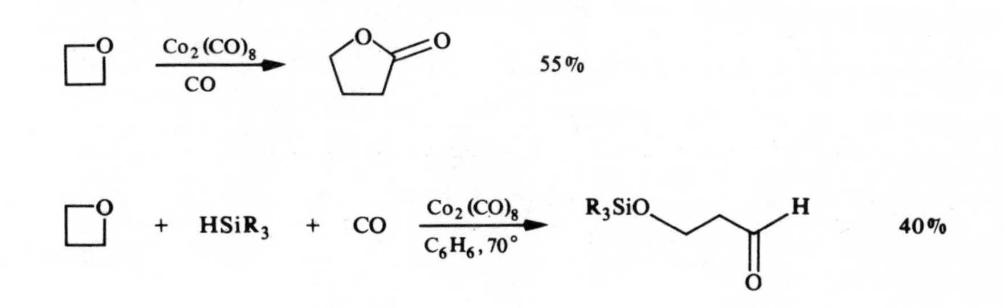

may occur in the presence of $HSiR_3$.[40]

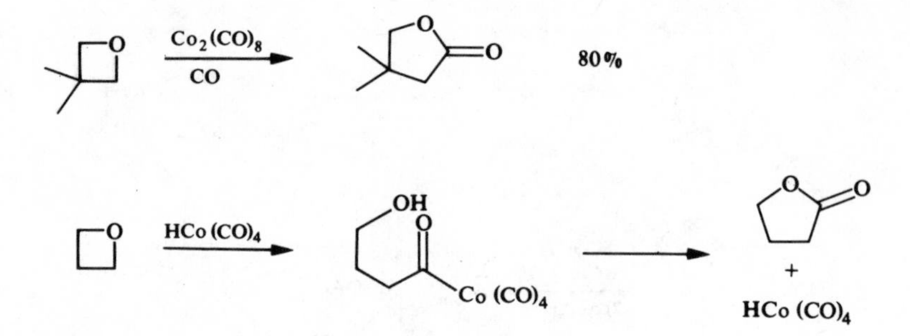

The reaction is an intramolecular analogue of the reaction of (acyl)$Co(CO)_4$ species with alcohols. The analogous reaction of epoxides to give β-lactones has not been observed. Larger ring lactones may be prepared from the appropriate chloro alcohols. The reaction is catalytic in the presence of hindered amine bases to reform $Co(CO)_4^-$ from $HCo(CO)_4$.

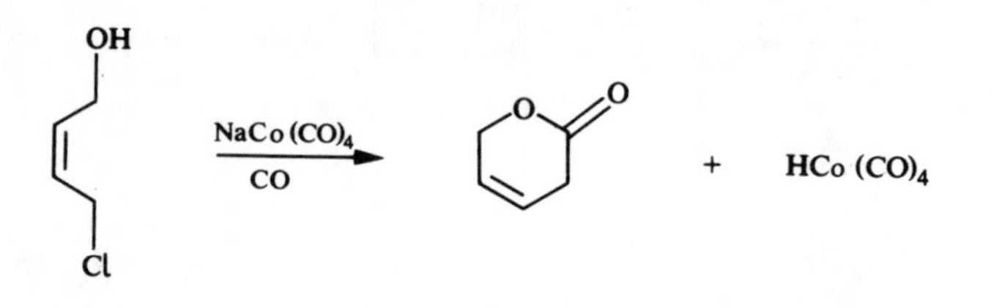

Lactones may also be prepared from unsaturated alcohols although once again mixtures are observed in many cases due to rearrangement reactions. β-Lactones are not formed.

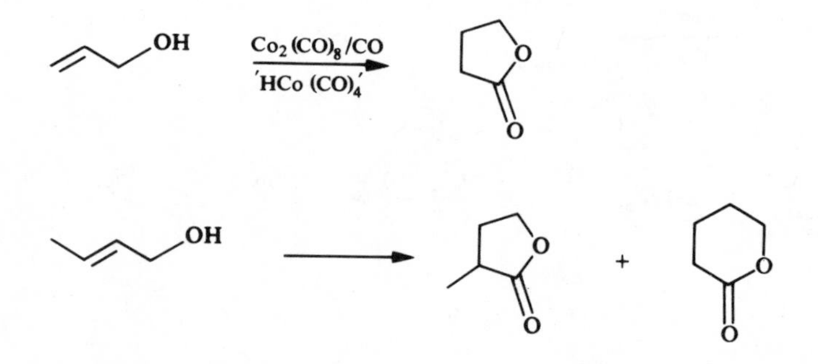

Unsaturated amines and amides give lactams and imides respectively, formation of the latter being preferred.[33,40a]

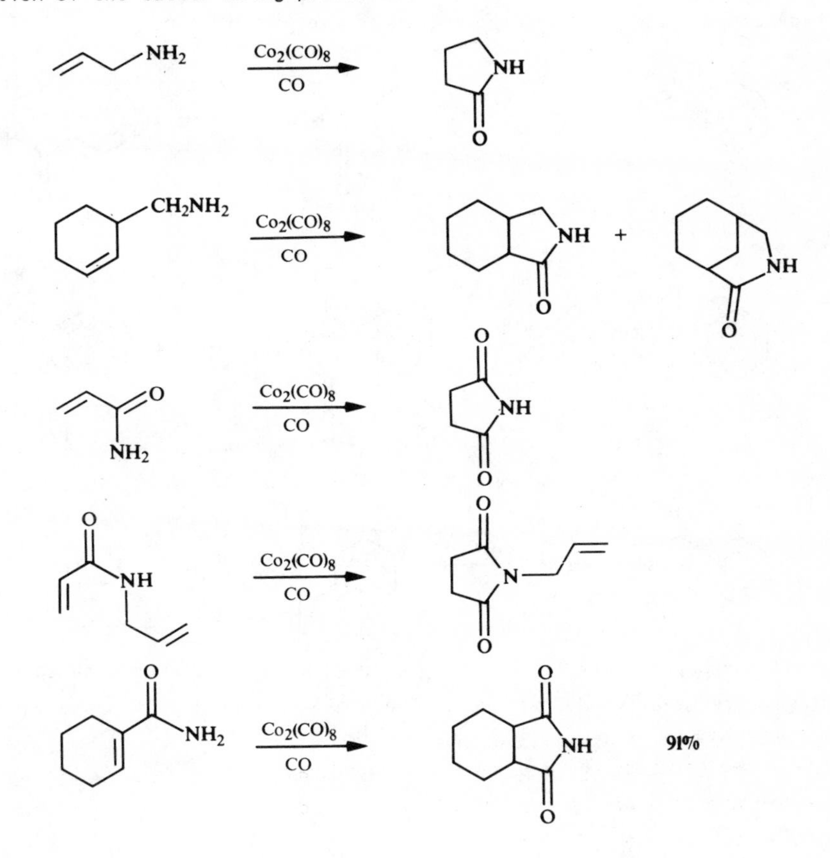

The unsaturation may be an aromatic ring. Many aromatic systems with a

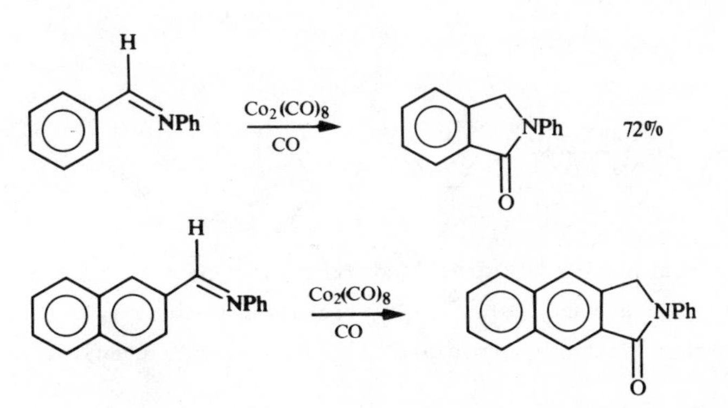

nitrogen substituent on the α-carbon undergo cyclocarbonylation reactions. Some examples are given for Schiff bases, azo compounds, oximes and phenyl hydrazones. In all cases the active species is believed to be $HCo(CO)_4$.[40b]

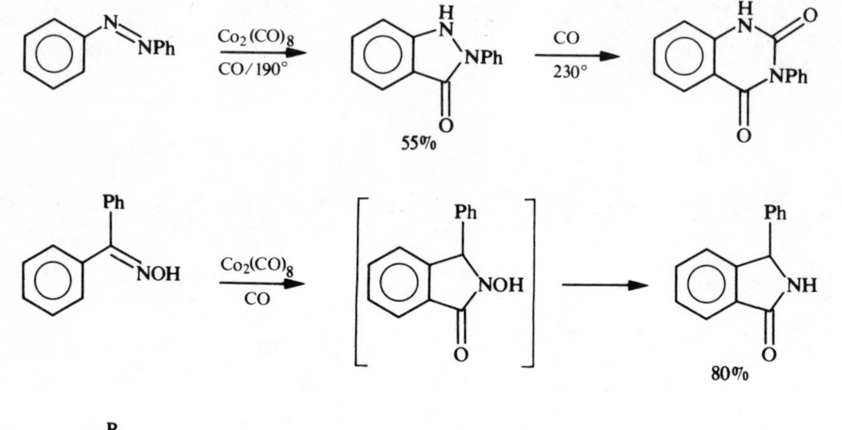

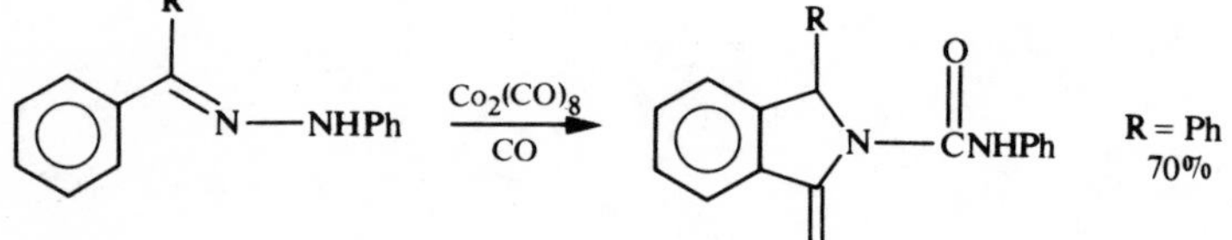

Alkyl and acyl cobalt tetracarbonyls react with 1,3-dienes with loss of **CO** to produce η^3-allyl derivatives. These η^3-allyl derivatives are decomposed by base to give 1-acyl 1,3-dienes.[33,41]

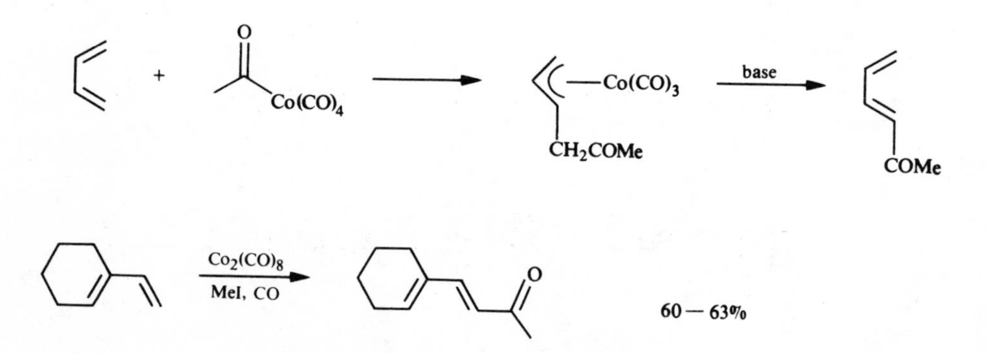

N-Acyl aminoacids **may be synthesised** from aldehydes and amides in the presence of CO and $Co_2(CO)_8$.[42] The mechanism presumably involves carbonylation of the intermediate **21**. For example phenyl acetaldehyde and acetamide give N-acetyl phenylalanine in 54% yield.

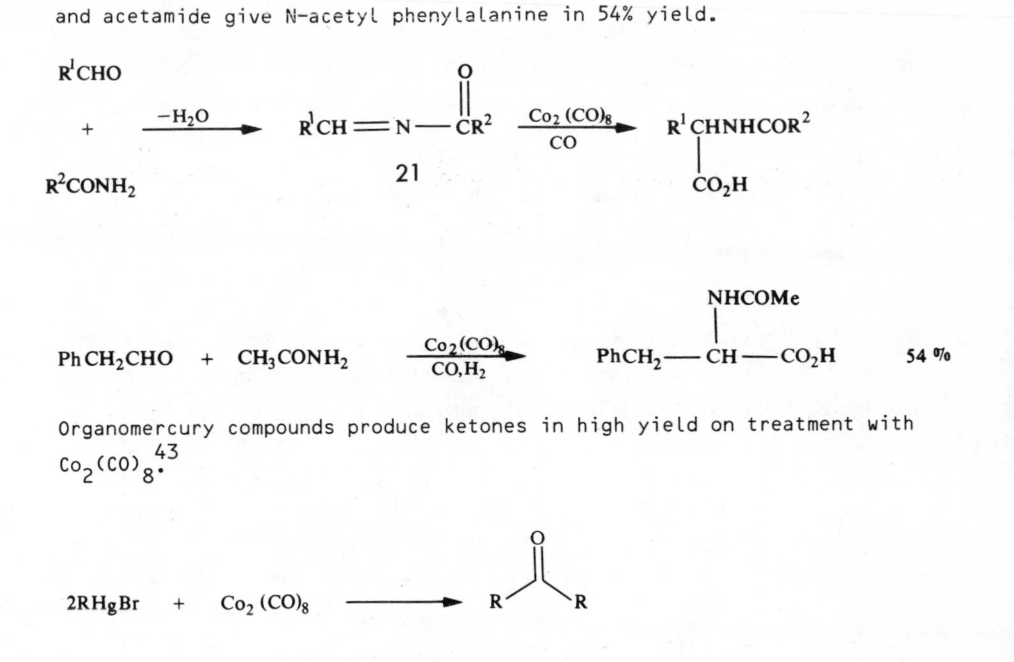

Organomercury compounds produce ketones in high yield on treatment with $Co_2(CO)_8$.[43]

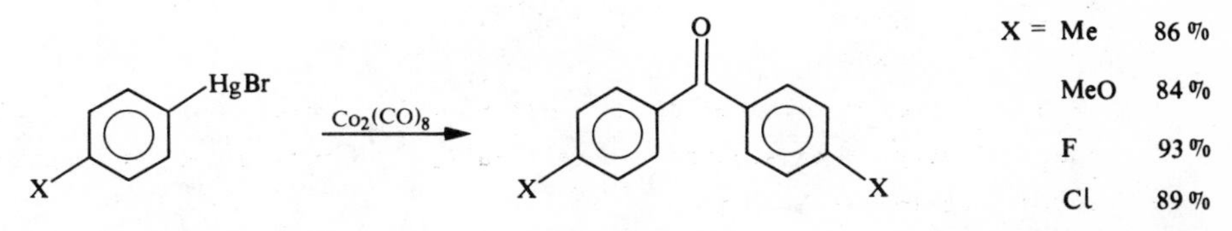

$[Rh(CO)_2Cl]_2$ reacts with strained rings to give carbonylated products.[44]

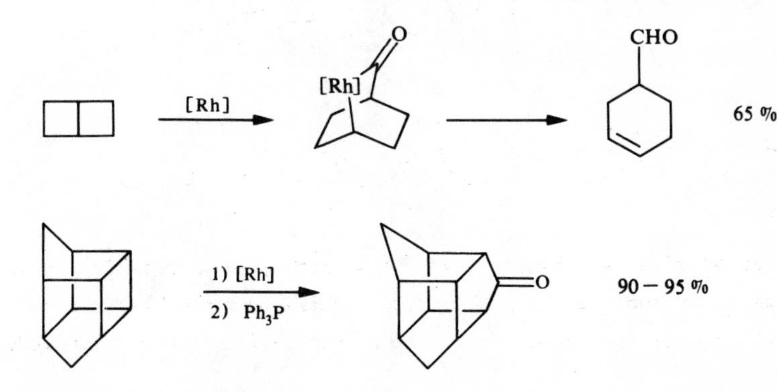

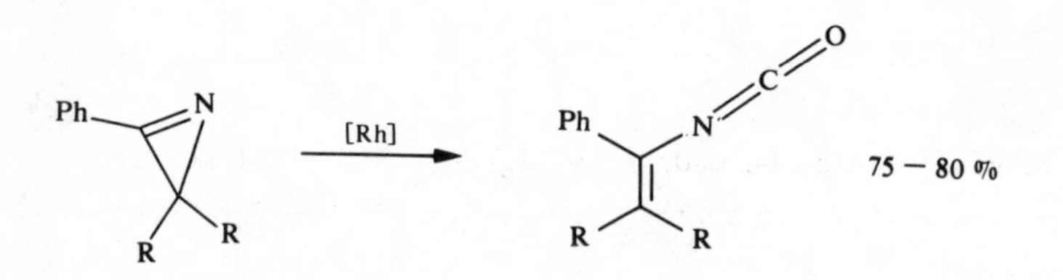

9.4 CARBONYLATION REACTIONS WITH Pd AND Ni COMPOUNDS

Like the carbonylation reactions of Zr, Fe and Co, the many CO insertion reactions involving the compounds of palladium can be regarded in terms of forming alkyl and acyl palladium intermediates, e.g.

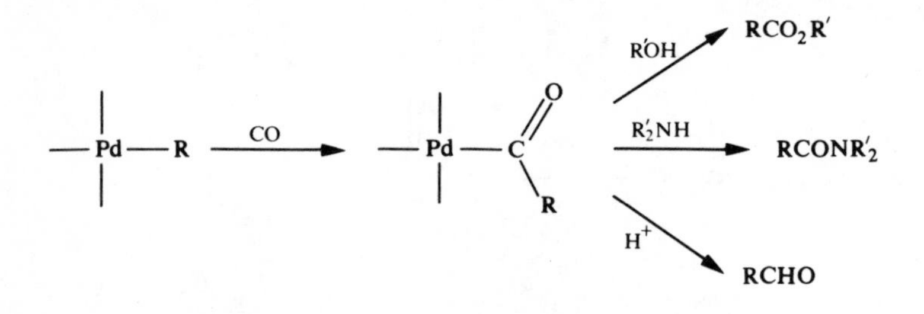

One method of preparing Pd-alkyl complexes is the S_N2 displacement of bromide from alkyl bromides with $Pd(PPh_3)_4$.[45] Conversion of the alkyl species into the acyl [Pd] occurs with retention of configuration at carbon and oxidative cleavage of the acyl in the presence of MeOH

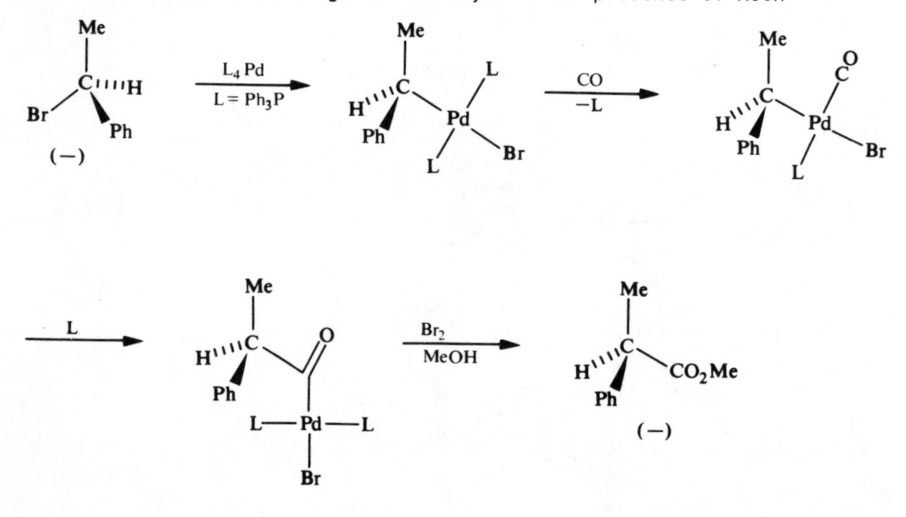

generates the methyl ester.

Palladium alkyls may also be made from Li_2PdCl_4 and alkyl mercurials. Treatment of the Pd–R species thus formed with CO and ROH likewise leads to ester formation.[46]

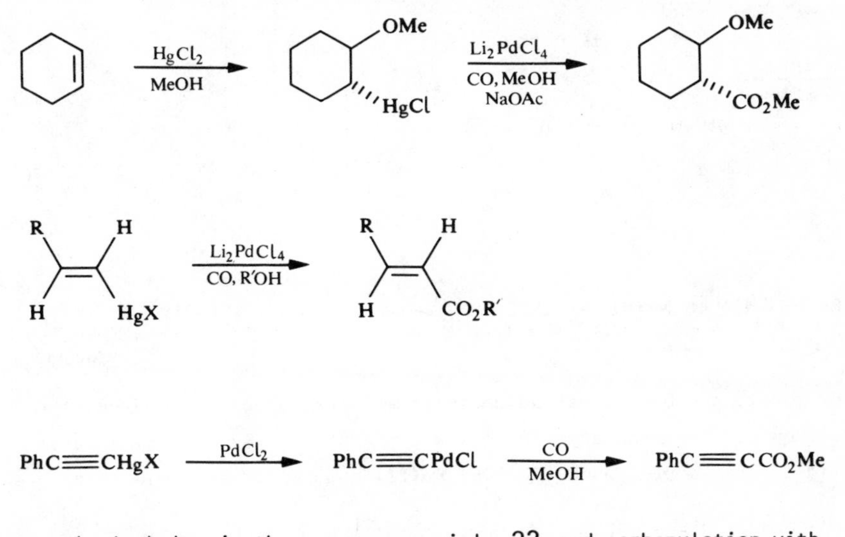

Propargyl alcohol, via the organomercurial 22 and carbonylation with Li_2PdCl_4/CO allows the synthesis of the important class of compounds, the butenolides 23.[47]

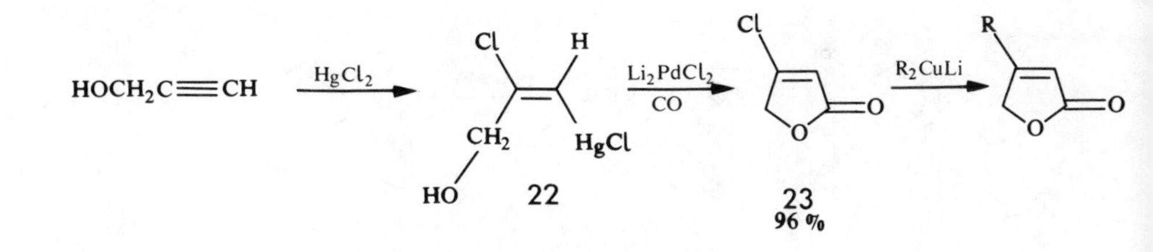

Palladium vinyl complexes may also be prepared directly from acetylenes.[48]

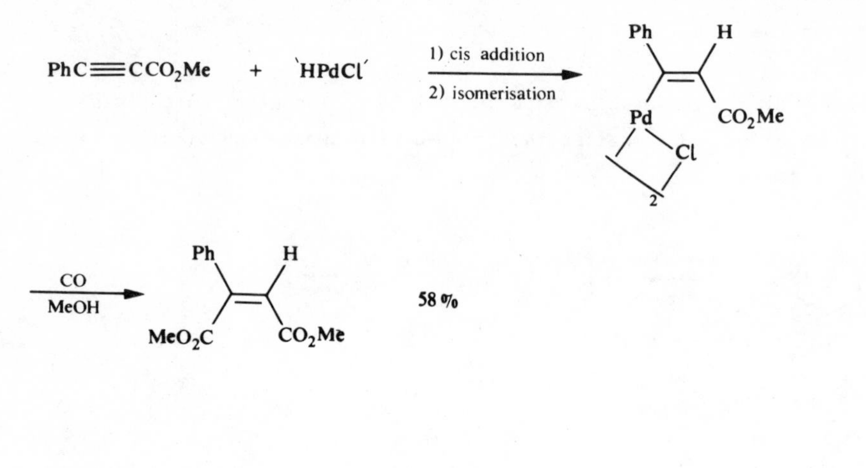

Vinyl,aryl, and benzyl palladium intermediates may be formed from the reaction of vinyl and aryl halides with L_2PdX_2. Treatment of an aryl halide, for example, with CO, ROH, Pd(II) and a tertiary amine base,

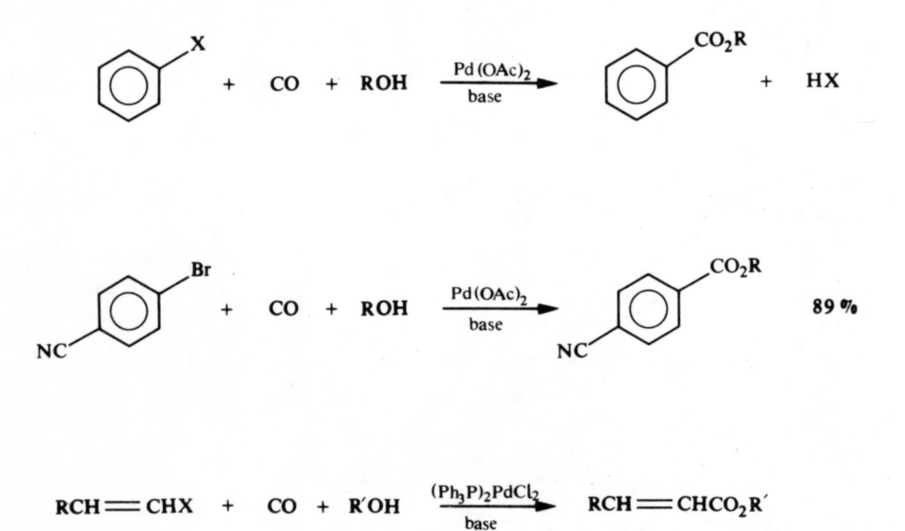

to remove the HX formed, leads to benzoate formation.[49]

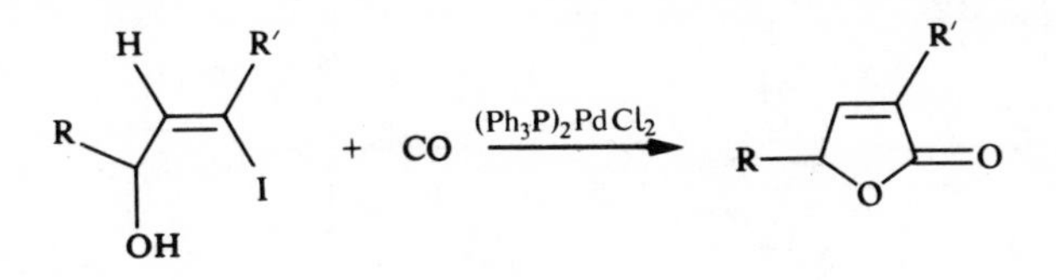

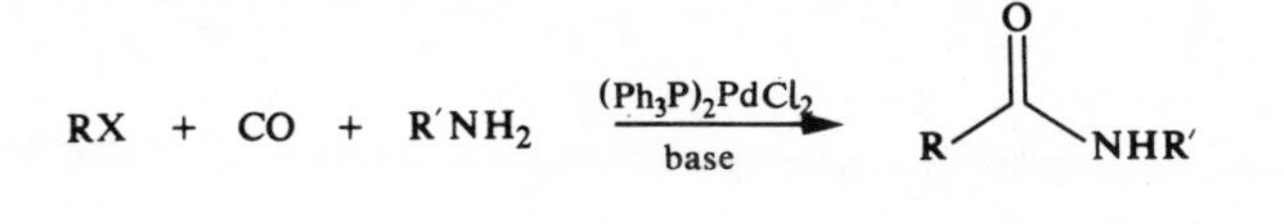

If the above reaction is carried out in the presence of a primary or secondary amine then the products are amides. The base used may be a tertiary amine.[50]

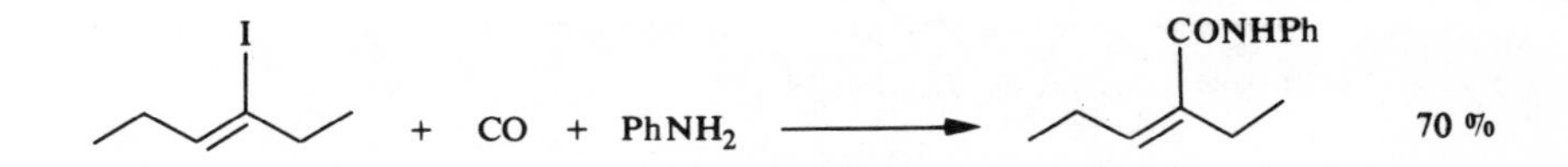

Cyclisation occurs when the aryl halide and the amine are in the same molecule.[51]

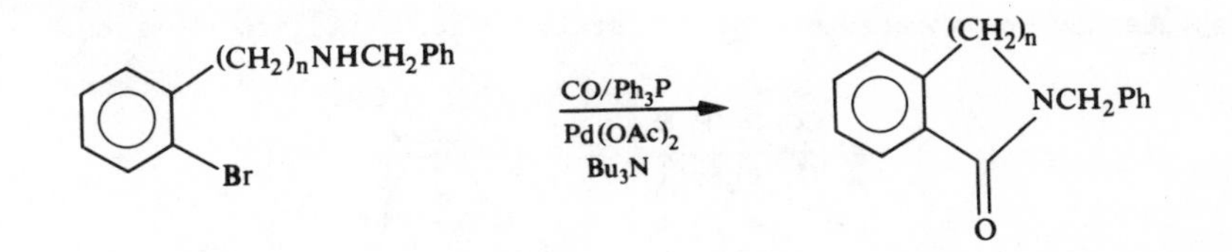

n = 1 (63 %), 2 (65 %), 3 (63 %)

This cyclisation has been employed as an essential step in the synthesis of the alkaloid sendaverine 24.[52]

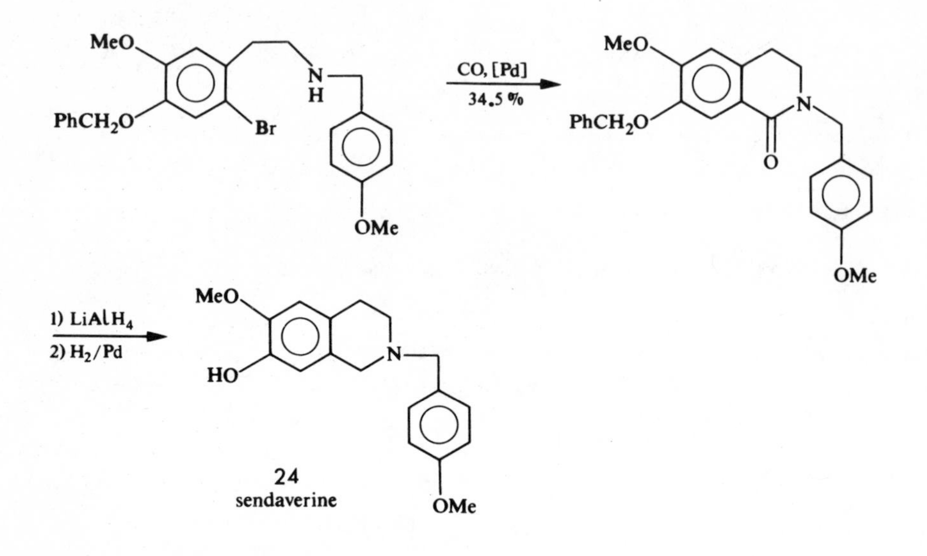

Cyclisation may also occur to give four membered rings and is illustrated by the following β-lactam synthesis.[53]

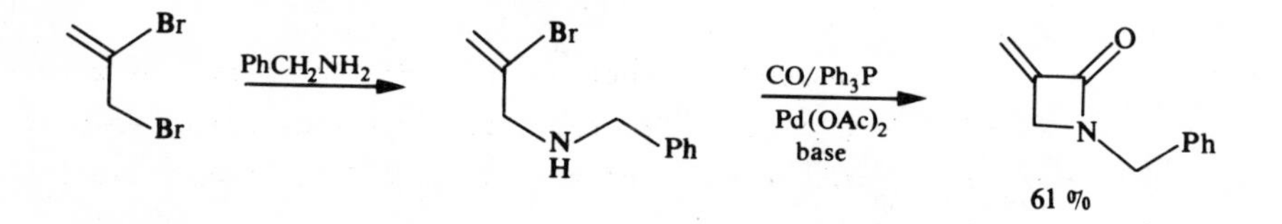

Aldehydes may be isolated if the Pd catalysed carbonylation reactions

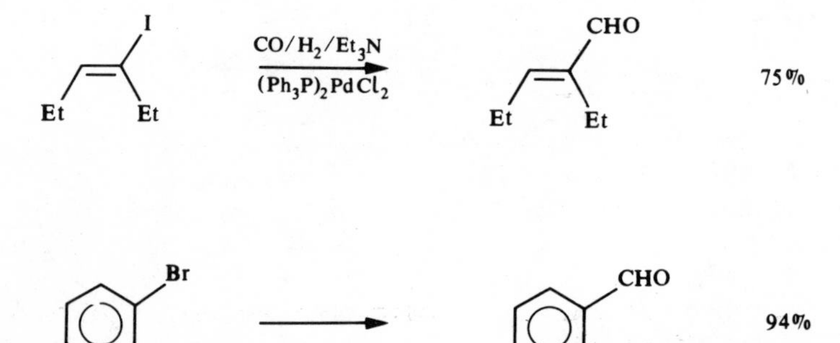

are performed with a 1:1 mixture of H_2 and CO.[54]

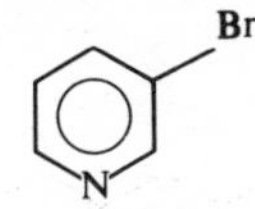

CHO
N
80%

Aryl compounds which contain a β-nitrogen atom that can coordinate to Pd undergo ortho insertion to give aryl-Pd complexes which readily undergo carbonylation with CO.[55]

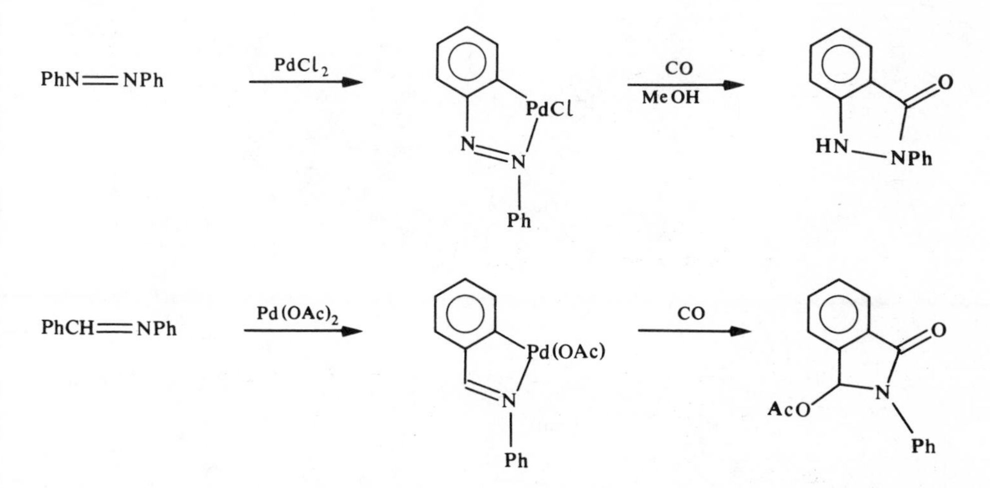

Palladium olefin complexes undergo nucleophilic addition reactions with a wide variety of nucleophiles. The products, α-alkyl palladium complexes, are readily carbonylated. Reaction of olefins with $PdCl_2$ and CO leads to 3-chloro acid chlorides. In the presence of MeOH and a tertiary amine base, α,β-unsaturated esters and 1,2-diesters are formed. The reaction

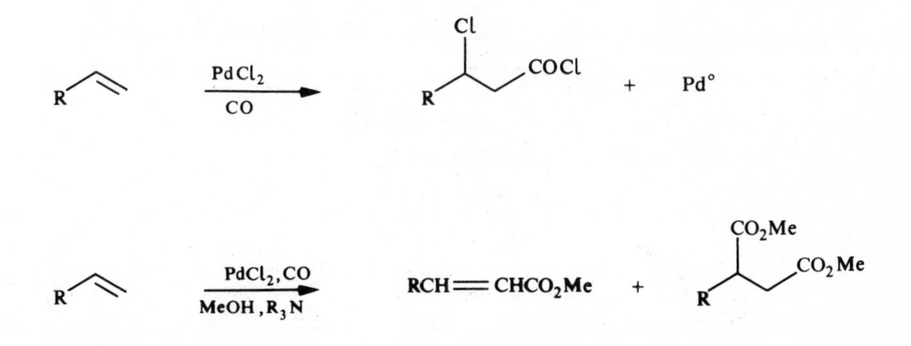

can be made catalytic in the presence of Cu^{II} and O_2 which reoxidises the Pd^0 to Pd^{II}.[56]

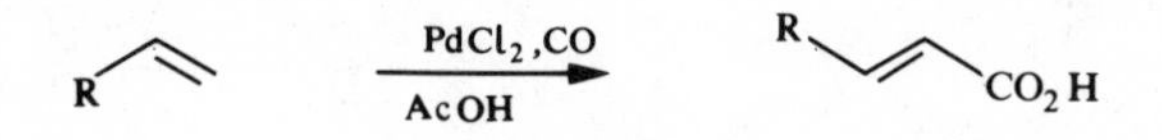

Depending on the reaction conditions and the nucleophiles present carboxylic acids, β-methoxy esters, β-amino esters or lactones may be produced.[57]

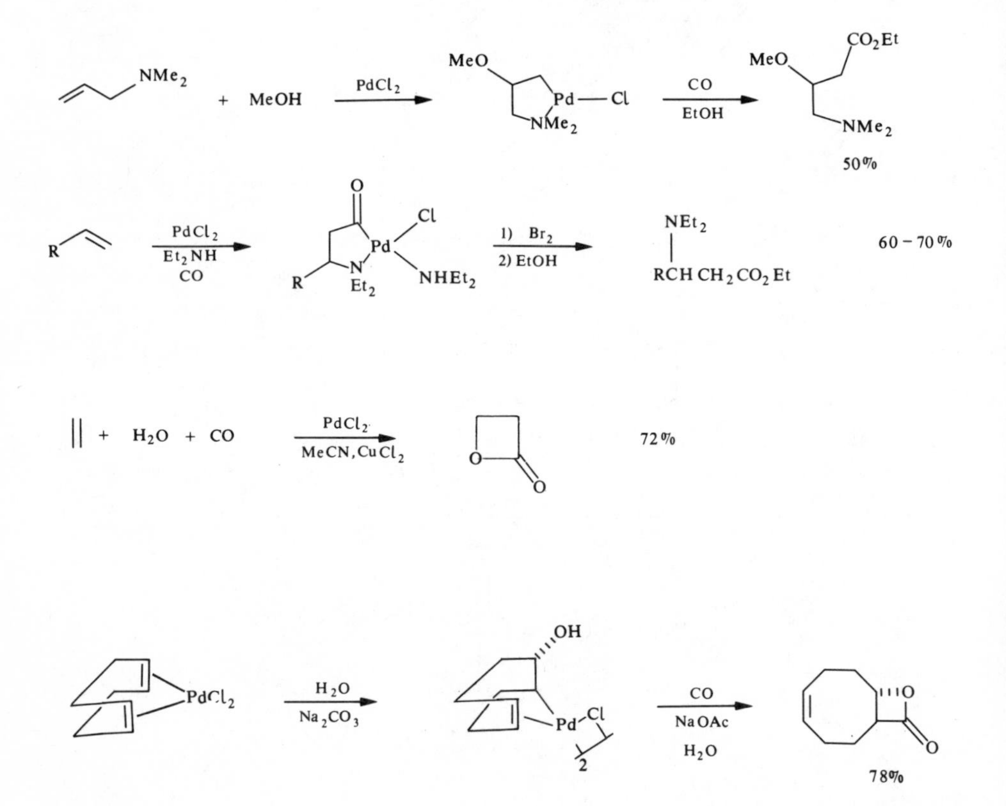

Treatment of the intermediate acyl-Pd species with H^+, or MeI or MeLi, leads to the formation of aldehydes and methyl ketones respectively.[58]

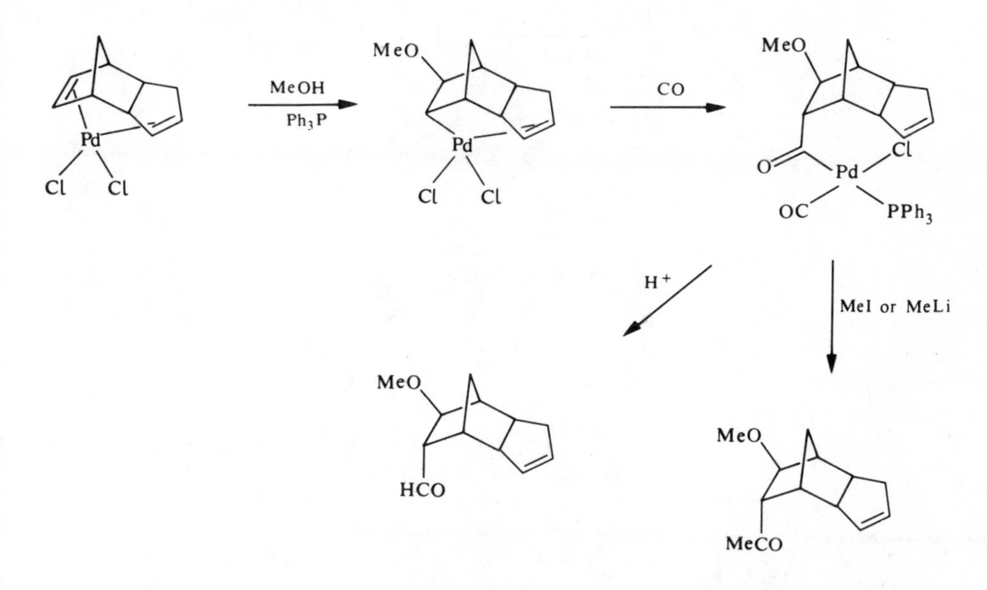

Phosphines are sometimes added to these reactions to coordinate to the Pd and thus stabilise the intermediates. With unsymmetrical double bonds mixtures of products are often obtained. The product mixture can be significantly altered, however, by changing the phosphine or solvent.[59]

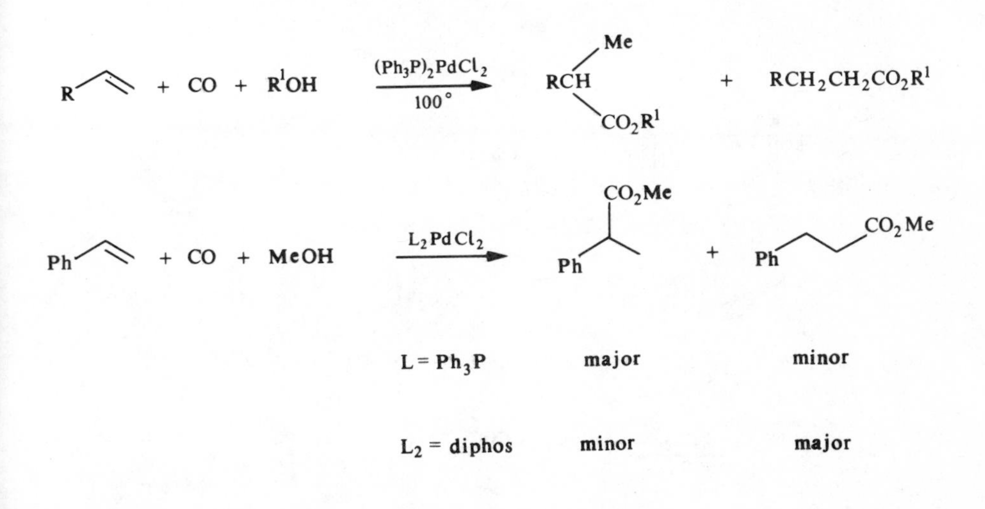

Palladium η^3-allyl complexes undergo carbonylation in the same way as σ-alkyl palladium species. The η^3-allyl complexes may be generated in a variety of ways, from dienes, allylic acetates (bromides), olefins, etc. (see section 2.7).[60]

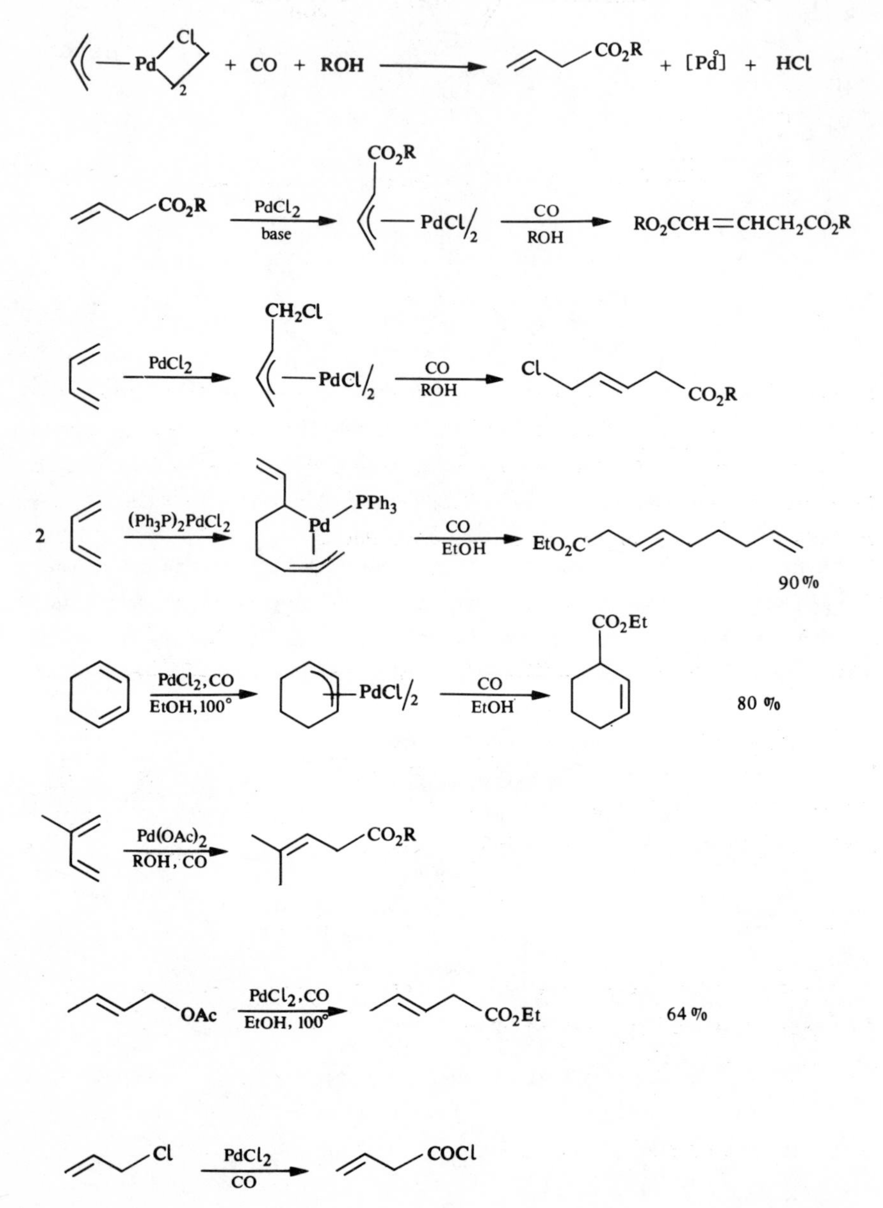

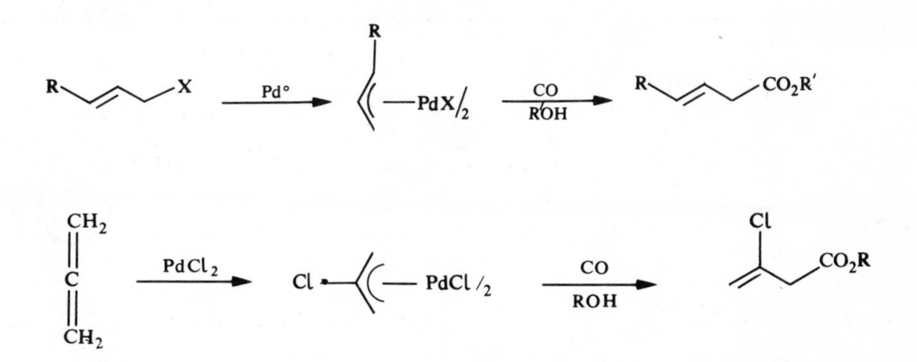

A second carbonylation mechanism has been observed for the formation of a Pd-carboxylate species and subsequent transfer of this group to a double or triple bond.[61] Pd carboxylate species may be generated from $PdCl_2$, CO, and MeOH in the presence of a base. Olefins react with CO in MeOH to form *cis* diesters in the presence of base and $PdCl_2$.

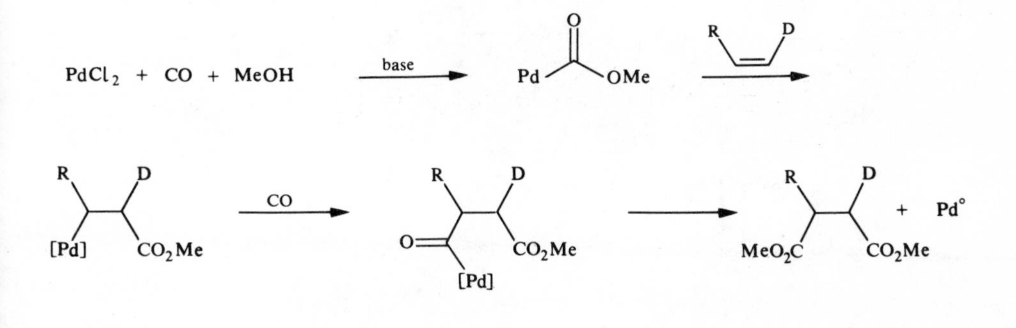

In the absence of base, β-methoxy esters are formed (see above). A carboxylate group may also be transferred to Pd from $XHgCO_2R$.[62]

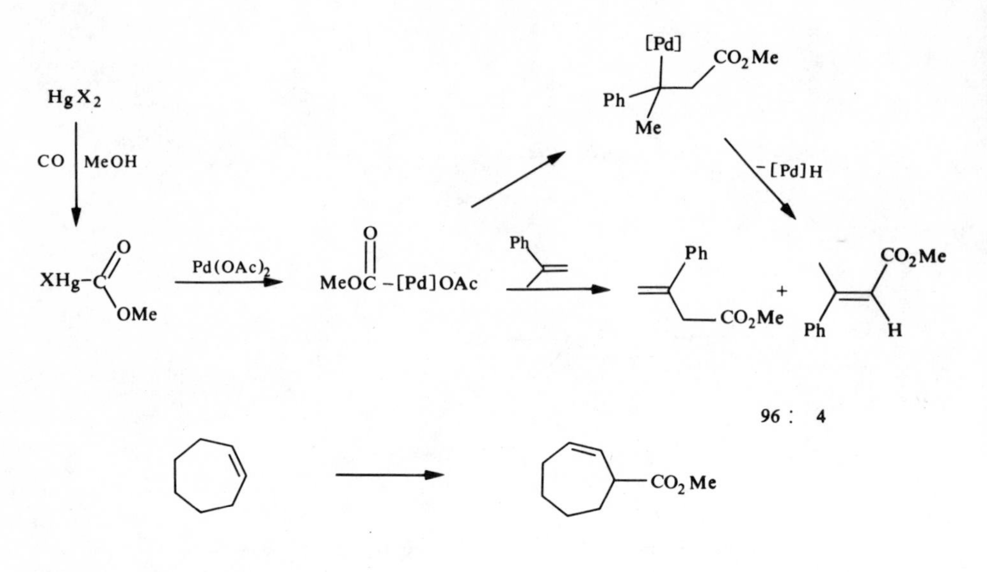

Acetylenic alcohols such as **24** lead to the formation of α-methylene γ-lactones in good yields.[63]

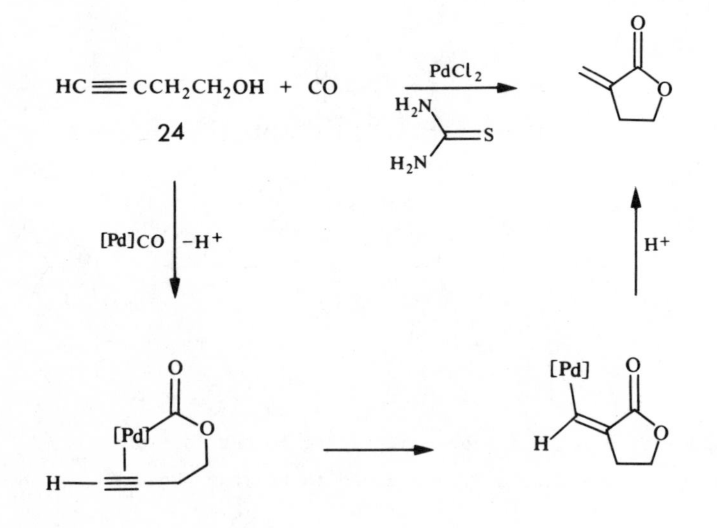

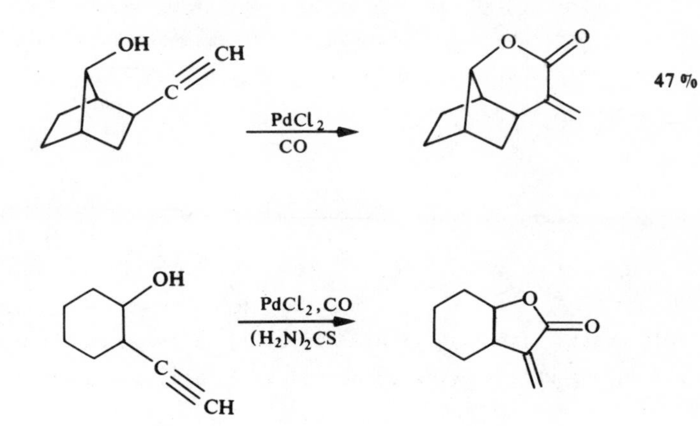

This reaction has been used for the synthesis of a venolepin derivative.[64]

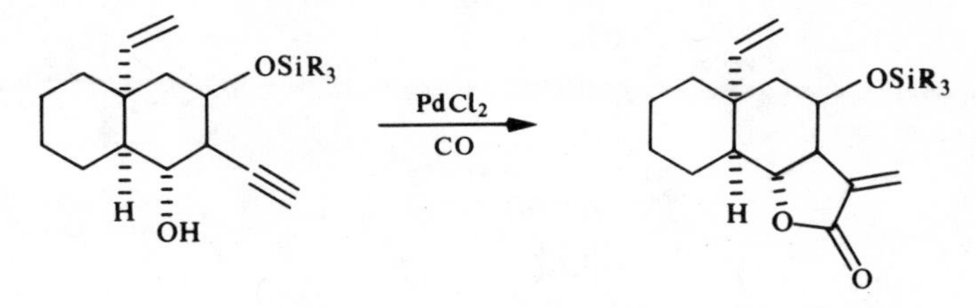

Propargyl alcohols lead to the formation of butenolides.[65]

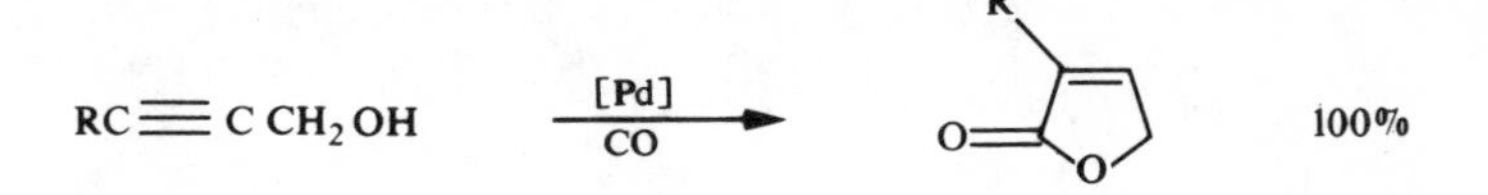

Pd compounds are also effective for the carbonylation of a variety of nitrogen containing groups.[66]

$$PhNO_2 + 3CO \xrightarrow{PdCl_2} PhNCO + 2CO_2$$

$$PhNO_2 + 3CO \xrightarrow[PhNH_2]{(Ph_3P)_2PdCl_2} PhNH.CO.NHPh$$

$$R_2NCl + CO \xrightarrow{PdCl_2} R_2NCOCl$$

Nickel complexes have been employed less frequently than Pd complexes although $Ni(CO)_4$ has been used to effect some interesting transformations. But-2-yne reacts stereoselectively with $(Ph_3P)_2NiPhBr$ followed by CO and MeOH to give **Z**-1,2-dimethylcinnamate.[67]

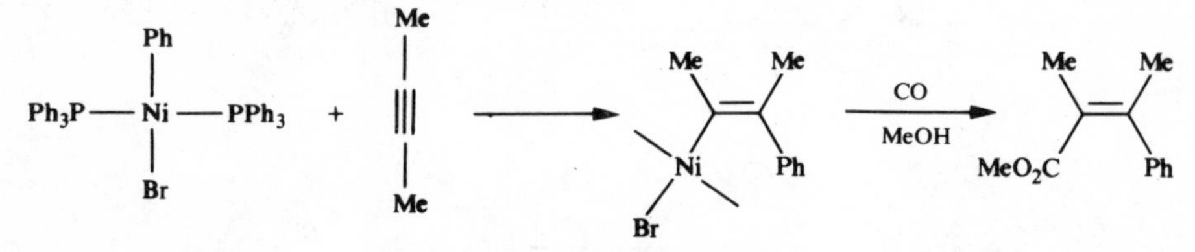

$Ni(CO)_4$ reacts with aryl lithium or Grignard reagents to generate $Ni(CO)_3$ (acyl) anions which may be converted to esters, benzoins or ketones.[68]

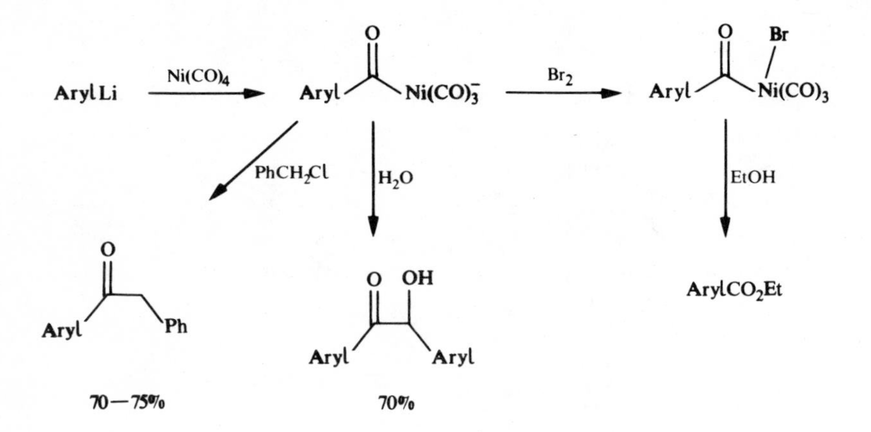

Organomercuric compounds react with $Ni(CO)_4$ and Ni-acyl complexes to generate ketones.[69]

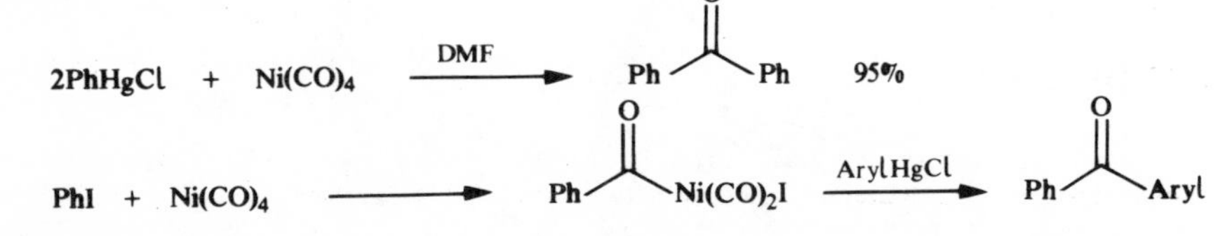

30—96%

Aromatic halides are converted to carboxylic acids on treatment with CO and $Ni(CO)_4$.[70]

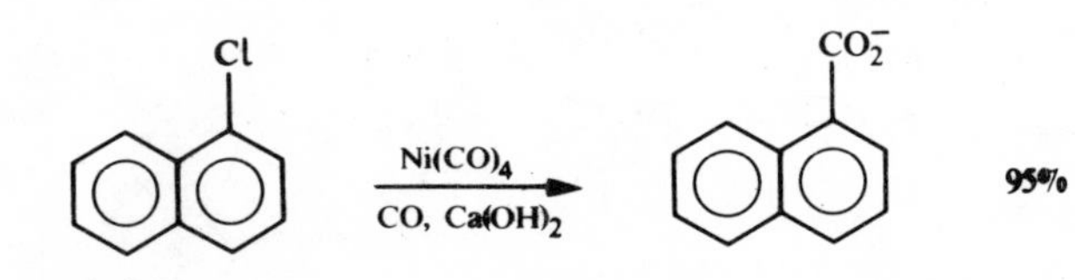

Acetylenes may be converted to cyclopentadienones by heating with $Ni(CO)_4$.[71] If conc. HCl is present then cyclopentenones are obtained.[72]

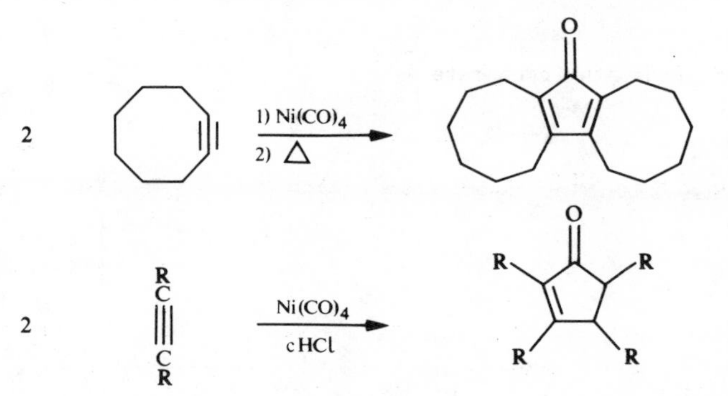

Carbonylative coupling with $Ni(CO)_4$ of allyl chlorides and acetylene in methanol leads to the formation of methyl hexa-2,5-dienates.[73]

Cl + HC≡CH $\xrightarrow[\text{MeOH}]{Ni(CO)_4}$ CO_2Me

Reaction of methallyl chloride, acetylene, methanol and $Ni(CO)_4$ leads to the formation of methyl 5-methyl-*trans*-hexa-2,4-dienate after treatment with base. The ester is readily converted to (±)-methyl-*trans*-chrysanthemate **25**.[74]

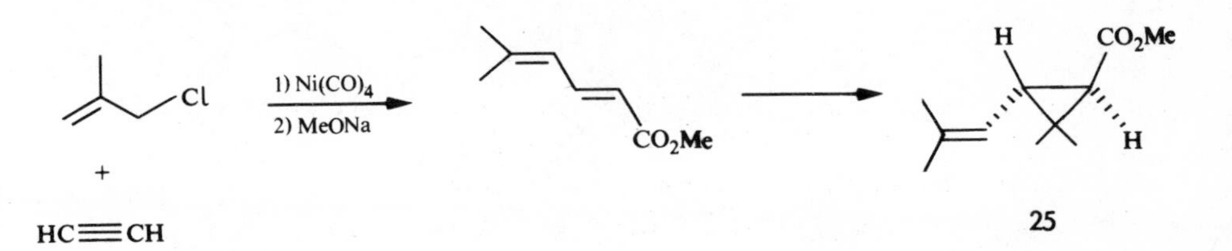

Carbonylative coupling is also observed in the reaction of allyl chlorides with olefins catalysed by $Ni(CO)_4$ in the presence of CO.[75]

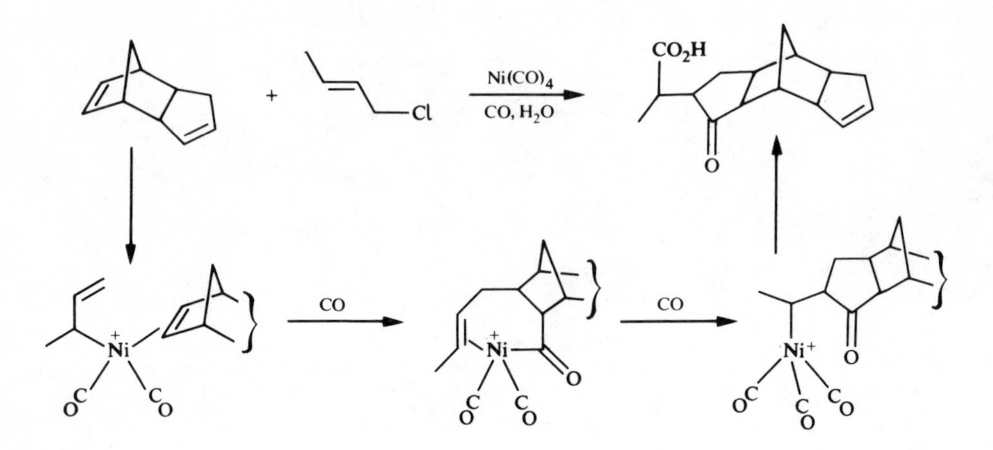

9.5 ASYMMETRIC CARBONYLATION REACTIONS

Several attempts to perform asymmetric hydroformylations and carbonylations using prochiral olefins and chiral phosphine ligands have been made. However the stereoselectivities observed are much less satisfactory than in asymmetric hydrogenation reactions.[76]

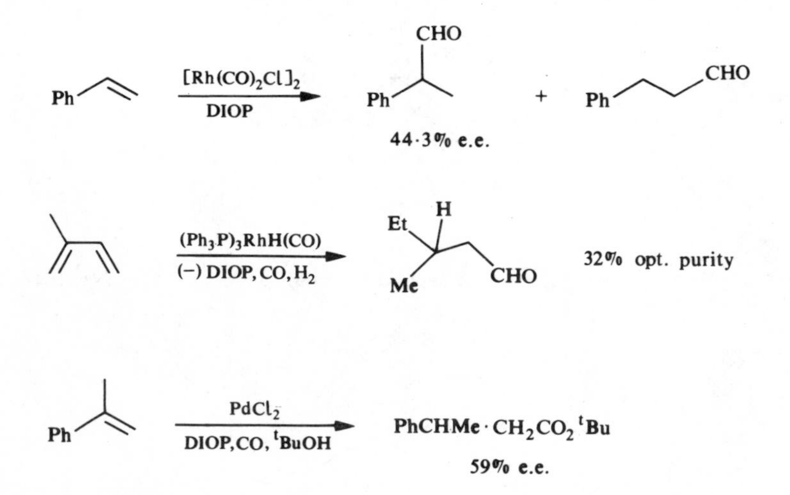

9.6 INSERTION OF CO_2

CO_2 reacts with bis-allyl nickel to give γ-lactones.[77]

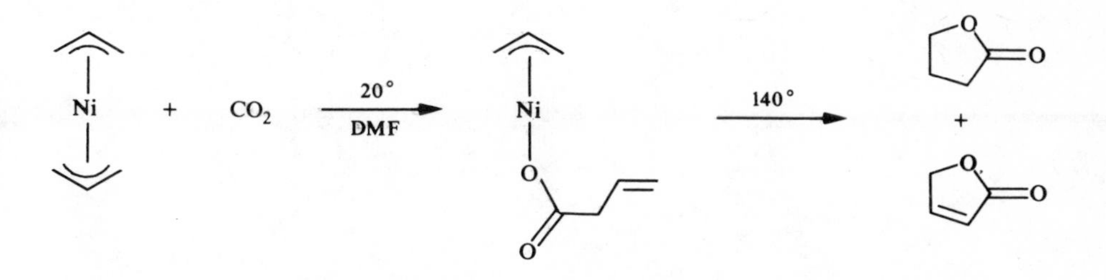

Butadiene reacts with palladium complexes in the presence of CO_2 to produce lactones.[78]

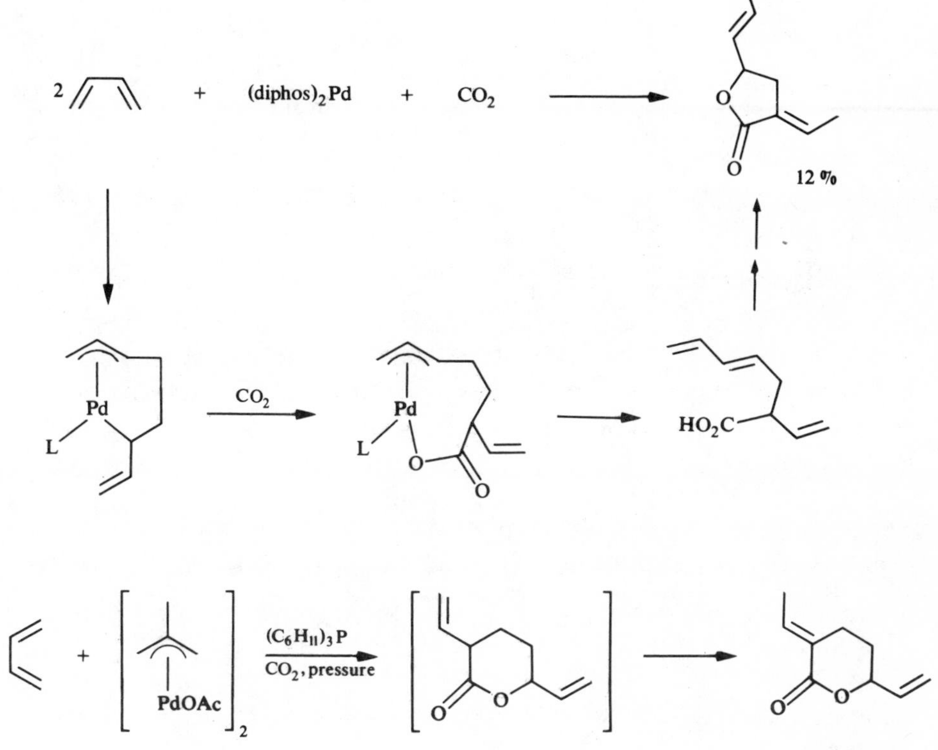

Butenolides are formed from alkylidene cyclopropanes, CO_2 and PdL_4.[79]

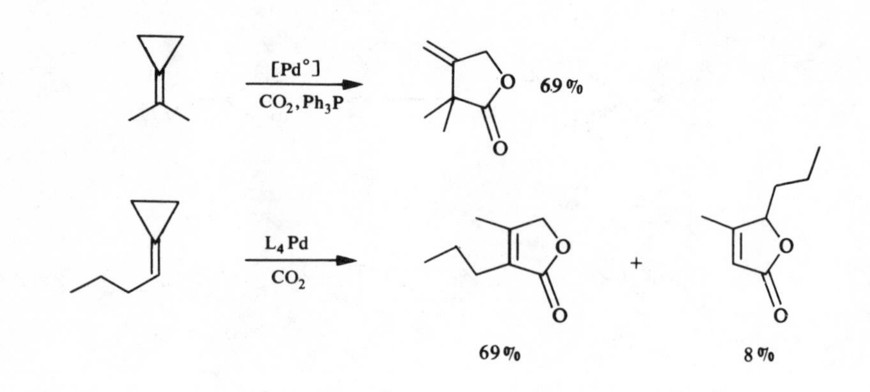

CO_2 can be inserted into epoxides to form ethylene carbonates.[80]

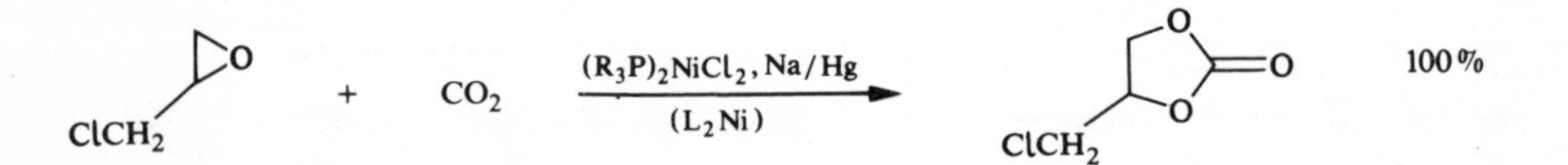

9.7 DECARBONYLATION REACTIONS

Many of the steps of the carbonylation reactions described above are reversible and this allows the reverse reaction, decarbonylation, to occur. There are relatively few examples available of this potentially extremely useful procedure.

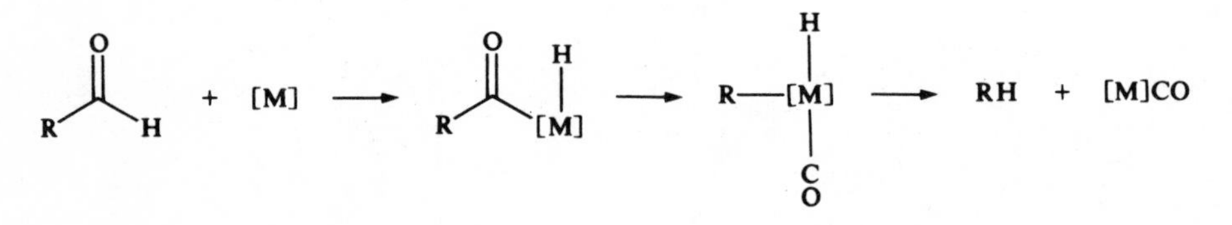

The most frequently used decarbonylation reagent is Wilkinson's catalyst $(Ph_3P)_3RhCl$ and the mechanism has been extensively studied for this complex.[81] The reaction has been shown to be intramolecular by deuterium labelling studies and the acyl to alkyl migration has been shown to be stereospecific with retention of configuration at carbon.

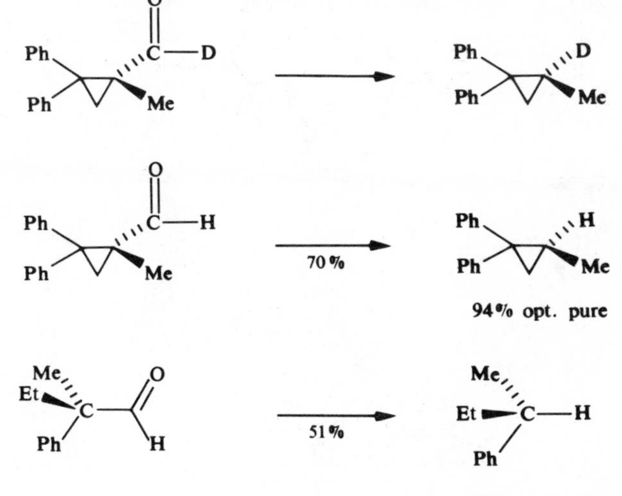

α,β-Unsaturated aldehydes may also be decarbonylated stereoselectively.[81,82]

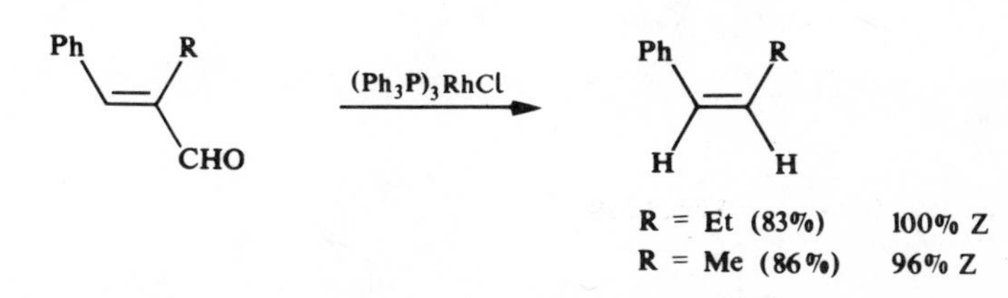

The aldehyde function in 26, readily prepared from cyclohexanone, is decarbonylated to a methyl group by $(Ph_3P)_3RhCl$.[83]

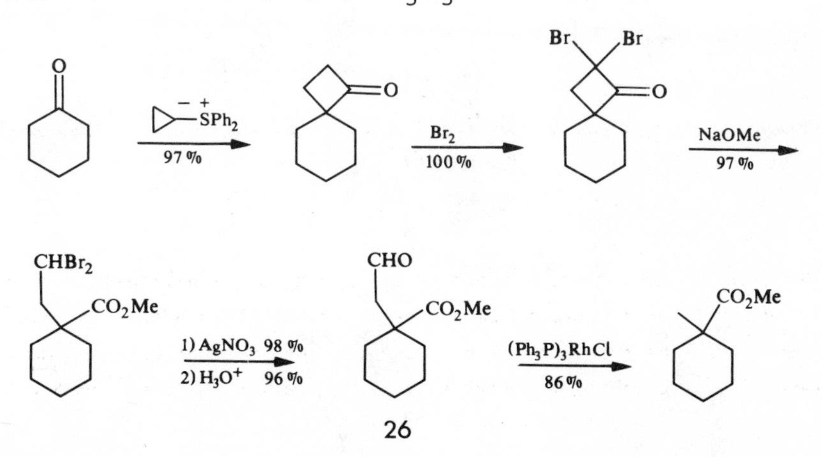

A similar geminal alkylation approach allowed the synthesis of methyl desoxypodocarpate 28 from the ketone 27.[84]

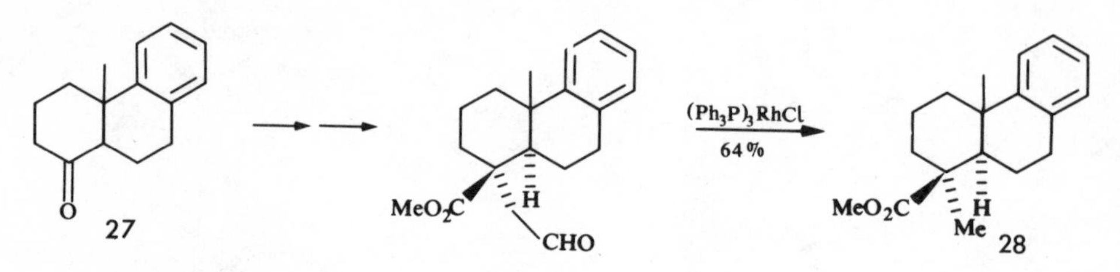

$(Ph_3P)_3RhCl$ decarbonylation of 29 and subsequent *cis* addition of HI to the triple bond of 30 by hydrozirconation form two essential steps in the synthesis of the vinyl iodide 31. This iodide is one of the key intermediates used in a total synthesis of the macrolide methynolide 32.[85]

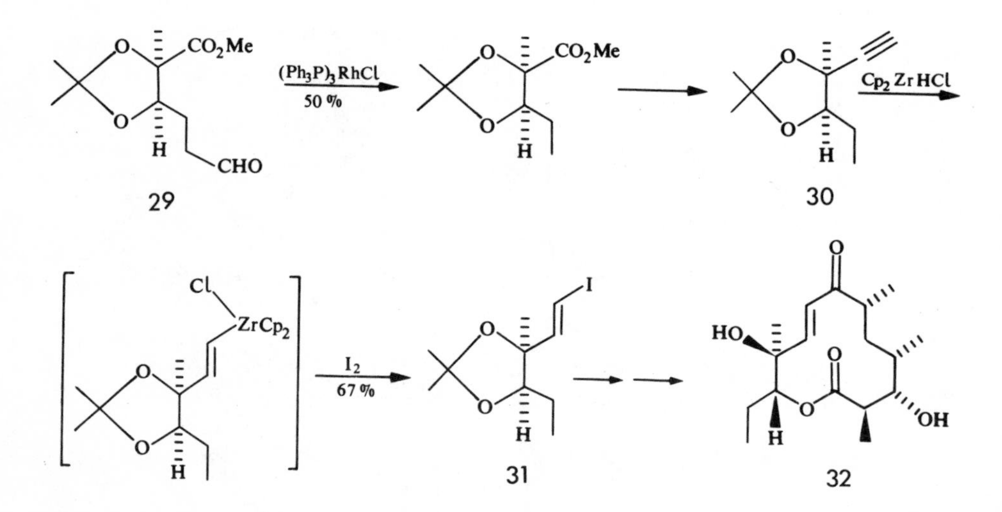

Angular methyl groups may be stereoselectively introduced by the addition first of a two carbon fragment to give an aldehyde followed by

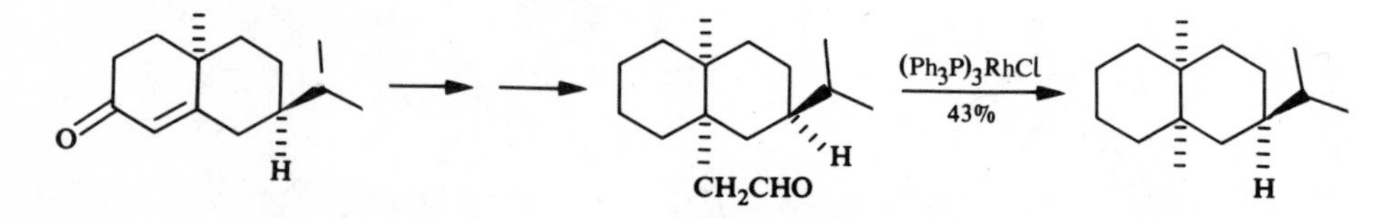

decarbonylation to the derived methyl group.[86,87]

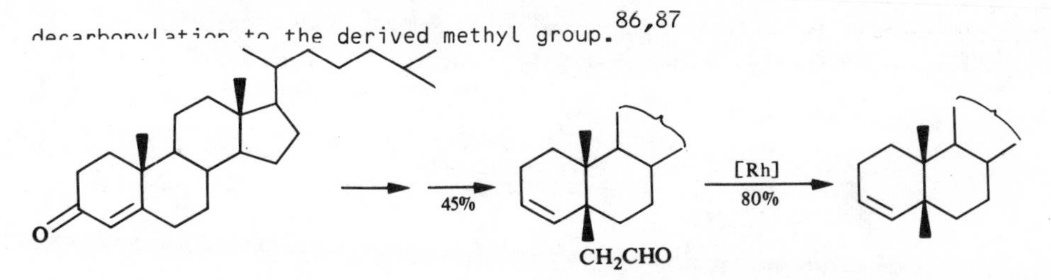

A 1,2-double bond can be introduced into 3-keto steroids by the addition of a 2-formyl group followed by oxidation and subsequent decarbonylation.[88]

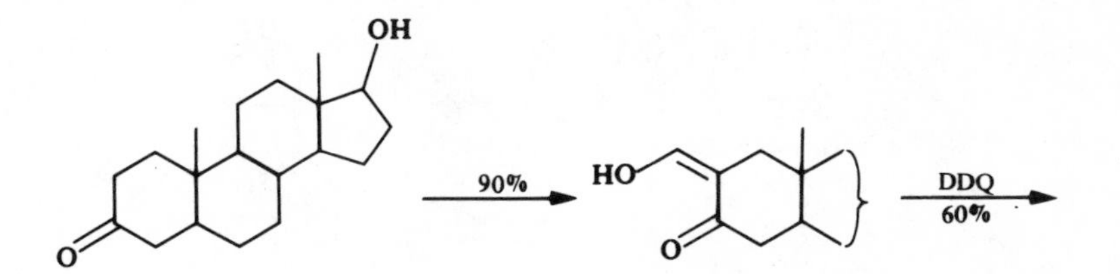

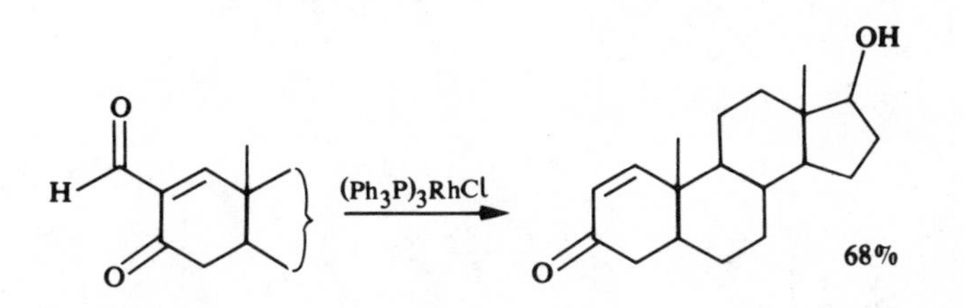

Decarbonylation of the disaccharide derivative 33 occurs in good yield with $(Ph_3P)_3RhCl$.[89]

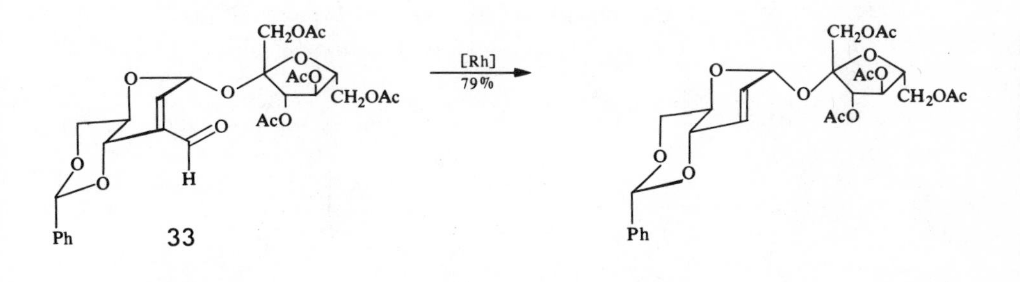

The aldehyde 34 is inert to $(Ph_3P)_3RhCl$ presumably for steric reasons. However, decarbonylation of 34 and 35 occurs with $(Ph_2MeP)_3RhCl$.[90]

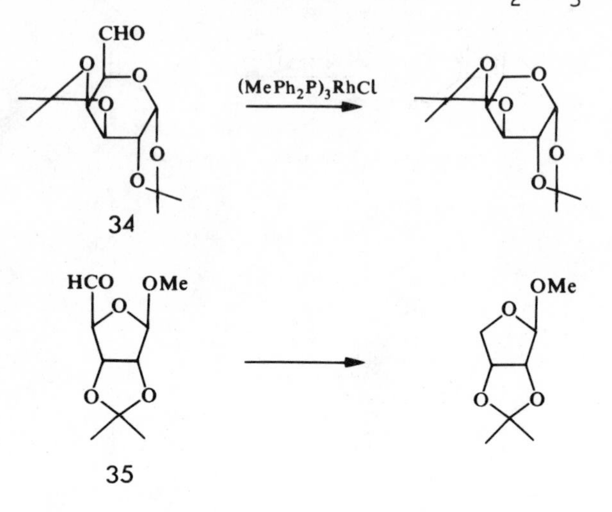

The chloro-aldehyde 36 is reductively decarbonylated by $Fe(CO)_5$ presumably via a vinyl radical mechanism.[91]

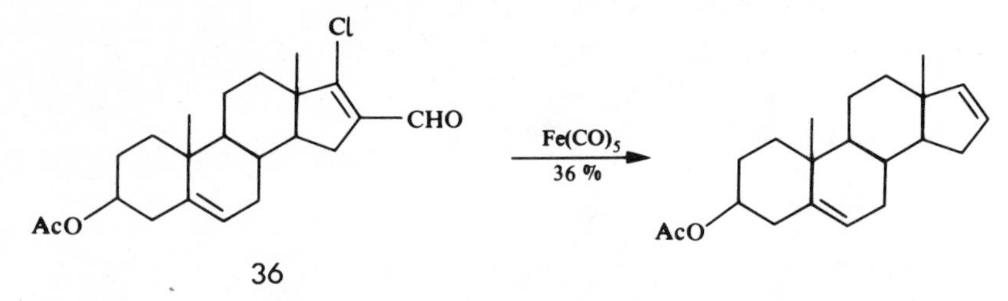

Acid chlorides are decarbonylated by $(Ph_3P)_3RhCl$ to give olefins. However, this reaction is not always useful as mixtures of isomerised olefins are produced, the rhodium complex being an isomerisation catalyst.

Both α and β diketones are readily decarbonylated to ketones by $(Ph_3P)_3RhCl$.

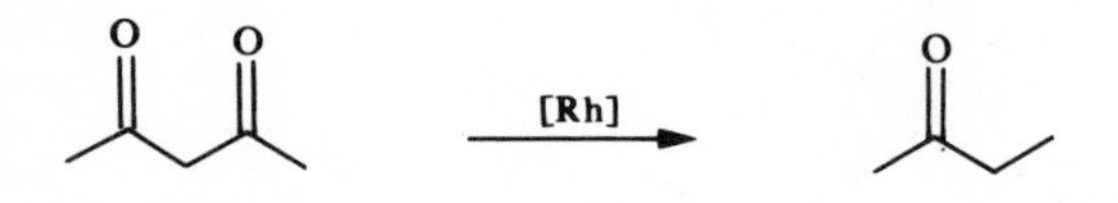

The reaction is catalytic.[92]

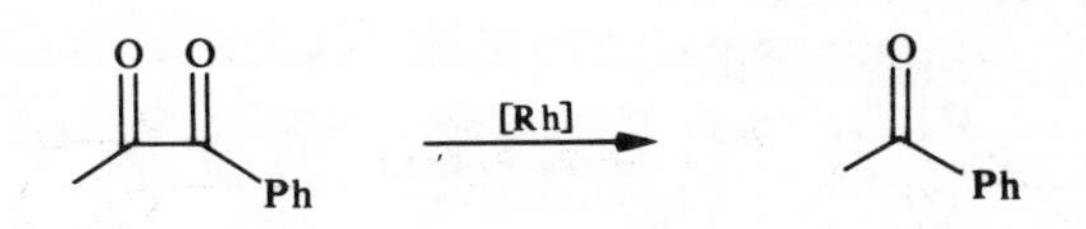

Conjugated diynes may be produced by decarbonylation of diethynyl ketones with $(Ph_3P)_3RhCl$.[93]

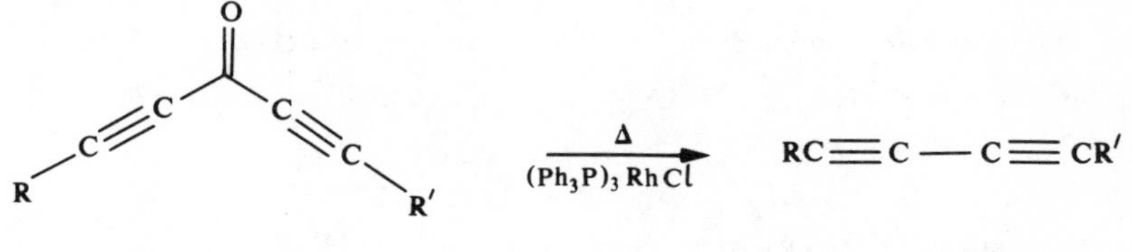

Catalytic decarbonylation of benzaldehyde has been achieved using $(diphos)_2RhCl$. High temperatures (115-180°C) are needed to effect decarbonylation of the intermediate cation $[(diphos)_2RhCO]^+Cl^-$.[94]

9.8. REFERENCES

1. P.L. Bock, D.J. Buschetto, J.R. Rasmussen, J. P. Demers and G.M. Whitesides, *J. Amer. Chem. Soc.*, 1974, *96*, 2814 and references therein.
2. J.A. Labinger, D.W. Hart, W.E. Seibert III, J. Schwartz, *J. Amer. Chem. Soc.*, 1975, *97*, 3851.
3. J. Schwartz and J.A. Labinger, *Ang. Chem. Int. Ed.*, 1976, *15*, 333.
4. C.A. Bertelo and J. Schwartz, *J. Amer. Chem. Soc.*, 1975, *97*, 228.
5. C.A. Bertelo and J. Schwartz, *J. Amer. Chem. Soc.*, 1976, *98*, 262.
6. I. Rhee, M. Ryang and S. Tsutsumi, *J. Organometal. Chem.*, 1967, *9*, 361.
7. A.F.M. Iqbal, *Helv. Chim. Acta*, 1976, *59*, 655.
8. E. Weiss and W. Hubel, *Chem. Ber.*, 1962, *95*, 1179.
9. R. Aumann and J. Knecht, *Chem. Ber.*, 1976, *109*, 174.
10. Y. Becker, A. Eisenstadt and Y. Shvo, *J. Organometal. Chem.*, 1978, *155*, 63; *Tetrahedron*, 1978, *34*, 799.
11. B.F.G. Johnson, J. Lewis and D.J. Thompson, *Tet. Letters*, 1974, 3789; E. Weissberger and P. Laszlo, *Acc. Chem. Res.*, 1976, *9*, 209.
12. B.F.G. Johnson, K.D. Karlin and J. Lewis, *J. Organometal. Chem.*, 1978, *145*, C23.
13. V. Heil, B.F.G. Johnson, J. Lewis and D.J. Thompson, *Chem. Comm.* 1974, 270.
14. P.A. Spanninger and J.L. von Rosenberg, *J. Org. Chem.*, 1969, *34*, 3658.
15. A. Stockis and E. Weissberger, *J. Amer. Chem. Soc.*, 1975, *97*, 4288.
16. R.R. Victor, R. Ben-Shoshan and S. Sarel, *Tet. Letters*, 1970, 4253; R. Aumann, *J. Amer. Chem. Soc.*, 1974, *96*, 2631.
17. G.D. Annis and S.V. Ley, *Chem. Comm.*, 1977, 581.

17a. G.D. Annis, E.M. Hebblethwaite and S.V. Ley, *Chem. Comm.*, 1980, 297.

18. R. Aumann and H. Ring, *Ang. Chem. Int. Ed.*, 1977, *16*, 50; R. Aumann, H. Ring, C. Kruger and R. Goddard, *Chem. Ber.*, 1979, *112*, 3644.
19. G.S. Giam and K. Ueno, *J. Amer. Chem. Soc.*, 1977, *99*, 3166.
20. M. Yamashita and Y. Watanabe, T. Mitsudo and Y. Takegami, *Tet. Letters*, 1976, 1585.
21. R.C. Cookson and G. Farquharson, *Tet. Letters*, 1979, 1255.
22. J.P. Collman, *Acc. Chem. Res.*, 1975, *8*, 342.

23. J.P. Collman, S.R. Winter and R.G. Komoto, *J. Amer. Chem. Soc.*, 1973, *95*, 249.

24. M.P. Cooke Jr., *J. Amer. Chem. Soc.*, 1970, *92*, 6080.

25. H. Masada, M. Mizuno, S. Suga, Y. Watanabe and Y. Takegami, *Bull. Chem. Soc. Jap.*, 1970, *43*, 3824.

26. Y. Takegami, Y. Watanabe, H. Masada and I. Kanaya, *Bull. Chem. Soc. Jap.*, 1967, *40*, 1456.

27. J.Y. Merour, J.L. Roustan, C. Charrier, J. Collin and J. Benaim, *J. Organometal. Chem.*, 1973, *51*, C24.

28. J.E. McMurry, A. Andrus, G.M. Ksander, J.H. Musser and M.A. Johnson, *J. Amer. Chem. Soc.*, 1979. *101*, 1330.

29. M. Rosenblum, *Acc. Chem. Res.*, 1974, *7*, 122 and references therein.

30. P.K. Wong, M. Madhavarao, D.F. Marten and M. Rosenblum, *J. Amer. Chem. Soc.*, 1977, *99*, 2823.

31. Ref. 9 in T. Aida, R. Legault, D. Dugat and T. Durst, *Tet. Letters*, 1979, 4993.

32. P. Lennon, A.M. Rosan and M. Rosenblum, *J. Amer. Chem. Soc.*, 1977, *99*, 8426.

33. I. Wender and P. Pino, Organic Syntheses via Metal Carbonyls, J. Wiley and Sons. Vol. 1, 1968, Vol. 2, 1977 and references therein.

34. R.P.A. Sneeden, *l'Actualité chimique*, 1979, 31.

35. C.W. Bird, *Chem. Rev.*, 1962, *62*, 283.

36. R.F. Heck and D.S. Breslow, *J. Amer. Chem. Soc.*, 1961, *83*, 4023.

37. A. Rosenthal, *Adv. Carbohydrate Chem.*, 1968, *23*, 59.

38. J.L. Eisenmann, *J. Org. Chem.*, 1962, *27*, 2706.

39. A. Rosenthal and G. Kan, *Carbohydrate Res.*, 1971, *19*, 145.

40. S. Murai and N. Sonoda, *Angew. Chem. Int. Ed.*, 1979, *18*, 837.

40a. J. Falbe and F. Korte, *Chem. Ber.*, 1965, *98*, 1928.

40b. A. Rosenthal and I. Wender in Organic Syntheses via Metal Carbonyls, Ed. I. Wender and P. Pino, Intersciences, London, 1968, *1*, 405.

41. H. Alper and J.K. Currie, *Tet. Letters*, 1979, 2665.

42. H. Wakamatsu, J. Uda and N. Yamakami, *Chem. Comm.*, 1971, 1540.

43. D. Seyferth and R.J. Spohn, *J. Amer. Chem. Soc.*, 1969, *91*, 3037.

44. M. Sohn, J. Blum and J. Halpern, *J. Amer. Chem. Soc.*, 1979, *101*, 2694; J. Blum, C. Zlotogorski and A. Zoran, *Tet. Letters*, 1975, 1117; T. Sakakibara and H. Alper, *Chem. Comm.*, 1979, 458.

45. K.S.Y. Lau, P.K. Wong and J.K. Stille, *J. Amer. Chem. Soc.*, 1976, *98*, 5832.

46. J.K. Stille and P.K. Wong, *J. Org. Chem.*, 1975, *40*, 335; R.C. Larock, *J. Org. Chem.*, 1975, *40*, 3237; R.C. Larock, *Ang. Chem. Int. Ed.*, 1978, *17*, 27, A. Kasahara, T. Izumi and A. Suzuki, *Bull. Chem. Soc. Japan*, 1977, *50*, 1639.

47. R.C. Larock and B. Riefling, *Tet. Letters*, 1976, 4661; R.C. Larock, B. Riefling and C.A. Fellows, *J. Org. Chem.*, 1978, *43*, 131.

48. A. Kasahara, T. Izumi and A. Suzuki, *Bull. Chem. Soc. Japan*, 1977, *50*, 1639; K. Mori, T. Mizoroki and A. Ozaki, *Chem.Letters*, 1975, 39, J.F. Krifton, *J. Mol. Catal.*, 1977, *2*, 293.

49. J.K. Stille and P.K. Wong, *J. Org. Chem.*, 1975, *40*, 532; Y. Fujiwara, I. Moritani, M. Matsuda and S. Teranishi, *Tet. Letters*, 1968, 633; A. Schoenberg, I. Bartoletti and R.F. Heck, *J. Org. Chem.*, 1974, *39*, 3318; M. Hidai, T. Hikita, Y. Wada, Y. Fujikura and Y. Uchida, *Bull. Chem. Soc. Japan*, 1975, *48*, 2075; T. Ito, K. Mori, T. Mizoroki and A. Ozaki, *Bull. Chem. Soc. Japan*, 1975, *48*, 2091; A. Cowell and J.K. Stille, *Tet. Letters*, 1979, 133; A. Cowell and J.K. Stille, *J. Amer. Chem. Soc.*, 1980, *102*, 4193.

50. A. Schoenberg and R.F. Heck, *J. Org. Chem.*, 1974, *39*, 3327; Y. Fujiwara, R. Asano and S. Feranishi, *Isr. J. Chem.*, 1977, *15*, 262.

51. M. Mori, K. Chiba Y. Ban, *J. Org. Chem.*, 1978, *43*, 1684; Y. Ban, T. Wakamatsu and M. Mori, *Heterocycles*, 1977, *6*, 1711; M. Mori, K. Chiba, N. Inotsume and Y. Ban, *Heterocycles*, 1979, *12*, 921.

52. M. Mori, K. Chiba and Y. Ban, *Heterocycles*, 1977, *6*, 1841.

53. M. Mori, K. Chiba, M. Okita and Y. Ban, *Chem. Comm.*, 1979, 698.

54. A. Schoenberg and R.F. Heck, *J. Amer. Chem. Soc.*, 1974, *96*, 7761; H. Yoshida, N. Sugita, K. Kudo and Y. Takezaki, *Bull. Chem. Soc. Japan*, 1976, *49*, 1681.

55. M. Takahashi and J. Tsuji, *J. Organometal. Chem.*, 1967, *10*, 511.

56. J. Tsuji, M. Morikawa and J. Kiji, *Tet. Letters*, 1963, 1061; *J. Amer. Chem. Soc.*, 1964, *86*, 4851; T. Yukawa and S. Tsutsumi, *J. Org. Chem.*, 1969, *34*, 738.

57. D.M. Fenton, *J. Org. Chem.*, 1973, *38*, 3192; D. Medema, R. van Helden and C.F. Kohll, *Inorg. Chim. Acta*, 1969, *3*, 255; L.S. Hegedus, O.P. Anderson, K. Zetterberg, G. Allen, K. Siirala-Hanseń, D.J. Olsen and A.B. Packard, *Inorg. Chem.*, 1977, *16*, 1887; J.K. Stille and R. Bivakaruni, *J. Amer. Chem. Soc.*, 1978, *100*, 1303;

J.K. Stille and D.E. James, *J. Organometal. Chem.*, 1976, *108*, 401; *J. Amer. Chem. Soc.*, 1975, *97*, 674; L.S. Hegedus and K. Siirala-Hanseń *J. Amer. Chem. Soc.*, 1975, *97*, 1184.

58. G. Carturan, M. Graziani, R. Ros and U. Belluco, *J.C.S. Dalton*, 1972, 262; L.F. Hines and J.K. Stille, *J. Amer. Chem. Soc.*, 1970, *92*, 1798; 1972, *94*, 485.

59. K. Bitter, N.V. Kutepow, K. Neubauer and H. Reis, *Ang. Chem.*, 1968, *80*, 352; J. Tsuji, M. Morikawa and J. Kiji, *Tet. Letters*, 1963, 1437; E.N. Frankel, *J. Am. Oil. Chem. Soc.*, 1973, *50*, 39; R. Bardi, A. Del Pra, A.M. Piazzesi and L. Toniolo, *Inorg. Chim. Acta*, 1979, *35*, L345.

60. J. Tsuji, J. Kiji and M. Morikawa, *Tet. Letters*, 1963, 1811; J. Tsuji, S. Imamura and J. Kiji, *J. Amer. Chem. Soc.*, 1964, *86*, 4491; J. Tsuji, J. Kiji and S. Hosaka, *Tet. Letters*, 1964, 605; J. Tsuji, Y. Mori and M. Hara, *Tetrahedron*, 1972, *28*, 3721; J. Tsuji, S. Hosaka, J. Kiji and T. Susuki, *Bull. Chem. Soc. Japan*, 1966, *39*, 141; J. Tsuji and H. Yasuda, *Bull. Chem. Soc. Japan*, 1977, *50*, 553; J. Tsuji, J. Kiji, S. Imamura and M. Morikawa, *J. Amer. Chem. Soc.*, 1964, *86*, 4350; W.T. Dent, R. Long and G.H. Whitfield, *J. Chem. Soc.*, 1964, 1588; J. Tsuji and T. Susuki, *Tet. Letters*, 1965, 3027.

61. J.K. Stille and R. Divakaruni, *J. Org. Chem.*, 1979, *44*, 3474.

62. R.F. Heck, *J. Amer. Chem. Soc.*, 1969, *91*, 6707; 1971, *93*, 6896.

63. G.P. Chiusoli, C. Venturello and S. Merzoni, *Chem. Ind.*, 1968, 977; T.F. Murray, V. Varma, J.R. Norton, *J. Amer. Chem. Soc.*, 1977, *99*, 8085; *J. Org. Chem.*, 1978, *43*, 353; T.F. Murray and J.R. Norton, *J. Amer. Chem. Soc.*, 1979, *101*, 4107; T.F. Murray, V. Varma and J.R. Norton, *Chem. Comm.*, 1976, 907.

64. C.G. Chavdarian, S.L. Woo, R.D. Clark and C.H. Heathcock, *Tet. Letters*, 1976, 1769.

65. A. Cowell and J.K. Stille, *Tet. Letters*, 1979, 133.

66. F.J. Weigert, *J. Org. Chem.*, 1973, *38*, 1316; H.A. Dieck, R.M.Laine and R.F. Heck, *J. Org. Chem.*, 1975, *40*, 2819; T. Saegusa, T. Tsuda, Y. Isegawa, *J. Org. Chem.*, 1971, *36*, 858.

67. S.J. Tremont and R.G. Bergman, *J. Organometal. Chem.*, 1977, *140*, C12.

68. M. Ryang, S. Kwang-Myeong, Y. Sawa and S. Tsutsumi, *J. Organometal. Chem.*, 1966, *5*, 305.

69. Y. Hirota, M. Ryang and S. Tsutsumi, *Tet. Letters*, 1971, 1531.

70. L. Cassar and M. Foà, *J. Organometal. Chem.*, 1973, *51*, 381.
71. G. Wittig and P. Fritze, *Ann.*, 1968, *712*, 79.
72. W. Best, B. Fell and G. Schmitt, *Chem. Ber.*, 1976, *109*, 2914.
73. G.P. Chiusoli and L. Cassar, *Ang. Chem. Int. Ed.*, 1967, *6*, 124.
74. E.J. Corey and M. Jautelat, *J. Amer. Chem. Soc.*, 1967, *89*, 3912.
75. G.P. Chiusoli, G. Cometti, G. Sacchelli, V. Bellotti, G.D. Andreetti, G. Bocelli and P. Sgarabotto, *J.C.S. Perkin II*, 1977, 389.
76. M. Tanaka, Y.I. Keda and I. Ogata, *Chem. Letters*, 1975, 1115; G. Consiglio and P. Pino, *Chimia*, 1976, *30*, 193; C. Botteghi, M. Branca and A. Saba, *J. Organometal. Chem.*, 1980, *184*, C17.
77. T. Tsuda, Y. Chujo and T. Saegusa, *Synth. Comm.*, 1979, *9*, 427.
78. Y. Sasaki, Y. Inoue and H. Hashimoto, *Chem. Comm.*, 1976, 605; A. Musco, C. Rerego and V. Tartiani, *Inorg. Chim. Acta*, 1978, *28*, L147.
79. Y. Inoue, T. Hibi, M. Satake and H. Hashimoto, *Chem. Comm.*, 1979, 982.
80. R.J. De Pasquale, *Chem. Comm.*, 1973, 157.
81. H.M. Walborsky and L.E. Allen, *J. Amer. Chem. Soc.*, 1971, *93*, 5465.
82. K. Ohno and J. Tsuji, *J. Amer. Chem. Soc.*, 1968, *90*, 99.
83. B.M. Trost and M.J. Bogdanowicz, *J. Amer. Chem. Soc.*, 1973, *95*, 2038.
84. B.M. Trost and M. Preckel , *J. Amer. Chem. Soc.*, 1973, *95*, 7862.
85. P.A. Grieco, Y. Ohfune, Y. Yokoyama and W. Owens, *J. Amer. Chem. Soc.*, 1979, *101*, 4749.
86. D.J. Dawson and R.E. Ireland, *Tet. Letters*, 1968, 1899.
87. R.E. Ireland and G. Pfister, *Tet. Letters*, 1969, 2145.
88. Y. Shimuzu, H. Mitsuhashi and E. Caspi, *Tet. Letters*, 1966, 4113.
89. D.E. Iley and B. Fraser-Reid, *J. Amer. Chem. Soc.*, 1975, *97*, 2563.
90. D.J. Ward, W.A. Szarket and J.K.N. Jones, *Chem. and Ind.*, 1976, 162.
91. S.J. Nelson, G. Detre and M. Tanabe, *Tet. Letters*, 1973, 447.
92. K. Kaneda, M. Azuma, M. Wayaku and S. Teranishi, *Chem. Letters*, 1974, 215.
93. E. Müller and A. Segnitz, *Ann.*, 1973, 1583.
94. D.H. Doughty and L.H. Pignolet, *J. Amer. Chem. Soc.*, 1978, *100*, 7083.

INDEX

Acetaldehyde
  from the Wacker Process 1,304-5
  to vinyl alcohol 109
Acetylenes
  benzyne 105
  carbonylation 388-390,391
  η^2-complexes 35
  coupling with aryl halides 222
  to cycloheptatrienone 354
  to diynes 399
  hydrozirconation 27,221
  hydrogenation of 343
  η^1-propargyl 50,192
  protection as $Co_2(CO)_8$ complexes 86-90
  to pyridines 260
  reaction with cyclobutadiene 102
  stabilisation of cycloalkynes 105
  structure of platinum complex 104
  trimerisation to arenes 255-259
Acetylide ligands
  as nucleophiles 42,193
Acidity
  benzoic acids 5,6
  dienyl acids 9
  phenols 7
Acyl complexes 348-392
  to carbene 40,41
  dissociation to acids 351-3
  dissociation to aldehydes 188,353
  dissociation to esters 31-2,350,352
  dissociation to ketones 350
  dissociation to lactams 133-4,355, 365
  intermediates in hydroformylation reactions 17,351-3
  intermediates in carbonylation reactions 348-392
  preparation 23,26,28-9,40,41,349
Adam's Catalyst 84
Adenine 141
Alkaloids
  cephalotaxinone 220
  codeine chromium tricarbonyl 211,70
  desethylibogamine 178,235
  ibogamine 178,235
  morphine chromium tricarbonyl 211
  muscaine alkaloids 245
  sendaverine 382
  thebaine iron tricarbonyl 55,65, 98,150
  tropane alkaloids 245
Allenes 88,192,272,387
Allyl acetates
  from allyl palladium 51,179
  to allyl palladium 173-4,176-80
  rearrangement 230,288
Allyl alcohols
  to allyl palladium complexes 175,233-4
  carbonylation 374,381
  epoxidation 313-6
  Felkin reaction 246-50
  preparation 53,319
  rearrangement 282
Allyl amides
  carbonylation 375
  rearrangement 289
Allyl amines
  carbonylation 375,382,384
  to olefin palladium complex 158-159
  preparation 177-9
  rearrangement 289-90
Allyl ethers
  to allyl complexes 47
  rearrangement 278,285-7
Allyl halides
  to η^1-allyl complexes 21
  to η^3-allyl complexes 46-7
  carbonylation 374
  coupling reactions 226,232,236-42, 391,2

Allyltrimethylsilane 88,89,142,145
Allylvinylethers 20
Amines
  addition to allyl complexes 174-5, 177-9
  addition to dienyl complexes 141, 143-4
  addition to olefin complexes 128, 132-4,160-1
  from arene chromium tricarbonyl 171
  arylation 111-2
  conversion to carbonyl 309
  directing groups 159
  protection 98
Amino acids
  from asymmetric reduction 328-9
  (±)-gabaculine 178
  phenylalanine 259
  protection of amine 98,290
  synthesis 375-7
Aphidicolin 363
Arene chromium tricarbonyl complexes
  aryl anions 202-8
  benzylic anions 8,203-4,209-13
  benzylic cations 8
  decomplexation 73
  electronic effects 5-9
  formation 69-71
  nucleophilic addition to 166-72
  stereochemical effects 10
  steric effects 12
Arene oxides 334
Aromatic nucleophilic substitution 6,152-3,154,170-2,206,307
Aryl halides
  addition to metal 24,218
  carbonylation 380-3,390
  η^6-complexes 68,152-4,167,170-2,206
  coupling reactions 219-20,222,231, 234,238-9
Asymmetric syntheses 11
  via η^3-allyl palladium 138,176-7
  chiral phosphine ligands 327
  coupling reactions 220,222-3,252-3
  cyclopropanes 43
  diene iron tricarbonyl 58
  epoxidations 316,318
  Felkin reaction 250
  hydroformylations 392
  hydrogenations 326-31
  isomerisations 289
Azide 66
Aziridines 299
Azirines 299,378
Back donation 5
Barbaralone 356
Benzocyclobutanes 258
Benzoic acids, acidity 5,6
Benzylidine acetone iron tricarbonyl 57
Benzyne 105
Biotin 286
Birch reduction 63,69,290
Bisbenzene titanium 336
Biscyclopentadienyl zirconium chloro complexes
  acyl 29,352-3
  alkyl 29,31,188,352-3
  hydride 27,353
Bonding
  Chatt-Dewar-Duncanson model 4,125
Butenolides 379,389,394
Cadmium alkyls 37,140,142,147
Camphor 242
Carbene ligands
  classification 3
  decomplexation 42-3,155-6
  deprotonation 201-3
  preparation 40-2
Carbohydrates 228,244
  amination 162,174
  carbonylation 369-70,372-3
  decarbonylation 397-8
  deoxygenation 333
  isomerisation of O-allyl protecting group 286
  olefin cations 35
Carbon dioxide
  insertion reactions 393-4
Carbonium ion stabilisation 8,10,86-91
Carveol 308
Carvone 277
Cephalotaxinone 220
Ceric salts 31,43,61,73,90-1,102,104,191, 203-4,272,358,364,366
Chiral complexes 10,58,328
Chiral phosphine ligands 11,327-9
Cholesta-1,3-diene 63
Cholestanone 173
Cholest-4-ene 49
Cholest-5-ene 49
4-Cholesten-3-one 323
Cholesterol 308,317
Chromanones 271

Chromium hexacarbonyl 3,19
  with arenes 69-70,72
  with dienes 34,71
  with dihydropyridines 107
  hydrogenation catalyst 322
  isomerisation of dienes 279-81
  with nucleophiles 40
  with pyrroles 72
Chrysanthemates 391
Citral 324
Claisen Rearrangement 287
Closed ligands 128
Cobalt hydridotetracarbonyl 17,366-376
Codeine chromium tricarbonyl 70,211
Coenzyme Q_1 242
Complexation reactions 19-82
Confertin 241
Coordination 13
Coprostanone 323
α-Cuparenone 242
α-Curcumene 222
Cyanide
  base 52,149
  decomplexation agent 40,274
  ligand 2,194
  nucleophile 25,141,146-7,180
Cycloalkynes 104-106
Cyclobutadienes 85,99-103,119
Cyclobutenes 293
Cycloheptatrienone 354
Cycloheptatrienyl anion; stabilisation 9
Cyclohexadienone iron tricarbonyl 110-111
trans-Cyclo-octene
  resolution 36,40
Cyclopentadienyl iron dicarbonyl
  acyl derivatives 29,41,364-6
  alkyl derivatives 20-22,25,29,31, 39,41,116,128-34,188,364-6
  allyl derivatives 28,30,38,47, 189-93
  anion 20-22,36-8,109,331-33
  carbene cations 41-2
  carbonylations 364-6
  deoxygenation of epoxides 331-33
  olefin cations 4,11,25,30,35-40, 83-6,101,109,116,124,128-34,189-93
Cyclopropanation 155-6,163-4
Damsin 271
Decarbonylation reactions
  aldehydes 394-8
  ketones 398-9
  metal carbonyls 29
Decomplexation
  of η^1-ligands 30
  of η^2-olefin ligand 40
  of carbene ligands 42-3
  of η^3-allyl 51-3
  of η^4-diene 60-2
  of η^5-dienyl 67
  of η^6-arene 73
Dehydrogenation 308-310
Desethylibogamine 178,235
Desoxypodocarpate 396
Deuteration
  of alkyls 31,230
  of allyl alcohols 283
  of anions 7,214
  arene reduction 153,326
  of carbenes 202
  of cyclobutadiene 101
  of esters 387
Diazocompounds 28,42,97,164
Diazomethane 28,42-3,180,204-5 207-8
Diborane 61,87,91,94,252
Dictyolene 237
Diels-Alder reactions 95-6,99
Diene iron tricarbonyl complexes
  carbonylation 354-6
  decomplexation 60-2
  to dienyl 62-6
  electronic effects 9,10
  with electrophiles 194-8
  formation 53-60
  in isomerisations 273-5,278-9, 293,296-7
  protecting group 91-8
  stabilising effect, 110-2
  stereochemical effects 10,11,58
  steric effects 12-13
Dihalocarbenes 94,95
Dihydrocarvone 270
Dihydrocoumarins 238
Dihydropyridines 61,106-108
Diimide 86
Diketene 51
DIOP
  176,220,250,327,329,330
Diphenylacetylene 90,104
Dissociation 13
Dodecene 306
β-Dolabrin 196

Diiron nonacarbonyl 19
with acetylene 354
with allylbromides 60
desulphurisation catalyst 339
with 1,1′-dibromoketones 242-5
with dienes 55,92,97,110
with dihydropyridines 108
with enones 56,57
with epoxides 296-8
isomerisation catalyst 300
with olefins 34,59,109
Eighteen electron rule 2-4
exceptions 2
Electronic effects of coordination 5
γ-Elemene 282
β-Elimination 14
Enol ethers 161,229
carbonylation 369-70
coupling 229
ligands 109,111,147-50
as nucleophiles 145
as electrophiles 161-2
synthesis 43,201,237,285-7
Enols see vinyl alcohols
Enones
alkylation 129-30,165,223,235 396-7
from carbonylation 357,376
complexation 34,56-8,97,165,198 229,245
from epoxides 295-6
Heck reaction 159
isomerisation 269-70,277
from ketones 310
preparation 67,168,320,334
reductions 323-4
Enyne complexes 86
Epiergosterol 12
epi-β-santalene 237
Episulphides 339
Epoxides
carbonylation 358,367,372-3
deoxygenation 36-7,331-6
preparation 311-8
rearrangement 294-8
Ergosterol 57,93,95,112,279-80
Estrone 70,259
Ethylene
carbonylation 17,351
coupling 224-5
ligand 23,33
nucleophilic addition to 25,116-7, 122-4,131-2,180
oxidation 1,304-6
reduction 15-16,26
Eugenol 85-6
Even **ligands 118**
Farnesol 221
Felkin reaction 246-50
Ferrocene 3,72
Fischer-Tropsch Synthesis 1
Fromaldehyde 101
Friedel-Crafts reaction 5,9,90,96 101,195-8,258
Fulvenes 194
Furans 243-5,298
Gabaculine 178
Geranyl acetate 239
Germacratriene 282
Grignard reagents 93,101,103
coupling 219-20,231
Felkin reaction 246-50
with metal halides 24-5
organometallic 21
as reducing agent 25,74
Haloarenes
carbonylation 380-2,390
chromium tricarbonyl complexes 6,169-72
coupling 219,220,231-3,238
iron cyclopentadienyl cations 154
manganese tricarbonyl cations 152-3
pericyclic addition to metal 24
reduction 341
Halopyridines 219,222,234,238,383
Halothiophenes 219,233
Hapto(η)number 3
Heck reaction 159,225,231,235
Hexafluorophosphate anion 20
Hibaene 249
Humulene 178,240,335
Hydorazulenes 191
Hydroboration 27,61,86,87,91,94,252
Hydroformylation 17,348-51,366-71
Hydrogenation 14,15-16,320-30
Hydrogen cyanide 180
Hydrogen peroxide 31,61,86,87,91,94,252,290,352-3
Hydrosilylations 14,16,324,331
Ibogamine 178,235
Imines 337,340
Indanone chromium tricarbonyl 12
Indoles 106,107,142-3,160,169,178, 223,232,235,238,299,310
α-Ionone 324

Iron pentacarbonyl 19
with benzylbromides 60
carbonylation catalyst 354,355,357,359-61
to cyclobutadiene complex 100,102
deoxygenation catalyst 334,337-8
with dienes 54,55,63,65,93,103,112,275,279
isomerisation catalyst 269,273,283,285,289,357
with KOH 340-1
with olefins 34,59
with vinyl epoxides 60,296-7,358
with vinyl halides 343
Isodamsin 271
Isomerisation
acetylenes 260-73
allyl derivatives 282-90
dienes 273-82
olefins 266-73
small ring hydrocarbons 291-3
small ring heterocycles 294-300
Isonitrile 2,40,194,380
Isoprene 135,139,140,254
Isoquinoline 260
Ketals 286
from olefins 306
Ketones
from alcohols 309
alkylation 89,104,178,211,220,240,248,366
from allyl alcohols 283-5
carbonylation 254,350,356-7,359-63,377,395
from Claisen rearrangement 287
complexation 41,70,92,97,111
coupling 256
decarbonylation 398-9
from epoxides 294,372
methylenation 200-1
from olefins 306-7
oxidation 318
preparation 90,93,96,101,103,131,135,142,145-9,164,168,195-7,199,206,233-4,237,245,252,258,340-1
reduction 12-13,92,101,325-6,330
Lactams
β-133-4,355,365-6
γ-232,375-6,381,383
δ-231,239,260,375,381-2
Lactones 159,180,228,238,271
isomerisation 299
macrolides 177,240
preparation from
alkoxycarbenes 203
anhydrides 325
Baeyer Villiger 244
carbon dioxide 393-4
carbon monoxide 358,373-4,379,384,388-9
coupling reactions 241
lactols 62
Lanalool 313
Lewis acids 2
Lewis bases 2
Ligands
non-hydrocarbon 2
hydrocarbon 3
Macrolides 177,239,396
Manool 249
MCPBA 32,53,160,312,314,319
Metathesis 251-2
Methallyl alcohol 283
Methylenation 200-201
Methynolide 396
Migration reactions
metal-ligand 14
Monoterpenes
see terpenes
Morphine chromium tricarbonyl 211
Muscaine alkaloids 245
Muscone 254
Myrcene
η^3-allyl 44,53
diene iron tricarbonyl 91,95-6,135,197
Naphthalene chromium tricarbonyl 70
reduction 326
Nezukone 245
Nickel tetracarbonyl 20
coupling with 236,239-40
7-Norbornadienone 103-4
Norbornene 227,230
Nornicotine 234
Ocimene 44
Odd ligands 118
Open ligands 118
Osmium tetroxide 94,244
Oxetanes
carbonylation 373-4
rearrangement 298-9

Oximes 337,376
Oxygen 31,43,67,360-1
Peptide synthesis 98
Perchlorate anion 20
Pericyclic addition 14-17
Pericyclic elimination 14-17
α-Phellandrene iron tricarbonyl 54
Phenol
  chromium tricarbonyl 7
  keto tautomer 110-11
Phenylalanine 259
Pheromones 89,233,272-3
Pimarene 249
Pinene 356-57
pKa changes on coordination
  benzoic acids 5,6
  cycloheptatriene 9
  dienylacids 9
  phenols 7
Precalciferol 112
η^1-propargyl 50,192
Prostaglandins 227-8,278,295,321
Protection
  acetylenes 83-6
  amines 86-90
  dienes 91-97
  olefins 98
Pseudouridine 245
Pterocarpin 227
Pyridines 107,219,222,234,238-9,
  260,359,383
2-Pyridones 260
Pyrimidines 161-2,228
α-Pyrone 99-100
Pyrroles 72,175,198,243,245,300
Queen Substance 233
o-Quinodimethane 60,101
Quinolines 232,310
Quinolones 231
Rearrangements
  see isomerisation
Regioselectivity rules 117-20
Resolution methods 10-11,36,40
Rhodium trichloride
  isomerisation with 269-71,274-77
Ruthenium trichloride 20,69
α-Santalene 237
Scopine 245
Sendaverine 382
Sesquiterpenes
  see terpenes
Showdomycin 245
Simmons-Smith cyclopropanation 94
Solvolysis reactions
  rate enhancement 8
  rate retardation 10
Spiro compounds 148-9,169
Stabilisation
  of benzyne 105
  of cycloalkynes 104
  of cyclobutadiene 99
  of dihydropyridines 106-8
  of enols 109
  of ketones 110
  of 7-norbornadienone 103
Steric effects 10,12
Steroids
  acetylene $Co_2(CO)_8$ 87
  allyl palladium 49,50,52-3,138
  165,173
  arene chromium tricarbonyl 70,71
  carbonylation 361
  decarbonylation 397-8
  deoxygenation 334
  diene iron tricarbonyl 12-3,57,
  58,61,63-4,93,94,112,146
  dienyl iron tricarbonyl cations
  146
  isomerisation 276,279,280
  olefin cations 39
  oxidation 308,315-23
Stilbene 334
$Tachysterol_2$ 112
Tebbe's Reagent 200-01
Terpenes
  mono- 44,49,54,58,91,95-6,135,137
  145,197,232,239,242-3,254,257,
  312-3,320,324,327,335,357
  sequi- 178,221,237,240-1,282
  di- 249
Testosterone 173,308
Tetrafluoroborate anion 20
Thebaine
  chromium tricarbonyl 70
  iron tricarbonyl 55,65,98,150
Thiophenes 233
Thiourea 388-9
Thujaplicin 196,245
Trapping agents 112
Tributyltin hydride 141-2,342

Triiron dodecacarbonyl
with dienes 278
Tropane alkaloids 245
Tropine 245
Tropone 196
Vanadium hexacarbonyl 19,73
Venolepin 389
Vinyl alcohols 109,110,305
Vinylcyclopropanes 357,391
Vinyl ethers 42-3,107,163,317
369-70
Vinyl halides
carbonylation 380-2
π-complexes 60,157
coupling 218-9,221-2,232
addition to metal 24,105
reduction 341
Vinylidene ligands
formation 42,193
Wacker Process 1,304-5,307
Wilkinson's catalyst
for coupling 235,255
for decarboxylations 394-9
for hydrogenations 1,14-16,320-1,
322,324
for hydrosilylations 14
for rearrangements 271,276,285-6,
289,290,294
Wittig reaction 63,65,93,163
Zeise's salt 33
Ziegler-Natta Polymerisation 1,
250-1